S0-AKJ-200

Practical Guide to Infrared Microspectroscopy

B/NA

PRACTICAL SPECTROSCOPY

A SERIES

Edited by Edward G. Brame, Jr.

The CECON Group
Wilmington, Delaware

1. Infrared and Raman Spectroscopy (in three parts), *edited by Edward G. Brame, Jr., and Jeanette G. Grasselli*
2. X-Ray Spectrometry, *edited by H. K. Herglotz and L. S. Birks*
3. Mass Spectrometry (in two parts), *edited by Charles Merritt, Jr., and Charles N. McEwen*
4. Infrared and Raman Spectroscopy of Polymers, *H. W. Siesler and K. Holland-Moritz*
5. NMR Spectroscopy Techniques, *edited by Cecil Dybowski and Rob- ert L. Lichter*
6. Infrared Microspectroscopy: Theory and Applications, *edited by Robert G. Messerschmidt and Matthew A. Harthcock*
7. Flow Injection Atomic Spectroscopy, *edited by Jose Luis Burguera*
8. Mass Spectrometry of Biological Materials, *edited by Charles N. McEwen and Barbara S. Larsen*
9. Field Desorption Mass Spectrometry, *László Prókai*
10. Chromatography/Fourier Transform Infrared Spectroscopy and Its Applications, *Robert White*
11. Modern NMR Techniques and Their Application in Chemistry, *edited by Alexander I. Popov and Klaas Hallenga*
12. Luminescence Techniques in Chemical and Biochemical Analysis, *edited by Willy R. G. Baeyens, Denis De Keukeleire, and Katherine Korkidis*
13. Handbook of Near-Infrared Analysis, *edited by Donald A. Burns and Emil W. Ciurczak*
14. Handbook of X-Ray Spectrometry: Methods and Techniques, *edited by René E. Van Grieken and Andrzej A. Markowicz*
15. Internal Reflection Spectroscopy: Theory and Applications, *edited by Francis M. Mirabella, Jr.*
16. Microscopic and Spectroscopic Imaging of the Chemical State, *edited by Michael D. Morris*
17. Mathematical Analysis of Spectral Orthogonality, *John H. Kalivas and Patrick M. Lang*
18. Laser Spectroscopy: Techniques and Applications, *E. Roland Menzel*

19. Practical Guide to Infrared Microspectroscopy, *edited by Howard J. Humecki*

ADDITIONAL VOLUMES IN PREPARATION

Quantitative X-ray Spectrometry, Second Edition, *Ron Jenkins, R. W. Gould, and Dale Gedcke*

Practical Guide to Infrared Microspectroscopy

edited by

Howard J. Humecki
McCrone Associates
Westmont, Illinois

Marcel Dekker, Inc. New York • Basel • Hong Kong

Library of Congress Cataloging-in-Publication Data

Practical guide to infrared microspectroscopy / edited by Howard J. Humecki.
p. cm. -- (Practical spectroscopy ; v. 19)
Includes bibliographical references and index.
ISBN 0-8247-9449-4 (alk. paper)
1. Infrared spectroscopy. 2. Fourier transform spectroscopy. I. Humecki, Howard J. II. Series.
QD96.I5P733 1995 94-42808
543'.08583--dc20 CIP

The publisher offers discounts on this book when ordered in bulk quantities. For more information, write to Special Sales/Professional Marketing at the address below.

This book is printed on acid-free paper.

MARCEL DEKKER, INC.
270 Madison Avenue, New York, New York 10016

Current printing (last digit):
10 9 8 7 6 5 4 3 2 1

PRINTED IN THE UNITED STATES OF AMERICA

Preface

Many new analytical techniques, when they first appear on the scene, fill a need of their inventors or developers. Those who share that need quickly adopt the new techniques and expand upon them. Sometimes these developers diffuse into a network that freely exchanges information regardless of their applications. This has not been the case with infrared microspectroscopy (IMS) to any significant degree. Of course, IMS is not a new technique; it is, in fact, an old one that has enjoyed a rebirth of sorts. But, because the lines of information within a specialty have been established for so long a time, there seems to have been little effort to communicate with others in diverse fields who may be exploring along similar avenues. For example, those in the packaging industry using IMS to examine multilayer thin films have little opportunity to discuss methodology with forensic scientists examining multilayered paint chips, and neither talks to art conservators even though there are considerable areas of overlap where each might profit from the others' experience.

To satisfy this need, resulting from a rapidly developing technology and expanding interest in this technique, required input from individuals having extensive experience in IMS and expertise in a wide range of specialties. I have attempted to bring together experts from various fields to discuss where and how they apply IMS in the solution of problems peculiar to their particular field or industry. Some have felt that it is important to preface their presentations with a discussion of theoretical considerations that show the limitations of IMS in regard to the problems under consideration. Rather than ask that these discussions be limited, I welcomed them as needed reminders that IMS has limitations that we must keep in mind when planning our approach to a difficult problem.

Likewise, an obvious extension of such limitation are techniques that complement or support evidence obtained by IMS. No method should be forced to stand alone if additional information can be obtained by other means. As my colleague Joe Barabe reminds me, "If you have a hammer, the only problem you can solve is a nail." Today, we have many tools at our disposal and, although this book is devoted to IMS, we must not fail to consider other means if they help us with needed information.

Robert Messerschmidt has been involved in the design and development of several microscopes for use with IMS. In Chapter 1, Bob describes theoretical considerations in instrument design that apply to the design of microscopes, and he discusses how they might affect performance. Some design features that have little importance in the design of standard instruments have enormous impact on the performance and the quality of the spectra generated for very small specimens.

Much to the chagrin of some spectroscopists, the transition from "standard" spectroscopy to microspectroscopy has not been an easy one. No one is better qualified to demonstrate the link between light microscopy and IMS than John Reffner, who with Pamela Martoglio demonstrates that light microscopy goes hand in hand with IMS.

K. Krishnan, Jay Powell, and Steve Hill discuss the principles of FT-IR microimaging, starting with a description of the essential components and their function and interaction. They describe applications of this technique on polymer, mineral, biological and semiconductor specimens.

The semiconductor industry has taken full advantage of IMS, not only to solve contamination problems but also to study materials and manufacturing processes. In Chapter 4, Kate Chess explores the use of reflectance techniques including ATR and grazing angle objectives in the study of surfaces, surface treatments, and processing problems in the electronics industry.

A study of thin films and laminates, and especially the problems associated with their manufacture, can be particularly difficult to solve. Richard Duerst and his associates describe how they have dealt with the variety of problems relating to surface defects, embedded particles, and diffusion of additives across an interface. They have included an example of analyzing a multilayer packaging film and of contour mapping of an interpenetrating network.

Probably no industry has exploited IMS more than law enforcement, where trace evidence plays a vital role in determining the success or failure of an investigation. Many of the techniques used by forensic scientists are also practiced by scientists in other fields. Scott Ryland describes in detail the steps involved in analyzing forensic paint evidence. Scott not only deals with the preparation of specimens, but also presents a detailed scheme for classifying and identifying the types of binders in single and multilayer applications.

Light microscopy and infrared spectroscopy are seen as the most useful techniques for identifying synthetic fibers. Edward Bartick, Mary Tungol, and Akbar Montaser compare sample preparation techniques, debating the pros and cons of flattened versus “as is.” They consider in great detail the effects of pressure, sheer force, and contamination on matching a questioned fiber to a suspected source. They present results of studies on the variations of peak area ratios for various copolymers and discuss results of studies of dichroism in single fibers.

Some of the problems encountered in art conservation bear certain similarities to those in forensics. Identification of protective coatings and fibers is a problem common to both, although sample preparation and the compositions may differ considerably as does the provenance of the articles and objectives of the analysis. Michele Derrick illustrates applications of IMS for the identification of coatings as well as some unique applications of IMS for the study of degradation of cultural artifacts, ancient and relatively modern.

The pharmaceutical industry has accepted IMS enthusiastically. In Chapter 9, Scott Aldrich and Mark Smith describe in detail the application of microscopical methods to problem solving in the pharmaceutical industry. They show in their examples not only the application and solution of problems, but in addition the need for detailed information concerning the history of the sample. The provider of the sample often has information that is vital in designing the analytical scheme.

The infrared microspectroscopy of mineral specimens presents unique problems, not the least of which is sample preparation. Very thin mineral specimens are difficult to polish, and often the thickness required to study one region of absorption is inappropriate for another. Anne Hofmeister discusses these problems and describes some techniques for preparing and carrying out both qualitative and quantitative analyses on small mineral specimens.

The previous chapters deal with applications and pertinent theory of IMS. It should be apparent that sample preparation is the single most important factor in determing the quality of information generated. Anna Teetsov, in Chapter 11, has carried sample preparation to a fine art describing techniques that some might find daunting. Methods for collecting a scattering of fine particles and droplets from surfaces and depositing them on a single spot on a salt plate are described, as are other simple but elegant techniques.

The last chapter deals with the identification of polymers in highly filled and pigmented systems through identification of their pyrolsis products. Nylons and polyesters can be subjected to controlled hydrolysis and their hydrolysis products may be recovered and identified by IMS.

I hope the reader will find these pages a handy guide to the basic theory of IMS and a ready source of information on techniques and applications. I would like to thank each of the authors for the thoughtfulness they have given to their topics, and the detail and professionalism in their presentation. My special thanks go to Bob Muggli, a friend, colleague, and early innovator in IMS, for his advice and editorial help.

Howard J. Humecki

Contents

Contributors

D. Scott Aldrich Trace Substance Analysis, The Upjohn Company, Kalamazoo, Michigan

Edward G. Bartick Forensic Science and Research Training Center (FSRTC), FBI Academy, Quantico, Virginia

William E. Breneman Analytical and Properties Research Laboratory, Corporate Research, 3M Center, St. Paul, Minnesota

Catherine A. Chess Chemistry and Materials Science, IBM Thomas J. Watson Research Center, Yorktown Heights, New York

Michele R. Derrick Scientific Program, The Getty Conservation Institute, Marina del Rey, California

Rebecca M. Dittmar Analytical and Properties Research Laboratory, Corporate Research, 3M Center, St. Paul, Minnesota

Marilyn D. Duerst Department of Chemistry, University of Wisconsin—River Falls, River Falls, Wisconsin

Richard W. Duerst Analytical and Properties Research Laboratory, Corporate Research, 3M Center, St. Paul, Minnesota

Stephen L. Hill Digilab Division, Bio-Rad Laboratories, Cambridge, Massachusetts

Anne M. Hofmeister Department of Earth and Planetary Science, Washington University, St. Louis, Missouri

Howard J. Humecki McCrone Associates, Inc., Westmont, Illinois

K. Krishnan Digilab Division, Bio-Rad Laboratories, Cambridge, Massachusetts

Gerald J. Lillquist Analytical and Properties Research Laboratory, Corporate Research, 3M Center, St. Paul, Minnesota

Pamela A. Martoglio Research Division, Spectra-Tech, Inc., Shelton, Connecticut

Robert G. Messerschmidt CIC Photonics, Inc., and Rio Grande Medical Technologies, Inc., Albuquerque, New Mexico

Akbar Montaser Department of Chemistry, The George Washington University, Washington, D.C.

Jay R. Powell Digilab Division, Bio-Rad Laboratories, Cambridge, Massachusetts

John A. Reffner Research Division, Spectra-Tech, Inc., Shelton, Connecticut

Scott G. Ryland Orlando Regional Crime Laboratory, Florida Department of Law Enforcement, Orlando, Florida

Mark A. Smith Trace Substance Analysis, The Upjohn Company, Kalamazoo, Michigan

Colleen K. Spicer Analytical and Properties Research Laboratory, Corporate Research, 3M Center, St. Paul, Minnesota

William L. Stebbings Analytical and Properties Research Laboratory, Corporate Research, 3M Center, St. Paul, Minnesota

Anna S. Teetsov McCrone Associates, Inc., Westmont, Illinois

Mary W. Tungol Hair and Fibers Unit, FBI Laboratory, Washington, D.C.

James W. Westburg Analytical and Properties Research Laboratory, Corporate Research, 3M Center, St. Paul, Minnesota

Practical Guide to Infrared Microspectroscopy

1
Minimizing Optical Nonlinearities in Infrared Microspectroscopy

Robert G. Messerschmidt CIC Photonics, Inc., and
Rio Grande Medical Technologies, Inc., Albuquerque, New Mexico

1. INTRODUCTION

There are several fundamental concepts that should be understood to perform FT-IR microspectroscopy successfully which are unimportant in macrospectroscopy. Many of these concepts are a result of optical principles which, to other than an optical physicist, may seem to contradict one's understanding of optics. In fact, some of the concepts necessary to describe the behavior of imaging in the infrared microscope *do* contradict classical geometric optics.

Most readers will be familiar with the concepts of the laws of reflection and refraction and the concept of magnification. These are the basis of geometric optics and can be summarized by one simple equation, Snell's law of refraction. This formula describes the bending of light as it passes through materials of differing refractive indices and past surfaces of various curvatures. A special case of this law for mirror surfaces states that upon the encounter of a "ray" of light with a reflective surface, the angle of incidence of that light ray with respect to the surface will equal the angle of reflection.

Many readers will also know about the common optical aberrations that can be predicted and proved through the application of geometric optical ray tracing. These, such as spherical aberration, chromatic aberration, and coma (offense against the sine condition), are controllable. It is the optical designer's responsibility to keep these aberrations small enough so that they are unobjectionable in a given application.

This is as far as one routinely needs to go in the design of instruments to be used in the visible or infrared region of the spectrum, at low magnification. However, to deal with the behavior of light when the eye is aided, for instance in a microscope system, the diffraction of light energy must be considered. Diffraction occurs everywhere, not only in microscopes. Insofar as the eye is an optical instrument, the image of everything one looks at is altered by diffraction. Fortunately, given the acceptance angle of light into the eye and the wavelengths involved, the diffraction effect is below the resolution limit. Therefore, one does not notice the effect.

The diffraction effect in infrared microspectroscopy manifests itself as a blurring of the image information which one obtains in order to measure the spectrum of a given sample. Primarily affected is the spatial resolution of the measurement. That is, the energy reaching the detector contains spectral information from a larger physical area than is expected or desired. This has its obvious consequences. Spurious energy and/or spectral peaks can be present in the resultant spectrum, inviting quantitative and qualitative misinterpretation. Unfortunately, diffraction is a physical phenomenon, and it is therefore not possible to eliminate the problem. But there are design considerations that can minimize the effect.

1.1. Rationale for an FT-IR Microscope

Infrared microscopes for dispersive spectrometers were developed in several academic laboratories in the late 1940s and early 1950s [1–3] but were used by only a handful a researchers. Even when Perkin-Elmer introduced a commercial unit in 1953 [4], not many people jumped on the bandwagon. A fairly in-depth search of the literature reveals few references to the infrared microscope in the 1960s and 1970s.

In the late 1970s, FT-IR spectrometers were starting to make their presence felt. A well-respected group of champions was at that time extolling the virtues of this type of instrument, such as higher resolution, lower noise, and better frequency accuracy than the ubiquitous dispersive instruments. Interestingly, an infrared microscope attached to an FT-IR spectrometer tends to restrict one of the inherent FT advantages: the throughput (Jacquinot) advantage. The optical throughput of an optical system is described by the equation

$$\Theta_x \Theta_s = A\Omega_s \qquad \mathrm{cm}^2 \cdot \mathrm{sr} \tag{1}$$

where A is the area and Ω_s is the solid angle subtended by the limiting aperture of the system (in the FT-IR microscope, this would be the area-defining aperture). Because a microscope is by definition used for small samples, the sample area contribution to the throughput calculation is reduced. On the other hand, one would like to increase the other factor, solid angle, as much as possible to

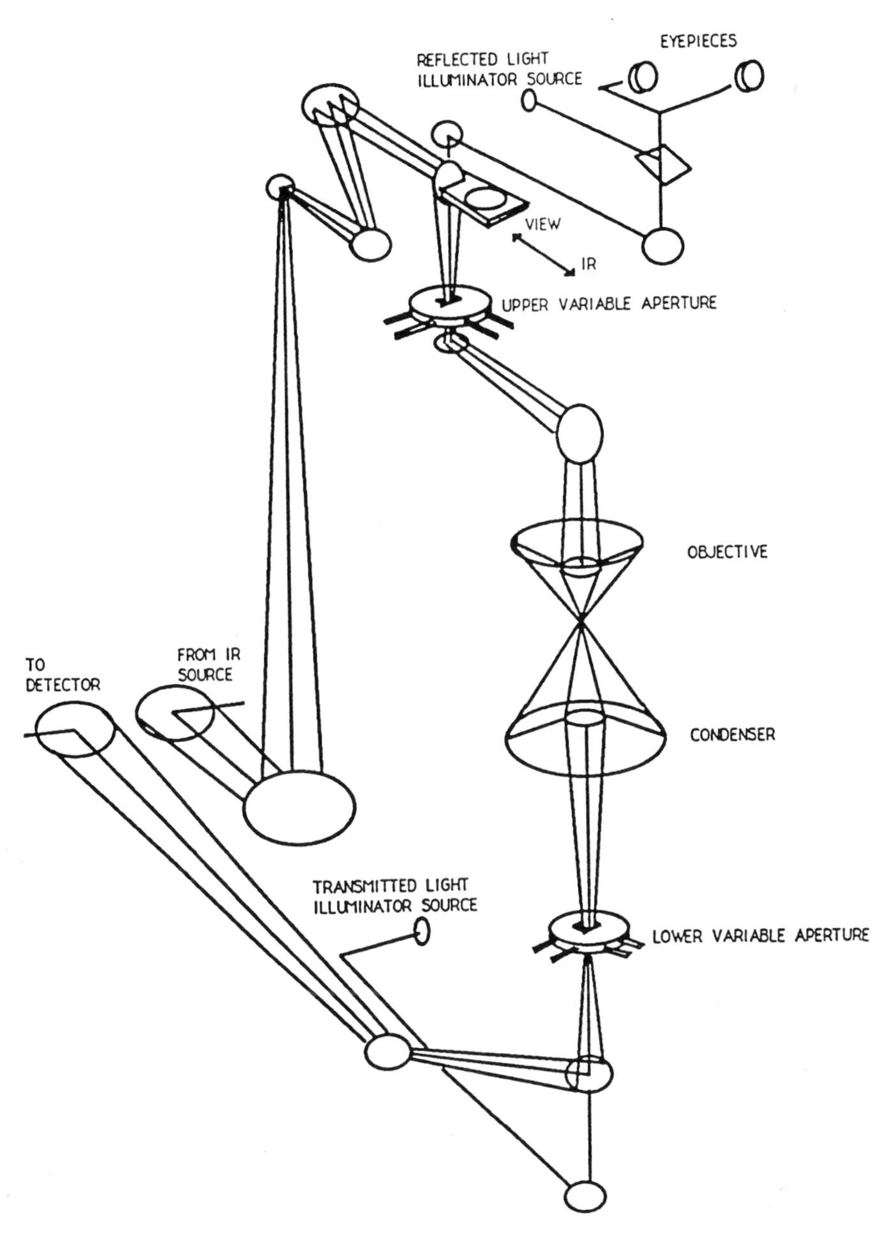

Figure 1. General layout of an FT-IR microscope.

improve signal *and* spatial resolution [5]. One normally still ends up in a region where more optical throughput than a dispersive spectrometer would allow is needed.

In addition, the other two fundamental advantages of the FT technique are still in full force: the multiplex (Felgett) and the frequency precision (Connes) advantages. The former results in a sensitivity benefit equal to the square root of the number of resolution elements. The latter relates to an improved ability to measure accurately the spectral frequency of absorption bands in a spectrum.

By 1982, the sensitivity of infrared spectroscopy had seemed to have developed to a point where low-energy and microscopic techniques were again being tried. A favorite demonstration on the floor of the Pittsburgh Conference in 1983 was obtaining the transmission spectrum of a brown paper bag: clearly, a low-energy situation. Down the aisle, another manufacturer was demonstrating the transmission spectrum of polystyrene through a 100-μm pinhole, without a beam condenser. Clearly, the stage was set for the reemergence of infrared microscopy. In the 1980s, FT-IR-based microscopes come into being, and the history of this development has been reviewed elsewhere [6]. Today, FT-IR microscopy is the chosen technique for a wide range of sampling problems, with about one-third of all FT-IR spectrometers being sold with a microscope attachment.

2. GENERAL CONSIDERATIONS

The general layout of an infrared microscope is shown in Figure 1. It consists of transfer optics to bring the infrared radiation from the interferometer, usually imaging both the source image and the pupil image (the beamsplitter image) through the microscope. The visible optical train is parfocal and collinear with the infrared radiation. This is achieved in setting up the microscope by viewing through a series of small pinholes and then aligning for infrared energy through the same pinholes. In actual use, the area of interest is brought to the center of the field of the microscope under visible (transmitted or reflected) light. Then the area to be analyzed is delineated with high-contrast apertures in the remote image planes of the sample. Preferably, these apertures are variable in size so that the area of interest may be "zeroed in on." Since the optical geometry is the same for visible evaluation and infrared detection, the diffraction effect is worse for the detection step, because the wavelength is longer. It follows that one can see the area one wants to measure more clearly than can actually be measured. For FT-IR microscopy, therefore, the spatial resolution is defined as the ability to measure the spectrum from an object delineated by the apertures without significant impurity radiation from neighboring objects. Since the infrared spectrum is quite complicated, consisting of thousands of data points, it is often difficult to say whether spurious energy has affected the result. If there is impurity, it can take the form of either spurious *narrowband impurity* from a nearby absorber, or spurious *broadband impurity* from a nearby hole. Obviously, this plays havoc with the quantitative accuracy of the measurement. Both quantitative accuracy and signal-to-noise ratio may be affected by the ability of a microscope to resolve the specimen spatially.

An interesting example is shown in Figure 2. An acrylic fiber 20 μm in diameter is embedded in polystyrene and then cross-sectioned to a 10-μm thickness. Three spectra are recorded, all with the same spectrometer conditions and

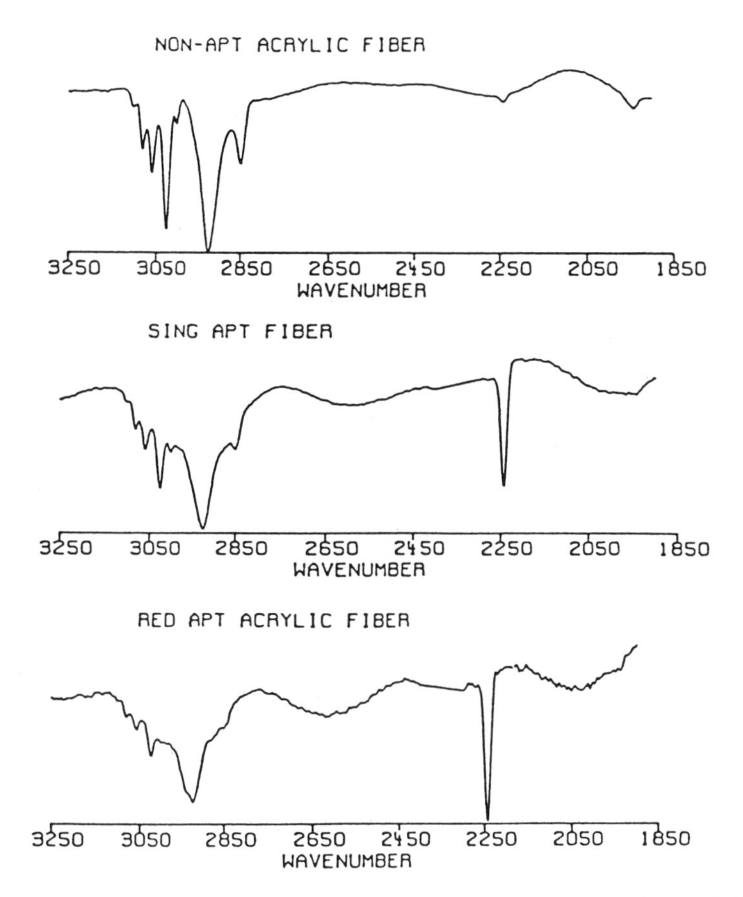

Figure 2. Demonstration of the problem of stray or spurious energy in an FT-IR microscope. The specimen in all three spectra is an acrylic fiber embedded in polystyrene. Top, wide apertures. Middle, single aperture defining the fiber area. Bottom, confocal apertures defining the fiber area.

for the same length of time. In the first spectrum, no particular effort is made to aperture out the surrounding embedding medium. In the second, one variable aperture is used to delineate the fiber. In the third spectrum, two apertures, one in the rear focal plane of the objective and one in the rear focal plane of the condenser, both delineate the same area of the sample. Quite obviously, the first spectrum has the best signal-to-noise ratio, but it is predominantly the spectrum of the embedding medium. It shows virtually no –C≡N absorption from the fiber and shows intense aromatic bands from the polystyrene. The singly apertured spectrum shows a stronger ν_{CN} vibration. But the result that best represents the fiber is the third one, even though it is noisier. Since the measured area is larger, a poorly spatially resolved spectrum appears to be better. But the ν_{CN} band at 2243 cm^{-1} is attributed to the fiber only. Therefore, the result where this band is strongest relative to the other bands is the most accurate result. This configuration using two apertures to derive the measured area is called the *confocal microscope.* This configuration is defined in the optical microscope literature as well.

3. CONFOCAL APERTURING

To describe the theory behind the operation of confocal aperturing, one must look at the diffraction theory of light. Specifically, the procedure is to ignore the imaging of the specimen in the microscope, and instead, talk about the imaging of the sharp-edged, high-contrast apertures which are used to delineate the sample. Of course, the sample can also play an important role in the performance of an infrared microscope, but for now, it will be ignored. The key to confocal aperturing is that the first aperture is actually defining the area of the sample that is illuminated. That is, the sample is nonluminous, and the system is only "lighting up" a small area, rather than illuminating the full field, as is done when viewing the sample. Therefore, one must know the shape of the diffraction-limited image of the aperture in the specimen plane.

3.1. Diffraction Imaging in the FT-IR Microscope

Remote image masking is used in an infrared microscope to provide spatial definition between two regions of a specimen. An object situated on the stage of a microscope is brought into focus and observed visually. The area of the specimen from which the infrared spectrum is desired is selected by means of an aperture at the back image plane of the microscope objective. As with any optical system, the imaging between the specimen and the image plane of the specimen is not ideal and is affected by diffraction. One can therefore think of the imaging of the remote aperture into the specimen plane as being "blurred" or modified by diffraction. The extent to which diffraction causes a problem in various optical systems has been studied extensively, and equations to quantitate the phenomenon have been available for a long time [7, pp. 370–455]. These equations deal with several special cases of imaging. First, for the limiting case where an optical system with a circular aperture stop is fed by light from a point source, the diffraction is described by

$$y = \left[\frac{2J_1(x)}{x}\right]^2 \tag{2}$$

where $J_1(x)$ is a first-order Bessel function and x is a function of numerical aperture (N.A.) and wavelength such that the first minimum occurs at 0.5λ/N.A. This result was first obtained by Airy [8] and the resulting pattern is known as an Airy disk (see Figure 18). Similar exact solutions are known for imaging involving optical systems with rectangular aperture stops [7, pp. 370–455].

In the infrared microscope, however, the slit or aperture is often larger than this. For instance, in the analysis of a 50-μm-diameter fiber stretched across the field of view, a remote aperture would be set to fit to the edges of the fiber such that no light could be seen around the fiber through the microscope. The imaging of this aperture in the specimen plane can then not be described by equation (2) since it is not infinitely narrow. There is a need to quantitate the percentage of infrared light that will miss the fiber due to diffraction, and contribute to spurious energy in the spectrum. In this paper, therefore, results are developed for the imaging of two parallel knife-edge apertures as they are brought closer together, approaching, in the limit, the infinitely narrow result.

Visible-light microscopes are often designed with a diffraction-limited back image plane not only behind the objective but also behind the condenser. The image plane behind the objective is imaged onto an eyepiece to provide additional magnification and to allow viewing of the sample. On the other hand, the condenser is used only to fully illuminate the field of view. The condenser often does possess a variable aperture in its back image plane, but this is used mainly as an aid to focus the condenser properly and hence provide uniform illumination of the full field. The optical quality of the condenser does not contribute to the quality of the image in a normal microscope as long as the field and the aperture of the objective are being filled uniformly [7, p. 522]. It is because of this fact that many early infrared microscopes were designed with nonimaging off-axis mirrors used as the condenser, since it was felt that imaging was not important here. Later it was realized that while the condenser quality does not contribute to visual spatial resolution per se, it does contribute to spatial definition if it is used to image another set of apertures into the specimen plane. The two superimposed patterns resulting from two sets of apertures of corresponding size in two different remote image planes causes the response function to be the product of the intensities produced by each set of apertures. It is then necessary to calculate the response function for a single set of aperture blades as the blades are brought closer together. The literature already provides us with the solution in the two limits [7, pp. 370–455]: (1) where the slit is infinitely narrow, and (2) where the aperture edge is so far removed from the other edge that no interference occurs. Unhappily, with the infrared microscope, one is generally working in the murky region in between, where the exact results have not been developed. An assumption must also be made about the coherence of the light being used to image the apertures, since the result depends on this. For the sake of simplicity, the assumption is made that the light is perfectly incoherent. In this chapter we then investigate the extent of the improvement in spatial definition expected through the application of dual remote image masking. This improvement, it will be shown, is not constant over all aperture widths.

The preceding discussion of the use of dual remote image masks to improve spatial definition would suggest that the two apertures, the source side aperture and the detector side aperture, are equivalent. From the standpoint of diffraction this is true. However, it is found in practice that the aperture before the specimen (the source-side aperture) is more important in achieving spatial definition in many situations. This is due to two factors. First, flooding the specimen plane with energy from an extended source potentially gives rise to glare, reflection, and scatter from areas of the specimen which are out of the field of view. This was recognized in the optics field in the 1950s and was named the Schwarzchild–Villiger effect [9]. Second, if the specimen is scattering or diffuse, the image information can be scrambled. If the aperture is after the specimen, this scrambling will not allow effective spatial definition. Since many samples induce some degree of scatter, if only one aperture is used, it should also be on the source side. This finding is counter to the practice of almost all previous microscope-photometers, but is consistent with the work of Schwarzchild and Villiger. In visible-light instruments, of course, sample morphology is more heavily controlled, so this constraint may be less important there.

It was mentioned earlier that the energy from an infinitely small point source, imaged through an optical system, cannot reimage to its original infinitely small dimensions because of the diffraction of light. The equation to describe the spreading out of this point source was given in equation (2). Using similar experimental treatment, it was discovered by Fresnel [10] that an infinitely narrow slit also possesses a characteristic diffraction pattern. In fact, every object in an optical system, especially those features that demarcate a large change in contrast, contribute a diffraction signal to the resultant image. Exact mathematical relationships exist, however, for only the simplest geometrical shapes in an optical system. The equation to describe the diffraction at an infinitely narrow slit is given by

$$y = \left[\frac{\sin x}{x}\right]^2 \tag{3}$$

where x is related to the N.A. and the wavelength in such a way as to place the first minimum of the function at 0.61λ/N.A. The graph of the point-source response and the slit response is shown in Figure 3. In both these cases, the area of the mask where light can transmit through the mask is assumed to be perfectly transmissive. The area where light is blocked by the mask is assumed to be totally opaque. In other words, the mask defines a large change in contrast, and this is where the magnitude of the diffraction phenomenon is largest. This is also the proper situation for an infrared microscope, where the masked-off region should be totally opaque to visible, and more important, infrared energy. These equations also assume that the slit or hole is infinitely thin. Obviously, it is physically impossible to have an infinitely thin mask that is also opaque. The phrases *infinitely thin* or *infinitely narrow* are used here to mean that the dimension in question is small relative to the least resolvable separation (LRS) of the optical system, which is defined as

$$\text{LRS} = \frac{0.61\lambda}{\text{N.A.}} \tag{4}$$

where λ is the wavelength of the energy and N.A. is the numerical aperture of the optical system at the position of the mask. The numerical aperture is defined

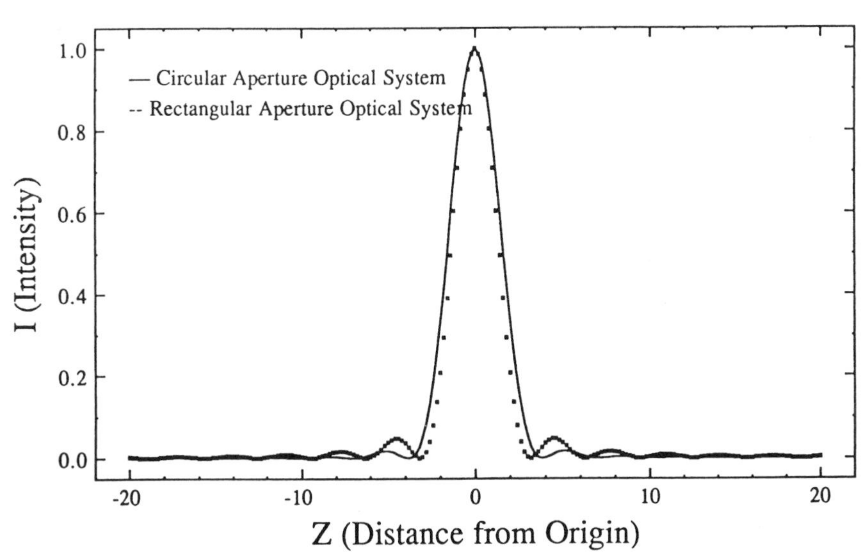

Figure 3. Point source and slit source response.

as

$$\text{N.A.} = n \sin u' \tag{5}$$

where n is the refractive index of the medium and u' is the semifield angle of the optical system at the position of the mask.

It is worth noting that in none of these equations regarding diffraction or throughput is the magnification of the system a direct factor, a point that has been stated incorrectly in the literature. Figure 4 shows that there is no dependence of magnification on the signal or throughput of the optical system. Numerical aperture is the variable that directly affects throughput, which would be predicted from equation (1) and which is shown in Figure 5. The misconception regarding the importance of magnification arises from the fact that light-microscope objectives of high magnification also have high N.A., in order to provide improved spatial resolution. The magnification of an FT-IR microscope objective should be chosen to be high enough to allow convenient visual masking of the desired area of the specimen at the remote field stop (15 or 32x is generally sufficient). On the other hand, N.A. should always be as large as practical.

A microscope system has now been developed that uses a unit magnification optical system for imaging the image masks into the specimen plane and uses

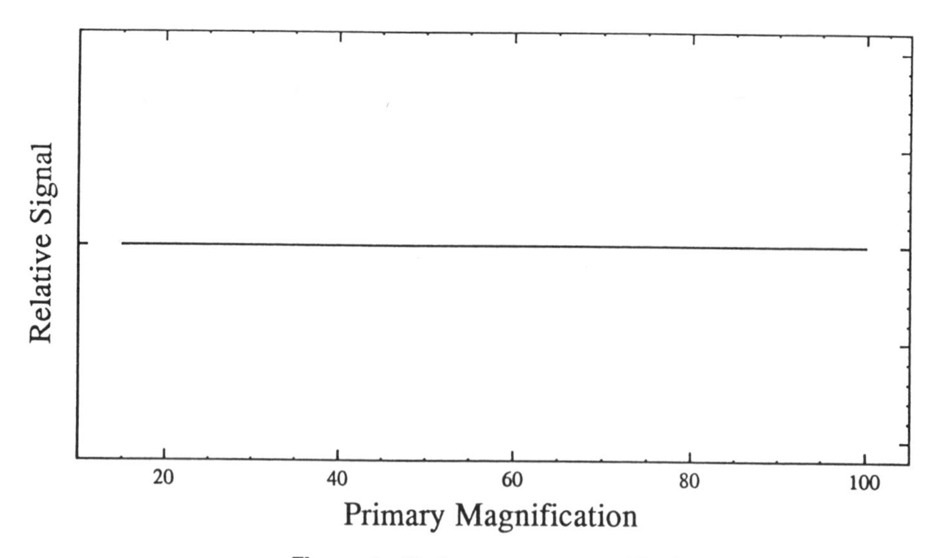

Figure 4. Performance vs. magnification.

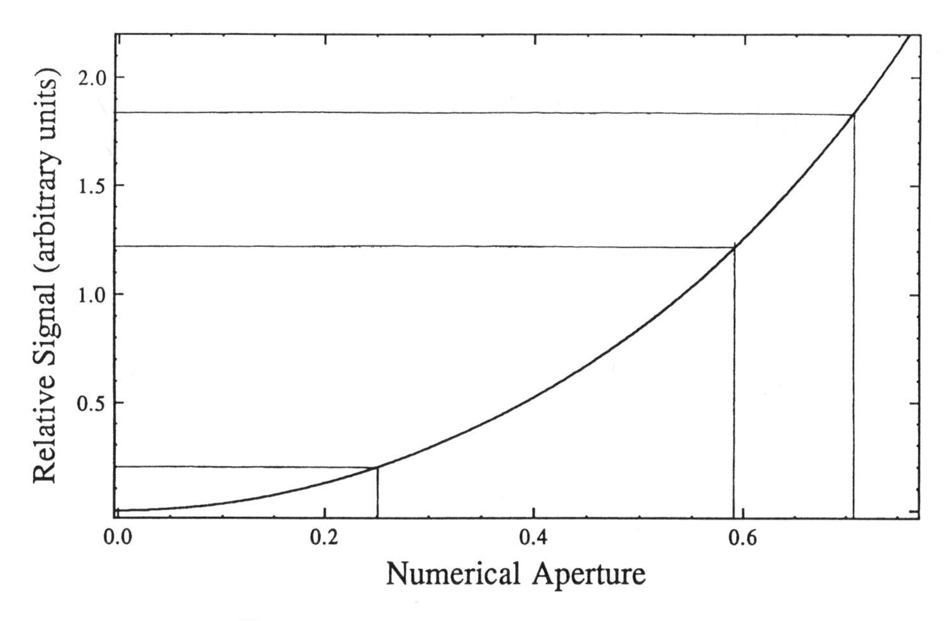

Figure 5. Performance vs. numerical aperture.

a higher-magnification refractive optical system to view the specimen and the mask. This system allows the use of inexpensive off-axis mirror pairs that can be diffraction limited for small field stops, and inexpensive refractive optics that work well with wide fields in the visible region. Another advantage of this system is that the numerical aperture can be made quite a bit larger than with Cassegrainian optics. As can be seen from the diffraction equations, and as discussed later, numerical aperture is an extremely important factor in the performance of an infrared microscope.

Also available in the literature is the mathematical treatment to describe the diffraction at a high-contrast straightedge [7, pp. 370–455]. This can be thought of as removing one blade of the infinitely narrow slit and allowing energy to flow past the remaining one high-contrast blade.

Several other mask and aperture shapes have been characterized in the literature, including toric apertures [11]. These will not be dealt with here, even though most FT-IR microscopes do have toric pupil stops. The effect of the toric pupil stop is to throw energy out of the central bright maximum of the Airy disk and into the wings, thereby reducing contrast and resolution.

As stated earlier, exact mathematical treatments are to be found in the literature for infinitely narrow slits and unobstructed high-contrast edges. Most spatial definition problems for FT-IR microscopy fall somewhere in between these two limits. Therefore, an analytical method has been developed to calculate the diffraction patterns for finite slit widths. This will then be useful in evaluating the expected limits of spatial definition as a function of slit width and wavelength, in the operating regions of the FT-IR microscope. Next, this methodology will be used to predict the improvement to be gained from dual-image masking.

The image of an unobstructed high-contrast edge is defined mathematically as

$$\int_0^\infty \left[\frac{\sin x}{x}\right]^2 dx \tag{6}$$

which is simply the integral of the expression for an infinitely narrow slit. Again, x is a function of wavelength and numerical aperture. It can be translated into such by knowing that the first minimum of this sine function occurs at $0.61\lambda/\text{N.A.}$

It is intuitive that one may achieve a graph for finite slit widths through a process of integration over a finite region, as depicted in Figure 6. This integration has been performed for a model 0.50 N.A. optical system, for slit widths ranging from $\frac{1}{2}$ a wavelength up to 5 wavelengths, and the curves are shown in Figure 7. These are all plotted on the same x,y-axis scales for simple comparison. The ideal square-wave function, which would prevail in the absence of diffrac-

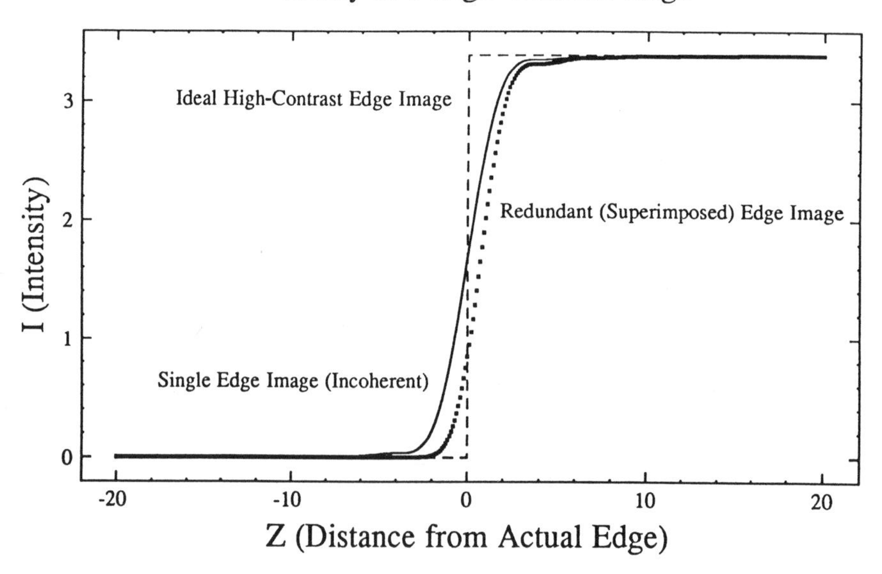

Figure 6. Two unobstructed high-contrast edge graph.

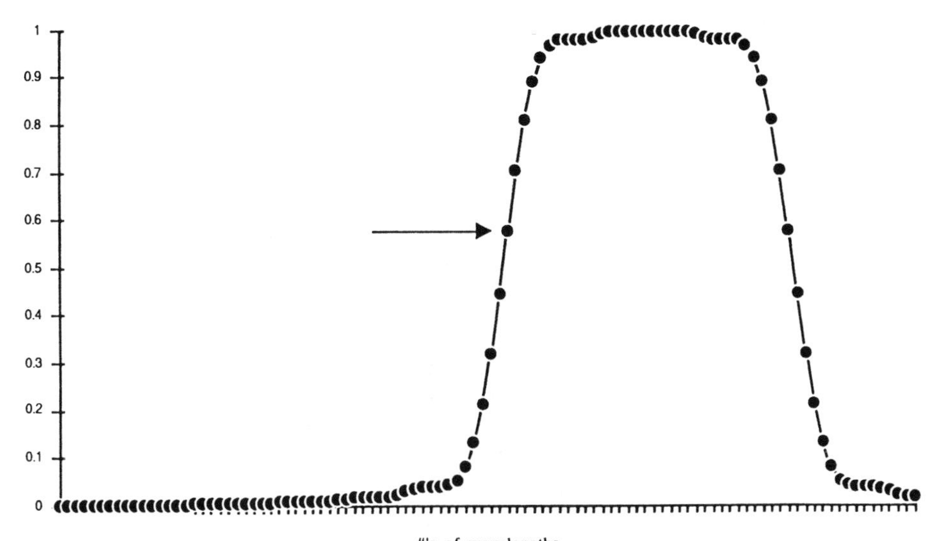

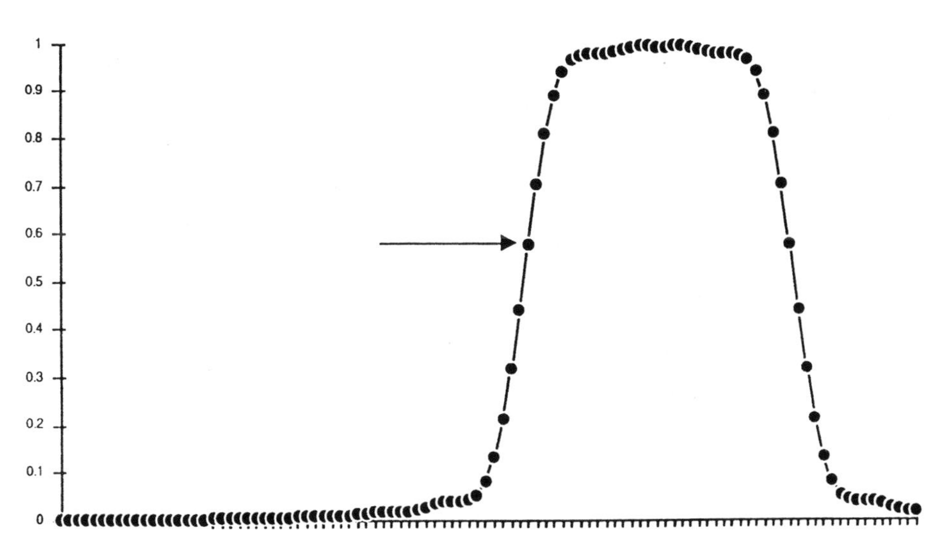

Figure 7. Various slit widths.

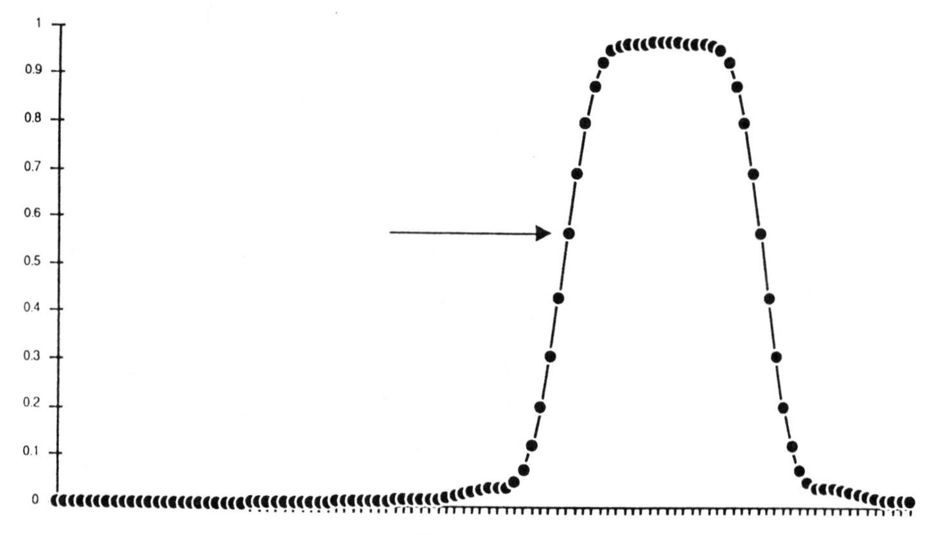

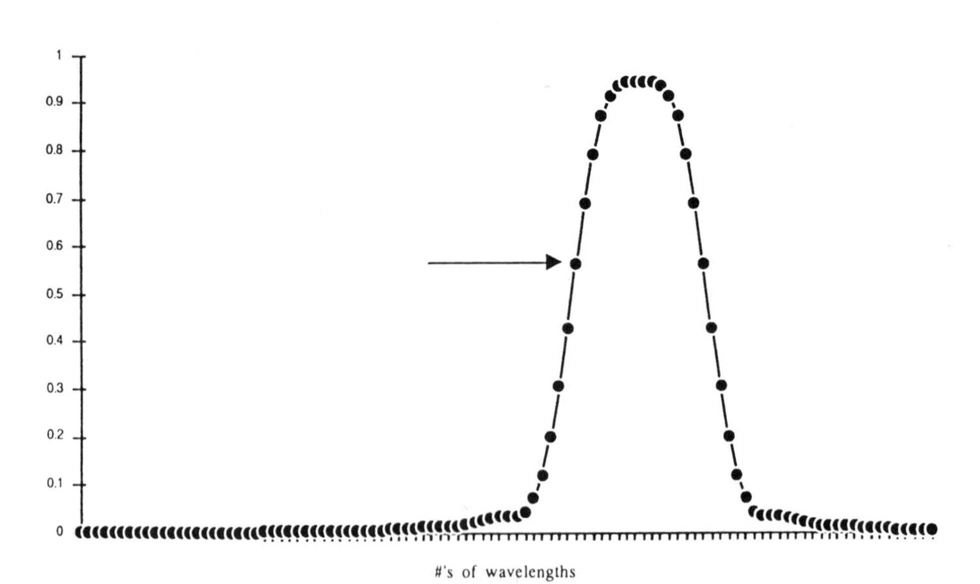

Figure 7. Various slit widths (*continued*).

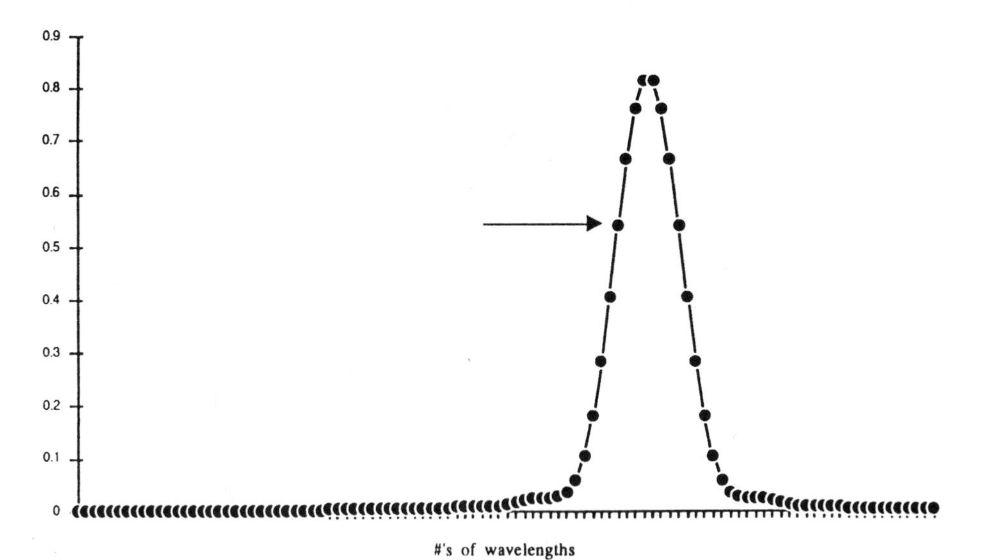
One Wave Slit Width—Single Aperture
0.9
0.8
0.7
0.6
0.5
0.4
0.3
0.2
0.1
0
#'s of wavelengths

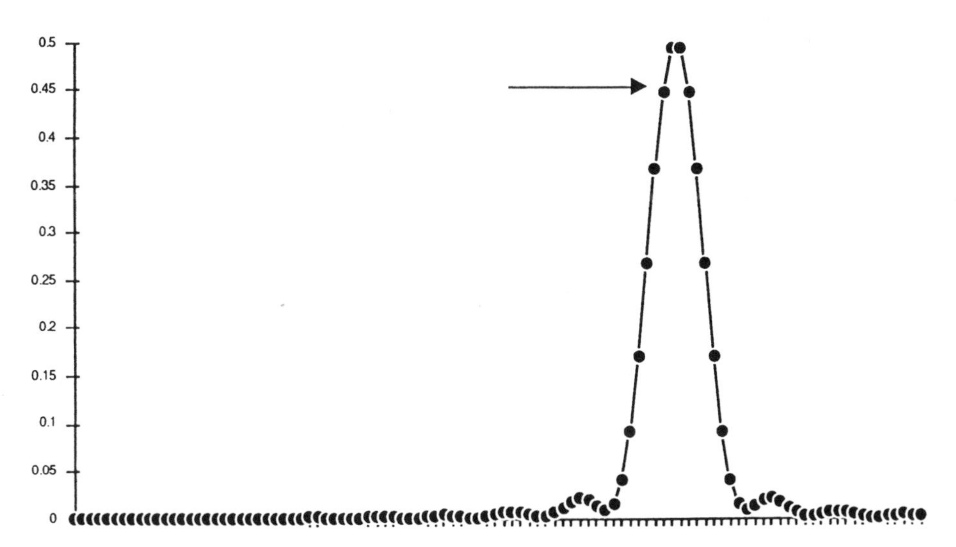
Half Wave Slit Width—Single Aperture
0.5
0.45
0.4
0.35
0.3
0.25
0.2
0.15
0.1
0.05
0
#'s of wavelengths

tion, has been plotted as well. This square wave is more like what would be seen looking through the microscope visually, since the diffraction is much less prevalent at the shorter visible wavelength.

For a 5-wavelength slit width, it is seen that at the center of the pattern, the energy has recovered from any shadowing from the aperture edges and is transmitting unit intensity. The falloff in energy at the edge is gradual, passing through the position of the actual edge at an intensity of about 0.6. Then there is a good deal of energy outside the ideal profile, gradually tending toward zero but actually having infinite extent. This pattern looks very similar to that for two unobstructed high-contrast edges. It is concluded that at a separation of 5 wavelengths, the two aperture blades have no effect on each other. At a 4-wavelength slit width, this also appears to be true. At a 3-wavelength slit width, the energy at the midpoint between the two blades is starting to fall away from unity. The severity of this effect continues to increase as the blades are brought closer together. At the same time, the percentage of energy falling outside the ideal square-wave function continues to increase. Remember that it is this energy which will impinge in the specimen plane of the microscope outside of where the area-defining mask edges have been placed. Upon reaching a $\frac{1}{2}$-wavelength slit width, the intensity through the system has fallen off severely, and the percentage of spurious energy is tending toward infinity. Graphs can now be drawn which show the percentage of spurious energy. To have these graphs make sense in a real-world sort of way, the results are plotted for two different optical systems, one having a numerical aperture of 0.25, and the other, 0.5. With these sorts of graphs, shown in Figure 8, the user can decide on an acceptable level of spurious energy for a given experiment, and immediately know how small an object can be analyzed while still achieving that level of spurious energy. Obviously, some experiments will call for greater assuredness of spatial definition than others. If one is trying to measure the spectrum of a weakly absorbing compound, and a strongly absorbing compound is just outside the field of view, only a 5% spurious energy may be acceptable. This would not be achieved for specimens smaller than about 5 wavelengths, or 50 to 100 μm. On the other hand, if there is no absorber next to the specimen and a little stray light in the measurement is acceptable, one can close the slits down to a fraction of a wavelength. But note also that the overall width of the pattern is not really decreasing below about 3 wavelengths wide. The maximum intensity is falling off, however. So for an isolated small object where one can live with stray light, the signal-to-noise ratio of the measurement will be best if the slits are left open to about 3 wavelengths (for this model 0.50-N.A. system).

An additional set of aperture blades at a remote image on the other side of the specimen, closed down to the same extent as the first set, further improves spatial definition. In imaging applications using light microscopes, this configuration using two apertures, called confocal imaging, was first proposed by

Minsky [12]. In the infrared microscope, the use of two apertures was first proposed by Messerchmidt [13] and first implemented by Spectra-Tech, Inc. In this chapter an attempt is made to quantitate this level of improvement through the use of what is called *dual remote image masking*, trademarked by Spectra-Tech as Redundant Aperturing. The mathematical treatment that is appropriate to deal with two superimposed sets of area-defining apertures is the product of the two single-aperture patterns. This makes sense intuitively if one thinks of the image of each set of apertures as setting up a gate through which energy from the source must travel. Note that for these graphs to have a basis in reality, all other optical aberrations in the system must be small enough that diffraction is the limiting cause of loss of spatial definition—the proverbial diffraction-limited system. It is interesting to note that most of the older microscope photometry literature in fact teaches against using two apertures closed down to the same extent, because of the decrease in energy [14]. But what was not noticed in the earlier work was that this loss of energy, which can be predicted from the following curves, is not heterogeneous across the field. A higher percentage loss occurs from the undesired regions than from the desired. Dual remote image masking describes the improved spatial definition which arises even when the pinholes or slits are not infinitely small or thin.

A set of curves similar to Figure 7 can be constructed, but for a dual masked system. This is accomplished in Figure 9. Note again that for a 5-wavelength slit width, it is seen that at the center of the pattern, the energy has recovered from any shadowing from the aperture edges and is transmitting unit intensity. The falloff in energy at the edge is again gradual, now passing through the position of the actual edge at an intensity of about 0.3, lower than the singly apertured system. There is still energy outside the square wave which defines the ideal response in the absence of diffraction, but it is much less than the single-apertured result. Again the maximum intensity starts falling off for about a 3-wavelength slit width. More dramatic difference between the singly and dual masked results occurs for the smallest slit widths. Note that for a given slit width, the dual masked system transmits less energy than does the singly masked system. To put this in perspective, Figure 8 again shows the percentage of spurious energy relative to the total energy for various slit widths, for both N.A. = 0.5 and N.A. = 0.25 systems. The curves for single and dual masked systems are presented on the same graph for simple comparison. The results are dramatic and agree with experimental findings from this laboratory. Also in Figure 8, lines have been drawn at the level of 95% confidence of exclusion of spurious energy in both single- and dual-masked situations. Comparing N.A. = 0.50 systems, the 95% confidence level for rejection of spurious energy occurs at a 2-wavelength-wide slit width for the dual-masked system and at a 5-wavelength-wide slit width for the single-masked system. Next, comparing dual-masked systems, the 95% confidence level for rejection of spurious energy oc-

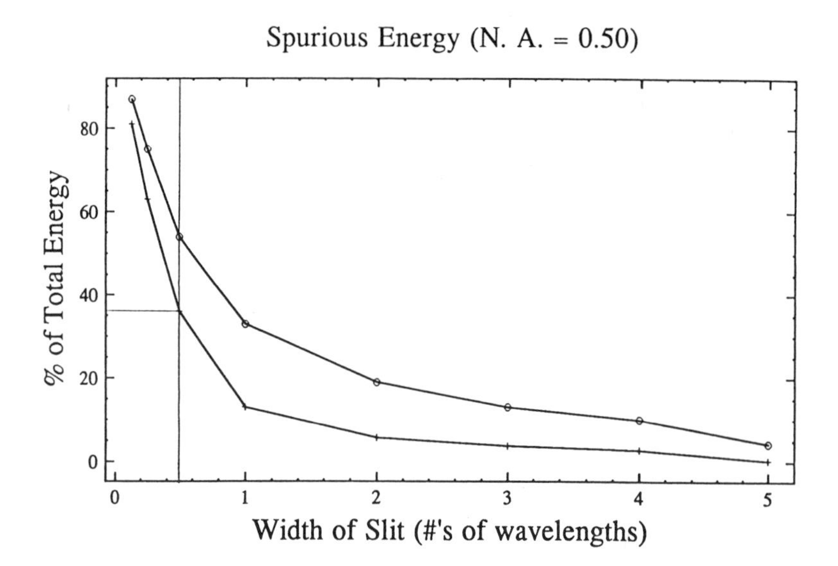

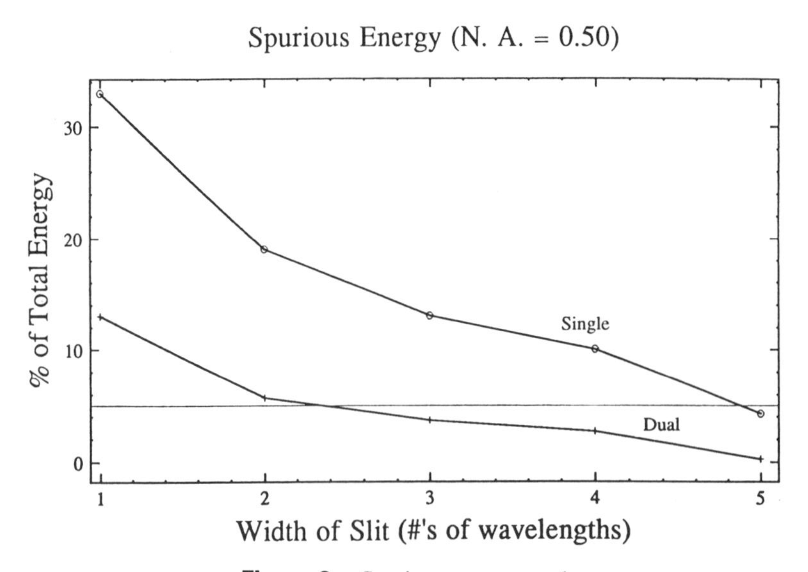

Figure 8. Spurious energy graphs.

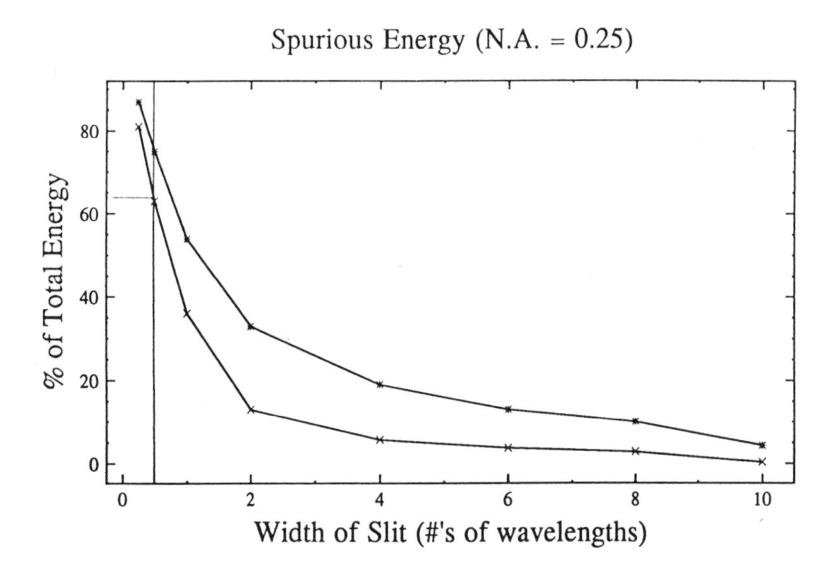
Spurious Energy (N.A. = 0.25)
% of Total Energy
80
60
40
20
0
0
2
4
6
8
10
Width of Slit (#'s of wavelengths)

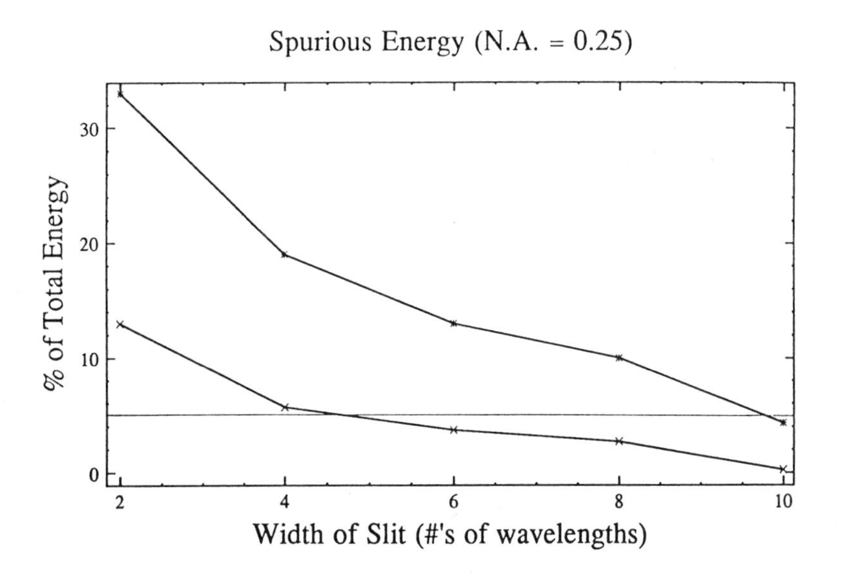
Spurious Energy (N.A. = 0.25)
% of Total Energy
30
20
10
0
2
4
6
8
10
Width of Slit (#'s of wavelengths)

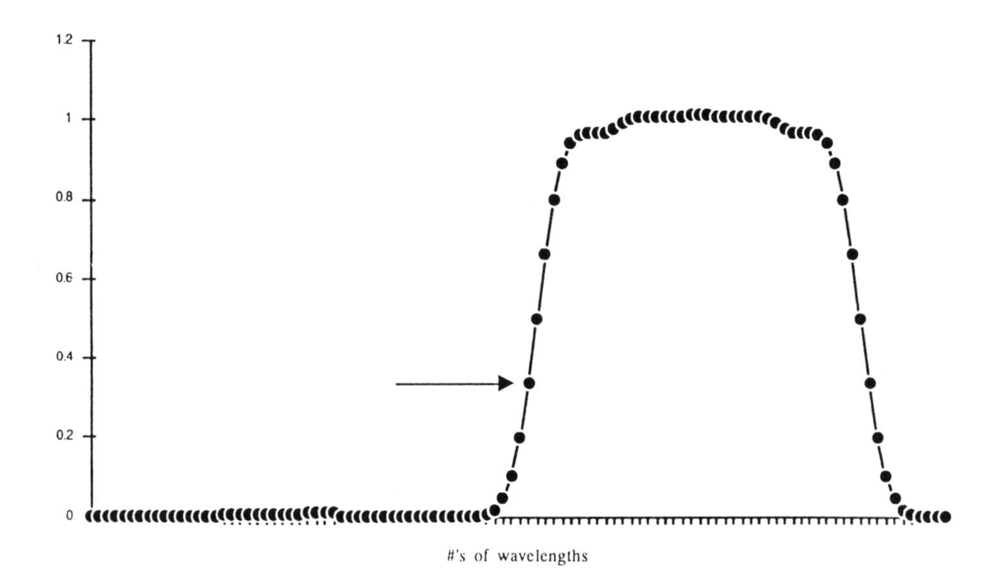

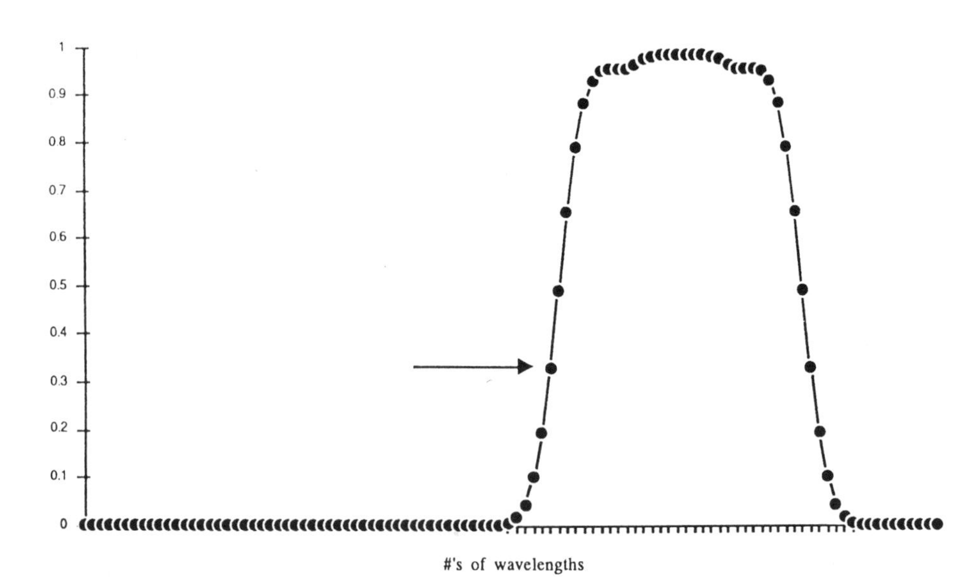

Figure 9. Slit widths—redundant.

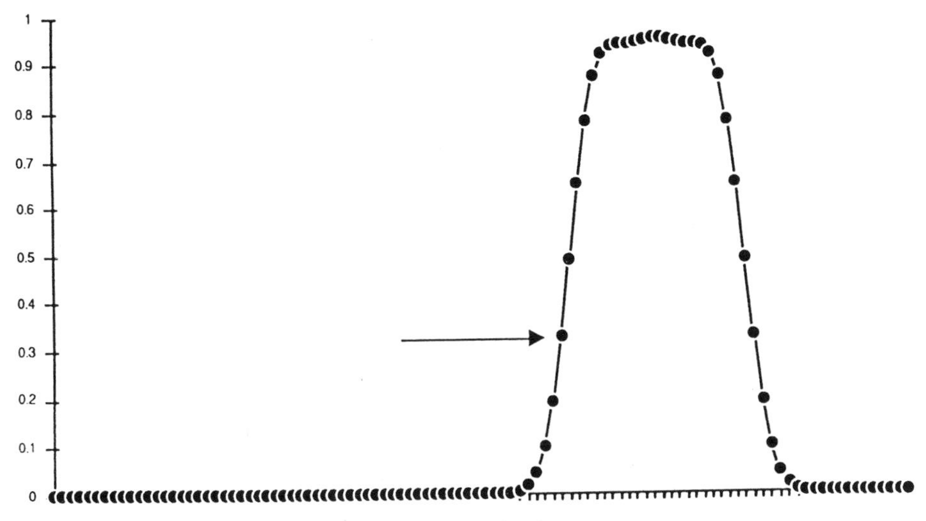
Three Wave Slit Width—Redundant Apertures
1
0.9
0.8
0.7
0.6
0.5
0.4
0.3
0.2
0.1
0
#'s of wavelengths

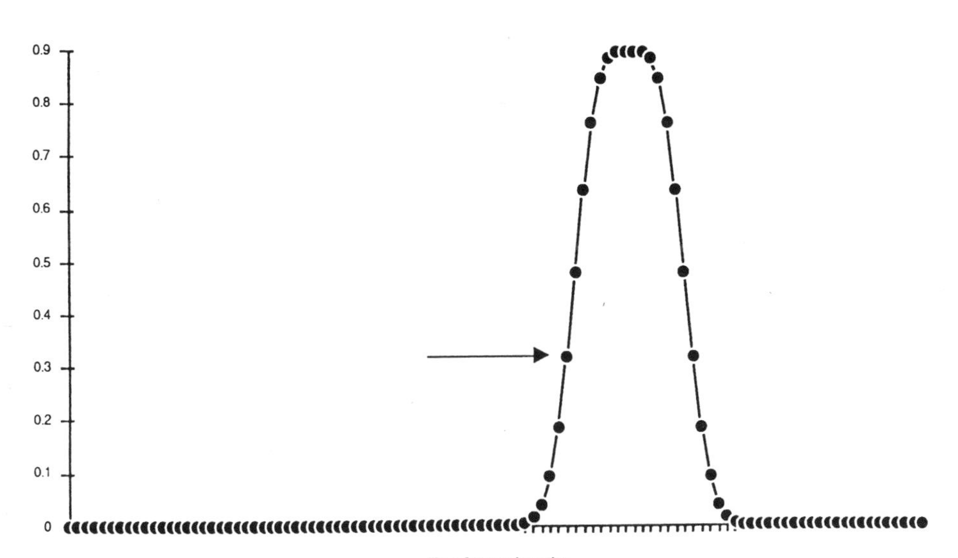
Two Wave Slit Width—Redundant Apertures
0.9
0.8
0.7
0.6
0.5
0.4
0.3
0.2
0.1
0
#'s of wavelengths

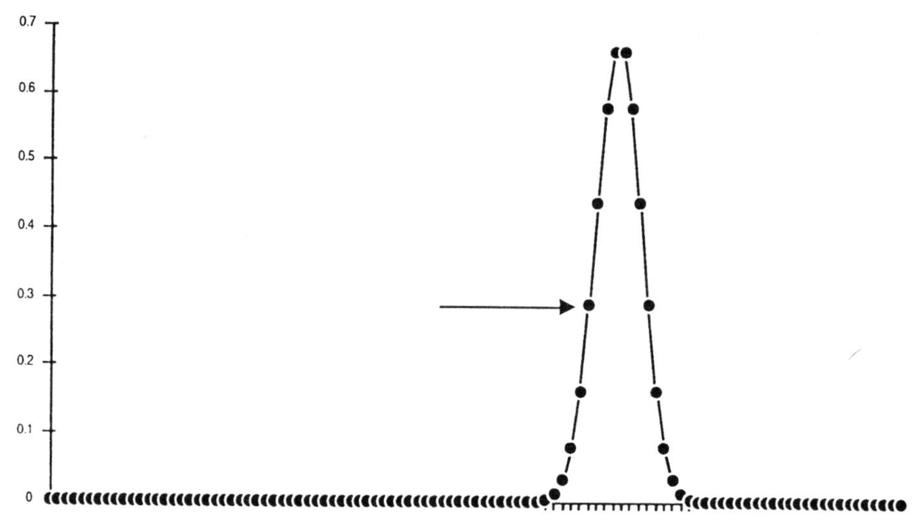

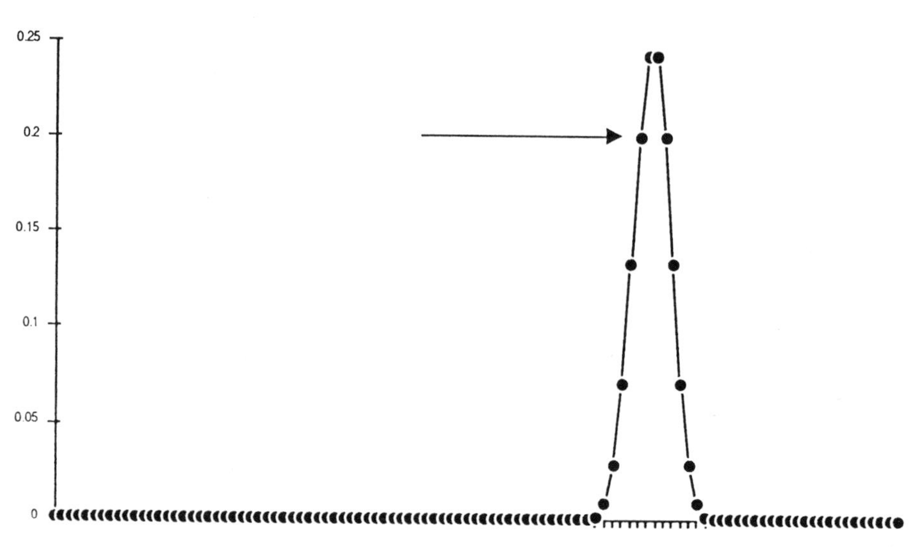

Figure 9. Slit widths—redundant (*continued*).

curs at a 2-wavelength-wide slit width for the N.A. = 0.50 system and at a 4-wavelength-wide slit width for the N.A. = 0.25 system. Also, we know from Figure 5 that the N.A. = 0.50 system lets through 6 times the energy of the N.A. = 0.25 system. So a dual-masked N.A. = 0.50 system can analyze samples five times smaller than a single-masked N.A. = 0.25 system with an equivalent level of spurious energy, and it can do so while allowing as much as six times the signal to get through. Note that if one tries to measure a 10-μm object—say, a single blood cell—with a single-remote-aperture N.A. = 0.25 system, then at the far end of the midinfrared, no spectral band from that cell will reach below 80% transmittance. This is obviously not a workable region for such a system. On the other hand, a dual-masked system employing a 0.5 N.A. objective would allow spectral features to reach 40% transmittance at the far end of the midinfrared. This would still not be a quantitatively accurate measurement but would be much better.

The ideal imaging of an edge in the absence of diffraction is a step function. Note that the energy which exists in the geometrical shadow of the edge is undesirable since it will illuminate an unwanted portion of the specimen. What is left to do is to superimpose a similar output function in the image plane after the specimen. The resultant energy profile at the detector is the product of these two functions (Figure 10). Of course, further assuredness of spatial purity can be obtained by overaperturing. This is accomplished by deliberately closing

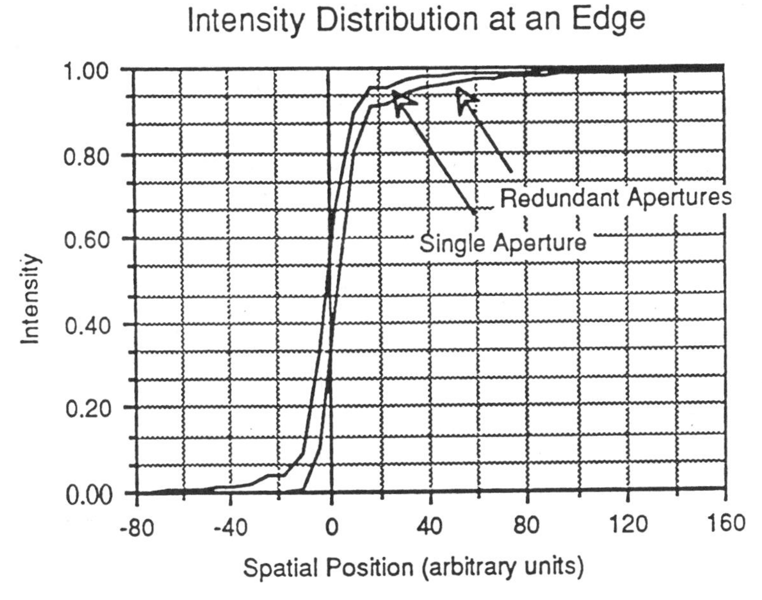

Figure 10. Intensity profile at a high contrast edge, single and dual image masking.

down the remote field apertures beyond the desired area of the specimen. In the size limit, however, one is often energy limited and can not afford to throw away energy. In this situation, it is much better to provide careful aperture redundancy to the edges of the area of interest. In some cases, for instance where the surrounding medium is known and can be subtracted, the second aperture may be neglected in the interest of speed.

Confocal apertures are, however, only one of many important considerations in the design of a quantitatively and qualitatively accurate, high-sensitivity instrument. Many of the other aspects were addressed in the many papers written about such instruments in the late 1940s and early 1950s [1–4].

4. NUMERICAL APERTURE

Chief among these details is the need for a high-numerical-aperture optical system which is also free of aberrations to below the diffraction limit, over a reasonable field. The numerical aperture of an optic is proportional to the sine of the angle of the most extreme ray accepted by the system. Typically, microscope objectives are highly corrected optical devices, employing combinations of many lenses in the design to cancel out optical aberrations such as spherical, chromatic, coma, and field curvature. Usually, field curvature is the least bothersome, since a microscope is normally used in a mode where an object of interest is brought to the center of the field of view. Of course, for use in both the visible region and the midinfrared, our choice of lens materials is extremely limited. Also, since the wavelengths of interest span a large range, the compensation for chromatic aberration becomes a more difficult problem. For these reasons and some others, it is desirable to consider the use of all-reflective optics in an FT-IR microscope.

The all-reflecting surface concept presents no major problems for the optical designer, except in the area of the objectives, and even here there is some historical information. Reflective surfaces have been *de rigueur* in the area of telescopy for many years [15]. The major concern with on-axis mirror systems is that the image falls in the path of the incoming energy, causing a partial obscuration of energy. Systems design must take this problem into account, minimizing the percentage obscuration. The result is a limitation on the number of good solutions to the design. As early as the beginning of the twentieth century, the all-reflective concept was applied to microscopy. Schwarzchild deduced a class of reflective microscope objective designs were both surfaces are spherical [16]. A number of years later, Burch [17] determined that for numerical apertures greater than 0.5, making one or both of the surfaces aspherical would improve performance. These designs still for some reason are colloquially called Schwarzchild configuration.

Designwise, an objective for photometric as well as visual use differs from a visual-only objective. In microscopy, the concept of "empty magnification" is used to explain the relationship between numerical aperture and magnification. For a given magnification, all available objectives from all manufacturers are virtually identical in their numerical aperture. Even though numerical aperture is proportional to the resolution of the system, resolution is also limited by the field angle subtended by the eye. Therefore, after a point, the increased resolution is "empty" or useless, and increasing the numerical aperture serves only to complicate the design. Similarly, a continued increase in magnification without increased aperture results in a larger image which does not contain more detail about the specimen. In photometric applications, however, increased aperture serves to increase the energy throughput of the system. This optical throughput, Θ, is directly related and proportional to the signal-to-noise ratio in the spectrum. Figure 11 shows the relationship between signal and numerical aperture for a microspectrophotometer system. Even for moderate magnification objectives, very high numerical apertures are desirable. A good value is N.A. = 0.5, which means that the most extreme ray is at 30° from the normal. A better value is 0.71, or 45°. Much beyond this, other problems occur, such as extremely small working distance, aberrations, and very small depth of field.

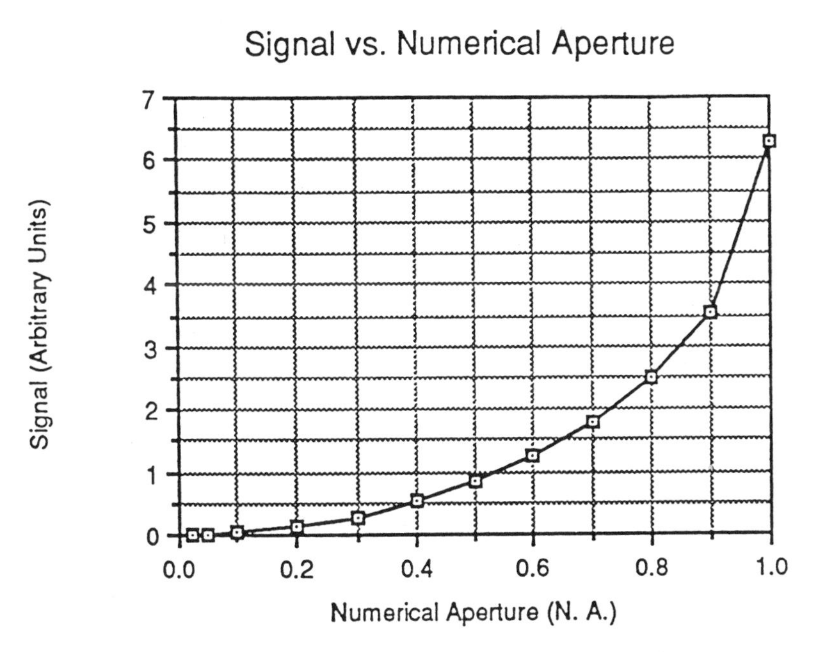

Figure 11. System performance as a function of numerical aperture.

In summary, the numerical aperture of the objective (and condenser) of the FT-IR microscope is very important to performance, for both energy throughput and spatial resolution.

Since high numerical aperture and confocal apertures contribute to improved spatial definition in similar ways, it is interesting to compare a medium-numerical-aperture system with two apertures to a high-numerical-aperture system with one aperture. Different microscopes on the market use these two approaches. There are trade-offs involved in using either two apertures or an extremely high numerical aperture. A two-aperture system is more difficult to use since the sample area must be defined with variable apertures twice (in a transmission mode). This effectively doubles the time required to set up a sample for analysis. Making a high-numerical-aperture system, on the other hand, tends to limit the working distance and impose restrictions on the thickness of the sample and substrate. High numerical aperture can be traded off for dual apertures. The choice of which is better is certainly a matter of personal preference. High-performance infrared microscopes are currently on the market that use both approaches.

5. SIGNAL-TO-NOISE RATIO

Numerical aperture as it applies to the attainment of the highest possible signal-to-noise ratio result needs further discussion. This field of discussion is called *radiometry*, the study of the collection, imaging, and detection of radiant energy. The signal-to-noise ratio of a radiometric optical system is determined by several factors, which have been formulated into an equation by Griffiths [18] and Griffiths and de Haseth [19] for the case of an FT-IR spectrometer that is not attached to a microscope device:

$$\text{SNR} = \frac{S}{N} = \frac{u_\nu (T)\Theta \ \Delta\nu \ t^{1/2} \ \xi}{\text{NEP}} \tag{7}$$

where

S = signal
N = noise
$u_\nu(T)$ = spectral energy density for a blackbody (derived by Planck)
Θ = limiting optical throughput, at either the detector or the interferometer
$\Delta\nu$ = resolution
t = measurement time
ξ = overall system efficiency
NEP = noise equivalent power of the detector

The spectral energy density, $u_\nu(T)$, is the quantity that describes the radiance of the source. The source in an FT-IR spectrometer is usually a nichrome wire or a silicon carbide material, both of which emit radiation that has essentially a blackbody profile. These sources are operated at different temperatures, depending on the manufacturer. Generally, though, the temperature is in the range 1273 to 1573 K. The spectral energy density at a given wavelength is given by the following equation by Planck:

$$u_\nu(T) = \frac{C_1 \nu^3}{\exp[C_2 \nu / T)] - 1} \tag{8}$$

where C_1 is a constant [1.191×10^{-12} W/cm^2 · sr (cm^{-1})^4], C_2 is a constant (1.439 K · cm), ν is the frequency (cm^{-1}), and T is the temperature of the blackbody (K). Griffiths and de Haseth, as well as others [20,21], have found that this equation fairly accurately correlates to the measured values for FT-IR instrumentation. This would indicate that the current hardware is working very near to its theoretical limit. This is not to say that the signal-to-noise ratio will not improve in the future. This can happen through improvements to detectors, sources, or system efficiency.

For the purpose of predicting the theoretical performance of an FT-IR microscope, two changes to this equation are needed. First, in the microscope, neither the detector nor the aperture usually limits the optical throughput. Instead, the specimen limits throughput in a general-purpose FT-IR microscope. One usually operates the FT-IR microscope in a mode where the specimen is brought to the center of the field of view, and then delineated by the variable apertures situated in remote image planes of the specimen. Normally, this specimen plane is imaged onto the detector, so that it is possible to customize the detector size for specific applications, and in that manner increase performance. This means that the throughput of the system changes based on the size of the specimen. There are two important consequences of this.

First, in the diffraction-limited region, the energy through the system is governed by the aperture size as well as the source profile. Therefore, it is imperative to measure both sample and background spectra through the same aperture size, else a sloped baseline will occur. The best way to avoid this is to measure the sample spectrum immediately after aperturing down on the specimen, then moving the specimen out of the way to record the background through the same aperture. Fortunately, this is the most convenient way to run a microscope anyway, allowing excellent compensation of water vapor, and minimizing problems from long-term instrument drift.

The second consequence is that the system can be throughput-matched for only one specimen size, and for smaller or larger specimens, the performance will not be ideal. The normal procedure is to throughput-match for a very small specimen. When a large specimen needs to be measured, the system is not optimized, but there is plenty of energy anyway.

The throughput at the specimen is the product of the specimen area (A), and the solid angle subtended by the objective from the specimen (Ω_s):

$$\Theta_s = A\Omega_s \qquad \mathrm{cm}^2 \cdot \mathrm{sr} \tag{9}$$

Let us assume that the specimen is self-luminous. The specimen can be thought of as being the source of the system. The solid angle is the portion of a sphere that a source of optical radiation radiates into when that source is at the center of the sphere. If the source radiates into the full sphere, this is equivalent to 4π steradians (sr). The solid angle may be evaluated in terms of the surface area of the portion of the sphere that is radiated into (S) and the radius of the sphere (R):

$$\Omega_s = \frac{S}{R^2} \tag{10}$$

It would be more convenient for our purpose, however, to express the solid angle in terms of the limiting numerical aperture of the microscope (usually the N.A. of the objective). Let us consider a sphere of unit radius, depicted in two dimensions in Figure 12.

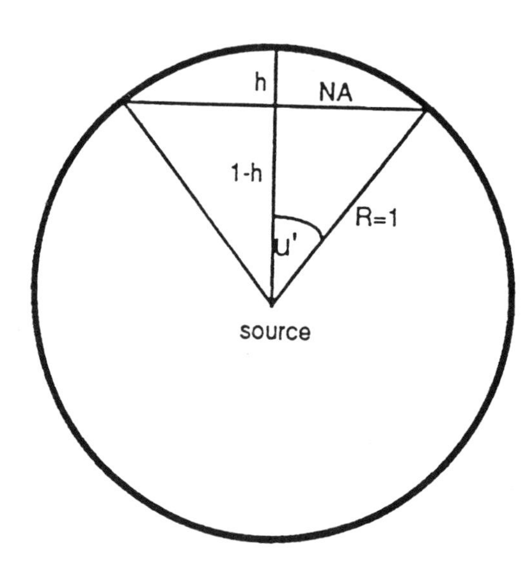

Figure 12. Construction of a unit radius sphere used to calculate solid angle in terms of the numerical aperture (see the text).

A source is placed at the center of the sphere, where it radiates into a u' semifield angle, and therefore a $2u'$ included angle. The quantity h is used in an equation which gives the surface area of the sphere irradiated by the source:

$$S = 2\pi Rh \tag{11}$$

Note that for the unit radius sphere, the value of the numerical aperture forms a side of a right triangle with a hypotenuse of unity and the other side equal to $(1 - h)$. This allows us to express h in terms of the numerical aperture by the Pythagorean relation:

$$(1 - h)^2 + (\text{N.A.})^2 = 1 \tag{12}$$

Rearranging yields

$$h = 1 - \sqrt{1 - (\text{N.A.})^2} \tag{13}$$

Therefore, Eqn. (8) can be written

$$\Omega_s = S = 2\pi\,[1 - \sqrt{1 - (\text{N.A.})^2}] \tag{14}$$

Next, a factor must be brought into the equation to predict the losses associated with the diffraction effect, because of the small sizes of specimens being observed. Smith [22] puts forth an equation that relates the irradiance at the center of the Airy disk pattern (H_0) to the total power in the Airy disk (P).

$$H_0 = 3.5P(\text{N.A.} \div \lambda)^2 \tag{15}$$

where λ is the wavelength. This equation predicts an efficiency due to diffraction of 1% for our system when the object being measured is small compared with the wavelength of light. The FT-IR microscope, however, is also used for larger samples where the diffraction effect still lowers system efficiency, but not to the extent predicted by Smith's equation.

Rather than developing the general equation for the efficiency due to diffraction for an FT-IR microscope, a graph of efficiency versus sample size is developed for rectangular samples in the IR-Plan microscope with the 15× magnification, N.A. = 0.58 objective, a 2000 cm^{-1}. Under incoherent illumination, the imaging of a high-contrast edge, such as the blades of the apertures used to delineate the sample in an FT-IR microscope, is described by the function:

$$I_{\text{im}}(x) = \int_0^\infty I_{\text{ob}}\,(\xi)\,\text{sinc}^2 x\,d\xi \tag{16}$$

where I_{im} and I_{ob} are the intensity at the image and object, respectively. The graph of this function for two parallel edges, as would be used to delineate a specimen, is shown in Figure 13. It is seen that the energy falling outside the

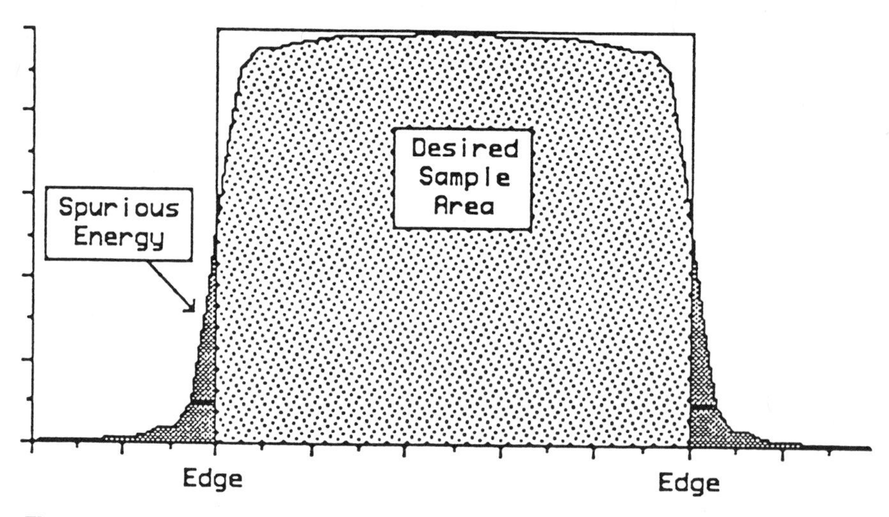

Figure 13. Imaging of two parallel high-contrast edges in the specimen plane of an FT-IR microscope.

rectangular area of the graph is lost due to diffraction. This energy is imaged outside the desired area, and therefore manifests itself as stray light. If redundant aperturing is used, much of it is effectively blocked by a second aperture. In either case, since this energy does not go through the part of the specimen that one wishes to sample, one should consider it as reducing the system efficiency. As the size of the specimen approaches the wavelength of light, this efficiency approaches zero. Figure 14, the graph of efficiency due to diffraction, can be constructed based on the percentage diffracted area demonstrated in the preceding graph.

Therefore, a new term is introduced into Griffith's equation δ, which is the efficiency of the system as influenced by diffraction. This value can be read from the graph Figure 7. So a complete equation for the theoretical signal-to-noise ratio in an FT-IR microscope can now be written:

$$\mathrm{SNR} = \frac{S}{N} = \frac{u_\nu(T)\Theta\ \Delta\nu\ t^{1/2}\xi\delta}{\mathrm{NEP}} \tag{17}$$

This equation can now be evaluated for any given set of conditions. Spectra-Tech uses a 100-μm-diameter object, 4 cm^{-1} resolution, and a measurement time of 2 minutes as a standard test of performance prior to shipping FT-IR microscopes. An average source temperature is 1273 K. The most commonly used detector is a 0.25-mm-square narrowband MCT detector, so an NEP of 5.5×10^{-13} is used, a typical value for this detector. For the overall system efficiency, Griffiths uses a value of 0.1. This must be reduced to take 10 extra

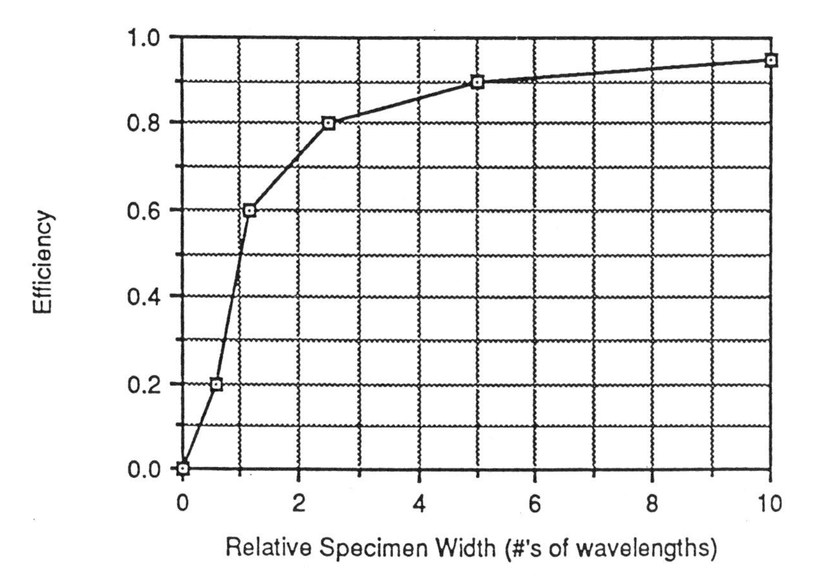

Figure 14. Efficiency due to diffraction.

mirrors into account, giving a value of 0.06. Since the sample width is 100 μm, the efficiency due to diffraction is 0.95. Given this set of data, the SNR calculates to a theoretical value of 73,000:1. This equation gives the rms noise value, which must be divided by 5 to give a peak-to-peak SNR of 14,000. This value is within a factor of 5 of the actual values measured for these conditions.

6. SAMPLE COMPENSATION

Another often overlooked problem is that the specimen to be examined under the microscope acts as an optical element itself. In the biological microscope, there is a very well-defined specimen geometry and most biological microscopes come already corrected for this. That is, the objective is designed with a 0.17-mm-thick coverglass expected and required for proper performance. The condenser is similarly corrected for the glass slide. Unfortunately, the specimen is much less well defined in the case of the FT-IR microscope. This microscope is expected to perform equally well for a cylindrical fiber suspended in air, and for that fiber squeezed between 2-mm-thick diamonds. The answer to the problem lies in the design of the Cassegrainian microscope objective and condenser [23]. This design has the nice feature that as the separation between the nearly concentric mirrors is decreased, the spherical aberration introduced by the specimen or support is compensated for (Figure 15). The chromatic aberration un-

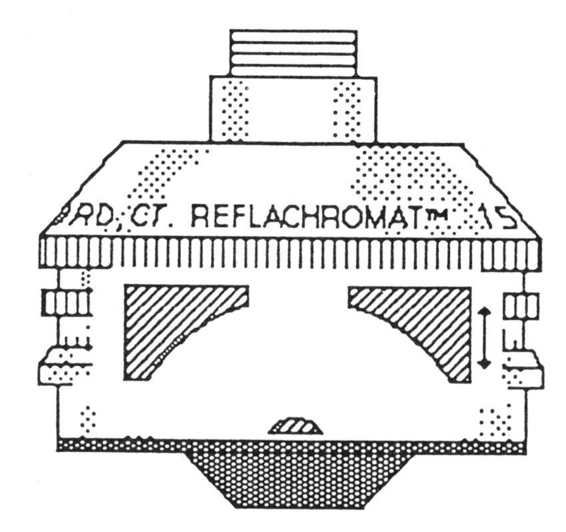

Figure 15. Optical configuration of the Schwarzchild-type Cassegrainian objective lens. The large primary mirror moves down to compensate for spherical aberrations.

fortunately remains, so it is still a good idea not to use thick or high-index specimen supports.

7. COHERENCE

The need to discuss the coherence of the light being used in the FT-IR microscope arises from the fact that image quality is affected by coherence, in situations where the diffraction limit is approached. This was realized as early as 1893 by Abbe [24], and illustrated experimentally in 1906 by Porter [25]. Rather than delving into the mathematical basis for this effect, an example would be in order. In a microscope measurement, the light is always at least partially coherent. That is, energy from neighboring points in the object (source), because of diffraction, tends to overlap and correlate. The visual result is that fringes (light and dark alternating features) are formed in the image plane of the microscope. These are considered undesirable from the standpoint of visual image quality, and an effort is made in a microscope system to minimize these features. One way to lessen the degree of coherence in a microscope is to place a diffusing glass after the source, although this is not extremely useful as a scrambler. Rotating diffusers are better, or even liquid diffusers such as milk solutions have been used. Liquid diffusers work because of Brownian motion in the milk suspension [26]. Time-averaged observation can also break down coherence, because of Brownian motion in the atmosphere. The degree of coherence is also related to the numerical aperture of the optical system and the mean wavelength of illumination [7]. The Smith–Helmholtz theorem gives rise to an equation that expresses the diameter of the coherently illuminated area of the exit pupil of a system (d'_{coh}) illuminated by an incoherent quasi-monochromatic uniform circular source:

$$d'_{coh} \sim 0.16\lambda_0 \div \text{N.A.} \tag{18}$$

where λ_0 is the mean wavelength and N.A. is the numerical aperture. Figure 16 evaluates this relationship for the IR-Plan FT-IR microscope operating in the midinfrared region. This shows that when dealing with objects at the limit of detection, that is, in the 5-μm-diameter region, the illumination in the IR-Plan microscope is fairly coherent. For larger specimens, the coherence interval falls off.

For the FT-IR microscope, one is again faced with a dichotomy. While incoherence is desired for the viewing step, an increased degree of coherence would give rise to a sharper edge function, slightly improving spatial resolution. Unfortunately, while there are several ways to destroy coherence, the only way to increase it is to lower the effective angular subtense of the optical system. Given the constraints of the type of measurement being made, this is only possible by decreasing the numerical aperture of the system. The small benefit

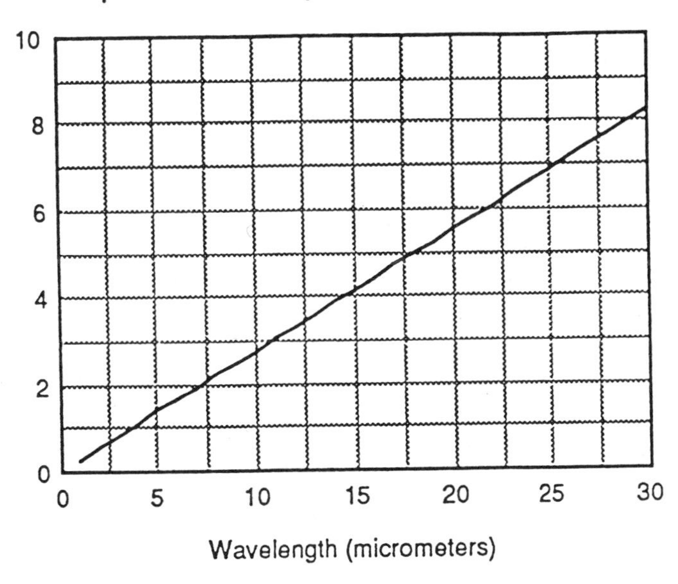

Figure 16. Coherence as it relates to the FT-IR microscope. For very small specimens, the system is almost totally coherent.

imparted would be more than offset by the increase in the size of the Airy disk and would also be undesirable from an energy-efficiency standpoint.

The temporal coherence of the illuminating energy does not affect the edge function unless chromaticity exists in the optical system (B. J. Thompson, personal communication). The implication here is that if our optical system has residual uncorrected chromatic aberration, an improved edge function for a narrower bandwidth source will be realized. Since most FT-IR microscopes are totally reflecting systems, chromatic aberration is not an issue, mirrors being totally achromatic.

8. PUPIL APODIZATION

A by-product of our need for a totally reflecting optical system for the FT-IR microscope is that the diffraction-limited Cassegrainian optics that must be used have a central obscuration (see Figure 15) which affects the diffraction imagery of the system. Figure 17 shows that the net effect of this central obscuration is to modify the Airy disk diffraction pattern, increasing the intensity in the first few lobes, and reducing the energy in the central bright maximum, while narrowing it. The Airy disk and its meaning to the imaging of a microscope system has been described previously, and the reader is referred here for more information [23].

9. CONCLUSIONS

A mathematical, graphical basis for the choice of design parameters in an FT-IR microscope has been presented, with the hope that an appreciation of these details will lead to a more successful use of the instrumentation. What is clear is that high numerical aperture and dual remote image masking are useful in obtaining good spatial definition in the measurement. Also clear is that spatial definition is not something that is important only for extremely small samples. The improvement of large numerical aperture and dual masking can be seen even for samples as large as several wavelengths in diameter, a good bit larger than what would normally be thought of as a diffraction-governed size realm.

Also dealt with in this chapter is the misconception that objective magnification is important to the energy throughput of an FT-IR microscope. Nothing could be further from the truth. There is a correlation between magnification and throughput in traditional microscope photometers because higher magnification objectives normally exhibit higher numerical aperture, which is the real arbiter of performance. It is therefore possible (and desirable) to construct all objectives for FT-IR microscope use with as high a numerical aperture as possible, regardless of magnification. Of course, a high-numerical-aperture objective

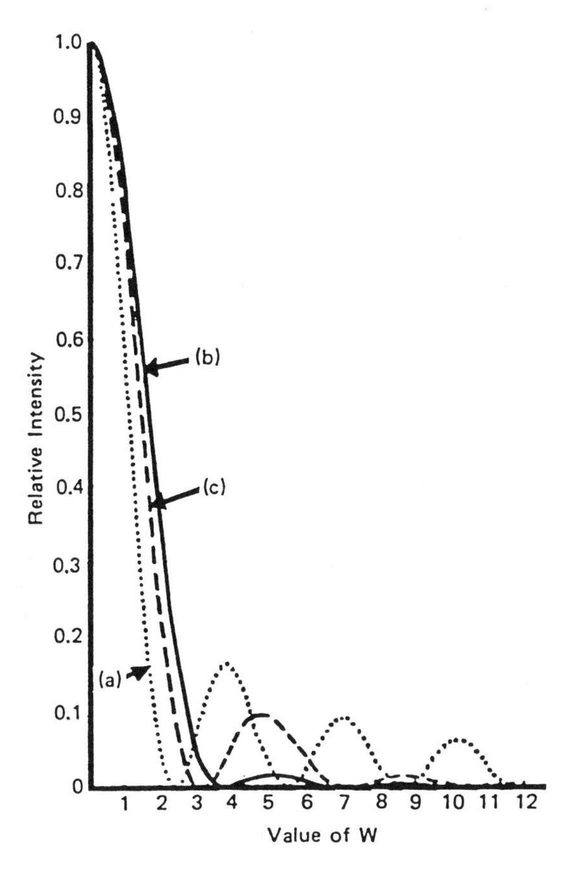

Figure 17. The Airy disk diffraction pattern is modified when the optical system contains a central obscuration, as is the case in most FT-IR microscopes. This figure shows the modification for various levels of central obstruction.

is much more difficult to produce, while maintaining diffraction-limited performance.

The author has made no statements regarding what is the smallest possible sample size measurable with an FT-IR microscope, because the data presented here suggest that there are too many variables and caveats to make such a statement. By way of explanation, a single red blood cell 7 μm in diameter can be measured successfully across the entire midinfrared region in a measurement time of 4 min [27]. At the far end of this spectrum there is considerable spurious energy, but there is no interfering neighboring material to contribute a spectrum. But a cross-sectioned 10-μm-diameter fiber embedded in polystyrene will always show polystyrene bands with the same optical system [15]. Spectral subtraction may be used to remove some but not all of the spurious spectral contribution.

The issues of spatial and temporal coherence have been dealt with in this chapter insofar as they affect diffraction-limited imagery. It has been put forth that temporal coherence—the spectral bandwidth—has no bearing on the imagery in the FT-IR microscope. On the other hand, spatial coherence does impinge on the data presented here. Total incoherence was assumed for the finite slit widths, which should be the worst case. This point may be investigated in the future.

It has also been noted here that given a choice of an aperture before the sample (on the source side) or after the sample (on the detector side), the aperture should be placed on the source side, especially if the specimen is a scatterer. In addition, the viewing apparatus is best placed on the source side also. Dual masking is still preferable to source-side masking from a spatial definition point of view, but it does require a more expensive apparatus, since high-quality imaging optics are required both above and below the specimen plane.

It is concluded that for many sampling situations, all else being equal, the use of dual remote image masks can allow the measurement of samples that are a factor or 2.5 smaller than a single-apertured system, with an equivalent level of spurious energy contamination. Also, the use of an N.A. = 0.50 objective will allow the measurement of samples that are a factor of 2 smaller than an N.A. = 0.25 system, with an equivalent level of spurious energy contamination. In addition to the spatial definition improvement attributed to the increased numerical aperture, the signal will be increased by a factor of up to 6 for an equivalent measurement. Recently, very high numerical aperture systems have been produced that allow excellent spatial definition with a single aperture. These systems approach the performance of medium-numerical-aperture dual-masked systems. It is safe to say that today, all of the commercially available FT-IR microscopes can give excellent results when used properly and kept in good alignment. Infrared microspectroscopy using the FT-IR microscope now stands as the best choice for identification of small amounts of organic materials and for the measurement of or correlation with a number of their physical properties, such as dichroic ratio or physical strength.

REFERENCES

1. R. Barer, A. R. H. Cole, and H. W. Thompson, *Nature, 163*: 198 (1949).
2. R. C. Gore, *Science, 110*: 710 (1949).
3. E. R. Blout, G. R. Bird, and D. S. Grey, *J. Opt. Soc. Am., 40*: 304 (1950).
4. V. J. Coates, A. Offner, and E. H. Siegler, Jr., *J. Opt. Soc. Am., 43*: 984 (1953).
5. R. G. Messerschmidt, in *Infrared Microspectroscopy: Theory and Applications*, R. G. Messerschmidt and M. A. Harthcock, eds., Marcel Dekker, New York, 1988, pp. 1–19.

6. R. G. Messerschmidt and M. A. Harthcock, eds., *Infrared Microspectroscopy: Theory and Applications*, Marcel Dekker, New York, 1988.
7. M. Born and E. Wolf, *Principles of Optics*, 6th ed., Pergamon Press, Oxford, 1980.
8. G. B. Airy, *Trans. Cambridge Philos. Soc*, 5: 283 (1835).
9. K. Schwarzschild and D. W. Villiger, *Astrophys. J*, *23*: 284 (1906).
10. A. Fresenel, *Ann. Chim. et Phys.*, (2), 1 (1916) 239.
11. C. J. R. Sheppard and T. Wilson, *Appl. Opt.*, *18*: 22 (1979).
12. M. Minsky, Microscopy Apparatus, U.S. Patent 3,013,467, Dec. 19, 1961.
13. R. G. Messerschmidt and D. W. Sting, Microscope Having Dual Remote Image Masking, U.S. Patent 4,877,960, Oct. 31, 1989.
14. H. Piller, *Microscope Photometry*, Springer-Verlag, Berlin, 1977, p. 167.
15. A. Bouwers, *Achievements in Optics*, Elsevier, New York, 1950.
16. K. Schwarzchild, *Astr. Mitt. Königl. Sternwarte Göttingen*, Göttingen, 1905.
17. C. R. Burch, *Proc. Phys. Soc.*, *59*: 41 (1947).
18. P. R. Griffiths, *Chemical Infrared Fourier Tranform Spectroscopy*, Wiley, New York, 1975.
19. P. R. Griffiths and J. A. de Haseth, *Fourier Transform Infrared Spectrometry*, Wiley, New York, 1986.
20. C. T. Foskett and T. Hirschfeld, *Appl. Spectrosc.*, *31*: 239 (1977).
21. D. R. Mattson, *Appl. Spectrosc.*, *32*: 335 (1978).
22. W. J. Smith, *Modern Optical Engineering*, McGraw-Hill, New York, 1966.
23. R. G. Messerschmidt, in *The Design, Sample Handling, and Applications of Infrared Microscopes*, P. B. Roush, ed., ASTM STP 949, American Society for Testing and Materials, Philadelphia, 1987.
24. E. Abbe, *Arch. Mikrosk. Anat.*, *9*: 413 (1893).
25. A. B. Porter, *Philos. Mag.*, *11*: 154 (1906).
26. B. J. Thompson, in *Progress in Optics*, E. Wolf, ed., Vol. 7, Elsevier North-Holland, New York, 1969.
27. A. Dong, R. G. Messerschmidt, J. A. Reffner, and W. S. Caughey, *Biochem. and Biophys. Resch. Comm.*, *156*: 752 (1988).

2
Uniting Microscopy and Spectroscopy

John A. Reffner and Pamela A. Martoglio Spectra-Tech, Inc., Shelton, Connecticut

1. INTRODUCTION

Infrared microspectroscopy (IMS) is the union of microscopy and spectroscopy for microanalysis. Microscopy is the science of creating, recording, and interpreting magnified images. Analytical spectroscopy is the science of emission, reflection, and absorption of radiant energy to determine the structure and composition of matter. Each science provides specific information about a material's composition and structure. Both sciences are rooted in optics and in the interaction of radiant energy with matter. Because of their common base, microscopy and spectroscopy are inseparable in IMS. The fullest application of IMS results from using proven techniques of both sciences.

Every aspect of IMS, from sample preparation to spectral interpretation, requires the understanding and application of the fundamental principles of both microscopy and spectroscopy. Microscopy plays a major role in selecting samples for analysis, assessing the best spectroscopic technique, observing the sample's microstructure, and defining the microscopic area for analysis. Also, many traditional sample-preparation techniques developed for microscopy are directly applicable or easily modified for IMS. Observing a sample's microstructure is a very important part of IMS analysis. A cotton fiber, a starch grain, a wood fiber, or pollen would produce infrared spectra of cellulose, but their microscopic shapes are readily distinguished from one another. Detailed microscopic examination can distinguish varieties of cotton, the origins of starches (e.g., corn, rice, potatoes), hard or soft wood fibers, and the nature of the plant that produced the pollen. However, spectroscopy alone can be used to determine the chemical composition in the absence of distinctive morphological features. Two colored fibers may appear similar under microscopic analysis but may actually be stained with two different dyes, which spectroscopy could detect and identify. A cross section of a polymer may show a uniform microstructure, but a linear-spatially-resolved spectral map of the carbonyl band's intensity can quantify the extent of photooxidative degradation (Figure 1). There is a symbiotic relationship between microscopy and spectroscopy that benefits each when they are united in IMS.

Today, IMS has become a versatile microanalysis technology, extending both microscopy and spectroscopy. Initially, IMS was used by spectroscopists only to analyze small samples. The first microscope accessories designed for FT-IR spectrometers could perform spectral analysis only in a transmission

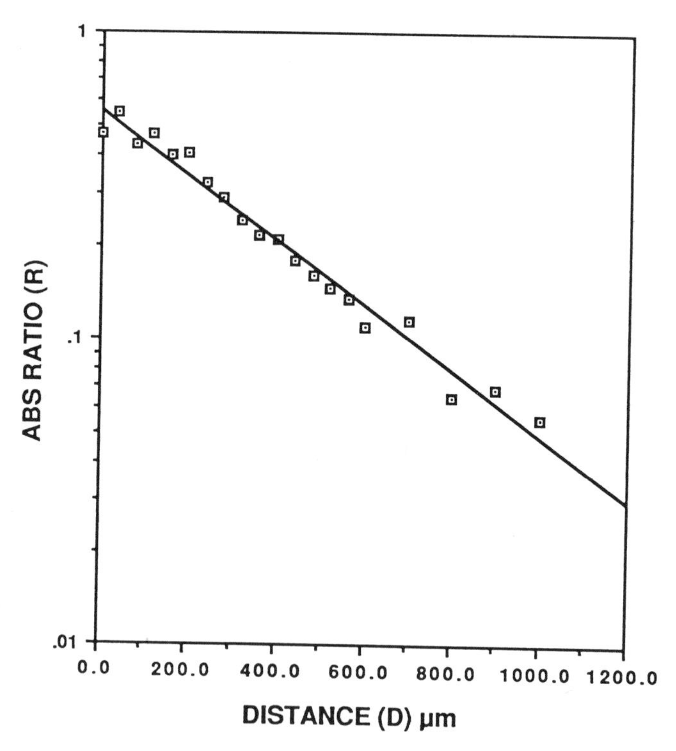

Figure 1. Photooxidative degradation of polypropylene is measured as the ratio of the carbonyl absorption band at 1735 cm^{-1} to the CH_2 band at 1240 cm^{-1} in the polymer. The photooxidative profile is plotted as a function of the distance from the outer surface of the polymer. The log of the absorbance ratio produced a linear relationship.

mode. IMS systems can now record specular, diffuse, internal, or grazing-incidence reflection spectra, in addition to conventional infrared-absorption spectra in the transmission mode. It is only the sample itself, not the microscope, that limits the choice of analysis technique. The analyst can quickly change IMS analysis modes, selecting the most appropriate technique for each sample. Automated sequential analysis of inhomogeneous materials has become a routine IMS technique. IMS is an all-purpose, versatile spectroscopic technology for the analysis of solid and liquid samples.

Applying IMS to materials analysis and characterization requires an understanding of the basic principles of microscopy, spectroscopy, and instrumental analysis. This chapter reviews the fundamentals of IMS. A systematic approach to analytical problem solving is presented, along with the relevant principles of instrumentation, microscopy, and spectroscopy. Specific examples of IMS analyses illustrate its versatility.

2. A UNIFIED APPROACH TO IMS ANALYSIS

Ever since infrared microspectroscopy became a practical analytical technique, the rush has been on to obtain spectra on samples with ever-increasing speed and ease. At one time or another, everyone has heard: "All you have to do is put your sample on the stage, take a spectrum, and you've got your answer!" Truly an oversimplification. Every sample is unique, and it is the job of the microspectroscopist to take a sample and decide the best way to analyze it.

When IMS is used to attack an analytical problem, there is a systematic process that, when followed, gives the highest probability of success. Basically, this scheme can be broken down into six steps. They are:

1. Look at the sample and decide on the appropriate IMS technique (i.e., design the experiment).
2. Perform the sample preparation necessary for the IMS technique selected.
3. Select an instrument and validate its performance.
4. Examine the microstructure and select the sample area for analysis.
5. Obtain spectra from the sample area.
6. Interpret the spectral results and relate the chemical analysis with the microstructure.

The first step (and one of the most important) when performing an IMS analysis is to look at the sample. Often, simply viewing it—preferably, with a low-power stereo microscope—tells the investigator which spectroscopic technique to use. Once the proper IMS technique has been chosen, analysts are well on their way to solving their analytical problem.

A contaminant on a plastic sheet is a sample that demonstrates the important role that microscopy plays in identifying samples and selecting the appropriate spectroscopic techniques for attacking a problem. A low-power microscopical examination can readily determine whether the contaminant is either on the surface or embedded within the plastic material. If it is on the surface, microscopical examination can reveal morphological details (such as smearing, elongation, roundness, or patterns) that may suggest the history of the contaminant. In fact, microscopical examination is one of the few methods that will actually allow the analyst to deduce a material's history. If the contaminant is on the surface, a reflection method could be the first spectroscopic technique to use in identifying the chemical nature of the contaminant. With the low-power microscope and an appropriate tool, it may also be possible to excise or remove a contaminant for analysis by transmission spectroscopy. The physical nature of the contaminant may be determined by probing with a sharp needle; for instance, is it rubbery or brittle? Depending on its physical nature, the sample may need to be pressed in a diamond cell or simply flattened with a probe in order to be analyzed by transmission spectroscopy. Whether thin sections of a sample can be prepared is determined by observing how the material behaves when being cut or sectioned by hand under the stereo microscope. Observations made with the stereo microscope are extremely useful for sample preparation and for selecting IMS methods, and provide useful information for solving problems.

Polarized-light microscopy provides additional insight into solving this contamination problem, as identifying the chemical nature of the contaminant does not provide a complete answer. There are two parts to this problem: what is it, and from what source did it originate? Spectroscopic analysis addresses the composition, and microscopy the origin, of the contaminant. Observation of the sample placed between crossed-polars can reveal whether or not the machine direction of the plastic sheet is aligned parallel with the contaminant. If a parallel orientation is observed, this indicates that the contamination occurred during the processing of the plastic sheet. A lack of alignment indicates that the contamination occurred after processing.

While each sample presents its own unique features, a systematic evaluation will lead to the selection of a best first approach. Table 1 contains a list of sample types, preparation restrictions, and recommendations to aid the analyst in selecting the appropriate IMS technique. To apply IMS successfully to analytical problems, it is necessary to understand the requirements of instrumentation, the principles of the methods, the technical details of the sample preparation, the proper spectral-data reduction, and spectral interpretation. Throughout this chapter, the various guidelines listed in Table 1 are described in detail. It is important that the analyst be equipped with as much information as possible concerning the IMS technique. Only with the most appropriate method and proper analytical technique may the highest-quality results be obtained.

Table 1. Guidelines for Selecting IMS Methods (Depending on Sample Type and Restrictions on Sample Preparation)

Sample type	Preparation Restriction[a]	IMS technique[b]					
			Reflection techniques[c]				
		Transmission	ATR	Diffuse	GRA	NRA	Specular
Biological tissue	1		P				
	2	P	S				
	3	P	S				
	4					P[d]	
Electronic device	1	S	P		S	S	P
	2	P	P		S	S	P
	3	P	P		P	S	P
	4		P		P	S	
Inclusion	1	S					
	2	P	P			S	S
	3	P	P	S		S	S
	4					P	
Isolated solid particle	1						
	2	S	P		S	P	S
	3	P		S			
	4		S		S	P[d]	
Liquid	1		P				S
	2	P	P				S
	3	P	P				S
	4		S		S	P[d]	
Multilayered laminate	1		P[e]				S[e]
	2	P	S				S
	3	P	S			S	S
	4						
Phase distribution	1		P				S
	2	P	S				S
	3	P	S				S
	4					P[d]	
Sub-μm films	1	S	P				
	2	S	P				P
	3						
	4				P		

Table 1. *(Continued)*

Sample type	Preparation Restriction[a]	IMS technique[b]					
			Reflection techniques[c]				
		Transmission	ATR	Diffuse	GRA	NRA	Specular
Surface defect	1		P	S			S
	2		P	S			P
	3	P	P	S		S	
	4				S	P	
Suspended particulates in liquids	1						
	2	P	S				
	3	P	S				
	4				S	P[f]	

[a]1, Nondestructive to specimen; 2, nondestructive to sample; 3, unrestricted; 4, thin film on metal substrate.
[b]P, primary method; S, secondary method.
[c]ATR = attenuated total reflection
GRA = grazing incidence reflection/absorption
NRA = normal incidence reflection/absorption
[d]On mirror mount.
[e]Edge must be accessible.
[f]Isolated on gold Nuclepore® filters. (Nuclepore is a trademark of Corning CoStar.)

3. INSTRUMENTATION

IMS is an instrumental technology, and the design and performance characteristics of infrared microspectrometers establish the limits for the application of IMS to analytical problems. It is not intended here to evaluate instrument designs but to present the basic instrumental requirements for microscopy and spectroscopy of microscopic samples. Many of these requirements are shared, while others are unique to one or the other disciplines.

3.1. Microscope Stand

While the fashion in microscope stands has changed over the years, the basic configuration of eyepiece, objective, condenser, stage, and focusing controls has remained constant. A stand must be sturdy, free of vibration, and large enough to support the microscope's optics. The sample stage area should accept samples of various sizes and shapes and should permit easy access to allow the analysis of large specimens. IMS microscopes should permit the observation of transparent or opaque materials. Transmitted light is supplied from below the transparent sample, while opaque materials are viewed using an illumination source from above the specimen (EPI illumination). The microscope should provide a means for polarizing the light prior to its entry into the sample and for positioning a second polarizer (analyzer) in the light path after the light has passed through the sample. The ability to examine the sample with polarized light provides a means for contrast enhancement as well as for assessing the optical properties of materials.

3.2. Imaging Optics and Quality

Achieving high-quality images is important in both microscopy and spectroscopy. The imaging optics of an IMS microscope consist of the light source(s), condenser, objective, and eyepiece(s), which must work together to form good images. While the magnification of a microscope and its resolution are determined primarily by the objective and eyepiece, uniform illumination is critical to the quality of images that the microscope will produce. For a uniformly illuminated field using a conventional, coiled, tungsten-wire-filament lamp, an optical arrangement referred to as Köhler illumination is the standard [1]. To achieve Köhler illumination, the lamp filament is imaged (via an appropriate optical system) into the back focal plane (also called the Fourier plane) of the objective.

To obtain high-quality images, it is also important to illuminate only the field of observation; otherwise, glare may occur. Glare (also referred to as the Schwarzschild–Villiger effect) is the reduction of image contrast caused by internally reflected or scattered light that is collected by the objective but is not part of the image [2]. Essentially, the reduced contrast is optical noise. Glare reduces the signal-to-noise (S/N) ratio in the image. From a spectroscopic perspective, glare has a similar effect, in that it represents radiant energy from outside the defined sample area which, when it reaches the detector, is perceived as having come from the sample. In the spectrum, glare produces extraneous spectral features from objects not in the defined sample area.

The objective and condenser are principal lenses in infrared microspectrometers, and various combinations of optical elements are used. (The various types of optical elements and their features that relate to IMS are summarized in Table 2.) For spectral measurements, reflecting optics are preferred, since they neither possess chromatic aberrations nor absorb radiant energy. They are also suitable for polarized-light measurements, as they do not introduce strain birefringents and cause little depolarization (due to the fact that incident radiation is reflected off a highly reflective metal surface at angles of incidence of less than 45°).

The design of the Schwarzschild objective and condenser (often erroneously called a Cassegrainian objective) is preferred because of its high image quality and efficiency (for an imaging optic) [3]. Schwarzschild lenses can contain a spherical-aberration correction mechanism to correct for refraction of samples and sample mounts. Conventional microscope objectives with glass refracting

Table 2. Summary of Objective and Condenser Factors in Infrared Microspectroscopy

Feature	Function: Microscopy	Function: Spectroscopy
Objectives		
Reflecting (Schwarzschild)	On-axis for highest correction; only minor coma	Best definition of sample area, with no chromatic defocus and no absorption of radiant energy
Reflecting (off-axis)	Not useful for imaging, since they are highly comatic	Poor definition of sampling area
Refracting	For viewing only; highest-quality images and contrast enhancement; variety of magnifications (4 to 100×)	Not used due to strong absorption
Numerical aperture	Defines resolution and depth of field for specific radiant energy	Limits sample definition and throughput while defining the acceptance angle for mounting cells and stages
Working distance	Limits space for sample access and manipulation	Limits space for mounting cells and special stages
Condensers		
Reflecting (Schwarzschild)	On-axis, highly corrected to control the illumination of the sample	High correction and efficiency for collecting radiant energy; images the field aperture to limit radiation to the sample area
Reflecting (off-axis)	Illuminates the sample	Floods the sample area with radiant energy (creating glare) and reduces sample definition
Numerical aperture	Defines resolution and depth of radiant energy for specific energy and for the angular aperture of the illumination system	Limits sample definition and throughput while defining the acceptance angle for sample-mounting cells and stages

lenses are useful for imaging and for special contrast-enhancement microscopy techniques.

3.3. Sample Definition and Performance

While the quality of a visual image is a good indicator of the microscope's performance, normally, the infrared image is not viewed; as a consequence, it is more difficult to assess the spectrometer's performance. To obtain high-quality IMS spectra, two performance criteria must be met: high through-put and an accurately defined sample area. Generally, the energy throughput of the spectrometer is considered a measure of its performance, but in IMS, the throughput can be misleading. Proper sample definition is more difficult in IMS. While it is easy to establish a sample area by viewing with visible radiation, the actual area to be analyzed is larger for infrared radiation because of diffraction. As the sample size decreases, it becomes even more important to define the sample area accurately.

One of the first attempts at microscopic sample definition was demonstrated by Shearer when he used small apertures for microanalysis [4]. By using pinholes in metal foils to define a sample area and then placing these samples in a beam condenser, sufficient energy could pass through the pinholes to record spectra. Placing a fixed-pinhole aperture at the specimen provides an ideal mask with which to define the sample area, but this is a very intractable method.

Infrared microspectrometers use images of masks superimposed on the specimen to define the sample area. In IMS, the image of the mask is the true definer of the sample area, so the quality of the image is critical. These masks are referred to as remote-image-plane masks; they may be positioned between the radiation source and the sample or between the sample and the radiation detector. To reduce glare, the area of illumination should contain only the area to be analyzed (i.e., the aperture should be placed between the source and the sample). A single aperture placed between the fully illuminated sample area and the detector cannot eliminate glare generated by radiation that is scattered by the sample, but the aperture will reduce diffraction effects.

Even with the use of remote-image-plane masks, the exact definition of the sample area is not easily measured. It can be estimated by moving a sharp interface systematically through the defined sample area and measuring the throughput (as energy reaching the spectrometer's detector) [5]. A poorly defined sample area is one in which the infrared sample area is larger than the visible sample area. A number of factors—such as glare, diffraction, and lens aberrations—can affect the quality of the sample definition. Any of these factors can be a source of extraneous radiation and will produce a higher throughput. Therefore, when the *S/N* evaluation is based on 100% line noise, an IMS spectrometer with a poorly defined sample area will produce high *S/N* values. In this case, a high *S/N* value is not necessarily indicative of good performance. The true performance test is measured by an absorbance band's peak-to-baseline *S/N* from the spectrum of a finite, small sample isolated in a continuous phase. In this test, both sample definition and energy throughput are used to assess the quality of the system. For IMS analysis, the spatial definition of the sample is critically important, so it is essential to minimize aberration, glare, and diffraction to achieve maximum performance.

3.4. Sample and Windows

The sample and its supporting windows (slide and coverplate) are elements in the microscope's optical system that affect image quality and system performance. Typical organic samples will have a refractive index of between 1.45 and 1.65. Since air has a refractive index of 1.0, there will be a significant change in the refractive index as the infrared beam passes from the air and into the sample. As is well known from Snell's law, a beam of light will refract when it passes from one refractive index medium into another. This law is illustrated by

$$\frac{\sin i}{\sin r} = \frac{n_2}{n_1} \tag{1}$$

where i and r are the angles of incidence and refraction measured from the normal to the surface and n is the refractive index. The subscripts 1 and 2 refer to the first and second media. Usually, n_1 is air and n_2 is the sample. In this case, this simplifies Snell's law to $\sin i/\sin r = n_2$.

Because of the refraction that occurs, the focus of the microscope's optical system will change. Since IMS samples are only a few micrometers thick, the beam displacement caused by the sample is not a major problem. However, if the sample is on a salt plate (e.g., KBr's refractive index = 1.524 at 1000 cm^{-1}), the beam displacement will be dramatic. As shown in Figure 2, the beam geometry going into and coming out of the sample does not change if the sample is placed on or between two salt windows, but the focal plane of the beam does change [6]. If the beam is not in focus at the sample, it cannot be focused correctly onto the detector, which means that little or no energy will reach the detector. For example, if a sample is placed first on a 2-mm-thick KCl plate and then on a 2-mm-thick ZnS plate, the focal plane will change by 400 μm. Since transmission samples are typically 1 to 30 μm thick, this means that the infrared beam will not be focused on the sample. To recover the energy, both the sample and condenser focus must be adjusted [26].

The use of windows also introduces spherical aberrations, which cause a loss of contrast in the image, poorer sample definition, and a reduction in energy. There are reflecting optics available (e.g., Spectra-Tech, Inc., Shelton, Connecticut) that can compensate for sample supports of various refractive indices and

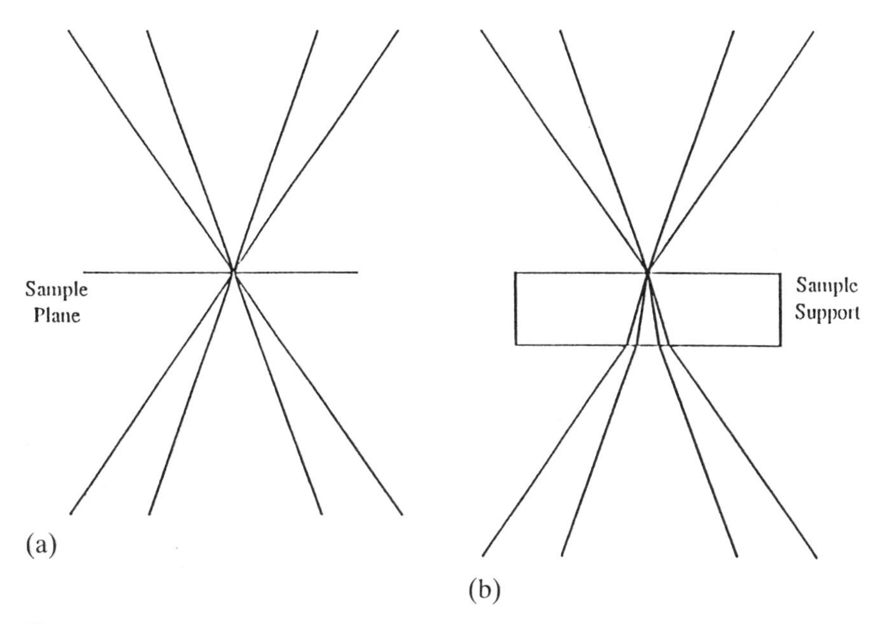

Figure 2. Change in focal plane with a window or cover plate: (a) focus of the beam on a thin, unsupported sample; (b) ray tracing for a sample supported on a window.

thicknesses. In Schwarzschild objectives, this task is accomplished by adjusting the distance between the primary and secondary mirrors. This operation can help eliminate spherical aberrations. The effect of spherical aberration correction is shown in Figure 3. The improved image quality is seen as a greater sharpness and higher contrast, which translates into improved sample definition and improved energy for spectral analysis.

4. BASIC REFLECTION TECHNIQUES AND PRINCIPLES OF IMS

The measurement of reflection spectra with the microspectrometer is the best method for analyzing surfaces, thin films on highly reflecting materials, or finely dispersed materials. Since reflection techniques require little or no sample preparation, they are the best choice for nondestructive analyses. An example of a type of sample that must be analyzed nondestructively is forensic trace evidence. Often, other analysts need to examine such a sample to confirm results. Because it may be the only piece of evidence available, the sample must be preserved. Fortunately, there are a number of ways in which samples can be analyzed nondestructively with IMS; the majority of the options are reflection methods. There are different types of reflection, each with its own advantages and dis-

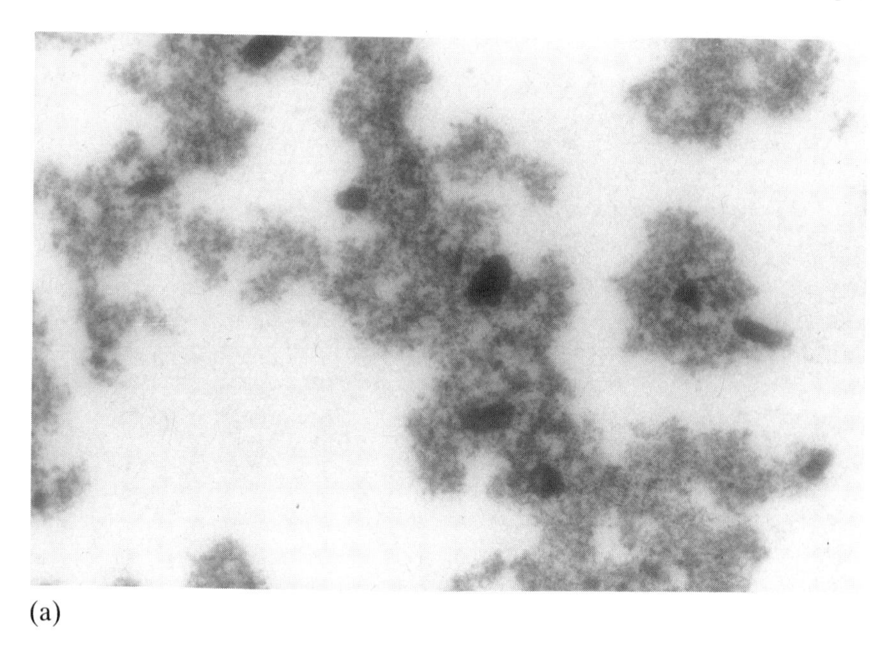

(a)

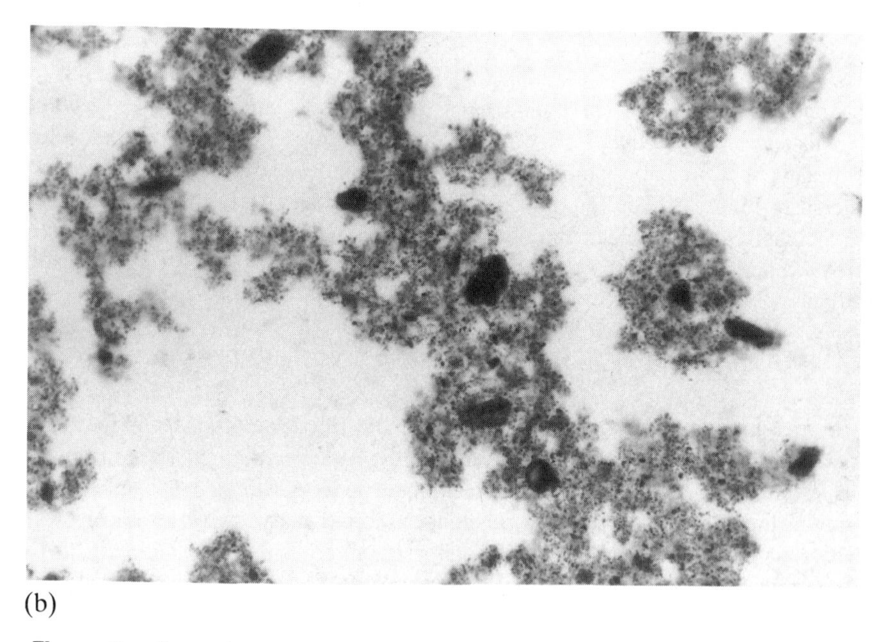

(b)

Figure 3. Comparison of image quality with and without spherical aberration. The sample seen in micrograph (a) was mounted with a 1.0-mm coverglass, while the objective used had its spherical aberration corrected for no coverglass. Micrograph (b) is the same sample with the objective's spherical correction for a 1.0-mm coverglass.

advantages. Basically, reflection techniques can be divided into four main categories: specular reflection (SR), diffuse reflection (DR or, more commonly, DRIFTS for diffuse-reflection infrared Fourier transform spectroscopy), internal (or attenuated total) reflection (ATR), and reflection/absorption (R/A).

4.1. Specular Reflection

Specular reflection is the front-surface (Fresnel) reflection from the exterior surface of a material. The simple rule for specular reflection is that the angle of reflection equals the angle of incidence. These angles are measured from the normal to the reflecting surface. Reflection spectra are generally measured at or near normal incidence. For most organic samples, only about 5 to 10% of the energy is reflected. Because specular reflection is a relatively weak process for dielectric materials, the sample should have a level, lustrous surface, and a large area should be analyzed to achieve high *S/N* spectra.

Specular reflectivities for different energies of incident radiation produce spectra that are different from normal transmission spectra. A specular-reflection spectrum of an ethylene/vinyl acetate copolymer is shown in Figure 4. The specular-reflection spectrum is distinguished by the presence of derivative-shaped bands. The changes in the reflectivity are the result of the superposition

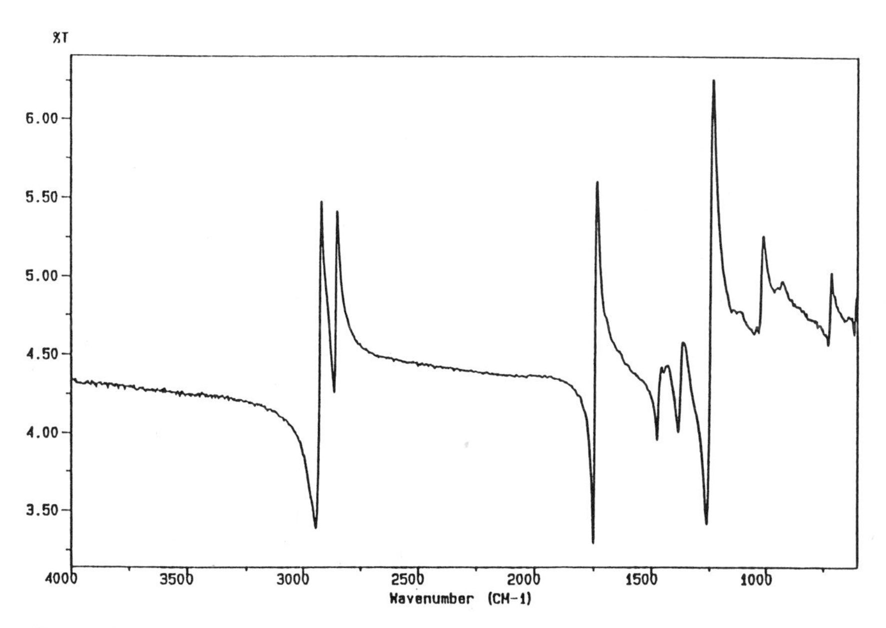

Figure 4. Specular reflectance spectrum of EVA. Note that the percent reflection is less than 10% for this polymer. This is typical of a dielectric material with weak absorption bands.

of two optical parameters: the refractive index and the extinction coefficient. In spectral regions of strong absorption, the refractive index of the sample will decrease at wavenumbers just higher than the maximum absorption and will then increase dramatically while passing through the band. This combination of the refractive-index component with the extinction coefficient k is what forms the distinctive specular-reflection spectrum. The mathematical relationship called the Kramers–Kronig (KK) transformation calculates the dispersion [$n(\nu')$] and the extinction coefficient spectrum [$k(\nu')$ from the specular data (Figure 5). This is expressed by

$$n(\nu') = n_\infty + \frac{2}{\pi} \int_{-\infty}^{+\infty} \frac{k(\nu')\ \nu}{\nu^2 - \nu'^2}\, d\nu \tag{2}$$

where n is the index of refraction at frequency ν', n_∞ the index of refraction at frequency $\rightarrow 0$, ν' the frequency, ν the frequency of maximum absorption, and k the extinction coefficient.

The KK transformation can now be calculated in seconds with most data-processing programs, making specular reflection a more desirable analytical measurement than it has been in the past. In specular-reflection spectra, the KK transform will be most accurate when the *S/N* is high, there is no diffuse re-

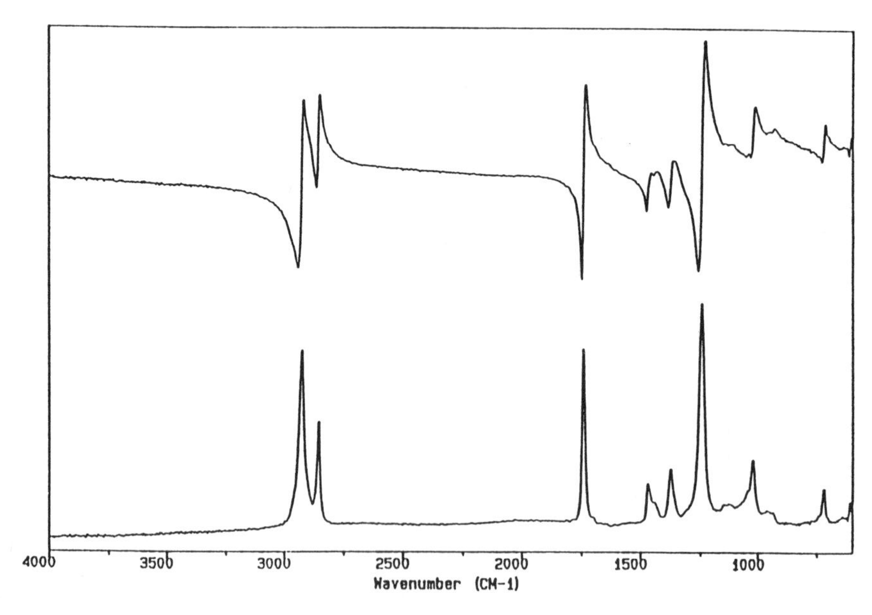

Figure 5. Refractive index dispersion (top) and extinction coefficient spectrum (bottom) calculated using the Kramers–Kronig relationship to transform the specular reflection spectrum of EVA, shown in Figure 4.

flection or scatter, and the interferences from water vapor and CO_2 are low. (An excellent discussion on specular reflection and other reflection techniques can be found in a book by Gustav Kortum [7].)

Ideal samples for specular reflection are level, lustrous, dielectric materials. Because reflection by dielectric materials is a low-efficiency process, the sample should be as large as possible (to improve *S/N*), level (so that the surface will be in focus and the reflected incident radiation will be collected by the objective), and smooth (so that most of the energy impinging on the sample will reflect specularly and not diffusely). Either a sample is naturally a good specular reflector or it is not; its ability to reflect is determined by its physical condition, refractive index, and absorption. Examples of good samples are plastic sheets, molded plastics, polished or microtomed sections of materials (preferably cut with a diamond knife), and liquids. After processing with the KK transformation, the correspondence of specular reflection spectrum with the sample's absorption spectrum is excellent, as shown in Figure 6. Pure specular reflection data can produce excellent analytical spectra.

Ensuring that the sample is level and normal to the microscope's optic axis is very important; this will ensure that the sample is in focus over the defined sample area. One simple way to obtain level samples is to use a leveling press. The sample is placed on top of a small amount of modeling clay fixed to a microscope slide. This arrangement is then placed in the leveling press and

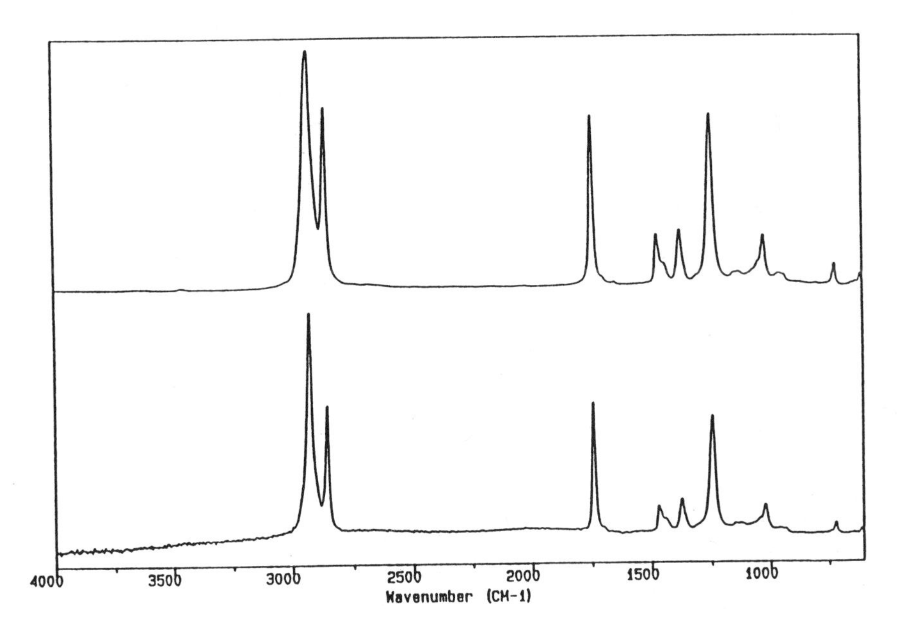

Figure 6. The absorbance spectrum of EVA, calculated from its specular-reflectance spectrum (bottom), and its absorption spectrum (top) showing their correspondence.

pressed between its parallel faces. The sample's surface will then become level with the microscope slide and will be ready for analysis. Rough sample surfaces can be "cleaned up" by polishing or removing the rough surface with an extremely sharp cutting utensil such as a diamond knife. However, when using a knife to obtain a fine surface, it is important to note that the surface may be left with cutting striations, which will decrease the efficiency of the specular reflection and introduce diffuse reflection and scatter.

When preparing materials for obtaining background spectra, the reference material should approximate the physical appearance of the sample. For example, for specular reflectance of highly reflective materials, the background single-beam spectrum should be obtained on something highly reflective and nonabsorbing. Good reference materials are polished metals (such as silver, aluminum, or gold) or aluminum or gold-coated mirrors.

4.2. Diffuse Reflection

Diffuse-reflection infrared Fourier transform spectroscopy (DRIFTS) is a useful IMS technique for specific samples when microstructural details are unimportant. Because the sample must be ground into small particles for DRIFTS, the sample's orientation and morphology are destroyed [8]. DRIFTS occurs when the incident radiation passes through an analyte particle, reflects off surfaces of a nonabsorbing matrix (such as KBr), and is then collected and detected. The best samples for diffuse-reflection spectroscopy are uniformly dispersed materials in nonabsorbing matrices. Spectra obtained by this method are similar to transmission spectra, but the intensity of the absorption bands is altered. Weak absorption bands increase in their absorbance intensity relative to stronger absorption bands. (For a detailed description of the theory of diffuse reflection, see Ref. 7). The Kubelka–Munk theory can be applied to the data to compensate for the distortions of band intensity (Figure 7). This theory is expressed by

$$f(R_\infty) = \frac{(1 - R_\infty)^2}{2R_\infty} = \frac{k}{s} \qquad (3)$$

where R_∞ is the absolute reflectance of the layer, k the molar absorption coefficient, and s the scattering coefficient.

Unfortunately, all incident radiation that interacts with the sample does not undergo diffuse reflection; most of the incident beam is lost in the sample or is specularly reflected (Figure 8). Specular reflections are a major interfering interaction, especially if the analyte is undiluted or is present in a high concentration in the nonabsorbing matrix. Since the particles are arranged more-or-less arbitrarily throughout the matrix, specular reflections from their surfaces will occur at all angles, making the total removal of the specular component impossible. Because of these potential interferences, diffuse reflection is a low-

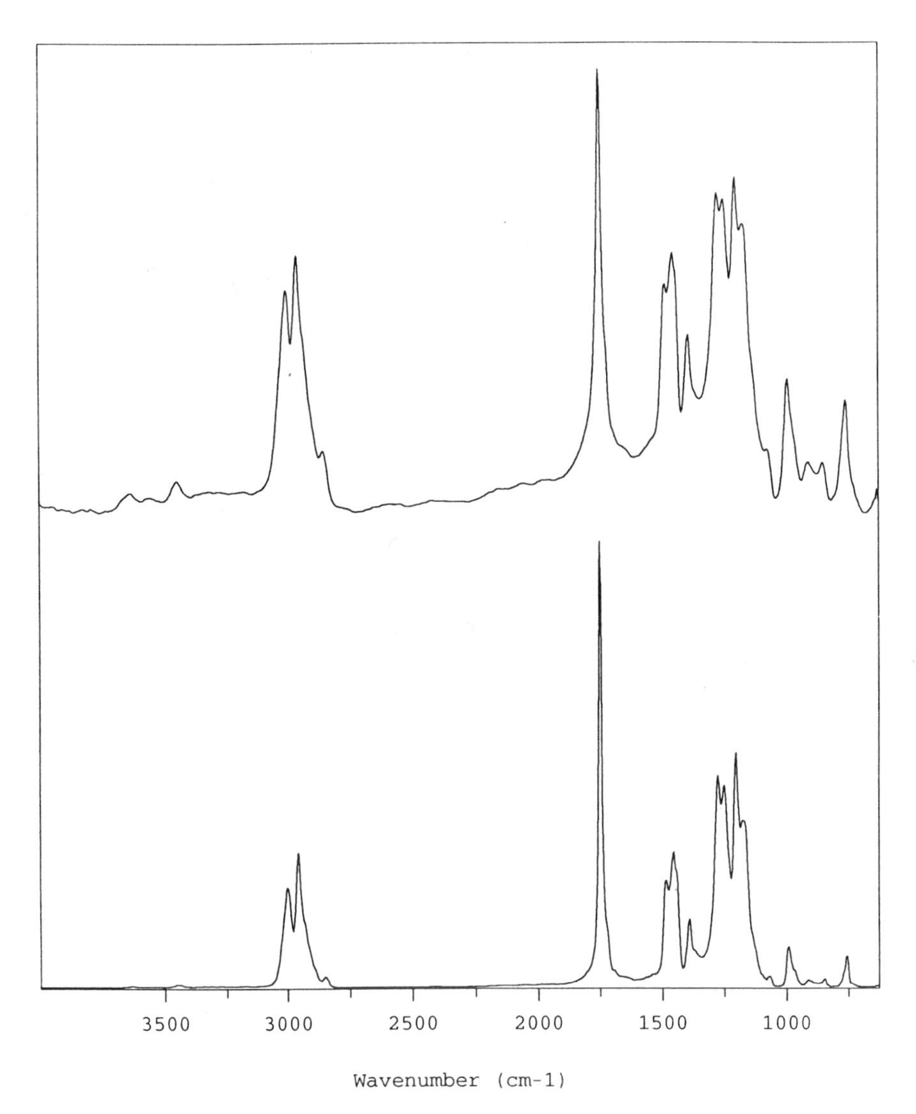

Figure 7. Diffuse-reflection spectra (top) from an abraded surface of poly(methyl methacrylate) (PMMA) and the same spectrum after (bottom) Kubelka–Munk transform of band intensities. Note that weak bands are relatively more intense in the experimental data than when the absorption is in Kubelka–Munk units.

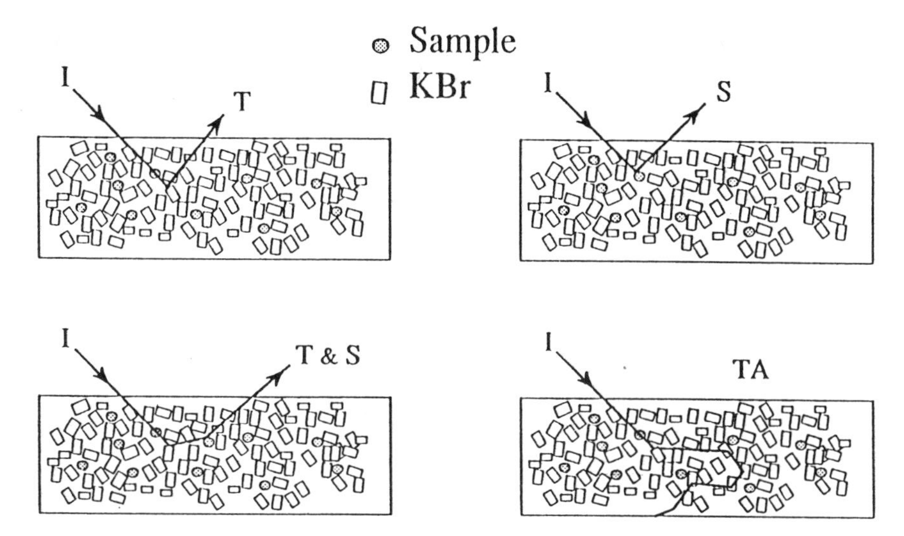

Figure 8. Mixed reflection modes in DRIFTS measurements. In these diagrams the possible paths of the incident beam (I) are shown that can produce transmission (T) spectra, specular reflection (S) spectra, and total absorption (TA).

efficiency technique (typically, 4 to 10%). Care must be taken to increase the percentage of diffuse-reflection interactions and to decrease the percentage of specular-reflection interactions.

As stated earlier, an ideal sample for diffuse reflection is a finely ground solid material or a liquid, dispersed in a nonabsorbing, highly reflecting matrix (KBr, KCl, etc.). However, there are a number of sample-preparation techniques that enable the user to analyze a wide variety of samples. For example, a dissolved solid in an aqueous solution can be analyzed by injecting the solution onto a small pile of sulfur crystals and allowing the water to evaporate. In this case, the background would be taken on a pile of pure sulfur crystals. Sulfur is not the conventional diluent used in diffuse-reflection but (in this instance) is used because, unlike KBr and KCl, sulfur is not water soluble.

For analyzing neat samples, the material's surface can be roughened with fine-grit silicon carbide or diamond abrasive paper (sandpaper). There are basically two choices: either the roughened sample surface or the abrasive paper can be analyzed. The background can be taken using KBr crystals (or a similar material), and the spectrum can be acquired from the sample's roughened surface. Conversely, the sample's spectrum can be obtained from the abrasive paper that had the ground sample dispersed on it, using a plain piece of the abrasive paper as the background. It is important to match the background and the sample, since the abrasive materials have their own absorption bands in the mid-IR region.

Since DRIFTS is a low-efficiency technique, the largest possible sample area should be analyzed. Grinding samples of DRIFTS analysis destroys the sample's morphology; hence the advantage of the microscope is reduced to that of requiring only a smaller quantity of material for analysis. Because of the many problems inherent in DRIFTS in the micro mode, other techniques should be considered first.

4.3. Internal (or Attenuated Total) Reflection

Internal reflection is an extremely useful method for IMS analysis, since it (1) requires little or no sample preparation, and (2) produces relatively distortion-free spectra that are comparable with transmission spectra [9–12]. Attenuated total reflection (ATR) occurs when a sample is brought into contact with an internal-reflection element (IRE) that has a higher refractive index than the sample. The IRE is also referred to as an ATR crystal. Typical IREs are ZnSe (which has a refractive index of 2.4) and Ge (which has a refractive index of 4.0). If radiation is brought through the IRE at an angle greater than the critical angle, the beam will be totally internally reflected (Figure 9).

The critical angle (ϕ_c) between two media with different indices of refraction (n_1 and n_2) is defined as the angle whose sin is n_1/n_2:

$$\phi_c = \arcsin \frac{n_1}{n_2} \tag{4}$$

For ATR spectroscopy, $n_1 < n_2$. A critical angle exists only for radiation traveling through a higher-index medium into a lower-index medium. At angles greater than the critical angle, the incident radiation is completely reflected, but there is an electromagnetic field that extends beyond the crystal surface. This field's

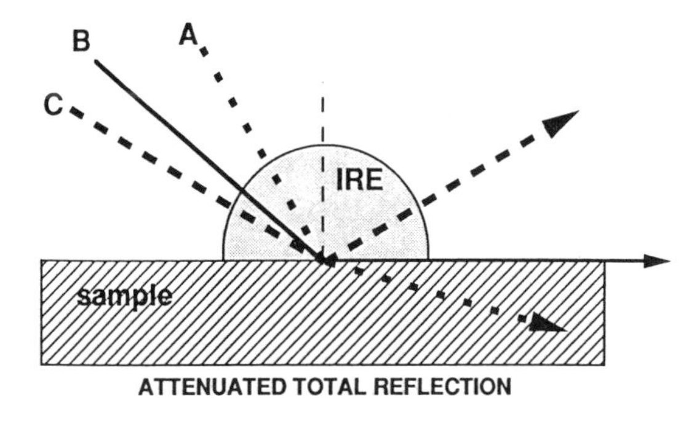

Figure 9. Diagram of rays through IRE. The critical angle ray (B) is refracted parallel to the interface between the IRE and the sample. Rays incident on the interface at angles smaller than the critical ray (A) are transmitted. Rays incident at greater angles (C) are internally reflected.

strength decreases as the distance from the reflecting surface increases; therefore, it is referred to as an evanescent wave. It follows that if an absorbing material is brought in contact with the IRE, the evanescent wave will be absorbed at wavelengths where the material has an absorption band, and the amount of energy reflected back through the IRE will be attenuated; hence the technique is called attenuated total reflection.

The distance that the evanescent wave extends past the crystal surface and into any sample in contact with this surface can be defined in terms of the depth of penetration. The depth of penetration (dp) is defined as

$$\mathrm{dp} = \frac{\lambda}{2\pi n_1 \, (n_1^2 \sin^2\phi \, - \, n^2)^{1/2}} \qquad (5)$$

where λ is the wavelength of radiant energy, n_1 the index of refraction of the IRE, n_2 the index of refraction of the sample, and ϕ the angle of incident radiation on the interface.

Since the electromagnetic field decays exponentially as it moves away from the crystal surface, perhaps a more useful definition is the sampling depth. The sampling depth is defined as the distance from the crystal–sample interface where the intensity of the evanescent wave decays to $1/e$ (approximately 37%) of its original value. A typical IRE has a sampling depth on the order of 0.5 to 3.5 μm, so the sample to be analyzed should be at least this thick; otherwise, materials beneath the sample of interest will be detected. An exception to this rule is if the thin sample is transferrable (i.e., it remains on the crystal surface after the IRE is removed from the sample). In this way, the ATR objective acts as a sampling probe, enabling a pure spectrum of the sample to be obtained without interference from the substrate. Oily or sticky materials are analyzed well by this technique.

An example illustrating the effect of sampling depth with ATR spectroscopy is shown in Figure 10. A thin film of polyethylene was analyzed by both transmission and ATR IMS. The transmission results indicate that the sample is mostly polyethylene with a small amount of fluorine. The ATR results show that the sample is polyethylene with a greater amount of fluorine than is seen in the transmission spectrum. When considered together, the two spectra indicate that the film has been fluorinated at the surface.

As is shown by equation (5), a variety of parameters can affect the depth of penetration, such as the refractive index of the sample and the crystal, the angle at which the radiation is brought into the crystal, and the wavelength of the radiation. It is obvious that as the wavelength of radiation increases (wavenumber decreases), the depth of penetration increases. When compared to transmission spectra, absorption bands in ATR spectra will be less intense at higher wavenumbers and more intense at lower wavenumbers. Correcting ATR spectra

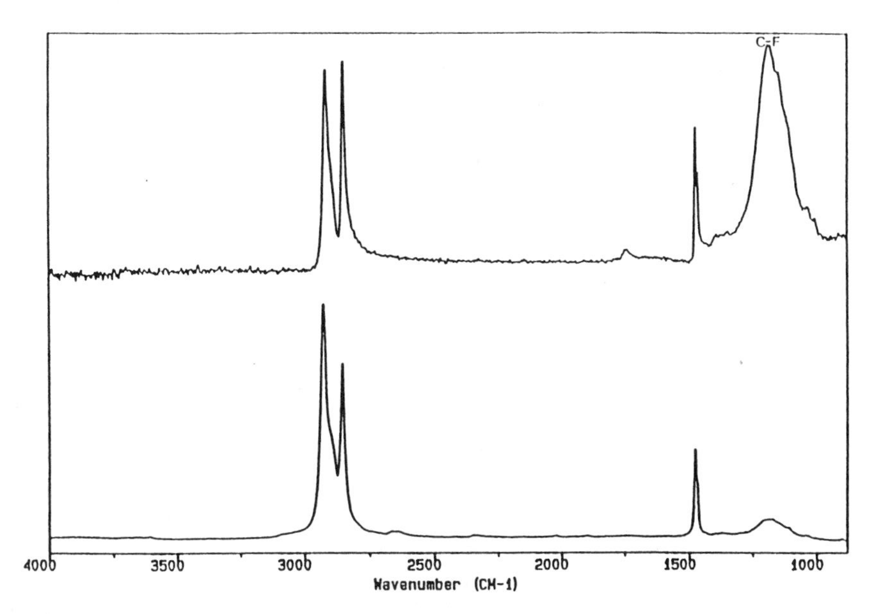

Figure 10. Comparison of transmission and ATR spectra of a fluorinated polymer. The 15-fold increase in the intensity of the C-F absorption (ca 1200 cm^{-1}) in the ATR spectrum indicates that the polymer has a higher level of fluorination at its surface.

for this difference in pathlength at different wavelengths can be accomplished by most data-processing systems.

What types of samples are ideal for ATR analysis? Basically, any sample that can come into contact with the IRE can be analyzed. One of the few limitations is that the sample's refractive index must be lower than that of the IRE; however, this is generally not a concern, since the crystal's refractive index is relatively high.

Achieving uniform sample–crystal contact is a greater concern. With traditional ATR spectroscopy (macro methods), it is desirable to achieve contact over the entire IRE surface (approximately 1 × 5 cm). While this requirement is easily achieved with liquid samples, obtaining good contact with solid samples (especially with irregularly shaped solids) can be difficult. Fortunately, micro-ATR makes analyzing solids much easier. With the ATR objective, contact is made with a very small area (50 × 50 μm or less) [13]. Because of the smaller area needed, finding a flat area for analysis is not as difficult for micro-ATR sampling as it is for macro-ATR sampling. Microscopically flat regions can be found on almost any sample; therefore, obtaining complete contact between the crystal and the sample is easy. With micro-ATR spectroscopy, simply touching a microscopic area of a sample is all that is needed to obtain a spectrum. Because the IRE will come into contact with the highest area of the sample first, the desired sample area should be at the highest position. This position can be determined by visually observing the change in focus at various sample areas. If the sample is irregularly shaped, or if it will not remain in the proper position, the user may need to mount it in modeling clay in order to hold it in place. For example, if the sample to be analyzed is the coating on a tablet, the tablet could be placed on a small piece of clay and pressed slightly so that the clay holds the tablet securely in place, with the area of interest at the top. The ATR spectra of the coating can then be measured.

Another important advantage that micro-ATR spectroscopy has over other IMS techniques is that due to the magnification factor of the IRE, even smaller sample areas can be examined than is usually possible. Since the IRE is a hemispherical lens, the beam cross-sectional area of the radiation passing through it will be magnified by the refractive index of the IRE. For example, for a ZnSe IRE (refractive index 2.4), an apparent 100×100 μm apertured area will actually be 42×42 μm at the sample. This magnification factor is a clear advantage of ATR microspectroscopy, because the conventional diffraction limit is reduced by the IRE index factor. Since the infrared radiation can only be absorbed by the material in contact with the IRE, the contact area defines the analytical sampling area, reducing the diffraction problem. To obtain a background, a spectrum is taken simply, with the crystal not in contact with anything. Essentially, the total internal reflection is the background.

4.4. Reflection/Absorption Spectroscopy

Reflection/absorption (R/A) measurements are divided into two classes, depending on the angle of incident radiation. At near-normal incidence (12 to 35°), samples $l/4$ or thicker are analyzed. (l is the wavelength of radiation being used, typically, 2 to 15 μm). For films of thicknesses less than $l/4$, grazing incidence (65 to 85°) is used. With the R/A method, the sample should be on a nonabsorbing reflective substrate, such as a polished piece of metal. The incident radiation passes through the sample, reflects off the substrate, and passes through the sample a second time, essentially performing a double-pass transmission operation.

R/A spectroscopy at near-normal incidence is a very efficient technique (about 80%). An example of a good sample for R/A is a contaminant on a circuit board, provided that the contaminant is 2.5 to 15 μm thick. However, if the sample of interest is not originally on a metallic substrate, it is usually quite simple to move the sample to an appropriate substrate. Soft, liquid-like samples can be transferred by picking up a portion with a probe and smearing the sample on a gold- or aluminium-coated slide. Fibers and soft solids can be rolled out to the proper thickness directly on the slide; the background would be taken on a clean portion of the slide. Microtomed sections can also be mounted onto metal-coated glass slides. These metal-coated slides are very useful for biological-tissue sections, since they can be handled by automatic tissue-processing equipment. The micrograph of artifacts in a spleen tissue is shown in Figure 11. These artifacts were analyzed by R/A spectroscopy (Figure 12). It is important that good optical contact be made between the sample and the metal. If not, interference fringes can appear in the spectra, due to the air gap between the sample and the metal.

Since near-normal R/A spectroscopy is a transmission-type measurement collected in a reflection operating mode, spectral distortions are few. If the sample is highly reflective, specular reflections from the sample's surface can distort strong bands in the absorption spectrum. (The effects of specular reflection were

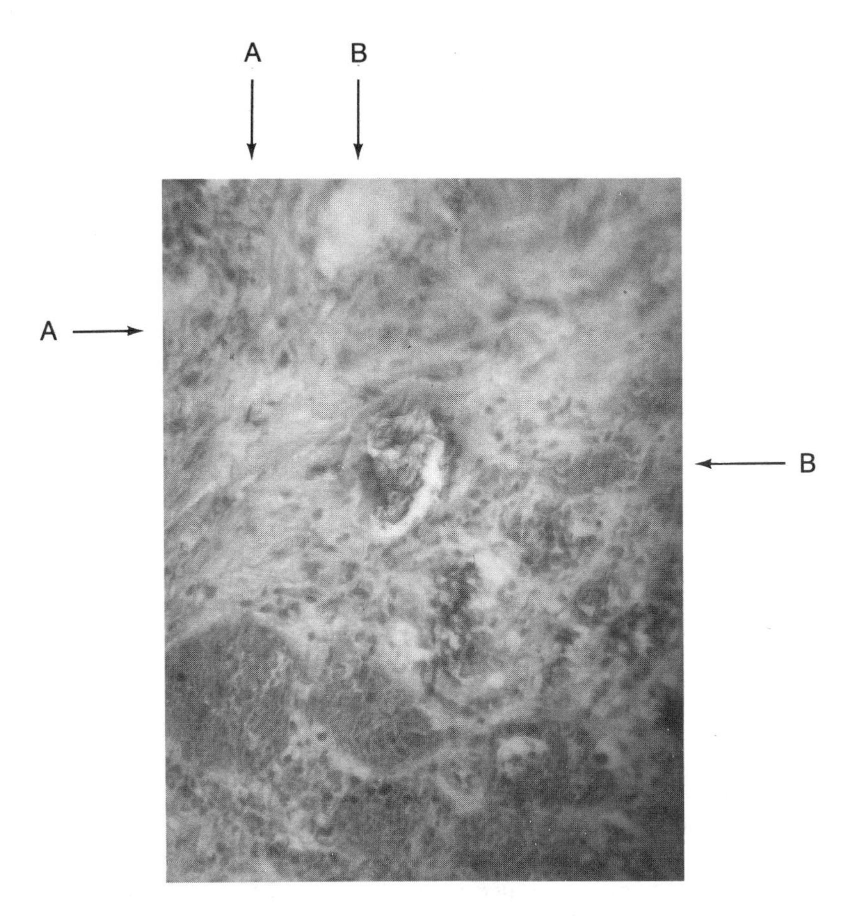

Figure 11. Micrograph of tissue mounted on a mirror slide, indicating areas selected for the spectral analysis shown in Figure 12.

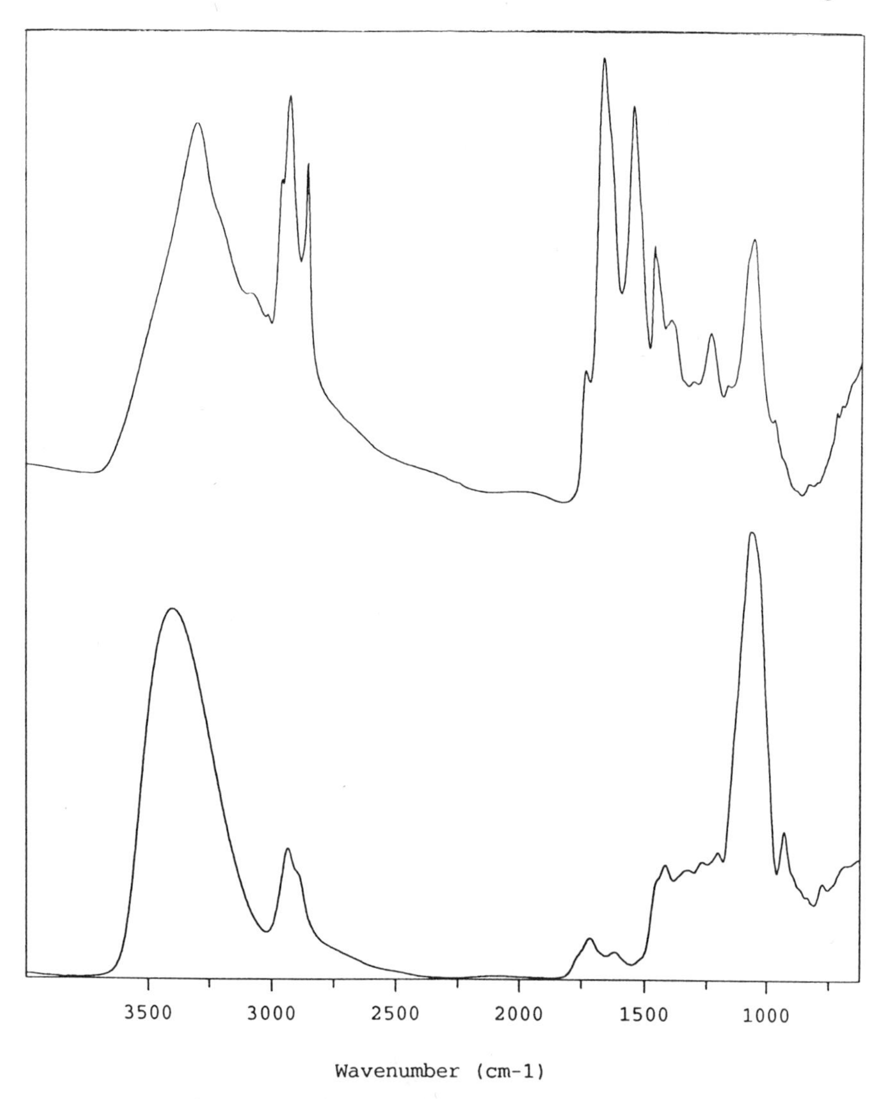

Figure 12. R/A spectrum of normal tissue (top) and an artifact (bottom) in a cross section of tissue (5 μm thick) mounted on an aluminized microscope slide.

described previously.) The best way to eliminate the specular component is to place a nonabsorbing material (such as a salt plate) on top of the sample. Since the refractive index of the salt plane (as opposed to air) is closer to that of the sample, the surface of the sample will appear to vanish. If (optically speaking) there is no surface, specular reflection cannot occur. In addition, since the top of the salt plate will be out of focus, specular reflection from its surface will not affect the spectral measurement of the sample. It is imperative that good optical contact be made between the salt plate and the sample surface; otherwise, interference fringes can appear. The best way to eliminate this possibility is to mount the sample in a compression cell. (We discuss the use of compression cells in more detail in Section 5.)

To analyze sub-μm-thick films, the incident beam of radiation must be brought in at higher angles of incidence (typically, 65 to 85° from normal). This type of R/A spectroscopy is called grazing-angle spectroscopy. Analyzing the sample near grazing angles results in an increase in the detectability of the thin film, due to the enhancement of the radiation that occurs [14].

Incident radiation's electric field can be resolved into two components: one component is parallel to the plane of incidence (*p*-polarized) and the other is perpendicular to the plane of incidence (*s*-polarized). The plane of incidence is the plane that contains the incident and reflected rays. Only the *p*-polarized component is enhanced in grazing-angle spectroscopy. Upon reflection of the *p*-polarized component, the incident and reflected beams constructively interfere with each other, producing a standing wave with a strong electric field oriented normal to the reflecting surface. Spectral absorption is greatly increased, provided that molecules on the surface have a vibrational component perpendicular to the surface. This effect is the main reason why grazing-angle spectroscopy can produce spectra on very thin films. Upon reflection of the *s*-polarized component, the incident and reflected beams destructively interfere with each other, forming a node at the surface. With no electric field, no interaction can occur between the infrared radiation molecules at the surface. For samples thinner than a quarter of a wavelength of the radiation, the *s*-polarized component is weakly absorbed and contributes little or no spectral information. Using linearly polarized incident radiation oriented parallel to the *p*-polarized component provides the greatest enhancement of the sample's absorption.

No sample preparation is required for grazing-angle samples. In many cases, grazing-angle microspectroscopy is a ''last-resort'' technique. It is advantageous only for sub-μm-thick films on metals, and it is of lower efficiency than most other techniques. However, if thicker samples cannot be obtained, or if molecular information (such as the orientation of bound monolayers at the surface) is desired, grazing-angle microspectroscopy is the only technique of choice. There is an objective available that enables grazing-angle spectroscopy to be performed on microsamples, which extends the usefulness of this technique to IMS [15].

A comparison between a grazing-angle reflection spectrum and an R/A spectrum is shown in Figure 13. The sample studied was a thin film of polyurethane on gold. The increase in intensity of the bands in the grazing-angle spectrum illustrates the greater sensitivity of this technique over the near-normal spectrum.

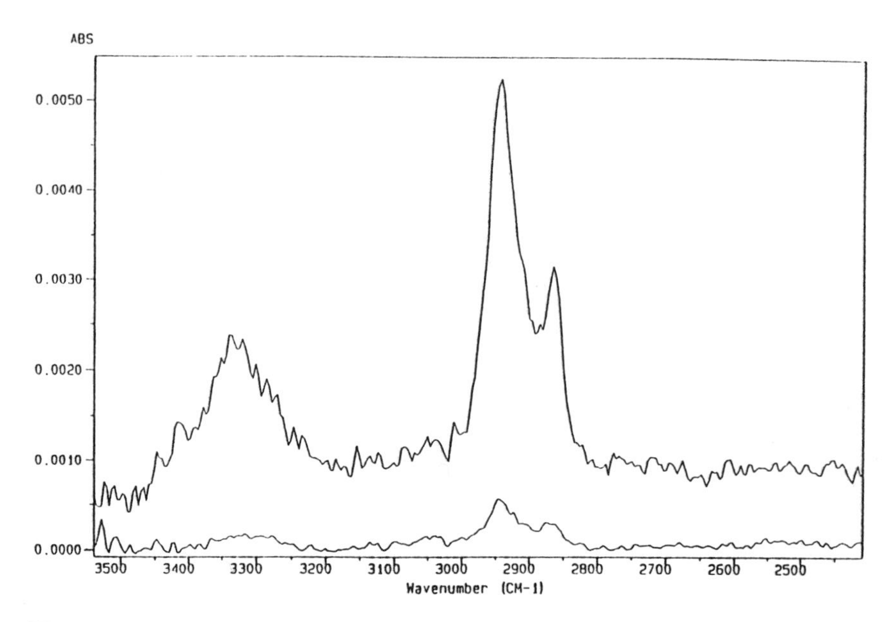

Figure 13. Grazing-angle spectrum of a thin film of polyurethane on a gold mirror (top) is compared to the near-normal incidence reflection-absorption spectrum of the same film (bottom).

5. BASIC TRANSMISSION TECHNIQUES AND PRINCIPLES OF IMS

The absorption of radiant energy as it travels through a material is the primary method for measuring absorption spectra. Analyzing samples by transmission is generally the preferred IMS method. Transmission spectroscopy can be photometrically accurate for quantitative analysis and is least likely to have spectral artifacts. A disadvantage of transmission spectroscopy is that it requires careful sample preparation. However, for problems that require the highest degree of spectral accuracy (e.g., quantitative studies, spectral subtractions), the time spent in sample preparation is well worth the effort.

There are many different techniques that are useful in the preparation of transmission samples, but they all have one common goal: obtaining thin samples. Depending on the sample's absorptivity, optimal sample thicknesses range from 0.5 to 30 μm. Basically, there are three broad categories of samples: those that can be sliced to the proper thickness, those that can be flattened to the proper thickness, and those that are formed out of smaller particles to obtain samples of the proper thickness.

5.1. Sectioning

Preparing thin (0.1 to 30 μm) sections is a routine microscopy procedure. Samples may be sectioned by hand-cutting, by using a microtome, or by grinding and polishing (called petrographic sectioning). For hand-sectioning, there are a variety of available tools and techniques. The simplest is to shave or slice portions off the bulk sample with a razor blade, diamond knife, or freshly fractured glass edge (depending on the hardness of the bulk material). The shaved fragments can then be sorted through under a low-power stereo microscope to find sufficiently thin sections. These tools are especially useful for sampling large materials, such as scraping the surface of a floor to analyze the wax or across a wood surface to analyze the varnish.

The second method—sectioning samples with a microtome—produces reproducible sample thicknesses, which is especially useful if any quantitative information on the sample is desired. However, if the sample is not self-supporting or is too small (so that it moves around when it is being cut), it must be mounted in a material that will hold it together. Various embedding materials (such as epoxy, wax, acrylic, gelatin, carboxymethyl cellulose, polyester, or fingernail polish) can be used. It is important to choose an embedding material that will not alter the sample chemically. The sample can also be physically clamped between two pieces of material to support it for sectioning. Plastic sheets, cork, soft metals (such as aluminum), and even raw carrots work well to hold samples for cutting. The samples and the embedding materials are sliced through together, and then the desired sample is taken out of the sliced fragments and analyzed. If the sample is permanently mounted in an embedding material, the entire sample/mount section is placed on the microscope stage for analysis. Care must be taken that no embedding material's spectral bands interfere with the sample's spectrum. This can usually be prevented by sampling far away from the embedding medium–sample interface or by spectrally subtracting the embedding medium's bands out of the sample spectrum.

An ideal candidate for microtoming is multilayer laminates. Cross sections of multilayer laminates will produce sections wherein each different layer can be analyzed, one layer at a time. In addition, the morphology of the sample will be preserved, as shown in Figure 14. Obviously, the wider the layers, the easier it is to detect them without interference from neighboring layers. Even if the layers are not sufficiently wide (around 15 to 40 μm), the sample can be prepared so that the layers will be wider. Figure 15 depicts the technique called oblique sectioning. If T is the original laminate layer thickness and S is the thickness of the cut section, cutting at an oblique angle (θ) will give a free width of the laminate layer that is neither above nor below any portion of a neighboring layer. This free width (FW) is defined as

$$\mathrm{FW} = T \sec\theta \; S \cot\theta \tag{6}$$

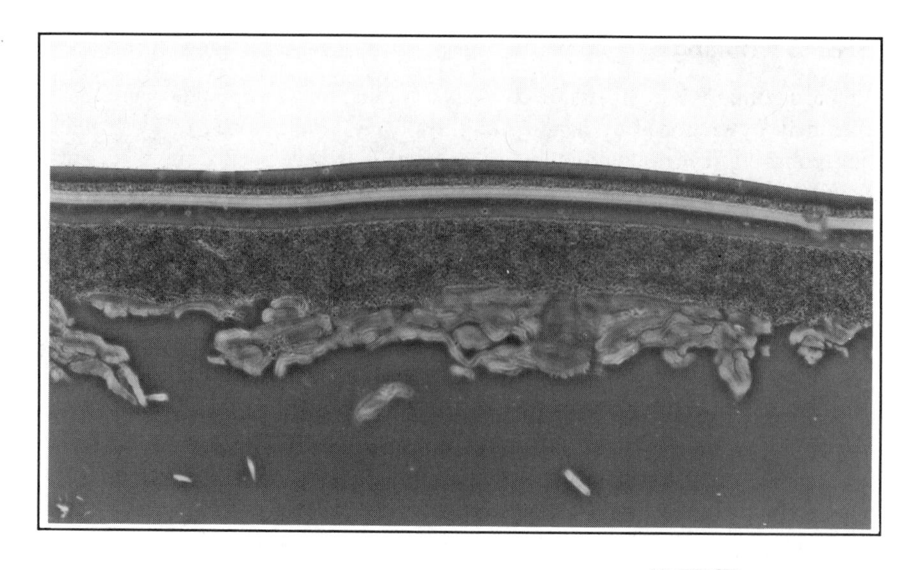

Figure 14. Photomicrograph of a multilayered laminate cross section. This sample is a multilayered coated photographic paper.

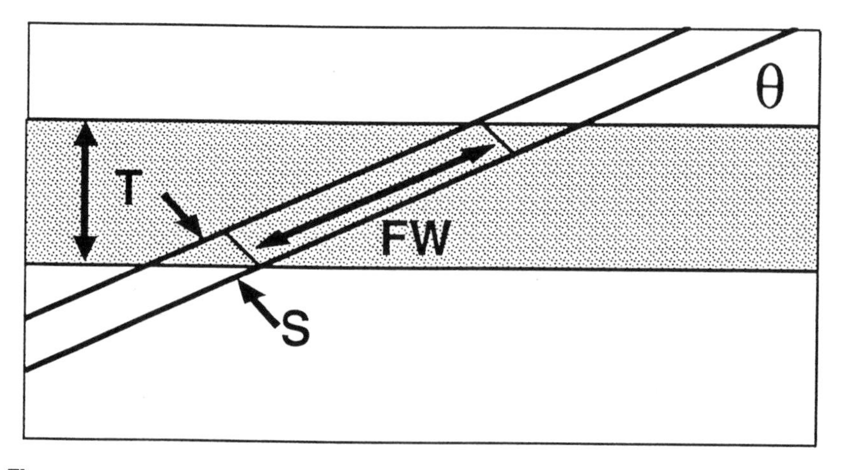

Figure 15. Diagram showing the geometry of oblique sectioning and the parameters that determine the free section width (FW).

Table 3 lists the free-width values resulting from a 1-μm-thick section of a 10-μm-thick layer at different angles. As shown by the table, the available sampling area can be increased by as much as 50 times the original area.

The third method used in preparing thin sections is grinding and polishing. Hard, brittle materials, which cannot be embedded and microtomed, can often be ground and polished to form thin sections. This technique is frequently used to prepare thin sections of rocks and minerals and is known as petrographic sectioning. Infrared studies of minerals and automotive brake linings have been reported using the petrographic thin-sectioning technique [16,17]. Fine grains of hard/brittle samples may be mounted in a suitable plastic embedding medium prior to grinding and polishing thin sections. For thin sections, the grinding-and-polishing operation is performed in two stages. The first involves grinding and polishing one surface of the material, while the second involves mounting this polished surface flat to a support material and then grinding and polishing the opposite side of the sample to its desired thickness.

Grinding is generally accomplished using silicon-carbide abrasive papers, ranging in abrasive-grit size from 60 to 600. Polishing is performed on a rotary lap using water or oil suspensions of finer and finer abrasives. Final polishing is generally achieved with $^{1}/_{4}$-μm diamond in kerosene or aqueous suspensions of 0.05-μm aluminum oxide or cerium oxide. Grinding and polishing samples thin enough for transmission infrared measurement is a tedious procedure. Before polishing the second surface, it is always advisable to examine the first (just-polished) surface to see if adequate data can be obtained by specular reflection. If so, this eliminates the need for preparing for thin section.

In the second step of the thin-section preparation technique, it is necessary to mount the polished surface onto a support. This should be done with a soluble adhesive so that the thin section can be removed from the support. If the sample

Table 3. Calculated FW Values for T = 10 μm and S = 1 μm

Angle	Free width (μm)
10	52.9
8	64.7
6	86.2
4	129.1
2	257.9
1	515.7
0	∞

itself is not affected by either acetone or methylene chloride, adhesives such as cyanoacrylate cements or certain hot-melt adhesives can be used.

5.2. Flattening

Flattening samples to the desired thickness is a feasible technique when it is not important to preserve the sample's morphology or orientation. Obviously, flattening techniques work best for samples that are soft and easily plastically deformed. Fibers, organic materials, soft minerals, plastics, and even tire rubber can be flattened with the appropriate tools. The process of flattening destroys sample morphology and orientation, which limits the use of flattening as a sample-preparation technique.

The simplest tool for flattening is a metal probe or roller. The sample is placed on a hard surface (such as a glass microscope slide), and the probe or roller is pressed over the sample. As the sample is flattened, it will often begin to appear transparent when it reaches the desired thickness. At this point, the sample can then be transferred to a sample holder for analysis. Flattening can also be performed on samples placed directly on transparent windows, but the windows are often damaged in the process. Polished aluminum plates or mirrors are good supports on which to flatten samples (in which case, spectra must be collected by reflection/absorption). This flattening technique works only for relatively soft samples that will retain their flattened shape once the probe is removed.

For harder, more-resilient samples, a diamond cell should be used. Unlike the high-pressure diamond cells currently on the market for analyzing samples under extreme pressures (like metallic hydrogen), these diamond cells are used to flatten and hold the samples in place. Since diamond windows are used, virtually any material can be flattened between them without damaging the windows.

5.3. Mini-KBr Disks

Small particulates (smaller than the diffraction limit of 10 μm) can easily be measured if they are mixed with a nonabsorbing matrix material (such as KBr) to form a micropellet [18]. While viewing under a low-power stereo microscope, the sample particulates and KBr crystals are crushed and mixed together with a fine-tipped metal probe, in approximately a 1:3 ratio. (A good probe to use is one made of tungsten wire; such a probe is easy to make.[1]) After the KBr and

[1]Tungsten wire (0.5 mm in diameter) heated in a flame until glowing will be etched to a very fine, sharp tip (2 to 5 μm) when placed into sodium nitrite or ammonium nitrite. The heat from the wire starts an exothermic reaction with the nitrite crystals, which etches away the outer surface of the tungsten wire. For details of this procedure, see ASTM E334-90 [19].

analyte crystals are mixed together, they are flattened with either a metal probe or a roller until a transparent micropellet is formed. This mixing-and-flattening procedure may need to be repeated once or twice to ensure that the crystals are mixed completely. There are a number of advantages to preparing samples in this manner. First, the sampling area is increased. Second, by dispersing the analyte crystals in the KBr, specular effects off the analyte particulates are diminished. Third, a sample flattened in KBr is usually much easier to pick up with a fine-tipped metal probe or razor blade than is the sample flattened by itself. Of course, crushing the sample destroys its morphology and orientation. For cystallinity studies, samples should not be prepared into micro-KBr pellets.

5.4. Optical Factors in Sample Mounting

Reflection and scatter are two problems inherent in IMS that can be minimized by proper sample mounting. If these problems are not addressed, spectral distortions may occur. A self-supported thin film will produce interference fringe patterns superimposed on the film's absorption spectrum. Optical inhomogeneities will scatter radiation from areas outside the desired sampling area, creating spectral interference. These optical factors must be understood and minimized if the user is to produce high-quality IMS spectra. Both reflection and scatter must be considered when performing either transmission measurements or reflection measurements, but the former are most affected.

5.4.1. Reflection

There are two types of unwanted reflections in transmission analyses: specular and internal. Since methods for eliminating unwanted specular reflection were discussed earlier, only internal reflection are discussed here. As an example of internal reflection, assume that the sample is a thin film suspended in air. While most of the radiation entering the film will be transmitted through, a fraction (around 4 to 8%) of the radiation will be reflected. Some of the radiation will reflect off the upper surface of the film. Some could reflect from the lower surface of the film, then reflect off the upper surface and back through the fiber a second time, and then on to the detector (Figure 16).

Since the film thickness should be around 0.5 to 30 μm and the infrared radiation passing through the sample has essentially the same wavelength (2 to 15 μm), the conditions are present for interference. Interference occurs between the transmitted beam and the doubly reflected beam, resulting in an interference fringe in the spectrum. For some qualitative work, the fringe may not be a problem, but it can interfere with library searching. When quantification is to be performed, the band intensities can be distorted by the superimposed interference fringe. These internal reflections can be eliminated if the sample is mounted in or on a material that has approximately the same refractive index

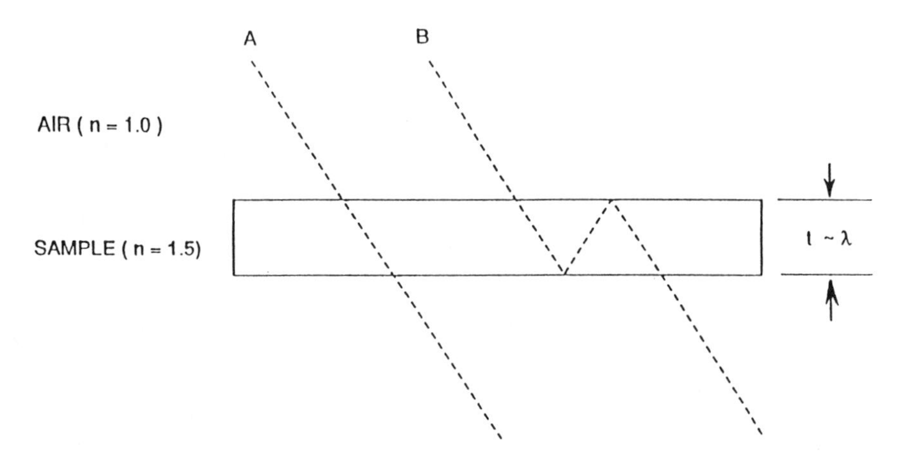

Figure 16. Diagram of internal reflection of a thin film, which leads to the production of interference fringes in spectra.

as the sample. The simplest way to accomplish this is to mount the sample in between or on top of salt plates. When mounting a sample between two windows, a crystal of KBr or another appropriate nonabsorbing material should also be placed between the windows to use for a background measurement. If the background is taken through the two windows in an area with no KBr in between them, the air gap between the windows will again produce an interference fringe. This time, the fringe is actually in the background single beam, not in the sample single-beam spectrum. Either way, the final ratioed spectrum will be distorted by the fringe. Taking the background through the KBr eliminates the fringe pattern in the ratioed sample spectrum.

Reflection and refraction at the sample edge can also affect spectral measurements. For example, if a section is placed on the stage and information is needed from both the interior and the edge of the section, the air–sample interface at the edge causes the beam to be scattered or reflected. The edge is visible because of the contrast created by the difference in refractive index between the sample and its surroundings. The eye or image detectors can only sense changes in brightness (intensity) or color (energy). If an intensity change results from differences in refractive index, the contrast is called phase contrast and the image is called a phase image. When absorption of radiation produces contrast, the image is called an absorption image. Obviously, only an absorption image is preferred for infrared measurements.

To obtain only an absorption image, the sample should be mounted in a refractive-index-matching fluid. If it is, refraction, scattering, and reflection will no longer occur at the sample edge; the phase image will be eliminated. Figure 17 shows an acrylic fiber mounted in air and in an oil that has an index of

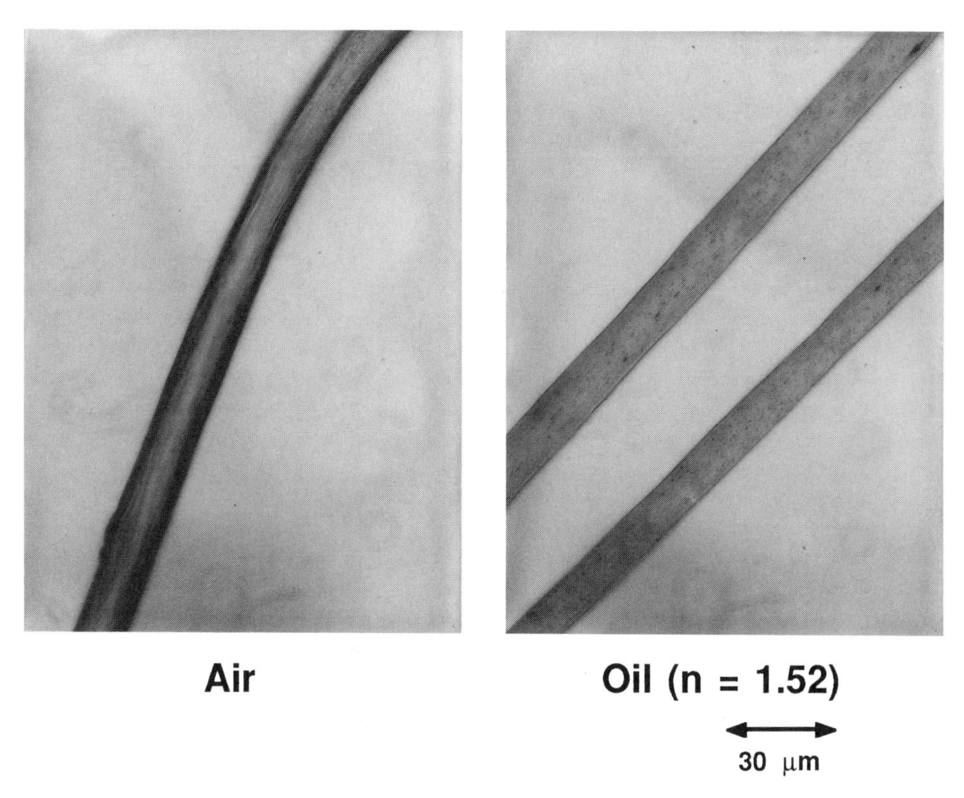

Figure 17. Micrographs of an acrylic fiber surrounded by air and similar fibers mounted in oil (n = 1.52). The high contrast of the fiber in air reduces its transmission and produces specular reflections that may distort spectral band shapes.

refraction similar to that of the fiber. It is obvious that in the air-mounted fiber, dark bands are apparent at the sample edges, due to the loss of light that occurs at the edge. When the same fiber is mounted in a liquid, these dark bands disappear, indicating that the light is no longer being lost at the edge. Preferably, the mounting medium should have few infrared absorption bands in addition to a refractive index medium close to that of the sample. Fully deuterated alkanes work well because their absorption bands are in spectral regions where those of typical organic samples do not occur. The spectrum of a single paper fiber mounted in deuterated hexadecane (Figure 18) illustrates the spectral quality that can be obtained with immersed samples. The infrared absorption bands due to this mounting medium can easily be subtracted. Of course, nonabsorbing materials (such as KBr, KI, or CsI) work extremely well as mounting media.

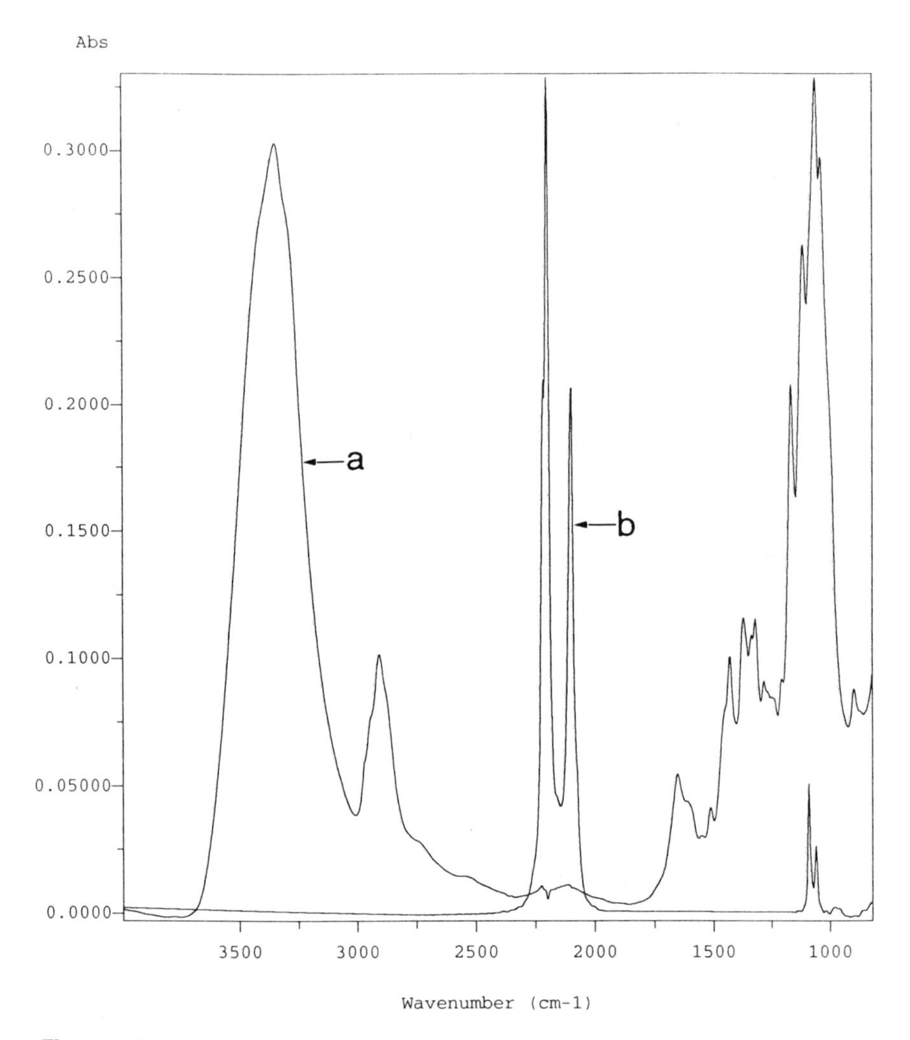

Figure 18. Deuterated alkane as a mounting medium. The spectrum of a single cellulose paper fiber (a) mounted in deuterated hexadecane, after subtraction of the deuterated hexadecane spectrum, and the spectrum of deuterated hexadecane (b).

5.4.2. Scatter

Radiation will be scattered if the particles have dimensions about the same as or smaller than the incident wavelengths, and if the particles are randomly distributed in a medium of refractive index different from their own [20]. There are two types of scatter: elastic and inelastic. In elastic scattering, no energy is gained or lost during the process. Elastic scattering, often called Rayleigh scattering, occurs when light scatters from a small particle dispersed in a matrix.

Scattering can also occur when small irregularities on the surface of a substance scatter light. The intensity of scattered light (I_s) is defined as

$$I_s = \frac{8\ \pi^4\ \alpha^2}{\lambda^4\ r^2}(1 + \cos^2\theta)I_0 \tag{7}$$

where α is the polarizability of the particle, λ the vacuum wavelength, θ the angle between the incident and the scattered rays, r the distance from the center of scattering to the detector, and I_0 the incident intensity.

The relationship between I_s and λ indicates that the scattering intensity will increase dramatically as the wavelength of light becomes shorter. As a practical example of this phenomenon, blue light (around 400 nm) is much more efficiently scattered than red light (around 600 nm). Dust particles and water vapor in the air will thus scatter blue light more effectively, resulting in the sky appearing to be blue.

When analyzing IMS samples that have a tendency to scatter, the spectra will be distorted due to the wavelength dependence of the scattering. For example, in a carbon-containing electrostatic copy toner sample, carbon-black particles scatter the light, giving a distorted baseline (see Figure 19). Fortunately, a baseline correction often makes the spectrum appear normal, so it can be interpreted.

5.5. Polarized Radiation in IMS

Polarized radiation adds a degree of order to radiation that, in turn, can detect order in materials on a molecular scale. In light microscopy, polarized light is used in the measurement of optical properties of crystals for chemical and mineralogical analysis. In spectroscopy, linearly polarized infrared radiation is used to supply the orientation of molecules in solids by recording the changes in the intensity of absorption bands as a function of the angle of polarization. IMS provides a unique ability to examine small crystals or areas in oriented materials via polarized infrared radiation. Furthermore, since crystals and oriented materials can be observed using polarized visible radiation, it is possible to be certain of the sample's orientation at the same time that its spectra are being measured with polarized infrared radiation. Because of this, polarized radiation is well suited to the study of molecular orientations and assists in vibrational-band assignments.

Polarized radiation can be used in transmission and internal-reflection spectral measurements. The resulting spectra are obtained for two orthogonal orientations of the polarized radiation. These spectra are referred to as dichroic spectra, and data are reported as dichroic ratios. Dichroic ratios have been defined in various forms [21]. The polarized IMS study of single fibers of polyacrylonitrile is an example of how polarized IMS is important for molecular-

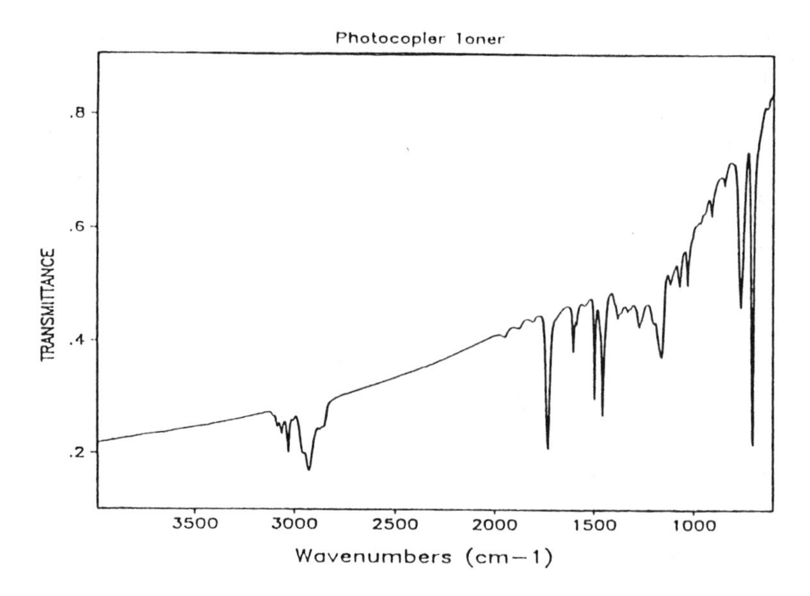

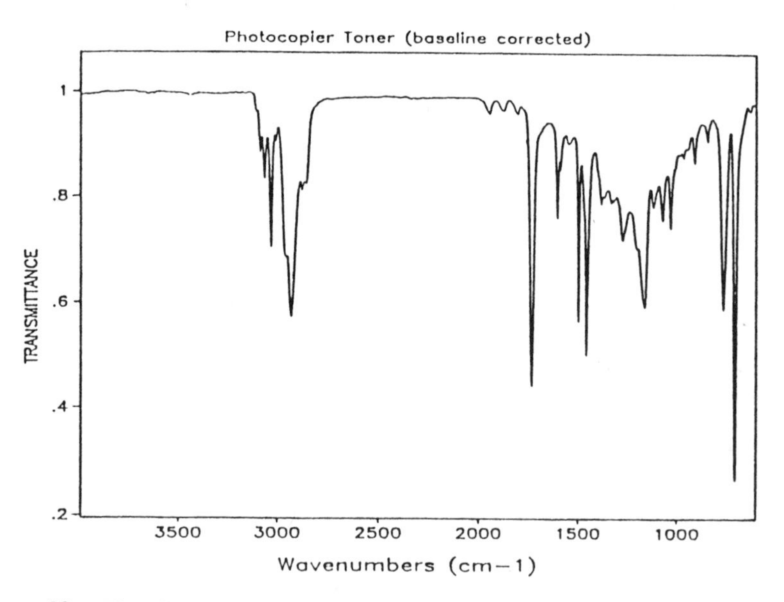

Figure 19. The effects of scatter are shown by the sloping background in the R/A spectrum of an electrostatic copier toner transferred to an aluminum mirror (top). The bottom spectrum is the is the same data after background correction.

orientation studies. In the fiber-manufacturing process, the polymer is extruded through a spinneret as a viscous solution (or a melt), the solvent is evaporated (or the melt is cooled), and then the fiber is stretched (or drawn) to induce orientation, which increases the mechanical strength of the fiber. Since the fiber's physical properties depend on the degree of orientation, it is important to be able to measure this property. Generally, optical birefringence is very useful in determining the degree of orientation. However, in fibers made from polyacrylamide, the nitrile groups are oriented perpendicularly to the carbon–polymer chain. Because of this, these fibers have unique optical properties. Even when highly oriented, their optical birefringence is nearly zero.

The polarized IMS spectra of single acrylic fibers are shown in Figure 20. The nitrile absorption band's intensity is greater when the electromagnetic-radiation field is perpendicular to the fiber axis than it is when the field is parallel to the axis. If the fiber were randomly oriented (or isotropic), there would be no difference between the absorption band for different orientations of the polarized infrared radiation. The fundamental work by Fraser in the mid-1950s provides a basis for interpreting polarized infrared spectra of oriented materials [22].

Polarized IMS studies of crystalline materials has not been a major application but is important to consider because of the effect of polarization on spectra. Generally, the infrared beam produced by an interferometer is partially polarized. While the degree of polarization may only be 15 to 20%, this can cause spectral variations that affect spectral interpretation. For example, if two single crystals were oriented differently when spectra were recorded, the crystals may show minor differences in their spectra. These differences may indicate that these two crystals represent different polymorphic forms. However, it may also be that since the crystals are arranged differently in the field, they are measured with different orientations of polarized radiation and that the spectral differences result from their geometrical arrangement.

In the critical analysis of single crystals, it is important to align the crystals using visible polarized light and to measure the infrared spectra with polarized radiation. If there is a distinct morphological feature, this can be used to align the materials prior to measurement. If no unique morphological feature exists, conoscopic observations of optical-interference figures can be used to determine optical orientation. When making spectral measurements on crystal and/or oriented materials, it is useful to have a centerable, rotating stage on the microscope. The rotating stage can be manipulated so that the axis of rotation is centered in the field of view. This allows samples to be rotated such that measurements can always be made with samples oriented in the same direction.

Polarized spectral measurement of single crystals is also useful in spectral-band assignments. The symmetry of the molecular vibration and of the lattice geometry can cause variations in dichroic ratios. These intensity changes with

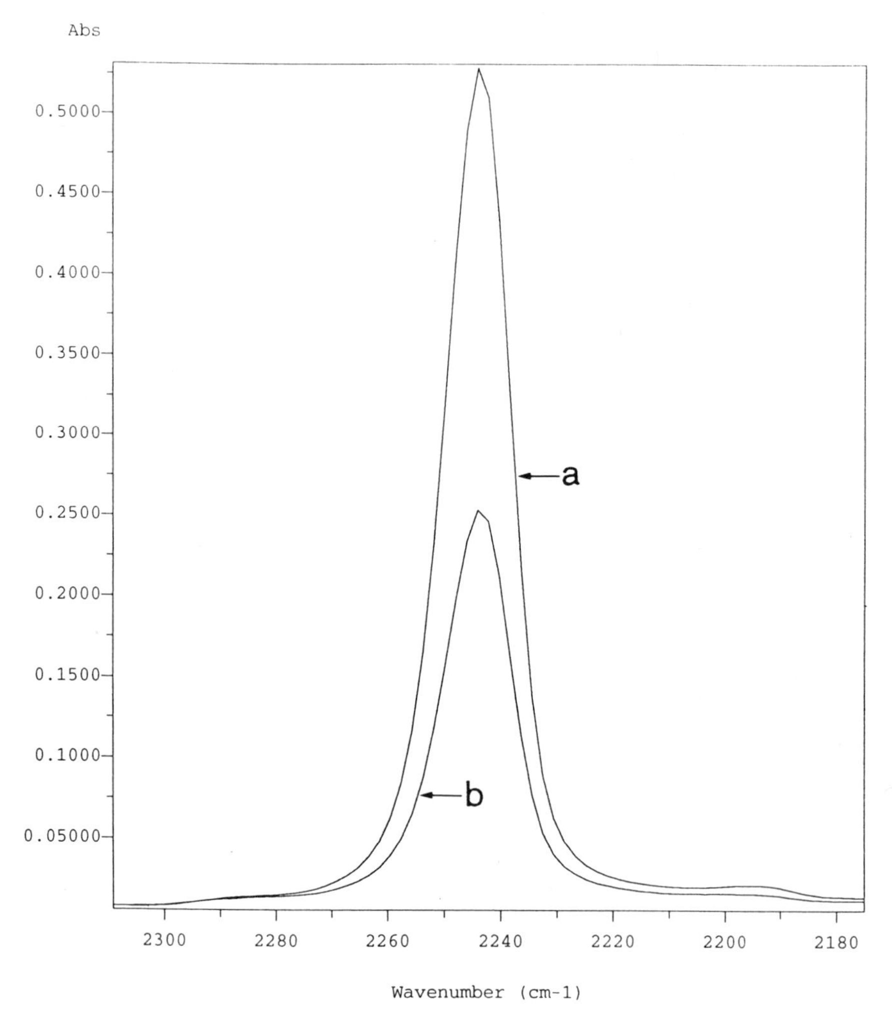

Figure 20. Polarized spectra of acrylic fiber wherein the highest absorption (a) of the nitrile band (2242.1 cm^{-1}) occurs when the polarized radiation field vector is perpendicular to the fiber axis and its lowest absorption (b) occurs for the parallel orientation. The dichroic ratio b/a = 0.497.

the relative orientation of polarized radiation can be used to deduce the spectral-band assignment [23,24]. For example, by taking polarized spectra of a crystal of tetronic acid that is oriented both parallel and perpendicularly to the infrared radiation, it is possible to distinguish between the OH in-plane bend and the CH_2 wag. Both vibrations absorb in the same spectral region, but the polarized spectra resolve them into two peaks at 1368 and 1358 cm^{-1} (Figure 21). Tetronic acid crystals, shown in Figure 22, were studied to determine their optical crystallographic properties. This study showed that tetronic acid was an orthorhombic crystal and that it could be orientated by its optical properties. This assured that polarized spectral measurements were made parallel to principal optical directions. The optical crystallographic data are presented in Figure 23.

Polarized radiation can also be used in grazing-incidence R/A to increase the sensitivity. In this technique, polarized radiation is used to eliminate the *s*-polarized component, since this component contains no spectral information for films less than one-fourth of a wavelength in thickness. It is important to note that it is not necessary to use polarized radiation in grazing-incidence R/A to detect molecular orientation; the reflection process itself causes an oriented electromagnetic field at the reflecting surface. Using *p*-polarized radiation in grazing-incidence R/A increases both the sensitivity and the photometric accuracy of the measurement of very thin films on metals.

Overall, polarized infrared radiation has been used in IMS studies of fiber orientation, single crystals, and grazing-incidence R/A. One disadvantage is that when a polarizer is inserted into the microscope, it reduces the energy throughput by a factor of at least 50%. In practice, 30 to 40% of the unpolarized energy emerges from wire-grid polarizers. This reduced intensity affects the *S/N* performance of the system. However, there is still energy sufficient to achieve high-quality spectra from microscopic samples with polarized IMS, and the additional structural information gained by using polarized radiation compensates for the reduced *S/N*.

Polarized light has a long history in aiding microscopists to extract more information from an image than its shape and form. Traditionally, the polarized-light microscope has been used to determine the optical properties of color, refractivity, optical signs, and conoscopic symmetries. With the addition of infrared absorption spectroscopy that IMS systems give, the microanalyst today has new power to analyze the microscopic world.

6. CONCLUSIONS

Throughout this chapter, the characteristics that both microscopy and spectroscopy contribute to IMS analysis were discussed. Each of these two parent disciplines has advantages that contribute to the success of IMS. For example:

Microscopy techniques are used to:

1. Provide the ability to view the sample and define its area for spectral analysis.
2. Provide morphological analysis to determine the origin of the sample (e.g., cornstarch vs. rice starch).

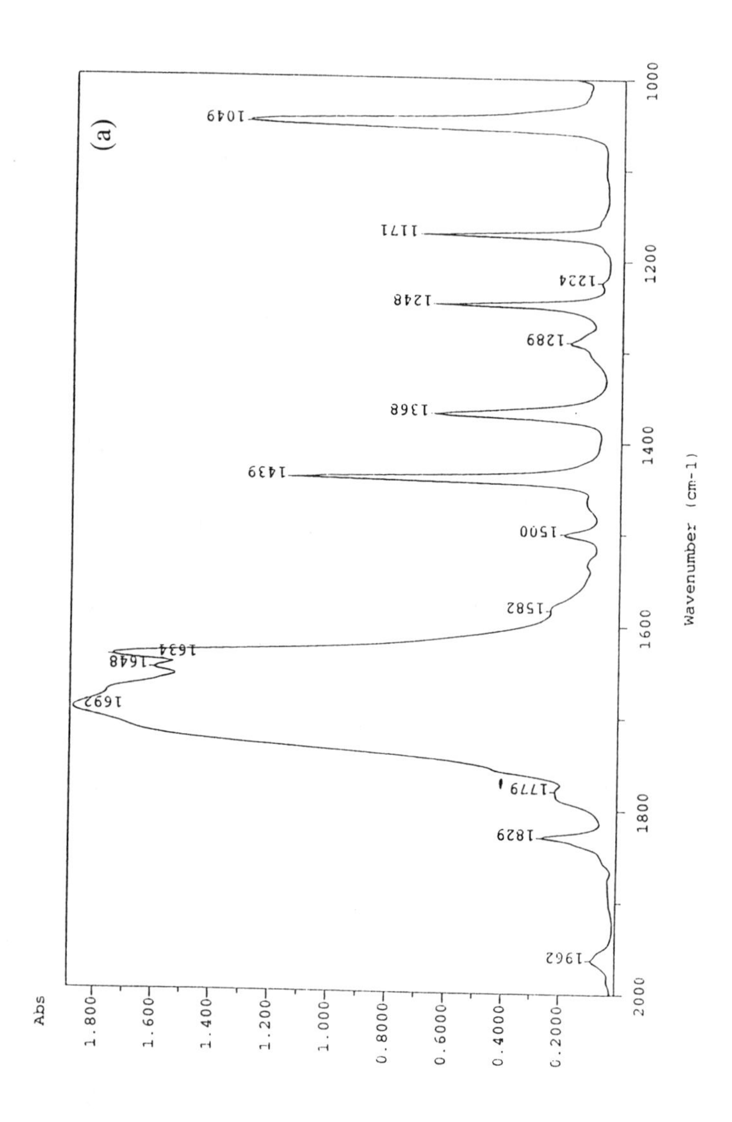
(a)
Abs
1.800
1.600
1.400
1.200
1.000
0.8000
0.6000
0.4000
0.2000
Wavenumber (cm-1)
2000
1800
1600
1400
1200
1000
1962
1829
1779
1692
1648
1634
1582
1500
1439
1368
1289
1248
1224
1171
1049

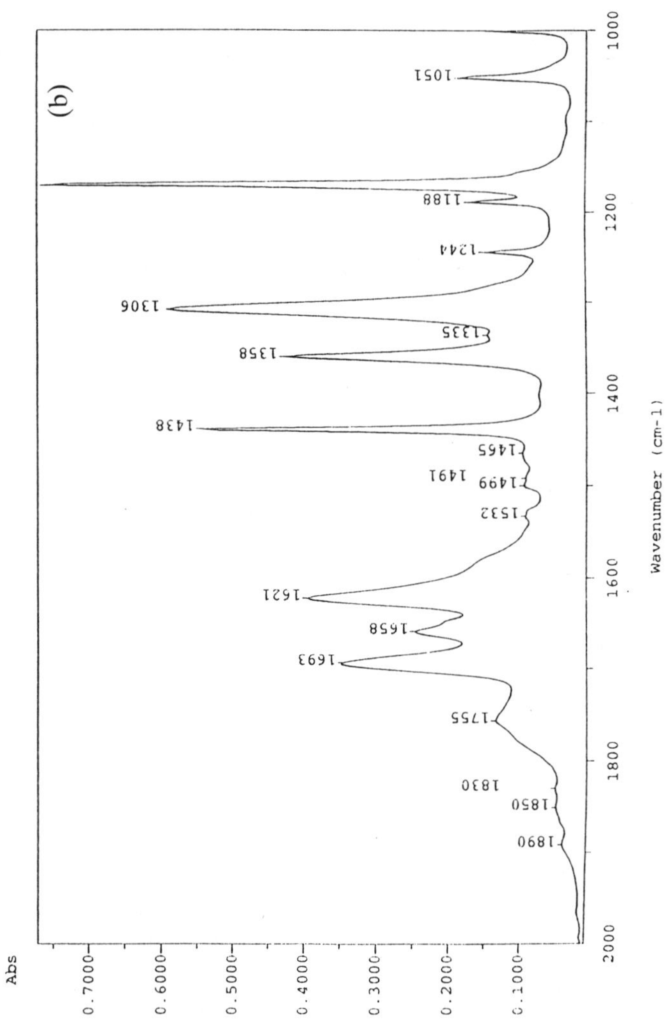

Figure 21. Polarized infrared spectra from a single crystal of tetronic acid showing the O—H in-plane bend at 1368 cm^{-1} (a) and the C—H_2 wag at 1358 cm^{-1} (b).

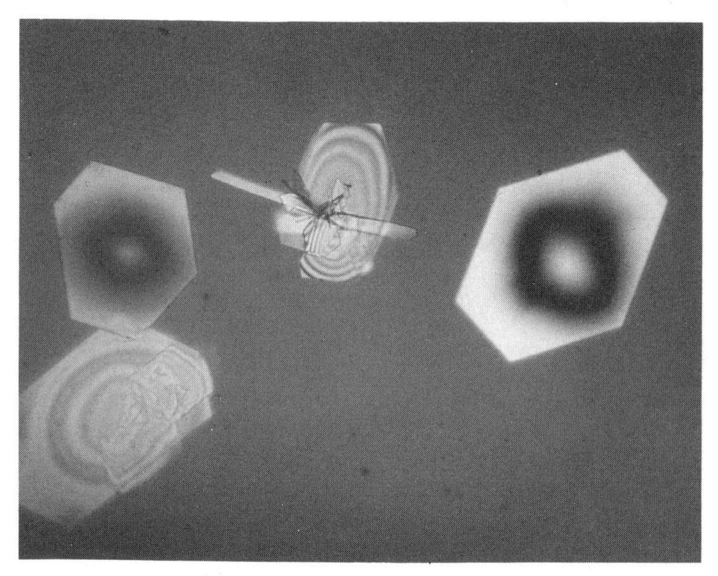

Figure 22. Micrograph of tectronic acid grown from water–glycerol solution.

3. Make use of polarized light to orient samples, measure optical properties, and enhance contrast to detect birefringent features.
4. Prepare samples for analysis (e.g., sectioning and mounting).

Spectroscopy techniques are used to:

1. Analyze the sample in a variety of ways (e.g., SR, DRIFTS, ATR, R/A, and transmission IMS).
2. Supply polarized radiation for orientation studies.
3. Determine the chemical composition of the sample.
4. Provide the means to relate the sample's chemistry to its microstructure.

Essentially, microscopy and spectroscopy are used together in IMS to solve the analytical problem at hand. The best way to achieve this goal is to ensure that the sample area is clearly defined, the instrumentation is free of aberrations, and the proper IMS technique is used. Only if these steps are taken can high-quality spectra be obtained. The lists above illustrate only a few of the valuable contributions of both techniques. The primary advantage in using two techniques is that where one of the disciplines may be lacking, the other may excel. For example, one problem with IMS is that a sample viewed with visible light is not representative of the area that will be analyzed by infrared light. By using techniques from both disciplines, this problem can be overcome. The sample can be prepared correctly to reduce scattering and unwanted reflections, and an

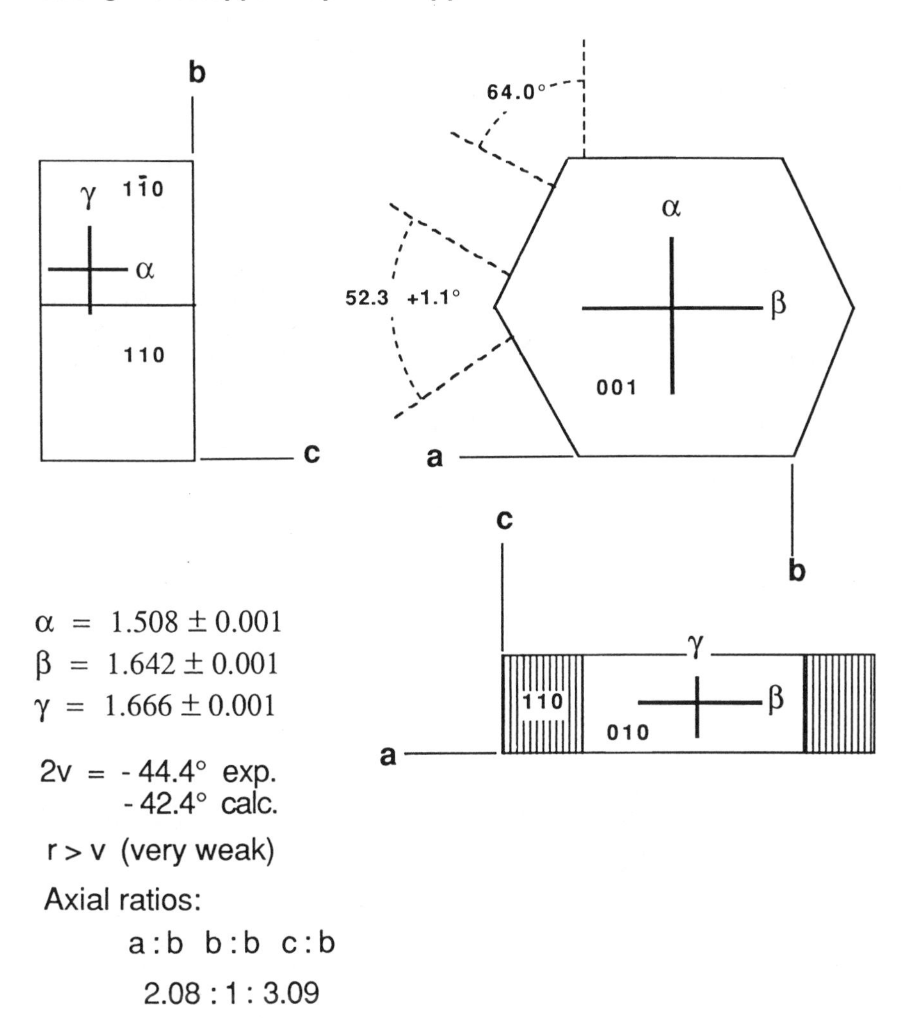

Figure 23. Optical crystallographic properties of tetronic acid determined by polarized-light microscopy.

instrument can be used that enables the user to reduce the amount of glare, diffraction, and refraction effects.

In conclusion, the intelligent user will recognize that IMS is truly a union of microscopy and spectroscopy. The purpose of this chapter was to provide the information that will guide the user in taking the best approach in analyzing IMS samples. Far too often, the advantages that one of the techniques can offer to obtain successful analyses are overlooked, since most of today's users are trained primarily as either a microscopist or a spectroscopist. IMS requires a new mind set, that of a microspectroscopist. In the words of Louis Pasteur: ''In the field of observation, chance favors only the prepared mind'' [25].

REFERENCES

1. Kohler, A. (1893). *Z. Wiss. Mikrosk., 10*: 433–440; translated (1993). *RMS Proc., 28*(4): 181–186.
2. Schwarzschild, K., and Villiger, W. (1907). *Astrophys. J., 40*: 317.
3. Schwarzschild, K. (1906). *Abhandlungen d. Koniglichen Gesellschaft d. Wissenshaften z. Gottingen, Mathematisch-Physikalische Klasse*, Band IV, pp. 1–54.
4. Cournoyer, R., Shearer, J. C., and Anderson, D. H. (1977). *Anal. Chem., 49*: 2275.
5. Reffner, J. A. (1992). *Proc. EMSA/MSA*, Boston, part II, pp. 1526–1527.
6. Smith, J. W. (1966). *Modern Optical Engineering*, McGraw-Hill, New York, p. 83.
7. Kortum, G. (1969). *Reflectance Spectroscopy: Principles, Methods, Applications*, Springer-Verlag, New York.
8. Wiley, R. R. (1976). *Appl. Spectrosc., 30*(6): 593–601.
9. Harrick, H. J. (1979). *Internal Reflection Spectroscopy*, Harrick Scientific Corp., Ossining, NY.
10. Reffner, J. A., Alexay, C. C., and Hornlein, R. W. (1991). *SPIE: 8th Int. Conf. on Fourier Transform Spectroscopy*, 1575, pp. 301–302.
11. Mirabella, F. M. (1993). *Internal Reflection Spectroscopy Theory and Applications*, Marcel Dekker, New York.
12. Fahrenfort, J. (1961). *Spectrochim. Acta., 17*: 698.
13. Sting, D. W. (1992). *ATR Objective and Method for Sample Analyzation Using an ATR Crystal*, U.S. patent 5,093,580 (assigned to Spectra-Tech, Inc., Shelton, CT).
14. Greenler, R. G. J. (1966). *J. Chem. Phys., 44*(1): 310–315.
15. Sting, D. W., Messerschmidt, R. G., and Reffner, J. A. (1991). *Optical System and Method for Sample Analyzation*, U.S. Patent 5,019,715 (assigned to Spectra-Tech, Inc., Shelton, CT).
16. Brenner, D. (1983). In *Chemistry and Characterization of Coal Macerals* (R. E. Winans and J. C. Crelling, eds.), *ACS Symposium Series* 252, pp. 4–64.
17. Nakashima, S., Ohki, S., and Ochiai, S. (1989). *Geochem. J., 23*: 57–64.
18. Sommer, A. J., and Katon, J. E. (1988). *Proc. 23rd Annual Conference of the Microbeam Analysis Society* (D. E. Newbury, ed.), San Francisco Press, San Francisco, pp. 207–214.
19. ASTM method E334-90, *Recommended Practices for General Techniques of Infrared Microanalysis*, ASTM, Philadelphia.
20. Strobel, H. A., and Heineman, W. R. (1989). *Chemical Instrumentation: A Systematic Approach*, 3rd ed., Wiley, New York, pp. 220–222.
21. Chase, B. (1988). In *Infrared Microspectroscopy* (R. G. Messerschmidt and M. A. Harthcock, eds.), Marcel Dekker, New York.
22. Fraser, R. D. B. (1953). *J. Chem. Phys., 21*: 1511–1515.
23. Hornig, D. F. (1948). *J. Chem. Phys., 16*(11): 1063–1076.
24. Krause, P. D., et al. (1977). *Appl. Spectrosc. 31*(2): 110–115.
25. Roberts, R. M. (1989). *Serendipity: Accidental Discoveries in Science*, Wiley, New York, p. 65.
26. Katon, J. E., and Sommer, A. J. (1992) *Anal. Chem., 64*: 931A.

3
Infrared Microimaging

K. Krishnan, Jay R. Powell, and Stephen L. Hill Bio-Rad Laboratories, Cambridge, Massachusetts

1. INTRODUCTION

The identification and study of microcontaminants or inhomogeneities in solid samples such as polymers, biological specimens, minerals, and semiconductor materials is an important application of analytical techniques. While many analytical techniques such as optical microscopy, electron microprobe, SIMS, and others can provide valuable information on these chemical systems with very high spatial resolution, infrared spectroscopy (and the complementary technique, Raman spectroscopy) has been shown to be a very useful tool as well. The infrared absorption spectrum of molecular compounds contains bands that can be associated with specific functional groups present, and a plot of the variation of such absorption band intensities over the entire area of the sample under study can be considered to be a chemical map of the sample. Examples of such infrared mapping of semiconductor samples using a laser scanning near-infrared microscope [1,2] have been published in the literature. Most such applications of chemical imaging by infrared and Raman spectroscopy have been reviewed by Treado and Morris [3]. More recently, such chemical mapping or functional group imaging on a microscopic level using Fourier transform infrared (FT-IR) spectroscopy has become popular, and this FT-IR microimaging technique is reviewed in this chapter.

Since its introduction in 1983, Fourier transform infrared (FT-IR) microspectroscopy has become a powerful tool for the analysis of microscopic contaminants in chemical systems and for the characterization of inhomogeneous materials on a microscopic scale [4–6]. Using commercially available FT-IR microspectrometers, it is easily possible to record the infrared transmission or reflectance spectra of samples with minimum dimensions on the order of 10×10 μm. Most of the initial experiments in this field consisted of moving the sample stage manually and recording a few of the infrared spectra over the sample area of interest. Most of the early work along these lines can be found in an earlier volume in this series [7]. By using a motorized *x–y* sample stage and appropriate software techniques, Harthcock et al. [8–10] have shown that it is possible to produce functional group maps or functional group images of the sample under study. This ability of FT-IR microimaging can thus be used to produce complementary information to go along with other established imaging techniques, such as electron microprobe [11,12], surface-enhanced Raman spectroscopy [13], and laser electron microscopy [14]. As mentioned above, the infrared and Raman spectroscopic imaging has also been reviewed by Treado and Morris [3].

Since the initial work of Harthcock et al. [8–10], a number of papers have been published in the literature using the FT-IR microimaging technique [15–24]. Attempts have also been made to enhance the infrared microimages by mathematical manipulations of the raw spectra data [25]. These and the other aspects of infrared microimaging are reviewed in this chapter.

2. EXPERIMENTAL DETAILS AND CONSIDERATIONS

2.1. Micro FT-IR Instrumentation

The design considerations and the operational principles of the microscopes used as FT-IR microsampling accessories have been well understood and described in the literature [6,7]. Such microscopes are available from FT-IR manufacturers such as Analect, Bio-Rad, Bruker, Horiba, Jasco, JEOL, Mattson, Nicolet, and Perkin-Elmer, and from FT-IR accessory suppliers such as Spectra-Tech. There are two categories of FT-IR microsampling systems currently available. In the first category the FT-IR sampling microscope is placed externally on to one side of a standard FT-IR instrument. These systems can then function as both general-purpose FT-IR instruments and FT-IR micro systems. In the second class of instruments the interferometer and related optics, together with the microscope, are located in one complete enclosure. Instruments in either category can be used to produce good-quality infrared microimages.

2.2. Principles of FT-IR Microimaging

Figure 1 shows the block diagram of a typical FT-IR microimaging experimental configuration. The essential components of the system are the FT-IR spectrom-

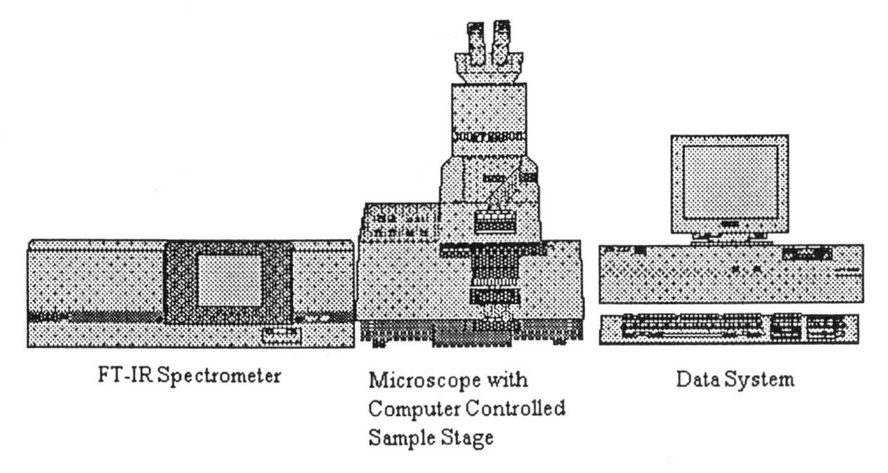

Figure 1. Typical FT-IR microimaging experimental configuration.

eter with its data system, and the microscope. The radiation from the FT-IR instrument is focused on the sample (in either the transmission or the reflection mode) located on a motorized *x–y* stage of the microscope. The *x–y* motion of the sample stage is controlled by the same computer that runs the FT-IR instrument. The radiation transmitted or reflected by the sample is passed on through the microscope optics to a high-sensitivity, small-area mercury–cadmium–telluride (MCT) detector. The variable aperture that is part of the microscope determines the area of the sample from which the FT-IR spectra are recorded (the individual cells in the grid patterns shown in Figure 2). With the sampling area thus predetermined by the use of the variable aperture, a series of infrared spectra of the sample under study are recorded by *x–y* translation of the sample stage. Typically, the sample stage movement is controlled in such a way that contiguous FT-IR spectra are recorded (Figure 2a). For instance, if a 20- × 20-μm sampling size is utilized in the experiment, the stage is moved in increments of 20 μm on the *x* and *y* axes. The effect of the stage step rate on the spatial resolution in the recorded spectra is discussed below. The computer that is part of the FT-IR data system keeps track of the spectra recorded and their corresponding *x–y* coordinates. It is a relatively simple matter, then, to produce the functional group images. These images are produced by displaying or plotting the infrared band intensities (above the optional local baselines) in the recorded spectra as a function of the sample *x–y* coordinates. The infrared bands under consideration can be selected based on their functional group assignments. Alternatively, images of infrared absorption band ratios, or the results of quantitative analysis based on the infrared spectra, could also be produced. These

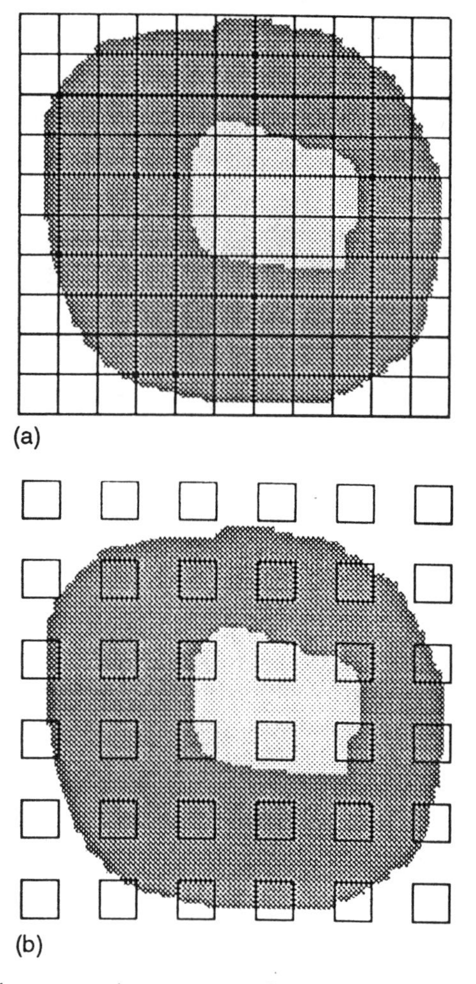

Figure 2. Variable aperture size versus sample stage movement: (a) stage movement same as aperture size; (b) stage movement greater than aperture size; (c) stage movement less than aperture size. In all cases, aperture size remains constant.

images can be presented in the form of three-dimensional wire frame or contour plots using commercially available software packages. Chemometric techniques such as factor analysis, principal component regression analysis, partial least squares analysis, and Gram–Schmidt orthogonalization procedure can be used to enhance the information content in the images. Numerous examples of the applications of the techniques above are reviewed in this chapter.

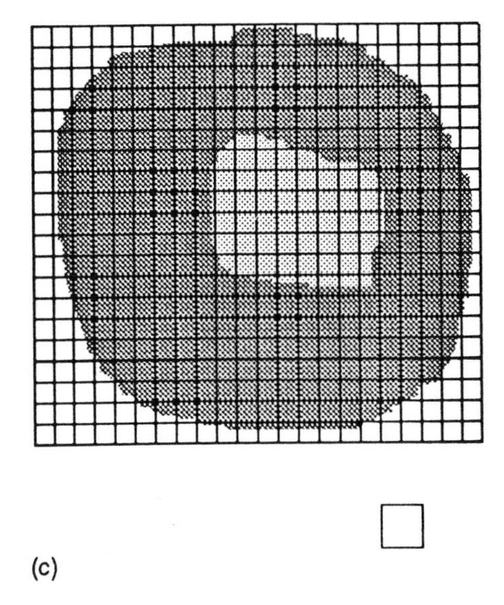

(c)

There has been considerable discussion in the literature regarding spatial resolution or stray light in FT-IR microspectroscopy. Messerschmidt [26] has claimed that "redundant aperturing" is necessary to guarantee true spatial resolution. However, some of the data presented [27] seem to indicate that the actual peak absorbances of a given sample change with the changes of the redundantly apertured sampling sizes. Perhaps more work is needed before one can understand the advantages of the redundant aperturing FT-IR microscopes applied to a wide range of chemical systems.

In the specific area of FT-IR microimaging, Harthcock et al. [27,28] have studied, for a given apertured FT-IR spectral sampling size, the effect of the sample x–y stage step rates on the spatial resolution of the microimages. They identify the three cases shown in Figure 2. The individual cells in the grids shown in this figure represent the areas of the sample from which FT-IR spectra are collected. The shaded area in the figure represents the x–y step used to map the sample.

a. *Step rate same as the spectral sampling size* (Figure 2a). The result will be to produce FT-IR spectra from contiguous areas of the sample and represent the best trade-off between spatial resolution and the experimental measurement time. The resultant microimage will be expected to have the same spatial resolution as that selected by the variable aperture.

b. *Step rate larger than the spectral sampling size* (Figure 2b). In this case there will be a considerable overlap of the spectral features due to the various imhomogeneities in the sample under study. The resultant microimage will be expected to have less spatial resolution than that selected by the variable aperture and may miss significant features from the sample under study. This condition should not be used for microimaging studies.

c. *Step rate smaller than the spectral sampling size* (Figure 2c). In this case there will be the guarantee that the spatial resolution selected by the variable aperture will be maintained in the microimage. However, the experiment under these conditions will take a considerably longer time. Indeed, Harthcock et al. [21] have shown that the microimages produced by the application of chemometric methods to the FT-IR data collected under this condition can be used to effectively increase the spatial resolution.

In summary, the experimenter will have to make the judicious choice from above based on his or her ultimate goals.

3. APPLICATIONS

The numerous publications in the literature attest to the usefulness of the FT-IR microimaging technique. Applications of the technique in the areas of polymers, biological specimens, minerals, and semiconductor materials are discussed in this section. As described elsewhere [6], micro FT-IR spectra of most solid materials can be recorded in the transmission or the reflection mode. For transmission studies the sample under consideration is microtomed into thin sections 5 to 15 μm in thickness. Thick samples can be studied in reflection. In this case, if the sample is a diffuse scatterer, the diffuse reflectance spectra of the sample will be recorded and can be analyzed directly. However, if the sample is a thick, smooth dielectric, the recorded reflectance spectra will contain first-derivative-like bands. A Kramers–Kronig analysis of such data will yield the extinction coefficient spectra, which can then be used for producing the microimages.

3.1. Polymer Systems

As mentioned in Section 1, Harthcock and co-workers [8–10] were the first to apply the technique of FT-IR microimaging, and they studied multilayer polymer laminates, imperfections in polymers, and polymer blends. Harthcock and Atkin [28] have reviewed some of these results as well as their imaging technique. In these initial experiments they used the Bio-Rad FTS 50 FT-IR instrument and the UMA-100 microscope equipped with a computer-controlled sample stage. The data collected were transferred over to a personal computer, and a software program written in the turbo Pascal language was used to produce color microimages. A commercially available plotting package was used to make three-dimensional wire frame and contour plots.

The simplest example of microimaging given by the above authors involved recording a series of FT-IR spectra across a microtomed five-layer laminate, including a layer of aluminum, and plotting the intensities of the infrared absorption at three different frequencies. From these one-dimensional images one can clearly identify the four infrared transparent layers. The authors also studied the two-dimensional mapping of an inclusion in a polyethylene film, and the presence of ''fish eyes'' on a painted, extruded polyurethane panel. The latter mapping was done using microreflectance spectroscopy.

Nishioka et al. [23] have analyzed the coated surface between two different polymers using the FT-IR microimaging technique. They prepared microtomed specimens of urethane paint coated on an ethylene–acrylic acid copolymer (EAA) and urethane paint coated on an ethylene ethyl acrylate copolymer with a partially hydrolyzed ethylester group (EAA/EEA). Using sampling sizes of 10×200, 10×100, and 10×50 μm (the long side being parallel to the paint substrate interface) and *x–y* steps of 10 μm, they recorded the microtransmission spectra over the interface. Figure 3 shows the microtransmission spectra of polyurethane and the EAA/EEA. Figure 4 shows the microimages created from the 1707/1464 and 1737/1464 cm^{-1} band ratios, respectively. In the EAA/EES system, the band at 1707 cm^{-1} arises due to the COOH group, the 1737-cm^{-1} band can be attributed to the C=O bond, and the 1464-cm^{-1} band is due to the scissoring vibration of the methylene groups. One can see clearly in Figure 4a that the $COOH/CH_2$ band ratio decreases at the interface. Such reduction in intensity is not seen for the 1737/1464 cm^{-1} band ratio. These results have been interpreted to indicate that there is a molecular interaction between the carboxyl group of EAA/EEA and the amine group of the urethane; however, there is no such interaction involving the carbonyl group.

Steger et al. [24] have reported on the one-dimensional local analysis across membranes of a flexible polyurethane foam. Polyurethane foams contain cells that are formed by stiff struts and thin membranes. By collecting a series of contiguous spectra, with a spatial resolution of 200 μm, across the membrane, the authors were able to monitor the thickness of the membrane, and the heterogeneity of its material. Zumbrum et al. [29] have published the FT-IR functional group images of multifunctional acrylate photopolymers. Various acrylate gels were prepared by ultraviolet photo-irradiation of the samples for various durations. From the FT-IR microimages of these samples, these authors have been able to draw conclusions regarding the effiency of cross-linking in the material as a function of the ultraviolet irradiation time.

The authors of the present chapter have studied the application of the FT-IR microimaging technique to the study of a glass epoxy composite. The sample showed the presence of gray-colored ''veins'' under optical microscopic examination, and was in the form of a 2-mm-thick polished plate. The sample was too thick to be used for transmission studies, and hence the FT-IR spectra

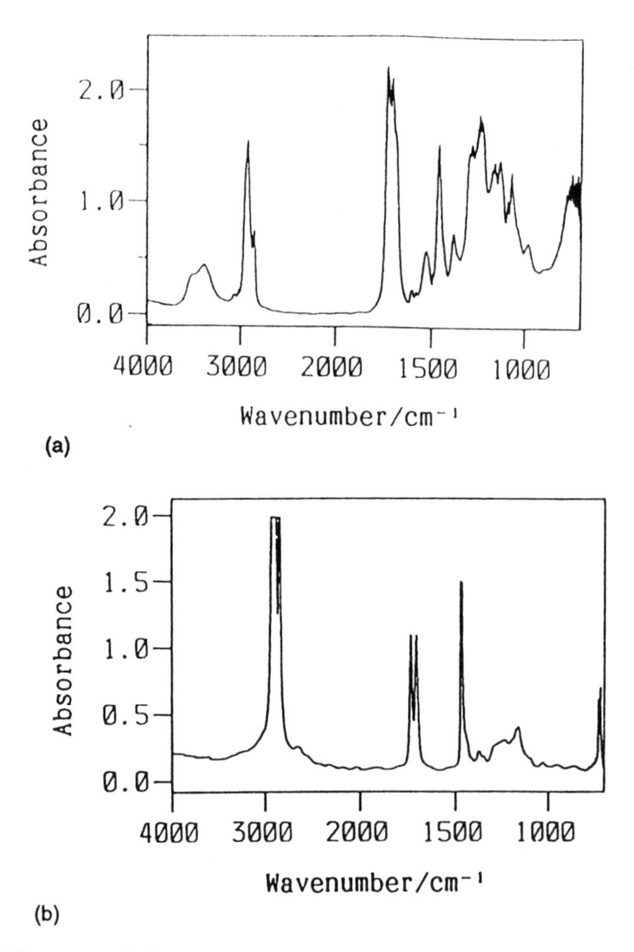

Figure 3. Microtransmission spectra of (a) polyurethane and (b) EAA/EEA copolymer. (From Ref. 23.)

were obtained by the microreflection technique. A Bio-Rad UMA 300A microscope was used to record the reflectance spectra over a 300- × 825-μm area of the sample. A 25- × 25-μm spatial resolution was used, and 256 scans were coadded to produce each spectrum. Figure 5 shows the typical microreflectance spectra from a normal region of the sample. Reflectance spectra show effects due to the dispersion of the refractive index, and these effects have been removed by the Kramers–Kronig transformation. These transformed spectra were used to produce the microimages of the sample. From Figure 6 one can easily see the broad silica band around 1100 cm^{-1}, and the epoxy band at around 1250 cm^{-1} in Figure 5. Figure 7 shows the functional group maps obtained by plotting

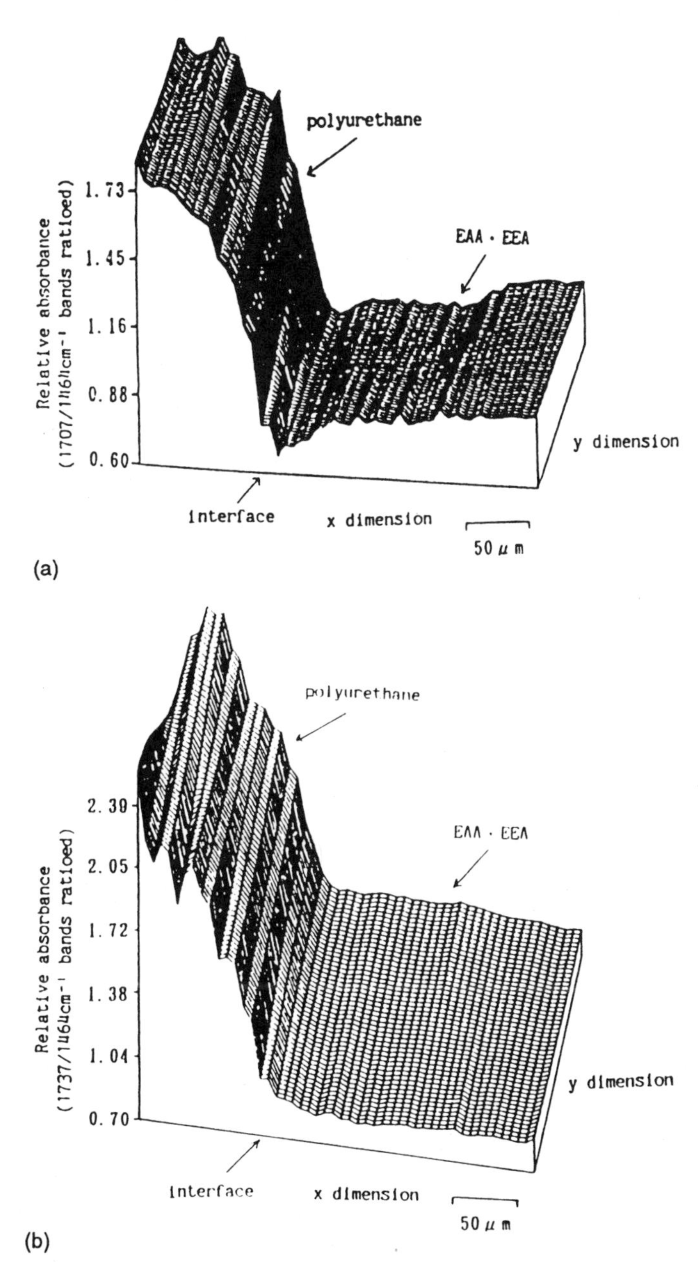

Figure 4. Functional group images in the EAA/EEA system of (a) 1707/1464 cm^{-1} band ratio and (b) 1737/1464 cm^{-1} band ratio. (From Ref. 23.)

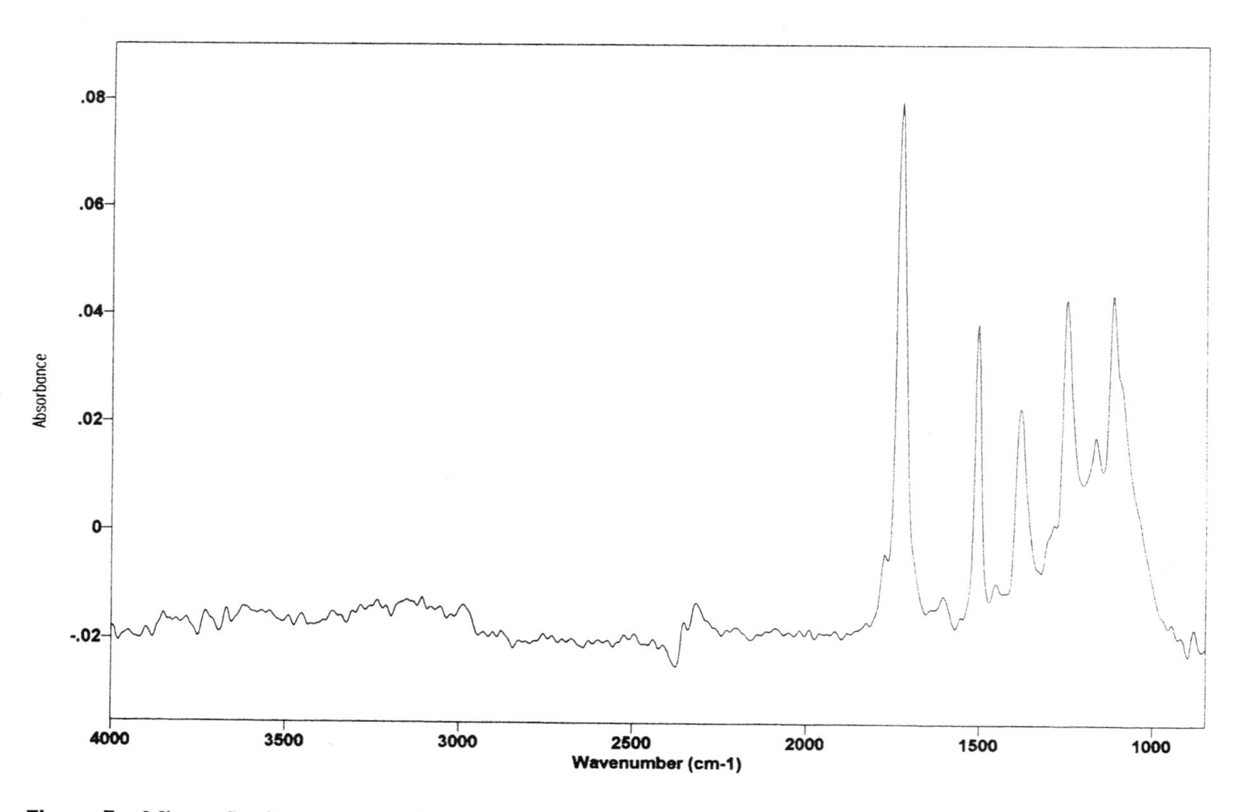

Figure 5. Microreflection spectrum of a normal region in a glass-epoxy composite, after Kramers-Kronig transformation.

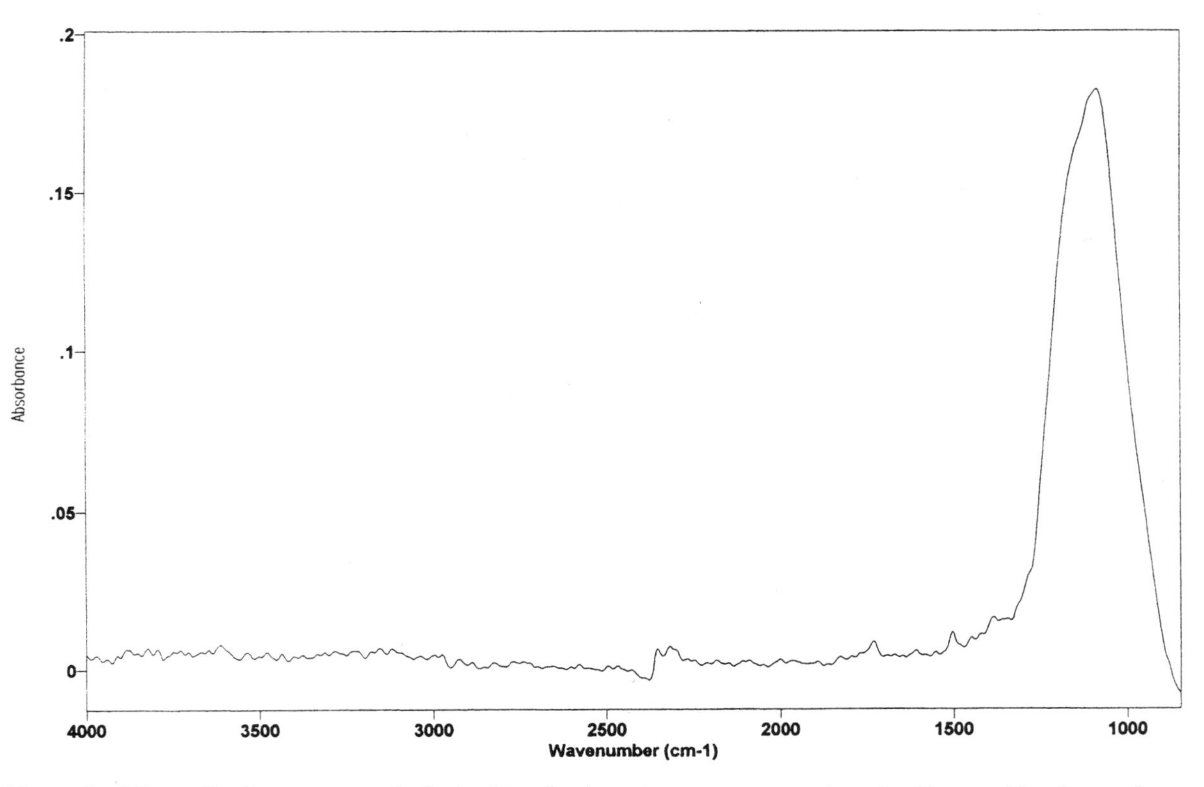

Figure 6. Microreflection spectrum of a "veined" region in a glass-epoxy composite, after Kramers-Kronig transformation.

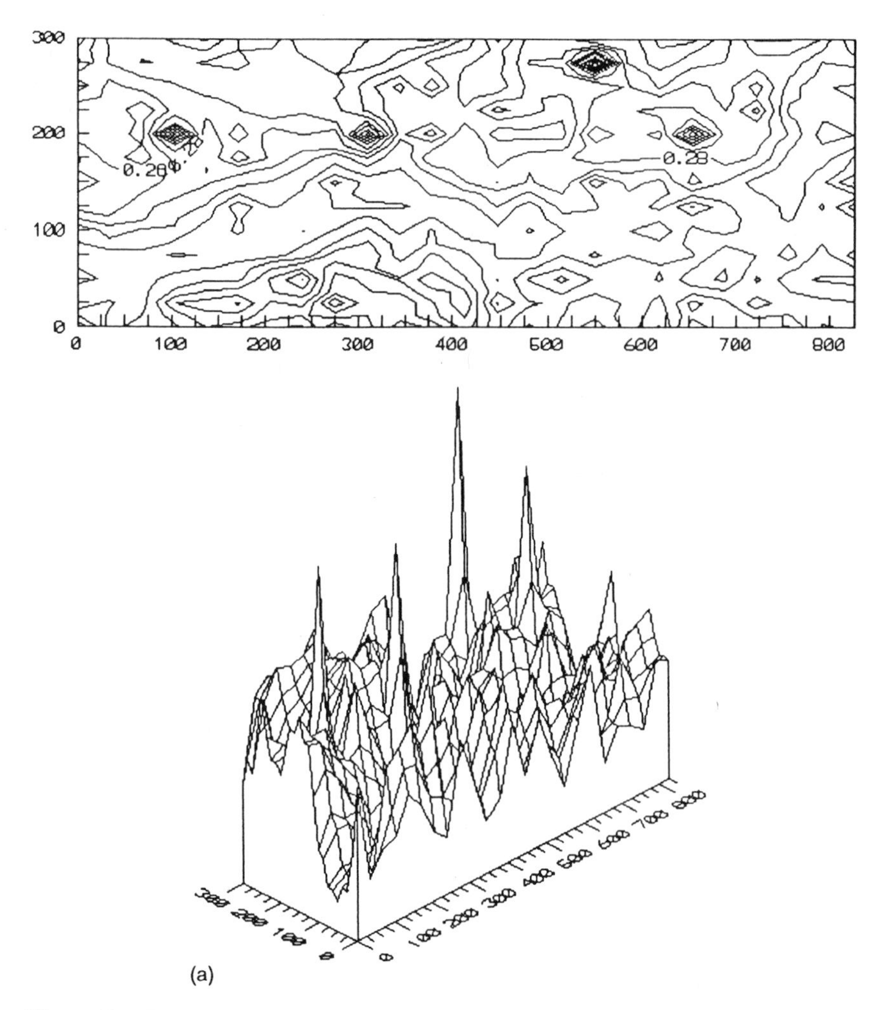

Figure 7. Function group images of the glass–epoxy composite: (a) 1036 cm^{-1} Si—O stretch; (b) 1250 cm^{-1} C—O epoxy stretch.

the intensities of the silica and the epoxy bands, respectively. Figure 7 shows that there are clear regions of glass and epoxy, thus providing an explanation for the appearance of the veins. The examples presented above indicate that the FT-IR microimaging technique can find many powerful and potential applications in the area of polymer analysis.

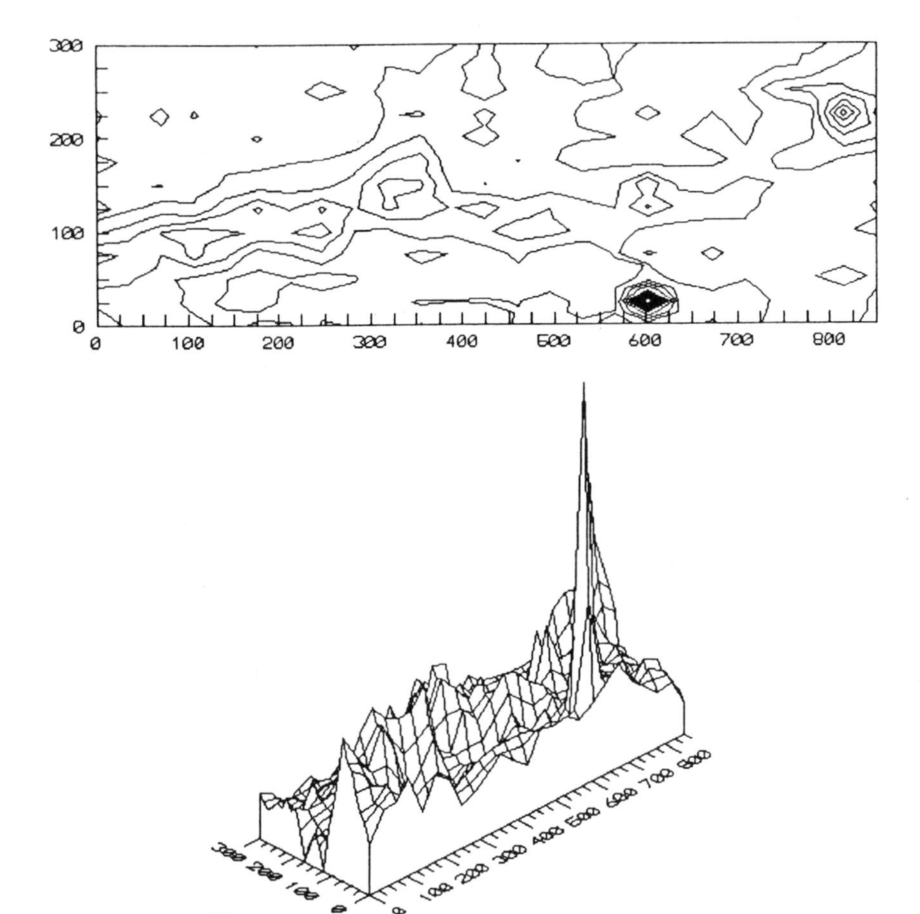

(b)

3.2. Minerals

The FT-IR microimaging technique in conjunction with reflectance spectroscopy should be a powerful tool for the mapping of mineral specimens. Surprisingly, however, little work along these lines seem to have been published in the literature. To illustrate the useful of the method for the analysis of mineral samples, the authors of this chapter have studied the microreflectance of the mineral, marble serpentine. The molecular formula of the mineral can be written as $Mg_3Si_2O_5(OH)_4$ plus $CaCO_3$. The sample, which had a mirror-polished surface, exhibited white, gray, green, yellow, and black crystallites. The microflectance spectra over 2 mm × 2 mm of the sample were recorded with a spatial resolution of 125 × 125 μm using a Bio-Rad UMA 300A microscope. A total of 256 scans at 8 cm^{-1} spectral resolution were coadded to produce each spectrum. The recorded spectra were all subjected to Kramers–Kronig transformation. Figure 8 (bottom to top) shows the final spectra so generated from the black, white, light green, and dark green areas of the sample. One can see absorption bands around 1100 cm^{-1} due to various Si—O stretching vibrations. In the lower-middle spectrum, that of the white crystallites, one can identify the familiar carbonate bands between 1400 and 1600 cm^{-1}. This spectrum could be identified to be that of $CaCO_3$, or marble. Figure 9 shows the functional group maps created by plotting the intensities of the bands between 1600 and 1400 cm^{-1}, and the Si—O stretching band, respectively. These plots clearly indicate the heterogeneity of the mineral sample.

3.3. Biological Materials

FT-IR microspectroscopy is ideally suited to the study of biological specimens, and a number of such applications have been described by Krishnan and Hill [6]. Since most biological specimen tend to be complex, functional groups im-

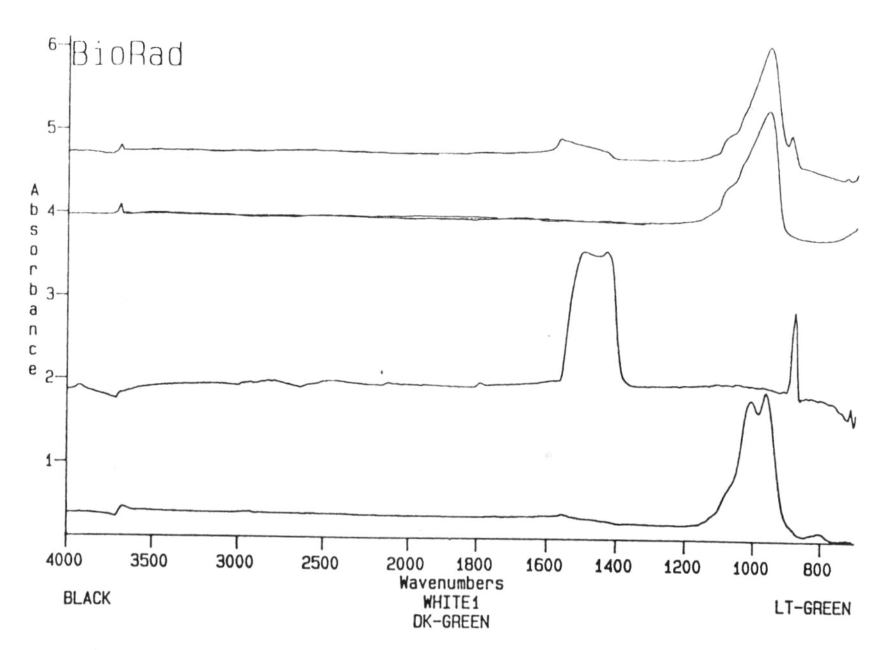

Figure 8. Microreflection spectra from different areas of a serpentine marble sample. Spectra correspond to (top to bottom) light green, dark green, white, and black areas from the sample. Spectra have been corrected by the Kramers–Kronig transformation.

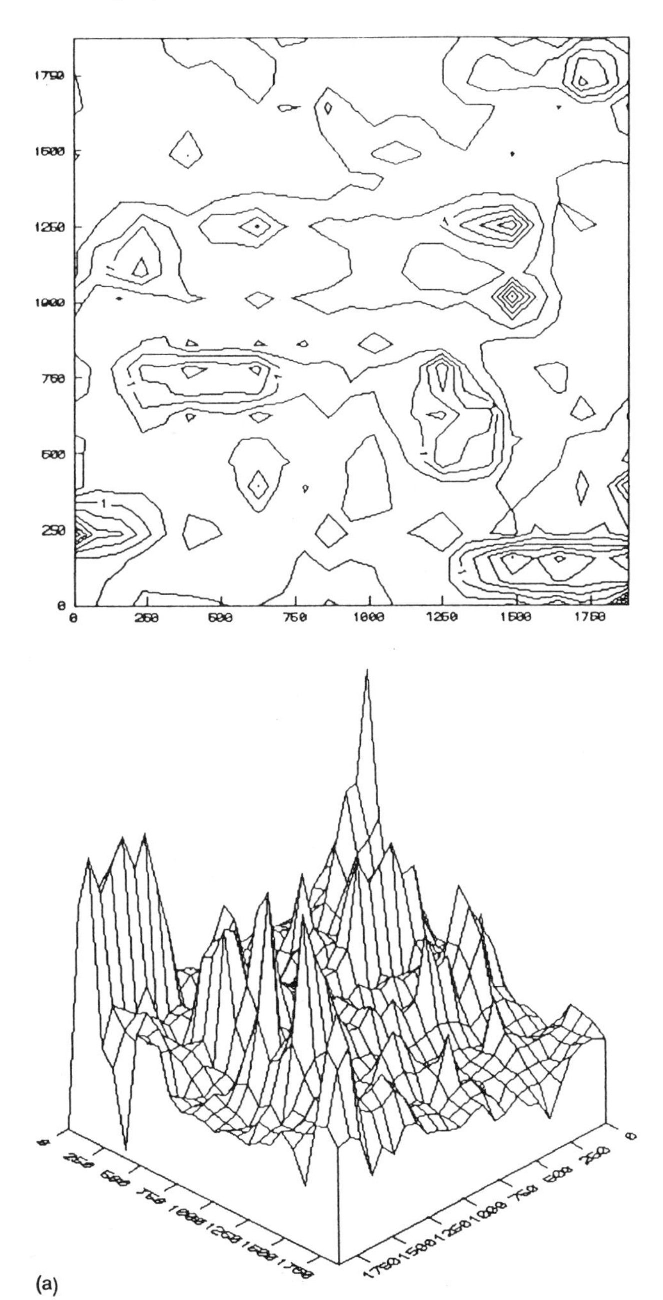

Figure 9. Functional group images from the serpentine marble sample of (a) Si—O stretch region, and (b) C—O carbonate stretch region.

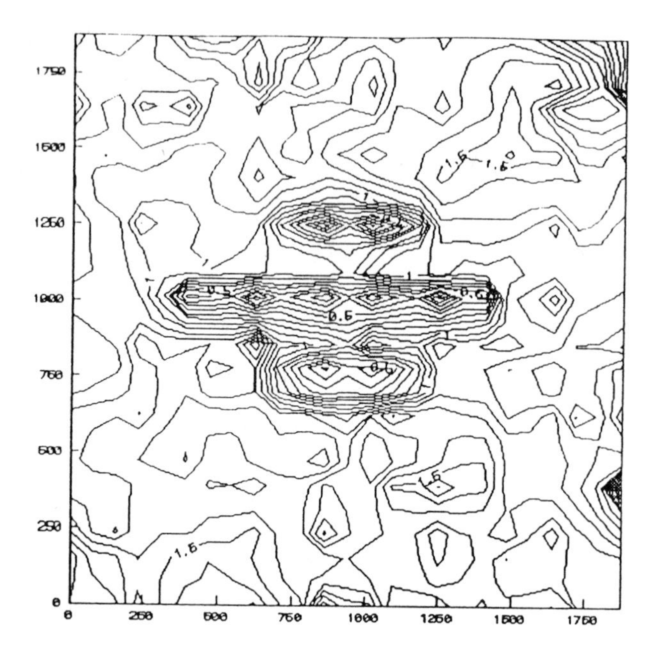

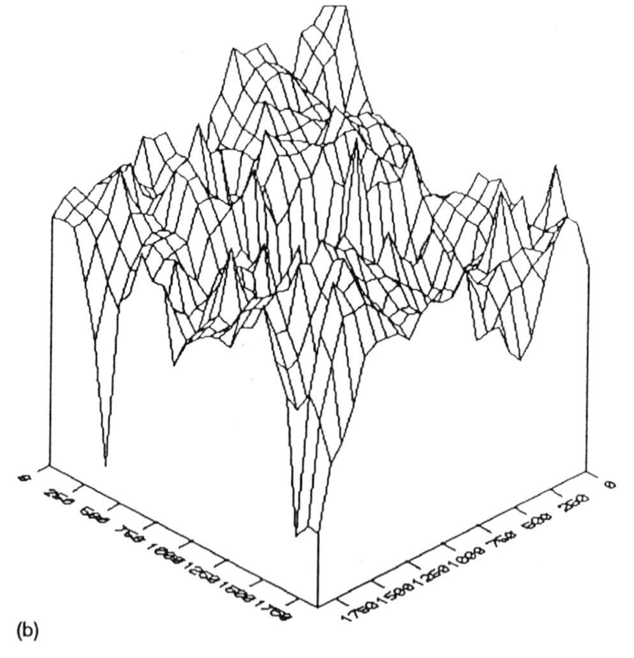

(b)

Figure 9. *Continued*

aging of these samples may be of great interest. Kodali et al. [22] have made a detailed FT-IR microimaging study of atherosclerotic arteries. They used 5- to 7.5-μm-thick arterial cross sections from a Watanabe Heritable Hyper Lipidemic (WHHL) rabbit, and a New Zealand White (NZW) rabbit in these studies. Figure 10 shows an optical micrograph of a 7.5-μm-thick section of a ballooned iliac of a cholesterol-fed NZW rabbit. Plaque deposits could clearly be seen in this micrograph. FT-IR microtransmission spectra over the whole arterial cross section using a spatial resolution of 100 × 100 μm were recorded by coadding 16 scans at 8 cm^{-1} resolution. A Bio-Rad UMA 300A FT-IR microscope was used to record the spectra. A comparison of a few of the recorded microtransmission spectra with pure lipid references, cholesterol, and cholesterol esters led to the identification of an absorption band at 1735 cm^{-1} as due to cholesterol esters or other ester-containing lipids. The amide I band could also be seen in most

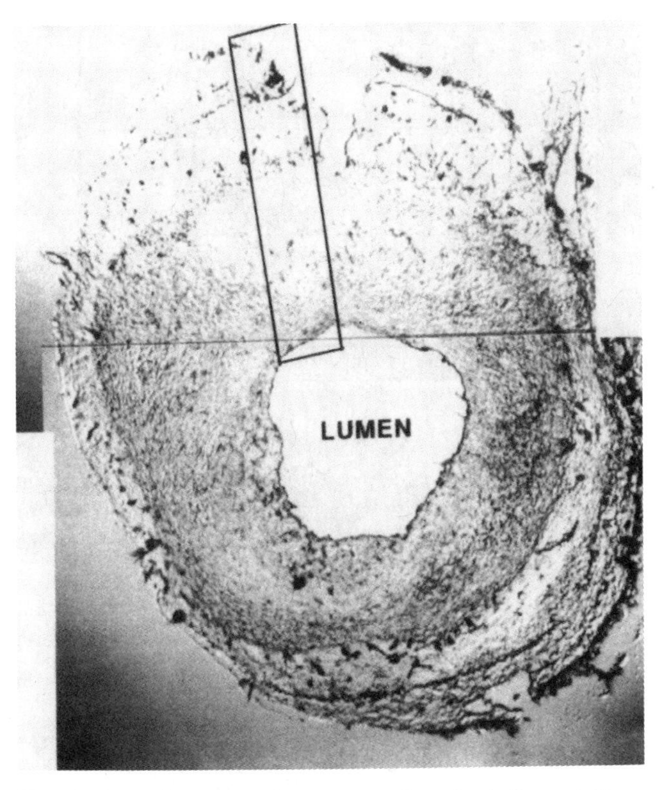

Figure 10. Optical micrograph of a cross section of a ballooned iliac artery from a cholesterol-fed New Zealand white rabbit. Area enclosed by the rectangle is imaged in detail in Figure 11. (From Ref. 22.)

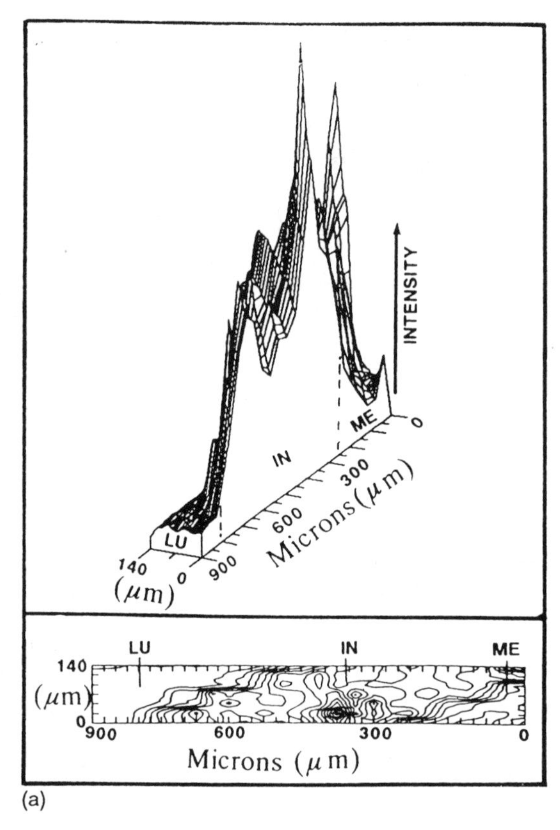

(a)

Figure 11. Functional group images from the enclosed area in Figure 10: (a) ester C=O stretch; (b) amide I band. LU, lumen; IN, intima; ME, media. (From Ref. 22.)

of the recorded spectra. Figure 11 shows the FT-IR functional group images due to the ester band and the amide I band, respectively. Both three-dimensional wire frame and contour plots are shown. One can see that the amide I band intensities are large in the media, and intima close to the lumen. The ester band intensity is strong in the media and in the internal edge of the intima. Figure 12 shows the polarized optical micrograph of a section of the artery, and the FT-IR microimage, in the form of a contour plot, of the ester band. From this figure one can see that lipid features seen as bright areas in the optical micrograph correspond to strong ester band intensities in the microimage. This figure shows the power of FT-IR microimaging in providing functional group information to complement that gained from conventional microscopy. More recently, Schiering et al. [30] have reported on the results of a similar microimaging study on a latex neuroanatomical probe in brain tissue. Their results reinforce

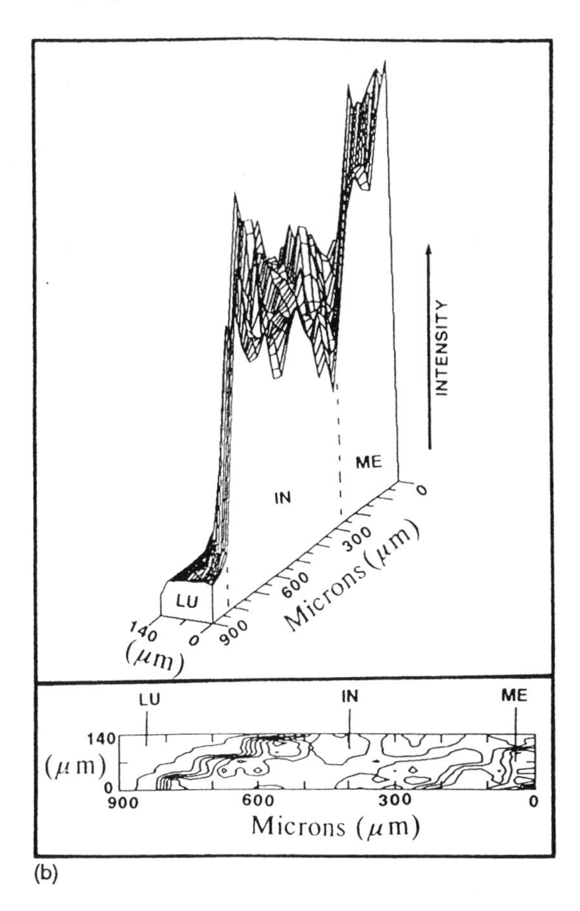

(b)

the conclusion of Kodali et al. [22] that the FT-IR microimaging technique is a powerful tool for the study of biological specimens.

Some interesting FT-IR microimaging studies have also been reported recently. Wetzel and Eilert [31] have examined a one-dimensional microimage of the diffusion of heavy water into a grain of wheat. Kalasinsky et al. [32] have used the one-dimensional FT-IR functional group mapping data to identify drugs of abuse in human hair. While conventional chemical extraction from hair, followed by a gas chromatography–mass spectroscopic analysis is the standard accepted method for such analysis, the method may suffer from the fact that passive environmental exposure to the drug may cause serious interferences with such analysis. On the other hand, if one examines the FT-IR spectrum of the central medulla of a microtomed section of the hair, such possible interferences can be eliminated.

a)

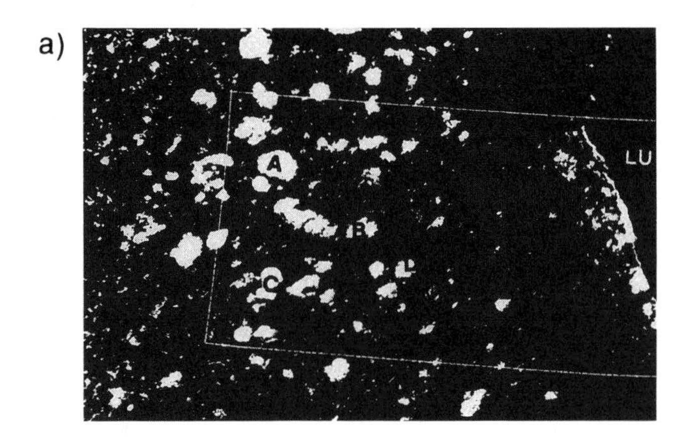

b)

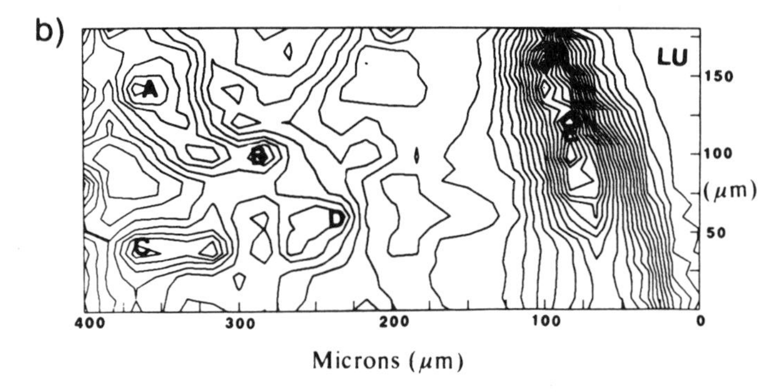

Figure 12. (a) Polarized optical micrograph of an area of the artery cross section shown in Figure 10. Bright areas indicate the presence of lipid droplets. (b) C=O stretch functional group image of the rectangular area marked in (a). Letters correspond to location of lipid droplets identified in (a). (From Ref. 22.)

3.4. Semiconductor Materials

FT-IR microspectroscopy has been proven to be very useful for the characterization of semiconductor materials. Whereas the most obvious use of the technique is for microcontamination analysis, the method could be used for study of the spatial distribution of the interstitial oxygen concentration in silicon and the determination of silicon epitaxial thicknesses over buried layers [5,6,32–34]. The theoretical and experimental basis of such measurements have been reviewed by Krishnan et al. [35]. If the FT-IR microimaging technique is applied to these systems, the images can be produced not only with the actual spectral data (qualitative images), but also from the results of quantitative analysis from the FT-IR spectral data.

Krishnan [32] had shown that precise measurements of silicon epitaxial thickness over buried layers could easily be made using the FT-IR microreflectance technique. Similar measurements could also be made on MCT (mercury–cadmium–telluride) epitaxial layers. MCT is a commonly used, high-sensitivity midinfrared detector. One of the ways of fabricating MCT is to deposit an epitaxial layer of the material on a cadmium telluride substrate. The thickness of the MCT layer will influence the response characteristics of the infrared detector made from the material. Therefore, it is very necessary to measure the thickness of this epitaxial layer precisely, and this can be achieved by FT-IR microreflectance spectroscopy as well. Furthermore, software routines are readily available that will actually provide the epitaxial thickness in units of microns. Thus one can actually produce microimages of the epitaxial thickness and visualize the intermediate microstructures during the MCT device manufacturing process. Figure 13 shows a wire frame plot of such an MCT structure. The sample in this case was a CdTe plate on which a number of "islands" of epitaxial MCT layers had been deposited. Microreflectance interferogram sets were collected from 20- × 20-μm sampling areas using a Bio-Rad UMA 300A microscope, over the entire area of the CdTe sample. Sixteen scans were coadded to produce each interferogram set. These interferogram sets were then analyzed using the standard Bio-Rad epitaxial thickness-determination software package,

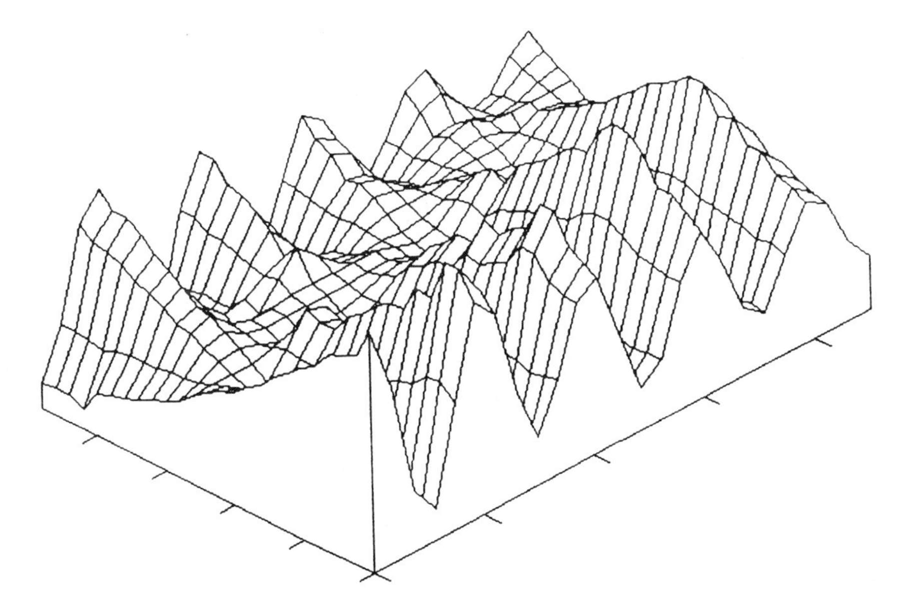

Figure 13. Wire frame representation of the epitaxial thickness of MCT on a CdTe substrate.

and each interferogram collected yielded one epitaxial thickness for the corresponding measurement spot. The contour plot from the same data is shown in Figure 14, and the numbers on the plot indicate the epitaxial thickness in microns.

4. ADVANCED DATA TREATMENT METHODS

All the examples of FT-IR microimaging were generated simply by using the raw spectral data (transmission measurements), spectral data after Kramers–Kronig transformation (reflectance measurements), or quantitative results. A number of papers have been published in the recent literature on the application of advanced software techniques such as Gram–Schmidt orthogonalization [36], and chemometric methods such as factor analysis, principal component regression analysis (PCR), or partial least squares (PLS) analysis as data treatment procedures prior to producing the FT-IR microimage. The basic idea behind the application of these chemometric methods is as follows. Since the ultimate resolution of the midinfrared microscope is limited, it is possible that any spectral features from sampling areas less than the resolution may appear superimposed on other spectral features. The chemometric methods offer the possibility of separating these spectral features, and thus providing effective resolution en-

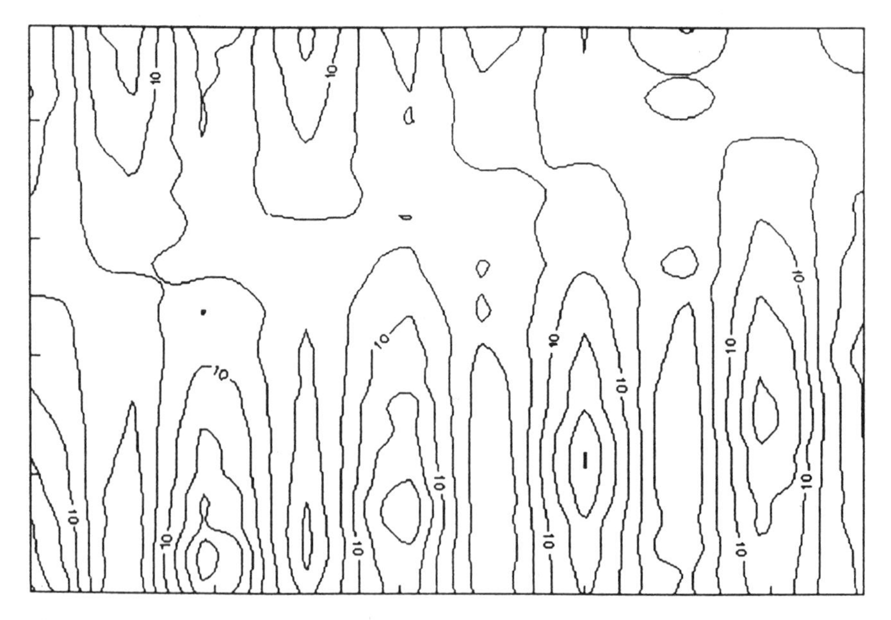

Figure 14. Contour plot of the epitaxial thickness (in microns) of the epitaxial thickness of MCT on a CdTe substrate.

hancement in the images created. When using the PCR method, for instance, one performs the calibration and creates a calibration file using all the recorded spectral data in the image area. Then the factor loading of each of the principal component vectors is calculated for each of the recorded spectra. These factor loadings are then plotted against the *x–y* coordinates of the recorded spectra. It is, however, beyond the scope of this chapter to describe the details of the chemometric methods, and the reader is referred to the article by Haaland [37].

Ward [25] has shown that under certain conditions, one can obtain greater clarity in the FT-IR microimage by pretreatment of the data by the Gram–Schmidt orthogonalization procedure. Carl and Weesner [38] have applied the above procedure to produce a map of silicon epitaxial thickness measurements over buried layers. Pell et al. [39] have reported on the possibility of enhancing the resolution of FT-IR microspectroscopic data by using chemometric techniques. As pointed out in Section 1, the practical minimum area from which a micro-infrared spectrum could be recorded is on the order of 10×10 μm. Pell et al. propose a computational method based on factor analysis which can effectively be used to obtain the spectra of sample dimensions smaller than 10×10 μm. The data to be input into their computational procedure are collected by moving the sample stage by increments that are much less than the sampling area defined by the variable aperture. Using the procedure above, McKelvy et al. [27] have managed to produce a recognizable infrared spectrum from a sub-10-μm-thick layer in a polymer laminate. Ward et al. [40] have also reported on the use of PCR for the enhancement of FT-IR microimages. Hill and Krishnan [41] have applied the PCR pretreatment to the FT-IR microreflectance data on a number of polymer systems—laminates, and samples with inclusions—and have shown that the results lead to enhanced microimages.

5. CONCLUSIONS

The technique of functional group imaging using FT-IR microspectroscopy has been described in this chapter with special emphasis on the application of the technique to polymer, biological, mineral, and semiconductor systems. The utility of the method in providing direct molecular information, and recent attempts to enhance the resolution of the images by a variety of chemometric data treatment methods, presage a bright future for the method.

REFERENCES

1. Laczik, Z., Booker, G. R., Bergholtz, W., and Falster, R. (1989). Investigation of oxide particles in Czochralski silicon heat treated for intrinsic gettering using scanning infrared microscopy, *Appl. Phys. Lett.*, *55*: 2625.

2. Kidd, P., Booker, G. R., and Stirland, D. J. (1987). IR laser scanning microscopy in transmission: a new high resolution technique for the study of inhomogeneities in bulk gallium arsenide.
3. Treado, P. J., and Morris, M. D. (1993). *Infrared and Raman Spectroscopic Imaging*, Practical Spectroscopy Series, Vol. 16 (M. D. Morris, ed.), Marcel Dekker, New York, p. 71.
4. Krishnan, K. (1984). Application of FT-IR microsampling techniques to some polymer systems, *Polym. Prepr. ACS Div. Polym. Chem., 25*: 182.
5. Krishnan, K., and Kuehl, D. (1984). A study of the spatial distribution of the oxygen content in silicon wafers using an infrared transmission microscope, *Semiconductor Processing*, ASTM Spec. Tech. Publ. 850, p. 325.
6. Krishnan, K., and Hill, S. L. (1990). FT-IR microsampling techniques, in *Practical Fourier Transform Spectroscopy* (J. R. Ferraro, and K. Krishnan, eds.), Academic Press, New York, p. 103.
7. Messerschmidt, R. G., and Harthcock, M. A. (1988). *Infrared Microspectroscopy: Theory and Applications*, Practical Spectroscopy Series, Vol. 6, Marcel Dekker, New York.
8. Harthcock, M. A., and Atkin, S. C. (1987). Compositional mapping with the use of functional group images obtained by infrared microprobe spectroscopy, in *Microbeam Analysis* (R. Geiss, ed.), San Francisco Press, San Francisco, p. 173.
9. Harthcock, M. A., and Atkin, S. C. (1988). Imaging with functional group maps using infrared microspectroscopy. *Appl. Spectrosc., 42*: 449.
10. Harthcock, M. A., Atkin, S. C., and Davis, B. L. (1988). Infrared microspectroscopy functional group imaging as a probe into compositional heterogeneity of polymer blends, in *Microbeam Analysis* (D. E. Newbury, ed.), San Francisco Press, San Francisco, p. 203.
11. Marienenko, R. B., Myklebust, R. L., Bright, D. S., and Newbury, D. E. (1985). In *Microbeam Analysis* (D. E. Newbury, ed.), San Francisco Press, San Francisco, p. 159.
12. Newbury, D. E. (1985). In *Microbeam Analysis* (D. E. Newbury, ed.), San Francisco Press, San Francisco, p. 204.
13. Van Duyne, R. P., Haller, K. L., and Altkorn, R. I. (1986). *Chem. Phys. Lett., 126*: 190.
14. Massey, G. A., Davis, J. A., Katnik, S. M., and Omon, E. (1985). *Appl. Opt., 24*: 1498.
15. Louden, J. D., and Kelly, J. (1990). Infrared mapping of deterrents (moderants) in nitrocellulose based propellent grains by Fourier transform infrared microscopy, in *Analytical Applied Spectroscopy*, Vol. 2 (A. M. C. Davies, and C. Creaser, eds.), Royal Society of Chemistry, London, p. 90.
16. Milledge, H. J., and Mendelssohn, M. (1988). Infrared microspectroscopy with special reference to computer controlled mapping of inhomogeneous specimens, *Proceedings of the International Conference on Analytical Applications of Spectroscopy—1987* (C. Creaser, and A. M. C. Davies, eds.), Royal Society of Chemistry, London, p. 217.
17. Nakashima, S., Ohki, S., and Ochiai, S. (1989). *Geochem. J., 23*: 57.
18. Nishioka, T. (1991). *Bunseki Kagaku, 40*: T21.

19. Nishioka, T., Nishikawa, T., Teremae, N., and Sawada, T. (1990). *Kobunshi Ronbunshu, 47*: 553.
20. Nishioka, T., and Teremae, N. (1991). Functional group imaging of a coated interface and polymer blends using Fourier transform infrared microspectroscopy, *Analytical Science 7, Supplement to Proceedings of the International Congress on Analytical Science—1991*, Pt. 2, p. 1633.
21. Reffner, J. A. (1990). Molecular microspectral mapping with FT-IR microscope, *Institute of Physics Conference Series, 98*: 97.
22. Kodali, D. R., Small, D. M., Powell, J., and Krishnan, K. (1991). Infrared microimaging of atherosclerotic arteries. *Appl. Spectrosc., 45*: 1310.
23. Nishioka, T., Nakano, T., and Teremae, N. (1992). Analysis of the coated interface using Fourier transform infrared microspectroscopy. *Appl. Spectrosc., 46*: 1904.
24. Steger, W. E., Machill, S., Herzog, K., Gerhards, R., Jussofie, I., and Schator, H. (1992). Local analysis by infrared microspectroscopy across membranes of a flexible polyurethane foam. *Fresenius J. Anal. Chem., 344*: 203.
25. Ward, K. J. (1989). Applications of image analysis for infrared microscopic detection of contaminants on microelectronic devices. *Proc. SPIE, 1145*: 212.
26. Messerschmidt, R. G. (1988). *Infrared Microspectroscopy: Theory and Applications*, Practical Spectroscopy Series, Vol. 6 (R. G. Messerschmidt and M. A. Harthcock, eds.), Marcel Dekker, New York.
27. McKelvy, M. L., Pell, R. J., and Harthcock, M. A. (1993). Effective resolution enhancement of infrared microspectroscopic data by multiresponse nonlinear optimization, paper presented at the *9th International Conference on Fourier Transform Spectroscopy*, Calgary, Alberta, Canada, Aug. 23–27.
28. Harthcock, M. A., and Atkin, S. C. (1988). Infrared microspectroscopy: development and applications of imaging capabilities, in *Infrared Microspectroscopy: Theory and Applications*, Practical Spectroscopy Series, Vol. 6 (R. G. Messerschmidt, and M. A. Harthcock, eds.), Marcel Dekker, New York, p. 21.
29. Zumbrum, M. A., Hellgath, J. W., Ward, T. C., and Carl, R. T. (1990). Evaluation of crosslinking efficiency of multifunctional acrylate photopolymers via infrared microspectroscopy, *Polym. Preparations, 31*: 393.
30. Schiering, D. W., Rapoza, D., Madison, R. D., Messerschmidt, R. G., and Kuehl, D. (1993). Infrared mapping microspectroscopy of a latex neuroanatomical probe in brain tissue, paper presented at the *9th International Conference on Fourier Transform Spectroscopy*, Calgary, Alberta, Canada, Aug. 23–27.
31. Wetzel, D. L., and Eilert, A. J. (1993). FT-IR microspectroscopic observation of the migration of water in individual grain kernels with tempering using D_2O, paper presented at the *9th International Conference on Fourier Transform Spectroscopy*, Calgary, Alberta, Canada, Aug. 23–27.
32. Krishnan, K. (1988). Characterization of semiconductor silicon using the FT-IR microsampling technique, in *Infrared Microspectroscopy: Theory and Applications*, Practical Spectroscopy Series, Vol. 6 (R. G. Messerschmidt, and M. A. Harthcock, eds.), Marcel Dekker, New York, p. 139.
33. Kim, K. M., and Smetana, P. (1986). Oxygen segregation in CZ silizon crystal growth on applying a high axial magnetic field, *J. Electrochem. Soc., 133*: 1682.

34. Yao, K. H. and Witt, A. F. (1987). Scanning Fourier transform infrared spectroscopy of carbon and oxygen microsegregation in silicon, *J. Cryst. Growth, 80*: 453.
35. Krishnan, K., Stout, P. J., and Watanabe, M. (1990). Characterization of semiconductor silicon using Fourier transform infrared spectrometry, in *Practical Fourier Transform Spectroscopy* (J. R. Ferraro, and K. Krishnan, eds.), Academic Press, New York, p. 285.
36. Hanna, D. A., Hangac, G., Hohne, B. A., Small, G. W., Niebolt, J. M., and Isenhour, T. L. (1979). A comparison of methods used for the reconstruction of GC-FT/IR chromatograms. *J. Chromatogr. Sci., 17*: 423.
37. Haaland, D. M. (1990). Multivariate calibration methods applied to quantitative FT-IR analyses, in *Practical Fourier Transform Spectroscopy,* (J. R. Ferraro and K. Krishnan, eds.), Academic Press, New York, p. 386.
38. Carl, R. T., and Weesner, F. J. (1992). Use of a Gram–Schmidt response function in infrared microscopic imaging, *Proceedings of the 80th Annual Meeting of the Electron Microscopy Society of America,* (G. W. Bailey, J. Bentley, and J. A. Small, eds.), San Francisco Press, San Francisco, p. 1524.
39. Pell, R. J., McKelvy, M. L. and Harthcock, M. A. (1993). Effective resolution enhancement of infrared microspectroscopic data by multiresponse nonlinear optimization, *Appl. Spectrosc., 47*: 634.
40. Ward, K. J., Reffner, J. A., and Martoglio, P. A. (1993). Applications of principal component analysis to multidimensional FT-IR microscopy data, paper presented at the *9th International Conference on Fourier Transform Spectroscopy*, Calgary, Alberta, Canada, Aug. 23–27.
41. Hill, S. L., and Krishnan, K. (1993). Enhanced infrared characterization in the microdomain by IR microreflectance principal component analysis, paper presented at the *9th International Conference on Fourier Transform Spectroscopy*, Calgary, Alberta, Canada, Aug. 23–27.

4

Applications of Reflectance Microspectroscopy in the Electronics Industry

Catherine A. Chess IBM, Thomas J. Watson Research Center, Yorktown Heights, New York

1. INTRODUCTION

Infrared microspectroscopy has been widely used in the semiconductor industry to many years [1,2]. The introduction of modern microscopes which include high-quality infrared and visible optics has extended the applicability of the technique. Fluorescence, polarized, and phase-contrast microscopy of the precise area analyzed spectroscopically is feasible. In addition, the reflectance objectives now available permit a wide range of analyses on a selected area with minimal sample preparation. Practical aspects of the technique have been discussed elsewhere [2,3].

Infrared microspectroscopy is employed within all areas of the electronics industry to study materials of interest, from semiconductors, packaging materials, ceramics, glasses, polymers [4], and adhesives, to magnetic media [5] and disk lubricants [6]. Semiconductor analyses include determination of dopants and contaminants such as oxygen in silicon [7,8], and characterization of amorphous and crystalline dielectrics.

Although failure analysis, contaminant identification, quality control, and quality assurance are still the major uses of infrared microspectroscopy, it is also widely employed in research and development. Characterization of selected areas is necessary to understand the impact of processing, cleaning, and etching on materials. Spectral mapping to verify composition, homogeneity, or thickness of films is facilitated by computer-controlled states. In situ heating studies yield epoxy curing data [9] or important materials degradation information.

Some samples presented special problems. Until recently, dark polymers, composite materials, thin films, and rough-surfaced materials were considered nearly impossible to characterize using infrared microspectroscopy [10]. The availability of the internal reflectance objective, however, now permits characterization of dark polymers and composites, and the grazing angle objective used with a polarizer permits characterization of thin films, especially on metals. In this chapter we discuss practical aspects and applications of these new methods.

2. EXPERIMENTAL PROCEDURES

Data were acquired on an IR-PLAN (Spectra-Tech, Inc.) research-grade infrared microscope, externally interfaced to a Nicolet 740 FTIR. The interface includes a small-element, midband mercury cadmium telluride (MCT) detector. The microscope is equipped with an internal reflection or attenuated total reflection (ATR) objective and a grazing-angle objective (GAO).[1] Both objectives are Schwarzchild Cassegrain objectives and consist of large concave primary mirror

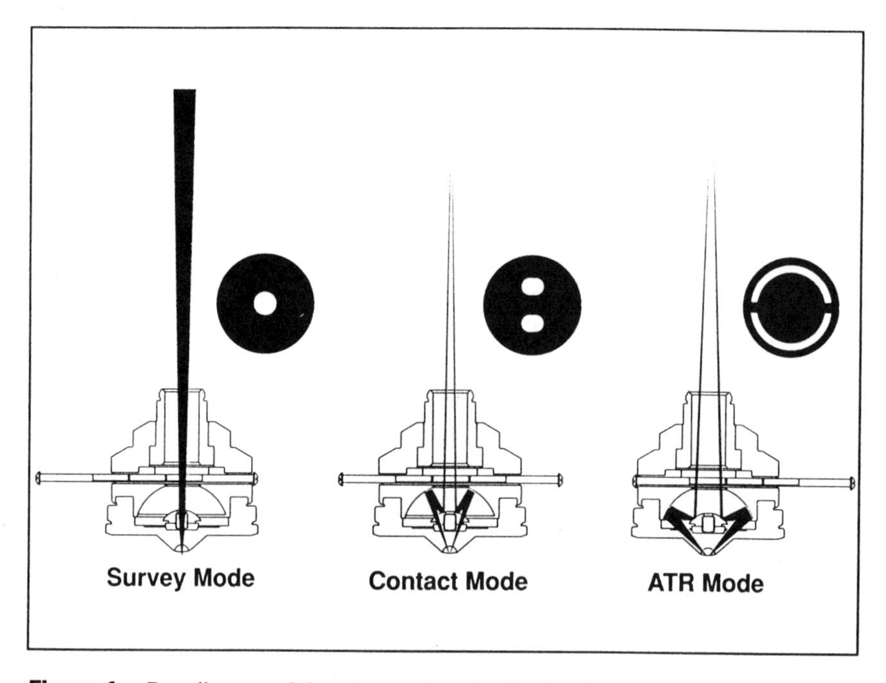

Figure 1. Ray diagram of the attenuated total reflection (ATR) objective (Courtesy of Spectra-Tech, Inc.)

[1]ATR objective and GAO are products covered by Spectra-Tech, Inc., patents.

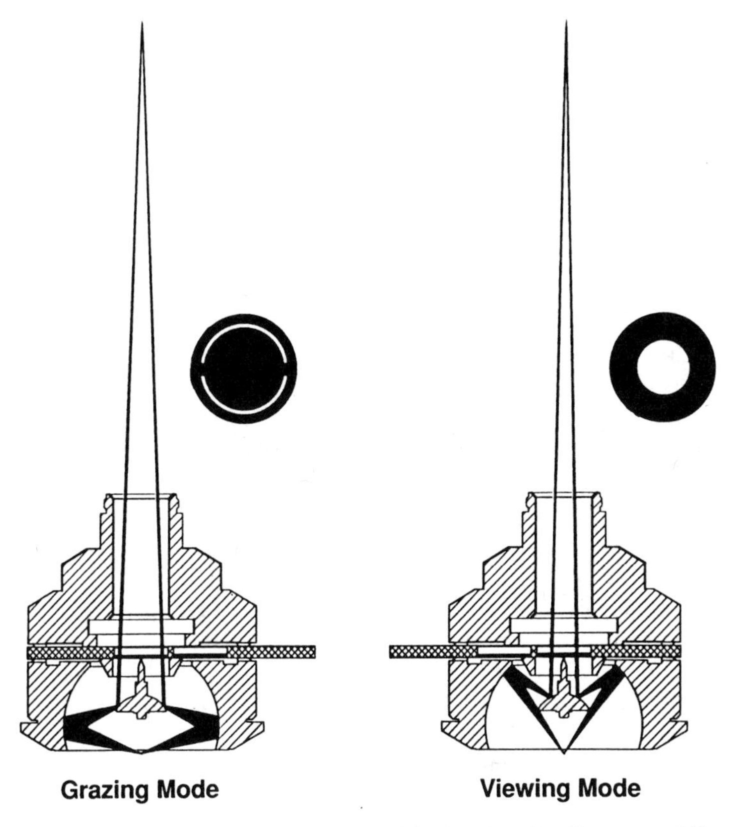

Figure 2. Ray diagram of the grazing-angle objective (GAO). (Courtesy of Spectra-Tech, Inc.)

and a small convex secondary mirror (see Figures 1 and 2). This design minimizes spherical abberrations. The beam passes through a central hole in the primary mirror, reflects off the secondary onto the primary, and is focused on the sample surface. These objectives have an objective aperture to restrict the incident angle of the light reaching the sample. The objective apertures are located in a slide at the top of the primary mirror. Data on the objectives are listed in Table 1, with data for the standard 15× objective for comparison. The microscope and objectives are described in detail elsewhere [11,12].

2.1. Microscopic Attenuated Total Reflection

The attenuated total reflection (ATR) objective equipped with a zinc selenide (ZnSe) crystal was used. Data were acquired at 4 cm^{-1} resolution. The ATR objective aperture slide contains three apertures, survey to examine the surface,

Table 1. Data for IR-Plan Objectives

Objective	Visible light	View angle	N.A.	IR	IR angle	N.A.
15×	15×	15.6–35.5	0.58	15×	15.6–35.5	0.58
ATR	15×	15.6–35.5	0.26	25×	45	0.87
GAO	30×	33–45	0.51	30×	65–85	0.996

Source: Data from Ref. 11 and Spectra-Tech, Inc.

contact to bring the crystal into contact with the sample, and ATR to collect data. In survey mode, the beam passes straight through the objective at normal incidence (Figure 1). The magnification in survey mode is 15× since the light does not reflect from the two Cassegrain mirrors. In this mode the working distance is 3 mm. In contact mode the light strikes the sample at about a 33° angle, and in ATR mode the light is at a 45° angle. The crystal must press against the sample in both contact and ATR modes, and the magnification is 25× in these modes.

The principles of internal reflection have been discussed in detail in the literature [13,14]. If light is passed into an optical crystal with an index of refraction greater than the medium that surrounds it and at an angle larger than the critical angle, the beam will be confined to the crystal. An evanescent wave will propagate a short distance into the surrounding medium at each point of refraction. The critical angle θ_c is calculated as

$$\theta_c = \sin^{-1} \frac{n_2}{n_1} \tag{1}$$

This technique is called attenuated total reflection because the beam of light will be selectively absorbed by the medium at each point of refraction. A pathlength of less than 10 μm is required for characterization of infrared-absorbing materials. FTIR spectra of strongly absorbing materials can be obtained by placing the sample in contact with a suitable optical crystal. Since the beam does not penetrate very far into the sample in this technique, extensive sample preparation can be avoided.

Small ATR crystal holders mounted in beam condensors permit nondestructive analysis of small specimens of "bulk" polymers and composites and have been in use for many years. However, these holders do not permit study of selected areas. Microscopic internal reflection using the ATR objective is better suited for semiconductor analysis than conventional ATR since it permits characterization of selected areas as small as 40 μm. In situ characterization of dielectrics for quality control purposes is now practical, as are developmental studies of the effects of processing on small areas of products. While this spot size is too large to study individual semiconductor devices, it is still quite useful to study uniformity and homogeneity across devices on a wafer.

Reflectance data of thick polymeric materials acquired at normal incidence will show film thickness effects as well as severe distortions known as reststrahlen effects in regions of strong absorptions. Microscopic ATR can easily be used for selective area characterization of relatively thick dielectrics used in the back end of the line (BEOL) while avoiding the problems associated with normal incidence reflectance measurements.

With ATR measurements it is important to bear in mind that the crystal is a fragile optical component of the system and is easily damaged. Once damaged, the objective performance will be diminished significantly. The objective is ideal for use with flat, flexible samples. It is not recommended for hard, rough-surfaced materials or irregular metal surfaces, due to the potential for crystal damage. For measurements of flat rigid samples it is best to place the sample on a thin foam pad on the sample stage prior to contact. Fully packaged chips require removal of metal leads prior to analysis to prevent damage to the crystal.

The depth of penetration of the light into the sample increases to longer wavelengths with internal reflection, which results in the characteristic increase in intensity of bands at longer wavelengths. Harrick [13] derived the following formula to calculate the depth of penetration (DP) for nonabsorbing materials:

$$DP = \frac{\lambda/n_1}{2\pi\,[\sin^2\theta - (n_2/n_1)^2]^{1/2}} \tag{2}$$

where n_1 is the index of refraction of the crystal, n_2 the index of refraction of the sample, and θ the angle of incidence.

The effective pathlength (EPL) is defined as the number of reflections in the crystal times the depth of penetration. This is a single-reflection case, so that EPL and DP are equivalent. Values for effective pathlength for a transparent sample with an index of refraction (n_2) of 1.5 are given in Table 2. For strongly absorbing materials, the pathlength will decrease much more rapidly. Switching to a crystal with an index of refraction close to that of the sample will increase the depth of penetration greatly. To probe different depths within the sample, data are collected using different crystal materials and different face angles.

Table 2. Effective Pathlength (μm) for Sample $n_2 = 1.5$

ATR	2.5 μm	5.0 μm	10 μm
ZnSe	0.50	1.0	2.0
Ge	0.17	0.33	0.66

Calculated using eq. 2.

While depth profiling is technically feasible, it is not yet practical on a microscopic scale.

Zinc selenide (ZnSe), diamond (C), and germanium (Ge) crystals are now available for the ojbective. Diamond has approximately the same index of refraction as ZnSe. Although it is much more expensive than ZnSe, it is much harder and therefore more difficult to scratch. Diamond is brittle and unfortunately, the crystal will break if struck. Germanium has an index of refraction of 4, which is necessary to study high-index materials, such as silicon, carbides, and diamondlike carbon materials. However, the angle of the germanium crystal for the ATR is below the critical angle required to characterize silicon. Germanium is useful for probing the surface of low-index materials such as polymers. Germanium is very brittle, slightly more expensive than ZnSe, and does not transmit visible radiation, so that it is more difficult to use than the other two materials.

2.2. Grazing-Angle Microspectroscopy

The grazing-angle objective (GAO) is illustrated in Figure 2. The GAO is a 7-cm-diameter 30× objective with a 1-mm working distance. Care must be taken to prevent contact with either mirror to avoid damage. There are two objective aperture slides for the GAO. The standard slide contains a view aperture and a grazing aperture. The other slide contains a view and p-polarizing aperture for performing polarized measurements. The p-polarizing objective aperture must be used with the polarizer whenever p-polarized light is desired.

At grazing angle, the penetration depth of unpolarized radiation into the sample is approximately one-tenth the wavelength. Surface species only interact with the *p*-polarized component of light. When *p*-polarized light is selected, the amount of stray light reaching the detector is reduced since the *s*-polarized component is absorbed. An 1800-line/mm wire grid on ZnSe infrared polarizer (Spectra-Tech, Inc.) was mounted in the exit beam of the microscope for polarized measurements. Although data can be acquired from relatively thick films without the polarizer, the objective is usually used with the polarizer mounted in the exit port of the microscope to improve sensitivity.

GAO is ideal for smooth, flat-surfaced samples, including fixed disks, wafers, and printed wiring boards prior to assembly. Samples with large changes in topography will not yield good results since a significant portion of the light will be scattered or blocked. The grazing-angle technique works best on evaporated metal surfaces, prior to wire bonding. Plated metals exhibit pitted surfaces which are often too rough for grazing-angle measurements. One drawback to the grazing-angle technique is that surface roughness will not only increase the spectral noise but may result in inadvertent sampling of a surrounding area. This is especially the case when the area to be analyzed is recessed. Some manufacturing processes, which include brushing or scrubbing with pumice prior to bonding, leave metal surfaces too rough for characterization at grazing angle.

To minimize the scattering, it is best to use the largest aperture possible. In some cases, data can be acquired at angles intermediate between grazing and near normal. This can be accomplished by acquiring while the objective is in view mode. Unfortunately, the numerical aperture of the objective in this position is only 0.51, so that by definition twice the sampling area is required in this mode than in grazing mode.

Although a clean gold first surface mirror is usually used as a reference in grazing-angle measurements, a control or known good sample is better since it will have about the same surface roughness and oxidation. If a control is not available, freshly prepared sputtered or evaporated metal films are recommended. It is important to use the same metal for a reference in order to compensate for metal oxidation. Aluminum first surface mirrors are not recommended since they oxidize rapidly. The aluminum oxide aborption on a first surface mirror is observed at about 860 cm^{-1} and is quite intense at grazing angle.

Macroscopic polarized grazing-angle spectroscopy has been used extensively to study monolayer adsorbates on metal surfaces [14–16]. Recently, grazing-angle measurements have been used to study polyimide surface modifications [17] and polyimide to metal bonding [18]. Microscopic grazing-angle measurements of the lube depletion on fixed disks [6], chemisorption of CO on Pt [19], and flux residues on gold pads [11] have already been demonstrated.

2.3. Practical Aspects of Reflection Microspectroscopy

There are several important considerations when adapting macroscopic techniques to infrared microspectroscopy, especially when developing quantitative methods. Diffraction at an aperture, sample diffraction and refraction, and substrate refraction [3] will greatly affect data. Katon et al. [12] have discussed diffraction-induced stray light in detail. Diffraction is the most significant factor to consider in reflectance measurements. In infrared microspectroscopy, an aperture is placed in the beam path to define the sampling area. Technically, this device is a field stop. In microspectroscopy it is normally referred to as the aperture, and for consistency that is the term we use throughout this discussion.

The IR-Plan design includes an aperture before and after the sample. In transmission measurement it is necessary to match the upper and lower apertures so that the area defined by the objective and condensor lens is the same. The light passes through the lower aperture and the condensor, which is focused on the sample. The objective collects the light and passes it to the detector through the upper aperture. For reflectance measurements only the upper aperture is required. The upper aperture is located between the eyepieces and objective. In reflectance mode, light travels through the aperture and the objective before reflecting from the surface of the sample. The reflected light is collected by the objective and passes back through the same aperture and to the detector.

The apertures may be fixed circular holes, an iris diaphram, or a pair of knife edges that can be adjusted to a rectangular shape. The apertures discussed here should not be confused with the objective aperture slide discussed earlier. Light will bend, or diffract, around the edges of an aperture. The light diffracted at the aperture forms an image that is a series of bright and dark lines known as a diffraction pattern. If the aperture is rectangular, the pattern will be a bright central rectangle with smaller, fainter rectangles to the sides separated by narrow dark lines. A circular aperture will generate a bright circular spot with concentric dark and light rings around it.

The formula to determine the location of the light maxima and minima at the focal plane of the objective [20] is

$$d = \frac{m\lambda}{\text{N.A.}} \tag{3}$$

where d is the distance between minima, N.A. the numerical aperture of the objective, and m is a variable. For a rectangular slit, m is an integer greater than or equal to 1 for the minima (dark rings). For a circular aperture, m is not an integer. The light intensity is a Bessel function of order 1. The values of m were calculated by Lommel [20] and are listed in Table 3. Note that the values for the circular rings are larger than for the rectangular slit. Light is diffracted across a larger area with a circular aperture than with a rectangular aperture of the same width.

Diffraction at the aperture determines the minimum spot size that is achievable. This spot also corresponds to the least resolvable separation of two ojbects within an optical system. The former is called the Fraunhofer diffraction limited spot size and the latter is defined by the Rayleigh criterion [21].

Table 3. Values of m for Slit and Circular Apertures[a]

Ring	Slit minimum	Circular minimum
First	1	1.220
Second	2	2.233
Third	3	3.238
Fourth	4	4.241
Fifth	5	5.243

Source: Data from Ref. 20.
[a]The value is for the Fraunhoefer minimum indicated in the ring column.

The formula to calculate the resolving power (D), or least resolvable separation, is

$$D = \frac{1.22\lambda}{\text{N.A.}} \tag{4}$$

This formula is valid when the desired sample area fills the aperture. The value calculated represents an ideal case, that is, the absolute minimum spot size achievable with the given objective. Table 4 contains the nominal spot size when using the fixed circular apertures supplied with the IR-Plan microscope, in the upper (reflectance) position. In Table 5, the diffraction limited spot size is listed for each objective at wavelengths of 2.5, 5.0, and 10 μm. Note that spatial resolution decreases at longer wavelengths. Since infrared data are acquired from 2.5 to 15 μm, an understanding of this phenomenon is key to correct interpretation of data.

The diffraction-limited spot is equivalent to the central bright spot. For a circular aperture this is known as the Airy disk, and it contains 84% of the radiation. The formula to calculate the resolving power is normally given to approximate the minimum aperture and minimum spot size. However, approximately 16% of the radiation falls outside this defined spot. This formula is satisfactory to calculate the resolving power of a visible microscope. However, it neglects contributions from the surrounding area, which can lead to substantial errors in infrared measurements.

A 40-μm × 240-μm aperture falls on a 64-μm-wide area [22]. The diffraction-limited spot size is 20 μm. Using equation (3), the width calculated using the third minimum is 52 μm, and for the fourth minimum, it is 69 μm. The third minimum of the circular Fraunhoefer diffraction pattern includes approximately 97% of the light. The fourth mimimum includes nearly 99% of the light.

Table 4. Nominal Spot Size of Fixed Circular Apertures (μm)[a]

Aperture	15×	32×	ATR	GA
1.0 mm[b]	67	31	40	33
1.5 mm[c]	100	47	60	50
3.2 mm	213	100	128	107
100 μm	6	3	4	3

[a]The size is calculated from the magnification of the objective. See Table 5 for the true spot sizes for each objective as a function of wavelength.
[b]Marked "lower."
[c]Marked "upper."

Table 5. Resolution of the IR-Plan Objectives (μm)[a]

	2.5 μm		5.0 μm		10 μm	
Objective	*D*	*D′*	*D*	*D′*	*D*	*D′*
15×	5	18	10	36	21	73
32×	3.8	13	8	26	15	53
ATR	3.5	11	7	24	14	49
GA	3	11	6	21	12	42

[a]The diffraction limited spot size (D), also known as the Airy disk defines the least resolveable separation. In reality, the minimum aperture diameter (in μm) for less than 2% stray light is given by *D′* (see Section 2.3).

To determine the area illuminated by approximately 98% of the radiation (*D′*), it is necessary to modify the calculation above so that it defines the fourth minimum of the Fraunhofer diffraction pattern of a circular aperture:

$$D' = \frac{4.241\lambda}{\text{N.A.}} \tag{5}$$

Calculated values of *D′* for each objective at wavelengths of 2.5, 5 and 10 μm, are listed in Table 5. Note that *D′* is nearly three times *D*. Using a 15× objective and a rectangular aperture in transmission, Sommer and Katon [22] empirically determined that bands due to a cellulose acetate film could be detected up to 56 μm away from the edge of the aperture. They observed that the 5.72-μm C=O stretch could be detected 32 μm away (36 μm from center), while the 9.5-μm absorption due to C—O—C asymmetric stretch could be detected 56 μm away (60 μm from center).

This increase in sampling area at longer wavelengths is very important to consider in reflectance measurements as well as in any type of quantitative determinations. Normally, it is best to select an aperture size smaller than the sample, known as inscribing the sample. For real-world samples where the area to be analyzed is embedded in an absorbing matrix, spectral contributions from the surroundings would be misleading. Spectral contributions are more likely at longer wavelength, and thus overaperturing is to be avoided since diffraction would increase the contribution from the surrounding area.

Use of the 1.0-mm aperture with each objective yields a nominal spot size that is outside the diffraction-limited spot size and is an ideal choice. The fixed circular aperture ensures reproducibility of spot size from measurement to measurement and is recommended for quantitative analyses instead of iris or rectangular (knife-edge) apertures. Variable apertures are difficult to reposition precisely, resulting in errors in quantitative measurements.

Although all calculations are shown for the IR-Plan microscope, they apply to other microscopes. The calculations for an off-axis, single-aperture system are more complex. Equation (5) will still yield a better approximation than the diffraction-limited spot size calculation. Calculations to determine the true sampling area for a rectangular aperture are similar to that shown for a circular aperture provided that the rectangle is slit-shaped. Square apertures involve more complex calculations. To a first approximation, the formula for the circular aperture can be used to estimate the minimum dimension of a rectangular aperture. The rectangular aperture is ideal for analysis of metal pads and lines on chips.

For reflection measurements, another important factor is depth of field. The axial depth of field (Z) of an objective represents the maximum depth from which the lens can collect light. Depth of field can be calculated as follows [23].

$$Z = \frac{4\lambda}{\text{N.A.}^2} \tag{6}$$

Values of Z for the 15× ojective are 30, 60, and 120 μm at 2.5-, 5-, and 10-μm wavelength. For the GAO, Z is 9, 20, and 40 μm, respectively. This calculation assumes that the sample has a smooth, mirrorlike surface and that the sample is metallic or absorbs infrared radiation. Most of the light will travel right through thin, nonabsorbing materials and will be lost unless the sample is placed on a clean, reflecting metal surface.

For reflection measurements of strongly absorbing materials it is often better to defocus slightly by lowering the sample stage. This reduces the depth of penetration into the sample and thus reduces the absorption. Each division on the fine-focus knob on the IR-Plan equals a 2-μm translation of the sample stage height. Sample and substrate refraction must also be considered. Refraction effects can be compensated for by adjusting the 15× objective and 10× condensor. This is required for transmission measurements, especially with high-index materials, such as silicon or germanium. Most samples are mounted on a substrate. If so, the condensor lens must be adjusted to compensate for the refraction that occurs when the radiation passes from the substrate to the sample into the air. This adjustment is described in the microscope manual and must be calculated based on the substrate thickness and index of refraction. Failure to perform this adjustment will result in an increase in stray light due to defocusing. For example, a 2-mm-thick silicon sample would shift the focus 1.5 mm [12].

3. RESULTS AND DISCUSSION

3.1. Internal Reflectance

3.1.1. Identification of Black Polymer

Black polymeric materials are difficult to characterize using near-normal reflectance techniques because they are totally absorbing. Since it probes only the near surface of a sample, the ATR objective enables rapid analysis of such material. A rigid black polymer used as a chemical waste holding tank liner was determined to be leaking. A thin sliver of the polymer was removed using a razor and a spectrum of the polymer was obtained using the ATR objective. A spectrum of carbon-black-filled PVC was obtained for comparison. Data are shown in Figure 3. It was determined that the liner was a carbon-black-filled PVC.

3.1.2. Determining Homogeneity and Cure in Epoxy Laminates

Epoxy laminates are used in the manufacture of printed wiring boards. The epoxy resin is mixed with hardener, catalyst, and solvents and is applied to woven silicate glass fabric mats that are pretreated with hydroxysilanes. Solvents

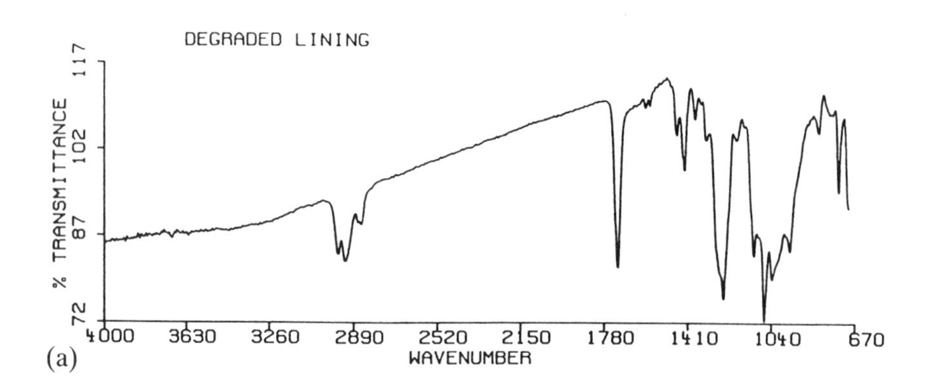

(a)

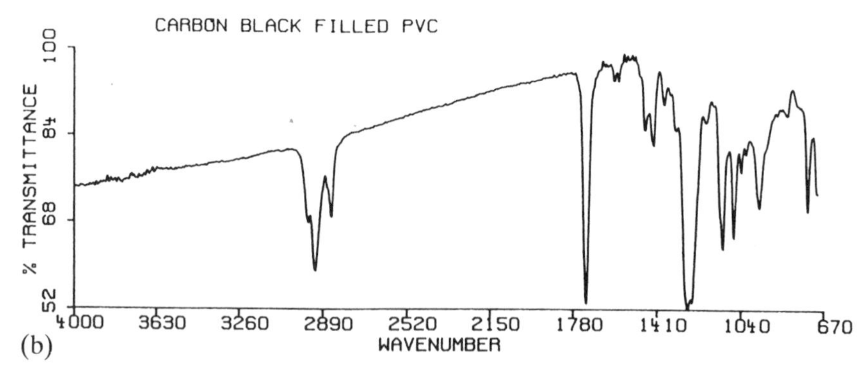

(b)

Figure 3. ATR objective spectra of unknown and PVC polymer. Spectrum of unknown black polymer (a) and PVC (b).

are removed by heating so that the epoxy is in a tack-free state, known as B-stage. B-stage materials are cut to size, stacked between sheets of copper foil, and then vacuum pressed at temperatures of up to 375°C for final epoxy cure. The epoxy mixture and B-stage materials react rapidly upon exposure to moisture and are processed, or characterized, shortly after preparation, One component of the epoxy mixture is a human sensitizer. The mixture and B-stage materials must be handled very carefully to avoid skin contact.

It has been shown that the reliability of the boards is directly proportional to the cure of the laminate [24]. Cure of the printed wiring boards is normally determined by differential scanning calorimetry (DSC) and FTIR. Unfortunately, DSC is an indirect method and is prone to errors since the measured value for the glass transition (T_g) is affected by the glass/epoxy ratio. FTIR determination of cure is performed by monitoring the intensity of the 910-cm^{-1} oxirane band intensity relative to the aromatic C═C band at 1600 cm^{-1} [25]. Coupons are cut from boards, and either a small piece of epoxy is excised with a sharp implement, or the copper is etched to expose the surface prior to analysis. The drawback to the former method is the effort involved in sample preparation, while the latter technique is prone to errors due to overlapping bands arising from silica glass fibers, which are strongly absorbing in the infrared.

Boards usually include mats of different thicknesses, so that individual layers may have significantly different ratios of epoxy to glass. Glass mat thicknesses are stated in mils (thousandth of an inch). In this case, samples were prepared by stacking five mats of the same thickness and curing them as described above. Specimens of each B-stage mat were also analyzed. The sample designation includes the last processing step and glass mat thickness. Data were acquired from several different thicknesses of mats and coupons cut from fully cured boards from the same mat lots. The B-stage and cure data are shown in Figure 4 for 142-mil glass and in Figure 5 for 280-mil glass. Note the presence of the C═N feature around 2180 cm^{-1} in the spectra of the B-stage samples. This band disappears upon exposure to air. Neither the oxirane absorption at 910 cm^{-1} nor absorptions due to the glass mat are observed in the spectra in repeated measurements across the surface of cured samples. The absence of the oxirane absorption indicates that the samples are fully cured.

In this example, the 280-mil mat is about twice as thick as the 142-mil mat, and the ratio of expoxy to glass is reduced nearly a factor of 2. Data acquired from specimens of thicker glass mats are identical to the examples shown here, indicating that the technique is useful even when the epoxy-to-glass ratio is very low. The cured boards are relatively rigid and are thus more difficult to characterize using this technique than the B-stage samples, however, the results obtained indicate that this technique is useful for determining cure, as well as verifying solvent removal and homogeneity.

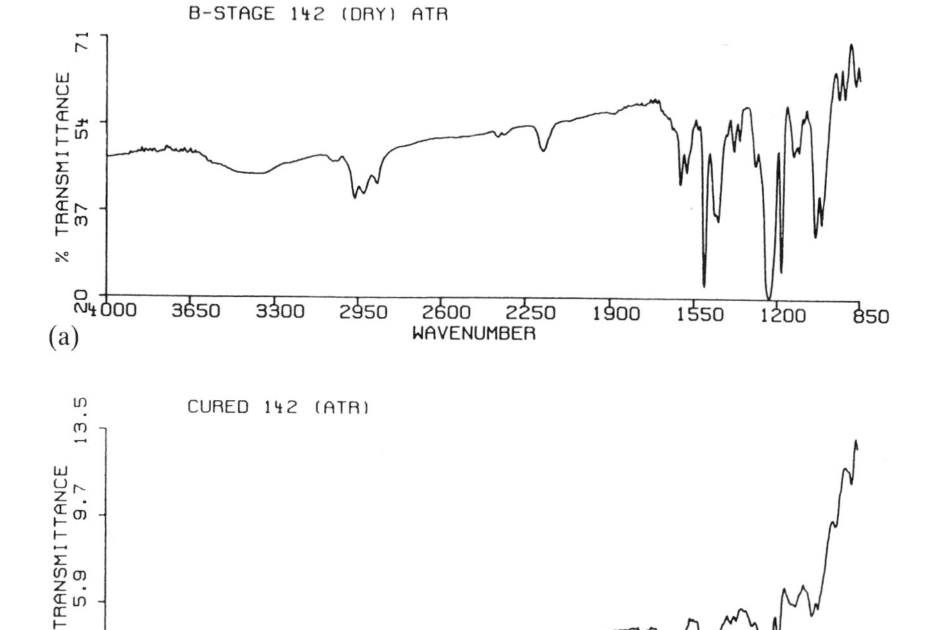

Figure 4. B-stage and cured epoxy spectra. Data acquired using ATR objective 100-μm aperture, 4-cm^{-1} resolution.

3.1.3. In Situ Characterization of Dielectric Materials

Data were acquired from a 1-μm-thick film of imidized 3,3′, 4,4′-biphenyltetracarboxylic acid dianhydride-*p*-diaminophenyl (BPDA-PDA) on silicon. The polyimide is produced by coating a clean silicon wafer with the polyamic acid form, which is heated in steps to 400°C under nitrogen. Water is a by-product of the imidization process. Both water and residual solvent are driven off by heat.

Figure 6a shows the reflectance spectrum of the film, and Figure 6b shows the internal reflection spectrum obtained with 100-μm aperture and ATR objective. The reflectance spectrum shows the sinusoidal baseline characteristic of a thin film, as well as distorted band shapes in the fingerprint region. In comparison, the spectrum obtained using the ATR objective is significantly easier to

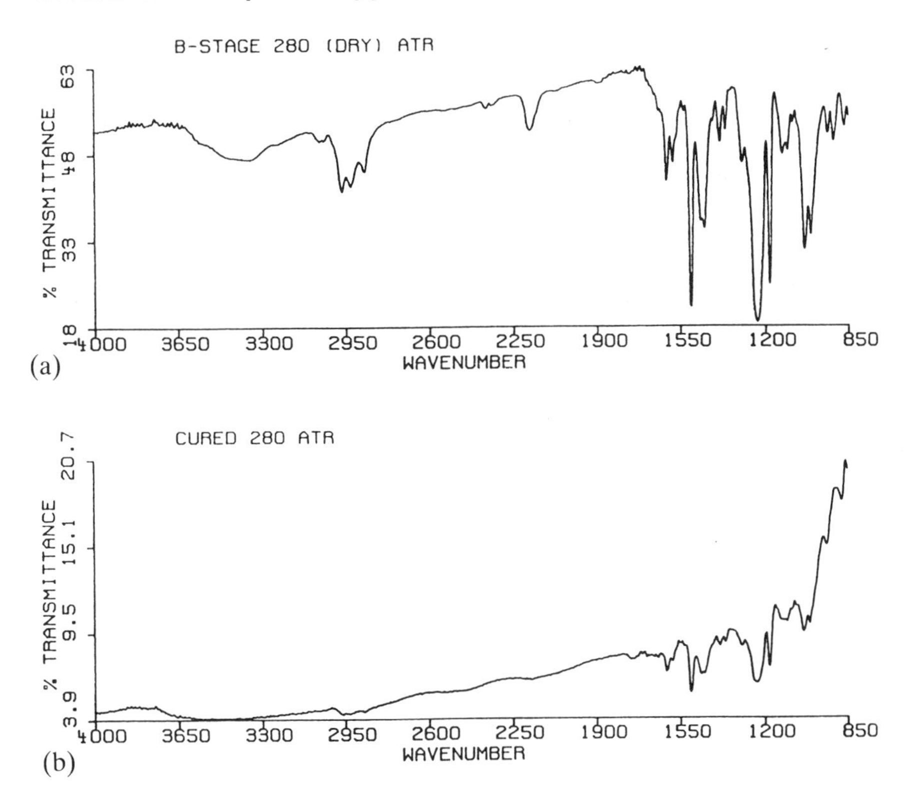

Figure 5. B-stage and cured epoxy spectra. ATR objective 100-μm aperture 4-cm^{-1} resolution.

interpret, even though it exhibits the characteristic sloping baseline at longer wavelengths.

To obtain an ATR spectrum, the sample must be pressed against the crystal. Figure 7 shows spectra obtained from the same BPDA-PDA film collected after the contact pressure is increased. As the pressure applied to the sample is increased, there is an increase in intensity of the bands. However, as the pressure is increased further, there is a 10-cm^{-1} shift in peak position. This is undoubtedly due to the compressive stress induced by pressing a flexible material that is bonded to an inflexible silicon substrate. This effect should be considered when interpreting data obtained from thin films or coatings on rigid substrates using this technique. It is important to apply the same pressure when performing quantitative analysis. Since one division of the fine-focus knob equals a 2-μm change, the fine-focus position may not be sufficient to determine the applied pressure precisely. To improve reproducibility of measurements, a piezoelectric

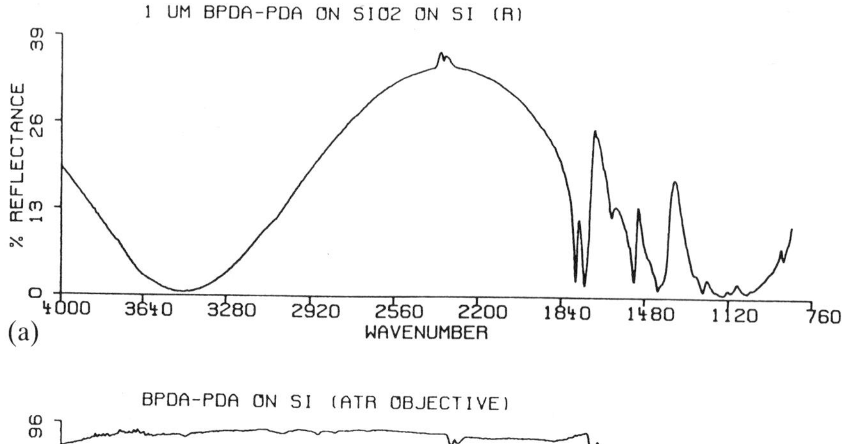

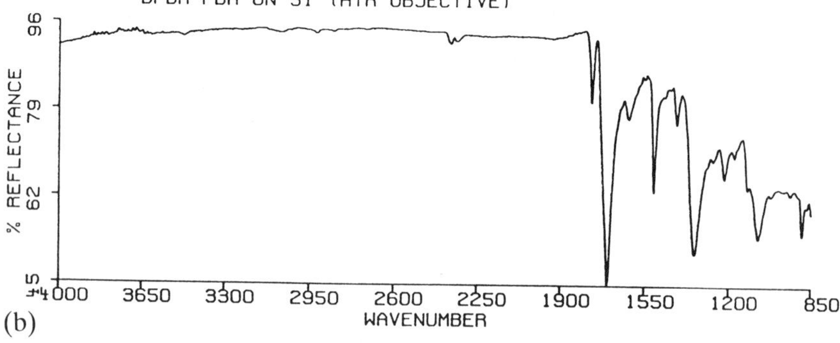

Figure 6. Reflection and ATR objective spectra of BPDA-PDA. Spectrum obtained with 15× (a) and ATR objective (b).

sensor such as Kynar film (Pennwalt) can be placed on the stage to measure the pressure at the point of contact.

3.1.4. Adhesive Failure Analysis

Identification of the type and locus of delamination is important in quality control and failure analysis. This is especially difficult when multiple layers of polymers, adhesives, organic coupling agents, and inorganic thin films are used. The standard peel test is used to verify proper adhesion of layers on patterened silicon wafers. Wafers that exhibited poor adhesion were examined to determine the mode of failure. Double-sided adhesive tape was used to "peel" the overlayers. The tape was first applied to a cleaned glass slide. Then the second tape surface was exposed and lightly pressed onto the surface of the wafer. The overlayers were then lifted, or peeled, from the wafer.

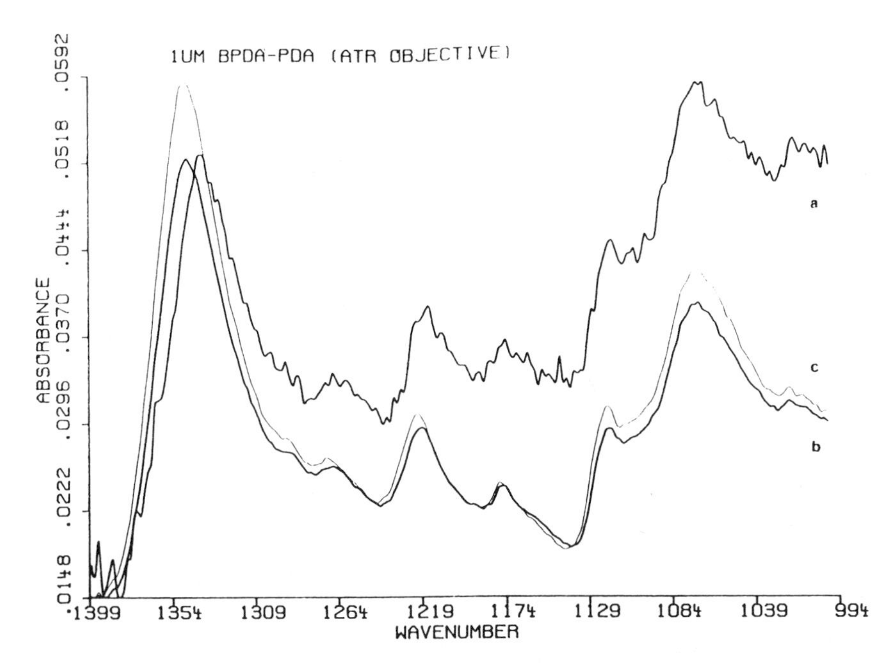

Figure 7. ATR objective spectra of BPDA-PDA film. Contact pressure low (a) highest for spectrum (c).

The exposed areas of the wafer were studied using visible microscopy and infrared microspectroscopy. It was evident from the examination that the failure was cohesive (Figure 8). This wafer consisted of approximately a 1-μ-thick layer of polyimide coated with about 600 Å of silicon dioxide, and a second 1-μm layer of polyimide and 600 Å of silicon dioxide. FTIR spectra obtained from both the peeled areas and the wafer show that the failure occurs in the lower oxide layer, underlying polyimide. Figure 9 shows the spectrum of the bottom of the peeled surface, with the ATR crystal just touching the specimen. The strong feature at about 1100 cm^{-1} corresponds to the Si—O—Si stretching mode. If the contact pressure is increased, strong features due to the polyimide are observed.

3.2. Polarized Grazing Angle

3.2.1. Contaminant Identification Relating to Bonding Failures and Corrosion

Fluorescence microscopy is used in the quality control process to ensure that there is no polyimide residue on metal surfaces. In some cases, chips that pass the fluorescence screening will fail during wire bonding. Infrared microspec-

Figure 8. Photomicrograph of peeled area on wafer. 90- × 90-μm pad with residue.

troscopy is used to identify the residue on the metal surface. The metal pads are surrounded by strongly absorbing material, such as polyimides. Often, contamination is a very thin film residue, too thin for near-normal reflection measurements. Chips that have failed in bonding will have large holes in the pads, which reduces the area available for sampling. Spectra were acquired from bonding failure pads at 4 cm^{-1} resolution. These pads are surrounded by 1-μm-thick polyimide. The pads have a depression in the surface due to the wire bonding test (Figure 10). A single beam spectrum taken from pads on a good chip was used as a reference. The rectangular aperture was adjusted to approximately 40 × 40 μm for these measurements.

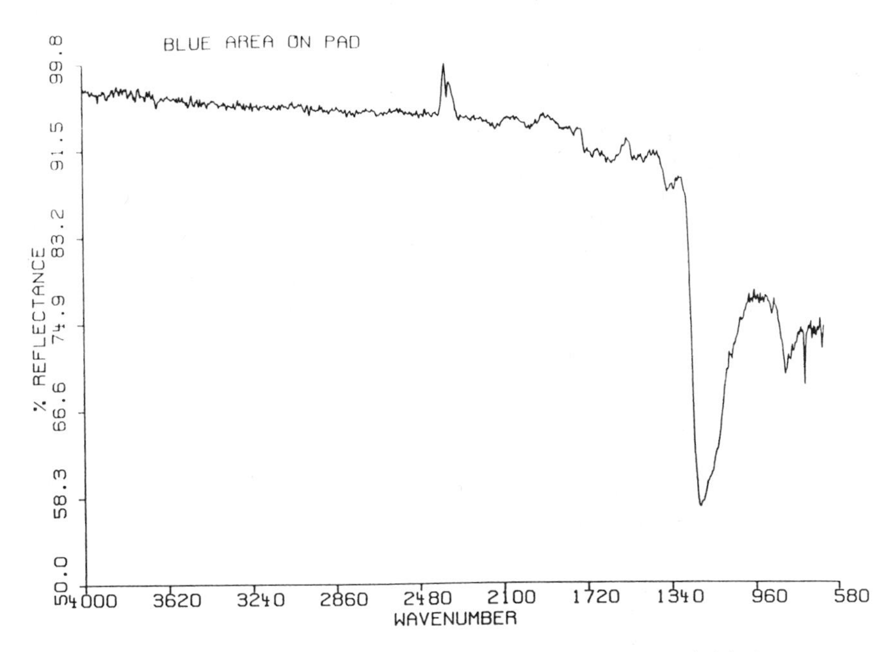

Figure 9. Spectrum of film peeled from wafer (ATR objective).

Data are presented in Figure 11. The bands at 1770, 1710, 1515, and 1370 cm^{-1} correspond to the imidized form of the polymer. The band at 1275 cm^{-1} is due to the aromatic ester and is characteristic of oxydianiline. These bands correspond to residual pyromellitic dianhydride–oxydianiline (PMDA-ODA) polyimide [26]. Since all of the polymer should be removed from the metal during processing, this indicates a failure in the removal process. Polarized measurements using the GAO with 1.0 mm aperture (33-μm spot size) have been used successfully to characterize similar samples. For many samples, however, the metal pads are too small or too rough to permit characterization using this technique.

3.2.2. Blue Spots on Copper

A wafer was examined to identify the cause of numerous circular brown and blue patches that appeared on exposed copper metal areas. The 200-μm-wide copper lines have 1-μm-thick polyimide along the sides (Figure 12). This wafer was stored in air for 2 weeks after reactive ion etching. Data were acquired at 8 cm^{-1} resolution using a 50-μm aperture and coaddition of 2000 scans. A clean evaporated copper film was used as a reference.

Figure 10. Photomicrograph of discolored pad on chip. 100- × 100-μm pad surrounded by polyimide.

Examination of the wafer revealed that the blue spots formed in the center of the dark reddish-brown areas on the wafer. Figure 13 shows the spectrum of a dark brown area on the wafer which exhibits a band at 605 cm^{-1} due to copper oxide. This finding is consistent with the discoloration observed. Figure 14 shows the spectrum of a dark blue spot on the wafer. This spectrum shows bands at 650, 610, and 605 cm^{-1}. The band at about 650 cm^{-1} is attributed to CF_2 molecule and the bands at around 600 cm^{-1} are due to copper oxides. The blue discoloration is consistent with formation of a copper salt. Normally, corrosion of copper proceeds from oxidation with or without hydrolysis, and subsequent attack from substances such as chlorine and fluorine to form salts

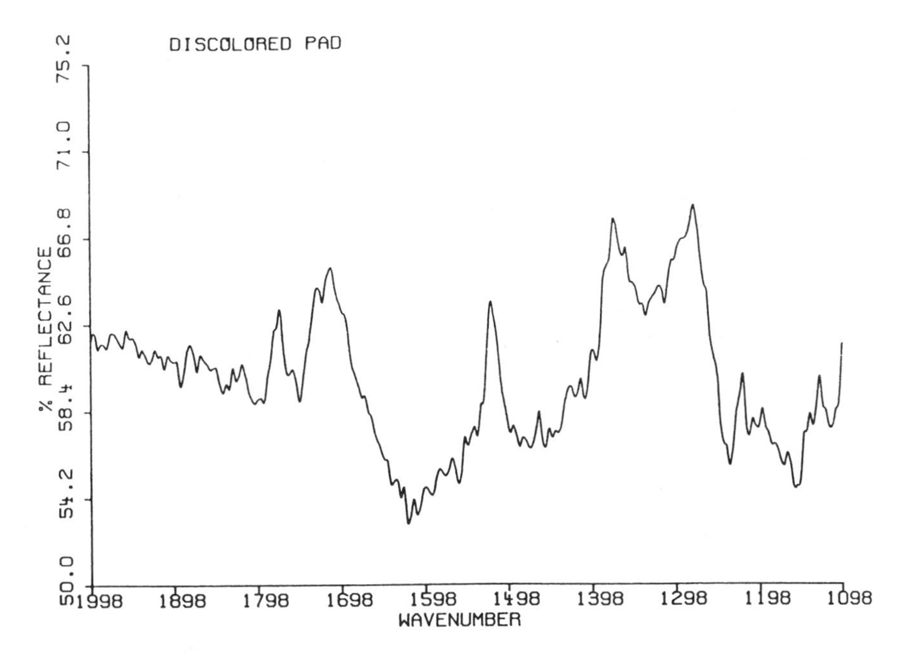

Figure 11. Polarized GAO spectrum of residue on metal pad.

[27,28]. This result was subsequently confirmed by secondary ion-mass spectroscopic (SIMS) imaging of the same sample. The CF_2 species was traced to fluorocarbon pump oil backstreaming into the process chamber.

3.2.3. Fixed Disk Media and Lubricants

Polarized grazing-angle FTIR is used extensively to characterize the thickness and distribution of the lubricant. As disks decrease in size, however, the microspectroscopical technique is useful in assuring even distribution, as well as in failure analysis. A hard disk that failed during testing was examined. This disk consisted of a metal disk coated with amorphous carbon to which a thin film of stearic acid lubricant was applied.

Spectra were obtained at 8 cm^{-1} resolution with a 100-μm nominal spot size, with a good disk as a reference. Figure 15 shows the spectrum obtained from the failed hard disk. It was observed that the lubricant was not detected, indicating either lube migration or improper application of the lube. The spectrum shows bands around 810 and 600 cm^{-1}, the first due to C—Cl and the second due to metal oxide. Since the carbon coating is used to prevent metal oxidation, the band due to oxidation was not anticipated. Pinholes in the carbon

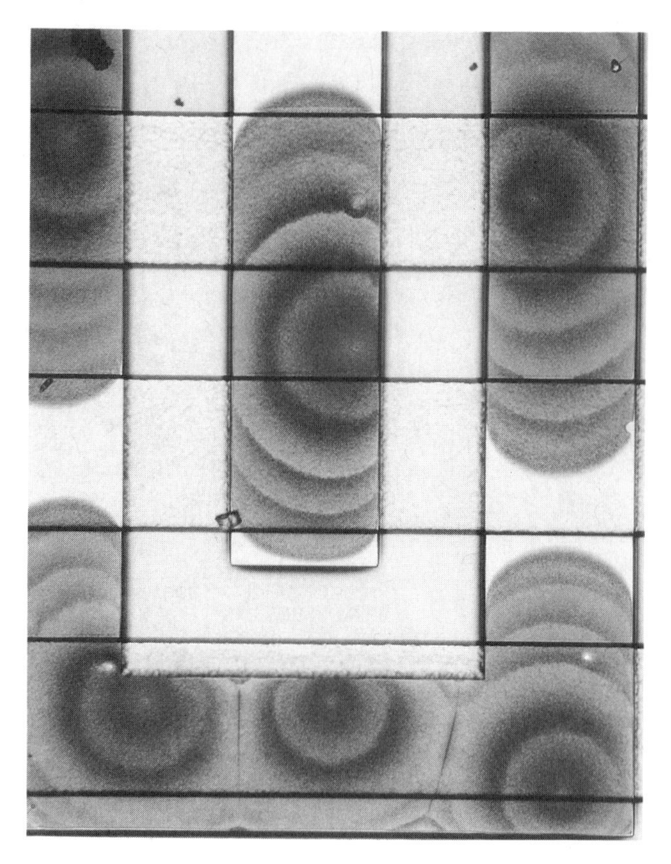

Figure 12. Photomicrograph of blue spots. 200-μm-wide copper line; white area is polymide.

coat were later detected with a scanning electron microscopy (SEM). The presence of oxide and chloride was confirmed with XPS analysis of disks from the same lot.

Unfortunately, exposure of the metal surface to air not only results in oxidation, but will lead to corrosion due to reaction with ambient chlorine species. The presence of chlorine contamination on the failed disk is not surprising. In this case, the failure results from improper carbon coating.

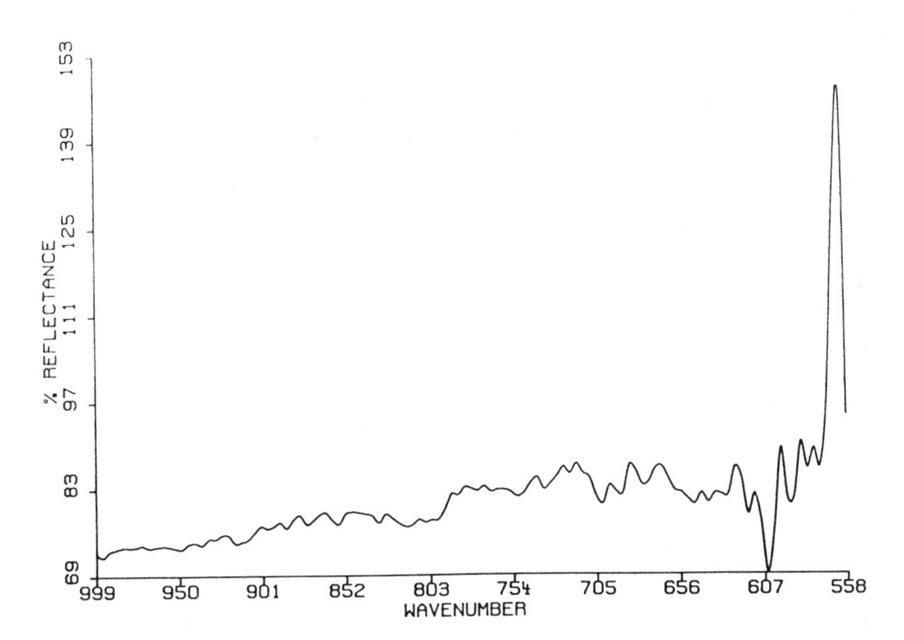

Figure 13. Polarized GAO spectrum of brown area on copper. 2000 scans, 8-cm^{-1} resolution, 50-μm aperture.

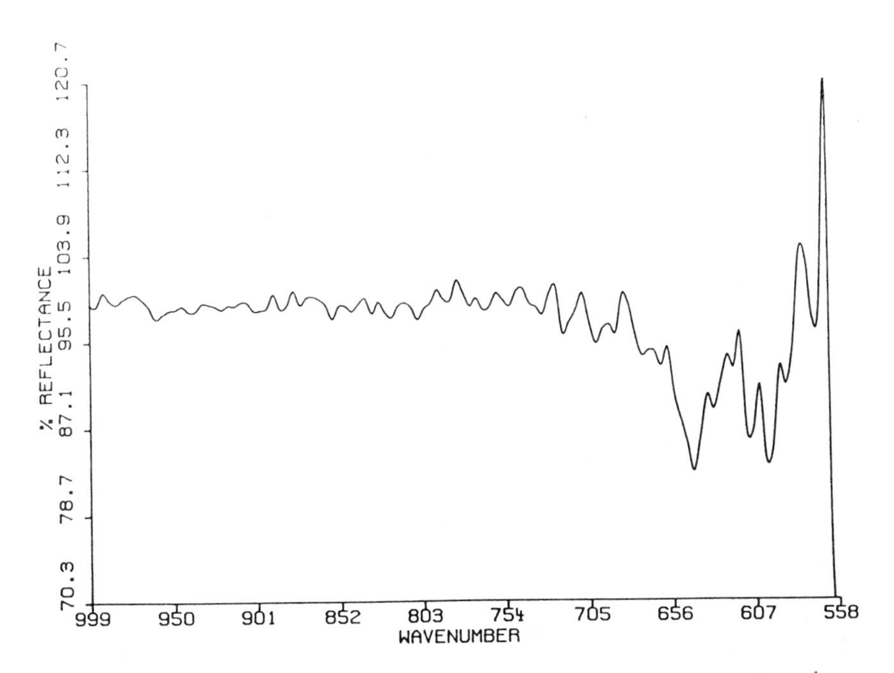

Figure 14. Polarized GAO spectrum of blue spot on copper. 2000 scans, 8-cm^{-1} resolution, 50-μm aperture.

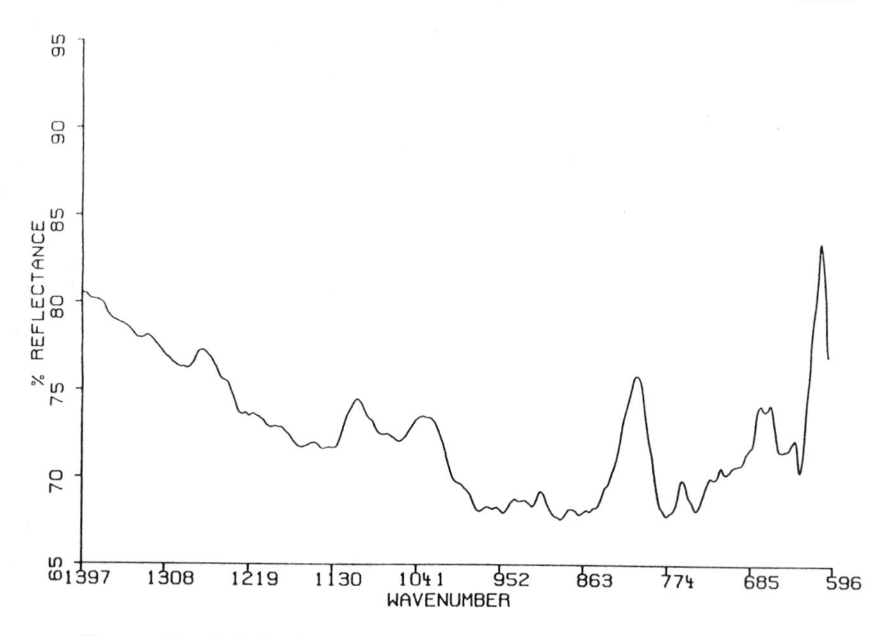

Figure 15. Polarized GAO spectrum of failed hard disk versus good disk.

4. CONCLUSIONS

Microinternal reflection objective and polarized GAO accessories to the IR-Plan are very useful for characterizing many of the challenging samples in the electronics industry. Care must be taken to select the proper aperture size to avoid spectral contamination from surroundings and to focus the objectives correctly to avoid false focus, spectral shifts, and increased noise. Due to the small area sampled, the probability that the data obtained are not representative of the whole is very high [12]. Quantitative analyses for AQ/QC must be designed thoughtfully to minimize false-positive and false-negative results.

The recent introduction of germanium and diamond internal reflectance crystals for the ATR objective improves the versatility of the technique. Although it is more difficult to work with a nontransparent crystal such as germanium, it is essential for characterization of high-refractive-index materials such as silicon. In addition, it is expected that polarized internal reflectance measurements would be advantageous in the study of surface modifications, molecular orientation within films, and stress in dielectric materials. Although there are still limitations to the techniques, they will be increasingly useful in characterizing materials within the electronics industry. Rough-surfaced materials still pose the greatest challenge to analysts and may be beyond the capabilities of either infrared or Raman microspectroscopy.

ACKNOWLEDGMENTS

The author would like to thank Michael Moore and Susan Garlock (IBM PC Company) for the B-stage and cured epoxy-laminate samples, and Paivikki Buchwalter for the cured BPDA-PDA on silicon. The author would also like to thank Paul Seidler for helpful discussions, and Robert Melcher, Trey Smith, and Bruce Scott for their support of this work.

REFERENCES

1. Ramsey, J. N., and Hausdorff, H. H. Application of small-area IR analysis to semiconductor processing problems, in *Microbeam Analysis*, R. D. Geiss, ed., San Francisco Press, San Francisco, 1981, p. 91.
2. Messerschmidt, R. G., and Harthcock, M. A. *Infrared Microspectroscopy: Theory and Applications*, Marcel Dekker, New York, 1988.
3. Katon, J. E., and Sommer, A. J. IR microspectroscopy routine IR sampling methods extended to the microscopic domain, *Anal. Chem.*, *64*: 931, 1992.
4. Young, P. H. Characterization of high-performance fibers using infrared microscopy, *Spectroscopy*, *3*: 24, 1988.
5. Webb, R. E., and Young, P. H. Identification of dropouts on magnetic media using infrared reflectance microspectroscopy, *Spectrosc. Lett.*, *23*: 679, 1990.
6. Mastrangelo, C. J., Von Schell, L., and Le, Y. IR microscopy applied to slider rail depletion of lube on a thin-film magnetic recording disk, *Appl. Spectrosc.*, *44*: 1415, 1990.
7. Borghesi, A., Geddo, M., Pivac, B., Sassella, A., and Stella, A. Quantitative determination of high-temperature oxygen microprecipitates in Czochralski silicon by micro-FTIR spectroscopy, *Appl. Phys. Lett.*, *58*: 2099, 1991.
8. Borghesi, A., Geddo, M., Pivac, B., Stella, A., and Lupano, P. Direct evidence of oxygen precipitates in epitaxial silicon obtained by micro-FTIR spectroscopy, *Appl. Phys. Lett.*, *58*: 2657, 1991.
9. Gaillard, F., Bonnard, A., and Hocquaux, H. Study of epoxy-bonded galvanized steels: applications of FTIR in situ analysis and microspectrometry, *Surface Interface Anal.*, *17*:537, 1991.
10. Compton, S., and Powell, J. Forensic applications of IR microscopy, *Am. Lab.*, *22*: 41, 1990.
11. Reffner, J. A., Wihlborg, W. T., and Strand, S. W. Chemical microscopy of surfaces by grazing angle and internal reflection FTIR microscopy, *Am. Lab.*, *23*: 46, 1991.
12. Katon, J. E., Sommer, A. J., and Lang, P. L. Infrared microspectroscopy, *Appl. Spectrosc. Rev.*, *25*: 173, 1989–1990.
13. Harrick, N. J. *Internal Reflection Spectroscopy*, Interscience, New York, 1967.

14. Ulman, A. *An Introduction to Ultrathin Organic Films from Langmuir-Blogett to Self-Assembly*, Academic Press, San Diego, 1991, p. 6.
15. Greenler, R. G. Infrared study of adsorbed molecules on metal surfaces by reflection techniques, *J. Chem. Phys.*, *44*: 310, 1966.
16. Yarwood, J. Fourier transform-infrared spectroscopy of ultrathin organic films, *Spectroscopy*, *5*: 34, 1990.
17. Lee, K.-W., Kowalczyk, S. P., and Shaw, J. M. Surface modification of BPDA-PDA polyimide, *Langmuir*, *7*: 2450. 1991.
18. Dunn, D. S., and Grant, J. L. Infrared spectroscopic study of Cr and Cu metallization of polyimide, *J. Vacuum Sci. Technol.*, *A7*: 253, 1989.
19. Self, V. A., and Sermon, P. A. In-situ FTIR-microspectrometry: spatial differentiation of CO during chemisorption and oxidation, *J. Chem. Commun.*, 834, 1990.
20. Jenkins, F. A., and White, H. E. *Fundamentals of Optics*, 4th ed. McGraw-Hill, New York, 1975, p. 315.
21. Meyer-Arendt, J. R. *Introduction to Classical and Modern Optics*. Prentice Hall, Englewood Cliffs, N.J., 1989, p. 253.
22. Sommer, A. J., and Katon, J. E. Diffraction-induced stray light in infrared microspectroscopy and its effect on spatial resolution, *Appl. Spectrosc.*, *45*: 1633, 1991.
23. Adar, F. Developments in Raman microanalysis, in *Microbeam Analysis*, R. H. Geiss, ed., San Francisco Press, San Francisco, 1981, p. 67.
24. Stone, F. E. Electroless copper, in *Printed Wiring Board Fabrication in Electroless Plating*, G. O. Mallory and J. B. Hadju, eds., 1990, p. 331.
25. Bower, D. I., and Maddams, W. F. *The Vibrational Spectroscopy of Polymers*, Cambridge University Press, New York, 1989.
26. Molis, S. E. Infrared characterization of molecular orientation in polyimide films, in *Polyimides: Materials, Chemistry, and Characterization*, C. Feger, M. M. Khojasteh, and J. E. McGrath, eds., Elsevier, New York, 1989.
27. Leidheiser, H., Jr., *The Corrosion of Copper, Tin and Their Alloys*, Electrochemical Society, R. E. Krieger Publishing, Huntington, NY, 1979.
28. Pecht, M. A model for moisture induced corrosion failures in microelectronics packages, *IEEE Trans. Components Hybrids Manuf. Technol.*, *13*: 383, 1990.

5

Depth Profiling and Defect Analysis of Films and Laminates: An Industrial Approach

Richard W. Duerst, William L. Stebbings, Gerald J. Lillquist, James W. Westberg, William E. Breneman, Colleen K. Spicer, and Rebecca M. Dittmar 3M Center, St. Paul, Minnesota

Marilyn D. Duerst University of Wisconsin–River Falls, River Falls, Wisconsin

John A. Reffner Spectra-Tech, Inc., Shelton, Connecticut

1. INTRODUCTION

The infrared microscope was introduced into the 3M Company about a decade ago. Since then, its use has spread rapidly throughout the company. To some extent use of infrared (IR) microspectroscopy has replaced standard infrared techniques, but in most cases the microscope added to the laboratories' workload because it could solve problems more quickly and efficiently than could other analytical methods. However, the microscope does not automatically allow unskilled or untrained personnel to solve problems with IR any more easily. In our experience, the technique of IR microscopy is more demanding in terms of the skill level needed, not less. The effective spectroscopists now must have microscopy skills in addition to the other sample preparation skills that he or she had before.

2. SAMPLING FACILITY REQUIREMENTS

As noted by many other authors, sample preparation techniques have a much greater effect on the quality of the analytical result than does any other factor in the analysis of defects and laminates using an FT-IR microscope. Infrared sample preparation requires a variety of specialized tools and equipment. We list the items and supplies in our sampling facility, together with the suppliers, as a starting point, realizing that other excellent suppliers are available for these items and that other laboratories may have alternative lists.

2.1. Preparative Microscope with Lighting Options

Probably the most critical piece of equipment in the sampling facility is the preparative microscope. It is essential that one's good binocular stereomicroscope with its lighting mechanism has a fairly large working space. Often, the success or failure in IR microscopy depends on one's skill in using this optical microscope. We use a Wild Heerbrugg M5A microscope, which is capable of magnification up to about 320×, although we use the 75 to 80× magnification range most frequently for sample preparation. The range of field diameters for this microscope is from 1.75 to 0.8 mm, with working distances from 91 to 25 mm.

For illumination of the sample, the options are incandescent lamps that illuminate the sample from above (sometimes referred to as incident illumination), transmission illumination, or a fiber optic system. For transmission illumination, a 6 V/10 W halogen bulb is available. For incident illumination, we have a low-voltage 6 V/20 W incident lamp with a color temperature of 3200 K for very uniform bright-field illumination, and a 6 V/10 W lamp for illumination of objects obliquely from above, provided with an inclinable or clampable lampholder. This method is useful for samples that need the increased relief that can be gained by sidelighting. We also have an Intralux 6000 source.

We frequently use a fiber optic ring illuminator for shadow-free cold light illumination of spatial objects. The intensity of a 90 W halogen bulb suffices for magnification up to at least 40×. We also use an 18-inch-long bifurcated gooseneck fiber optic light guide (Volpe, Inc.), which can be directed at almost any oblique angle at a specific part of a sample. A number of objects display their features best when illuminated with a combination of lighting techniques. Semitransparent objects with uneven surface features, such as fibers, for example, are best viewed by using both transmitted and incident light.

2.2. Static Control Devices

Static can be a very serious problem when transferring very small particles, especially from or in the presence of films such as polyethylene terephthalate.

One method of reducing static effects is to place a static control bar, which uses Polonium-210 (Nuclear Products Company) near the work area. Another static control product is Zerostat 3, a product made in England but distributed in this country as ''Discwasher'' for removing debris from phonograph records.

2.3. Small Tools

Tools needed for microsampling include scalpels and scalpel blades, tweezers, and various types of needles. Most, if not all, IR suppliers furnish some type of toolkit with their instruments. Generally, these are fairly useful for particles in the range >50 μm. With a lot of care, one can use them for smaller particles. However, we have created our own toolkit for microsampling, which includes:

1. *Scalpel blades and handles.* We use scalpels for particle transfer, flattening, and small-scale microtoming or slicing. We have found that scalpel blades vary greatly in quality and utility. Some have serrated edges, which are not good for microtomy work. Recently, we have been using Feather Brand surgical scalpels with Bard-Parker scalpel blade handles, which seem to be quite good in quality. The size we use the most is number 15.

2. *Tweezers.* SPI Miracle Tip tweezers (size 3) from Structure Probe, Inc. are useful for transferring particles.

3. *Needles/probes.* Common sewing needles with some type of handle attached and tungsten needles are extremely useful for probing the samples as received. The sewing needles may also be used to flatten fibers by rolling the needles over the samples to prepare them for transmission IR analysis. Tungsten needles are available commercially, or you can make your own by heating a 20-mil tungsten wire to ''red'' hot and poking it into a block of $NaNO_2$. This is an ''exciting'' procedure because the chemical reaction is quite exothermic, etching the tungsten and forging a point. The tungsten needles can be made to have extremely sharp points and are used mainly to transfer extremely small microtomed samples. Holders for the tungsten needles are obtained from THI Company and X-Auto.

2.4. Microtome

An AO/Reichert FC4 Reichert-Jung Cryo-Ultramicrotome for preparing thin sections of samples is essential in many cases. The specimen temperature can be kept as low as −190°C. Thicknesses between 2 and 10 μm are cut with cross sections typically 0.2 × 0.2 mm. In most cases, more art than science is required in microtoming such tiny samples, especially for low-temperature cryomicrotomy. Additional difficulties arise when two layers with different T_g values are layered next to each other.

2.5. Other Useful Items

Other items that should be available in the sample preparation facility include microscope slides, KBr plates, wick sticks from Harshaw, Inc., and safety razors. A diamond anvil cell may be quite useful in some sample preparations.

3. EXPERIMENTAL

3.1. Infrared Analysis Methodologies for Specific Problems

In the particular experience of our laboratory in the 3M Company, the two main categories of samples are laminated or multilayered films and surface or embedded defects on dielectric or metallic substrates. The nature of the sample itself and the purpose for the analysis usually dictates what sample preparation is required, if any, and what IR technique is most amenable to the analysis.

3.1.1. Surface Defects

Surface defects (specks from 10 to 500 μm in diameter) may respond to three experimental aproaches: attenuated total reflectance (ATR), specular reflectance or reflection-absorption (RAIRS), or removal of the particle from the sample via the techniques described above to a salt plate for IR transmission measurements.

ATR-IR

The attenuated total reflectance (ATR) method is most useful for surface particles as small as 25 μm in diameter. The defect particle needs to be thicker than 0.2 μm or the IR spectrometer will also probe the substrate material and the analyst may have difficulty distinguishing between the absorbances due to the sample and those of the substrate.

The internal reflectance element (IRE) we usually use is made of ZnSe, which has a penetration depth of 1 to 5 μm (note that penetration depth is wavelength dependent). In our experience, the softer the particle, the better the results, due to better contact between the sample and the crystal. Also, the ZnSe is relatively soft, so it is easily damaged by contact with a hard material. One main advantage of the ZnSe crystal is that it is quite clear visually, and employing it in an analysis is fairly simple.

Another available ATR objective at 3M has a diamond crystal. This material is very hard, very expensive, and has about the same penetration depth as ZnSe. The diamond crystal is much less transparent visually (because the edges of the crystal are facets), and positioning the sample is consequently much more difficult. The advantage of this type of crystal is that it is much more robust and can be used on harder materials without it being damaged, assuming that one uses proper microscope techniques.

A third IRE option is germanium, which is hard, expensive, and difficult to use. The Ge crystal has a much higher index of refraction (4.0) than the ZnSe or the diamond crystal, and consequently, the normal depth of sampling is much less, generally less than 1 μm. The Ge crystal, however, is visually opaque, and consequently much harder to use. It is the opinion of one of the authors (J.W.W.) that there is a good deal of art required in the use of this crystal, due to its fragile nature.

In our experience it is useful to confirm ATR results, when feasible, by removing the speck using appropriate tools described above, and then transferring the speck to a salt plate for transmission IR analysis. ATR spectra look somewhat different from transmission spectra in that the relative peak intensities are different and the peaks may be shifted somewhat to lower wavelengths. Modern software packages for IR data analysis incorporate programs that correct for the intensity distortion inherent in the ATR method, and this is quite helpful in the quest to match spectra obtained from defects against commercial and company libraries of spectra of known materials.

Specular Reflectance and Reflection-Absorption

Specular reflectance and reflection-absorption infrared spectroscopy (RAIRS) basically utilize the same physical setup, in which the infrared beam is impinged on the sample surface. For specular reflectance, the beam is reflected off the surface of the particle, a method that only occasionally works well for specks on polymeric substrates. For RAIRS analysis, the defect is usually received on a metallic substrate or is transferred to a metal surface (mirror). The method seems to work well on particles at least 15 μm in diameter that are fairly thin (5 to 15 μm). A new side-arm attachment for reflectance measurements is very valuable for analyzing surfaces of large samples that cannot be destroyed.

Transmission IR

The preferred method for obtaining good, interpretable spectra without artifacts or cross-contamination remains transmission IR. It is definitely worthwhile to go to the trouble of removing the defect with an implement (such as one of those described above) and transferring it to a salt plate. Note that this absolutely requires a good stereomicroscope, which exhibits a flat field, excellent image brightness, sharp image, contrast mechanisms, and variable magnification. In addition, high-quality scalpel blades, sharpened tungsten needles, static control devices, a lot of practice, a steady hand, and even good luck are also required.

3.1.2. Embedded Particles

If a particle is embedded in the substrate rather than lying on its surface, a thin section through the particle may be prepared with a scalpel or a microtome as described above. Transmission IR analysis with double aperturing will usually furnish an acceptable spectrum of the particle and exclude effects of the surrounding material.

3.1.3. Multilayered Films

The general procedure in our laboratory for handling multilayered films consists of obtaining ATR spectra of both sides of the film as well as a transmission IR spectrum of the bulk of the material when the sample is sufficiently thin. One then slices off one thin section at a time with a scalpel, while viewing under a stereomicroscope. ATR spectra of the layers are obtained as soon as they are exposed. This method has proved to be very useful since ATR analysis assists in isolating the components of one layer from those of another layer, although ATR examines only the surface that is exposed after each scraping.

Sometimes the scrapings from a layer can be transferred to a salt plate or pressed into a pellet for transmission analysis. Also, the layers can sometimes be separated if soaked for a period of time in a solvent or in liquid nitrogen, or placed on a dry ice block. Subsequent squeezing between steel plates may break apart the bonding between the layers.

Another method we are using more frequently is the microtoming of cross sections of the multilayered film and then transferring the cross sections to IR transparent plates. Each layer of the sample is then analyzed by transmission IR using the instrument's apertures to isolate one layer from another. We have found that the dual or redundant apertures are very useful for this technique. We have successfully analyzed layers as thin as 3 to 4 μm, using a rectangular aperture set to a width of 3 to 4 μm and a length of at least 10 μm and have avoided most of the diffraction problems. A rectangularly shaped aperture with one dimension fixed by the thickness of the layer of interest allows more energy to reach the detector than does a square aperture while providing a larger "view" of the same layer. Our experience has shown that a cryogenic microtome, as described above, can be very useful for this type of analysis.

Occasionally, top layers of multilayered samples can be analyzed by specular reflectance, but this works less than 10% of the time on our samples. Spectral artifacts and bandshape changes are problems. Because of the mix of types of interactions taking place, including reflection, absorption, and transmission, the Kramers–Kronig transform rarely improves the spectrum to a level that is considered acceptable. For the transform to work properly, the beam reaching the detector must be from reflection alone [1].

3.2. Microscopy Optics

Our laboratory uses an IRμs molecular microspectroscopy system, which is shown in Figure 1. We have a motorized stage to position the sample conveniently, an optical and CCD camera, and a special bench for stabilization of the infrared microscope. The motorized stage exhibits spatial resolution of 0.1 μm, stepping increments of 1 μm, and a digital readout to 0.1 μm.

Since the infrared microscopy technique is usually severely taxed because of the very limited sample available for analysis, the technique requires a good optical match between the optics of the microscope and the optics of the infrared instrument. Each of the two modes of operation, the viewing mode and the infrared mode, use a different method of sample illumination. Figure 1 illustrates the image-forming paths in both the viewing and infrared mode.

The spectroscopist should have a basic knowledge of the optics of both systems in order to obtain the maximum spectral information from the instrumentation. It should be noted that some elements in the viewing mode optical path are also used in the infrared mode optical path.

3.2.1. Viewing Mode

The visible microscopes at 3M, such as those described above, generally use the Kohler illumination method [2, p. 317 ff]. The IRμs system also incorporates this type of sample illumination when the sample is being positioned for examination (i.e., we are in the ''viewing'' mode). The illuminating rays from a tungsten light source focus on the aperture diaphragm (stop), the objective back focal plane, and the eyepoint. This system provides uniform illumination of a large field of view so that the sample and/or defect is readily found.

In the viewing mode, the aperture stop is that opening or diaphragm that limits the amount of light (energy) from the source that can be collected or transmitted through the rest of the instrumental optics. This stop controls the illumination level, so that the observer is comfortable with the energy impinged on his/her eye, as well as controls the resolving power. In contrast, the image-forming ray paths focus on the field diaphragm, the specimen plane, the entrance pupil of the eyepiece, and the film plane (if the image is being photographed) or the retina (if the specimen is being viewed by the scientist). Just as aperture stops limit the energy throughput, so do field stops limit (determine) the field of view. Note that although the image-forming paths and the illuminating paths may be treated as separate, in reality, they are not independent of each other.

3.2.2. Infrared Mode

After a small sample (less than 100 μm) has been positioned visibly and its infrared spectrum is to be obtained by the infrared microscope, one switches to

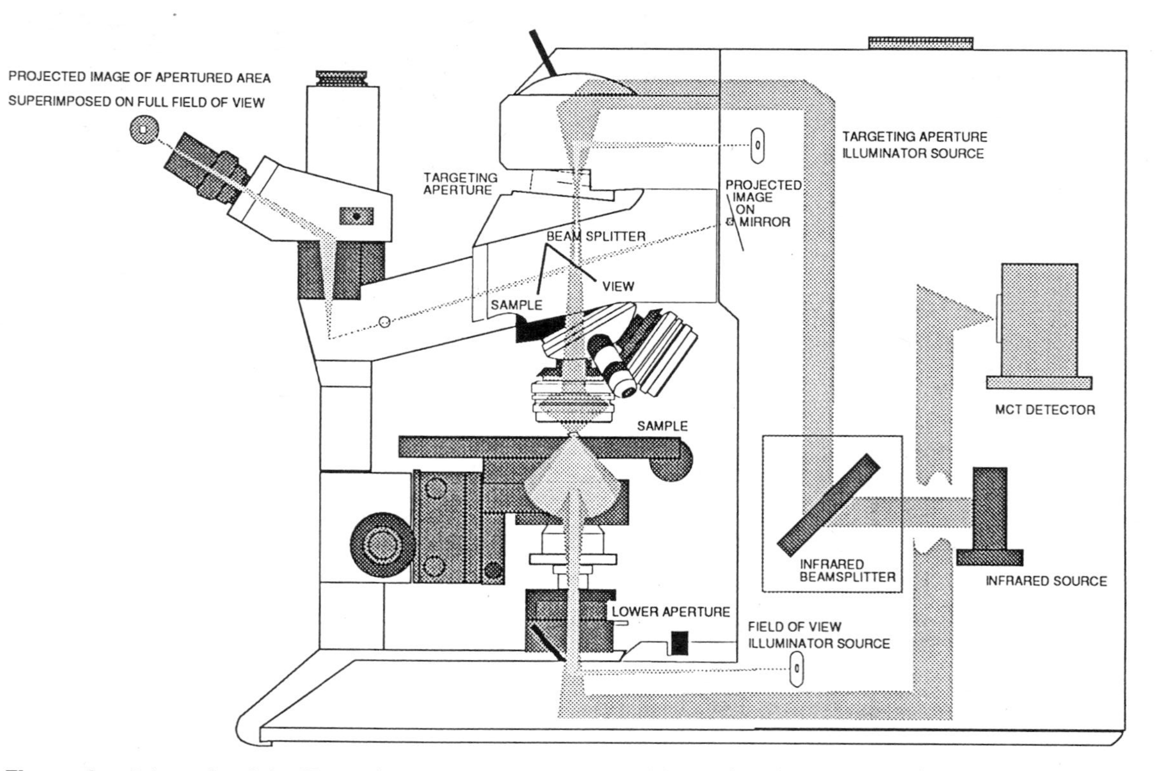

Figure 1. Schematic of the IRμs microspectroscopy system and image-forming ray paths. (Courtesy of Spectro-Tech, Inc.)

the "infrared mode." In this mode, the Nelsonian (or critical [2, p. 315 ff]) rather than the Kohler illumination system is employed, so as to maximize the energy reaching the detector. In this case, the stops for the image-forming path and the illuminating path are one and the same. The aperture and field stops for the infrared path of the system are usually determined by the amount of sample available.

Typically, the detector is a square about 250 μm on a side. Most samples are considerably smaller than this. What is the minimum sample size necessary for obtaining useful spectral data in the transmission mode? The answer to this question must refer both to the thickness of the sample as well as to its cross-sectional area. A minimum practical thickness is estimated to be 0.05 μm (500 Å) for a 25-μm cross section, based on a typical absorbance and *S/N* ratio of 10:1.

What is the smallest cross-sectional area that we can examine? The answer to this question is not as simple. For a well-designed system and a fixed set of operating conditions, the minimum sample area is inversely proportional to the source brightness, the transmission efficiency, the detector detectivity (D^*), and the square of the effective numerical aperture of the microscope objective [3]. These relationships imply that a smaller sample requires a greater numerical aperture (see the discussion below), a brighter source, and a greater detectivity. For any instrument, the minimum sample area is also directly proportional to the square root of the amplifier band pass and the *S/N* ratio, and inversely proportional to the spectral band width [3]. In practice, 10 μm is probably near the limit of cross-sectional area that we can examine today, although the technique of ultramicroscopy, which we describe in more detail in Section 3.5, is able to examine specks less than 10 μm in diameter.

3.3. IR Spectral Optimization

The fundamental factors to be considered in attempting to optimize infrared spectra of small samples using IR microscopy revolve around the same three accessible variables as should be considered for the usual infrared analyses: namely, the desired spectral resolution, the desired signal-to-noise level (*S/N* for a single scan), and the time available for the analysis [4, p. 254 ff].

The key question in optimizing any system is to identify the relevant variables that influence spectral quality (i.e., resolution and *S/N*) and then to optimize their interrelationships for optimum performance, or, in spectroscopy parlance, to balance the "trade-off." Although we intend to discuss the *optical variables* in this chapter, the electronic variables (such as the A/D converter and the preamplifier gain) and the mathematical variables (such as apodization and phase corrections) are also important. Even if the optics have been optimized, spectral quality may not be optimized unless the electronic adjustments as well as the algorithms have been applied properly. In fact, today we can easily choose the algorithm we wish to use to massage the data.

3.3.1. *S/N* Factors

It is generally understood that it is critical that the *S/N* ratio be high enough for detecting small amounts of material and/or for using subtraction spectra to determine reliably if two samples differ in composition. A practical example of where such an analysis may be important in industry is in the analysis of "good" and "bad/defective" samples from a given production run.

For a spectrum measured with a Michelson interferometer, the relationship between the relevant *S/N* level and a number of other variables is given by

$$S/N = \frac{U_\nu\,(T)\,E\,\Delta\nu\,t^{1/2}\,\xi\,D^*\,(0.1e^{-0.2\Delta\nu})}{A_D^{1/2}}$$

$U_\nu\,(T)$ is the radiance of the source (in $W/cm^2 \cdot sr \cdot cm^{-1}$) for a wavenumber range from a blackbody source at temperature T, $\Delta\nu$ is the instrument spectral resolution, ξ is the energy transmission efficiency, E is the etendue, A_D is the area of the detector with specific detectivity D^*, and t is the collection measurement time [4, p. 249]. The factor 0.1 $\exp(-0.2\Delta\nu)$ has been included to approximate the fact that a band must be resolved in order to have an analyte signal. An excellent discussion concerning the errors in the absorbance measurement has been written by Anderson and Griffiths [5].

3.3.2. Infrared Source Intensity/Cooling

In any discussion of the actual energy reaching the detector, we should first consider the intensity of the source itself. One way to improve the *S/N* level is by increasing the temperature of the source, thus increasing the spectral energy density, $U_\nu(T)$. In theory, a good source for mid-IR work would be one whose emission peaks at approximately 1000 cm^{-1}. However, such a source would be cooler than room temperature, about 290 K. Higher-temperature sources will produce more intense beams with higher infrared energy, and according to the *S/N* equation above, a higher-energy beam would also increase the *S/N* ratio. Therefore, in reality, practical infrared sources operate at temperatures of 1200 to 1800 K [6, p. 287].

A number of spectrometers such as the IRμs cool the source compartment to eliminate heating of other spectrometer elements. In Figure 2 we compare our best estimates of curves for a water-cooled source and an air-cooled source. Our water-cooled source, which is operated at a higher temperature than our air-cooled source, delivers more radiance with its peak at about 1750 cm^{-1}. How-

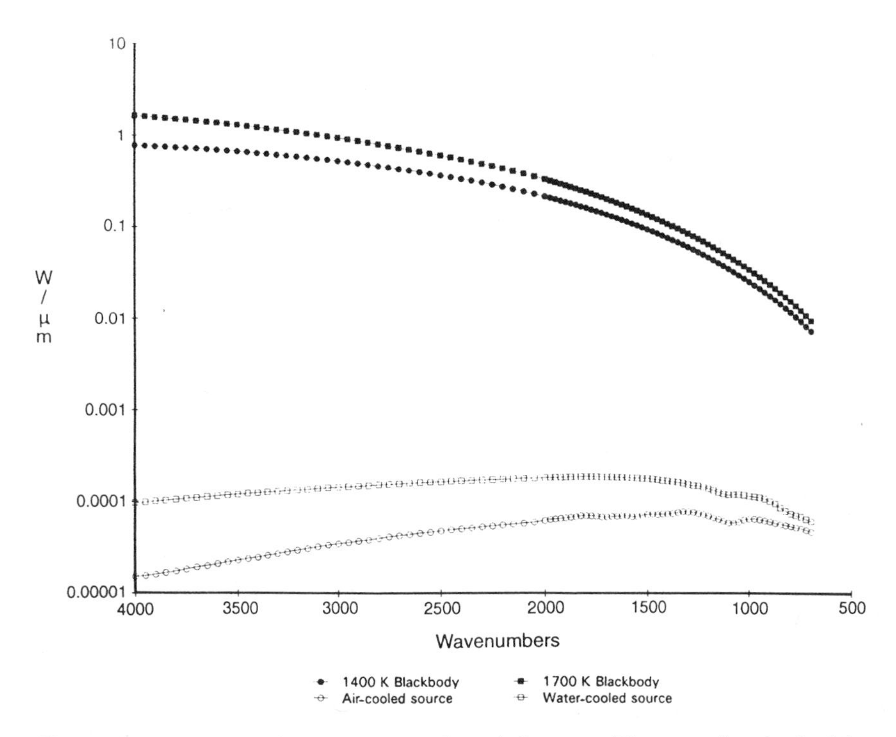

Figure 2. Water-cooled source versus air-cooled source. The power/μm emitted by blackbody sources into the spectrometer employing a 1-cm^2 aperture and a solid collection angle of 0.38 sr are plotted versus wavenumber. An estimate of the power/μm impinging the detector after passing through the interferometer are represented in the lower curves for the two sources.

ever, its relative output is more affected by environmental temperature variations such as changes in the cooling-water temperature. If a background scan and a sample collection scan are taken using an unstable system, when components vary in temperature, undesired spectral artifacts are created, which are difficult, at times, to interpret.

A new IR source, with two to eight times the brightness of conventional sources, is being studied [7]. If and when commercially available, such a source should improve the *S/N* ratio. (See also the current efforts of John Reffner, Section 3.5.)

3.3.3. Optical Throughput (Etendue)

Recall from the discussion above that image-forming beams and illuminating beams are essentially the same for the infrared spectral mode of an IRμs microscope. This illuminating system (Nelsonian) is employed in the infrared system, as we have stated above, to maximize the energy passing through the sample and reaching the detector. Each collection point on the collecting mirror gathers light from the source IR lamp and contributes to the illumination of a region of the specimen. The energy reaching the detector through the illuminating path depends on the radiance and the optical throughput, or etendue.

This infrared energy throughput is a key issue. Why? Because the energy throughput directly affects the efficiency of the instrument and ultimately determines the *S/N* level of the spectrum. For the infrared system, the sample cross section defines the field of view. This sample cross section also dictates the energy in the illuminating path, since the cross section of the sample (which is also the aperture stop) cannot be varied, as we do with large samples. This transmitted energy is characterized by the throughput and is defined [8, p. 337] by

$$E = \Omega A$$

Ω is the solid angle subtended by a point on, for example, the sample (e.g., the apex of the solid angle is at the sample and defines a cone which just contains the detector cross section), and A is the area of the sample. It may be useful to visualize the sample as a collection of point sources, each of which has the apex of the cone defined by the detector (i.e., from the sample "viewpoint" each point on the sample subtends a solid angle of radiation onto the detector). See Figure 3 for an illustration of these relationships.

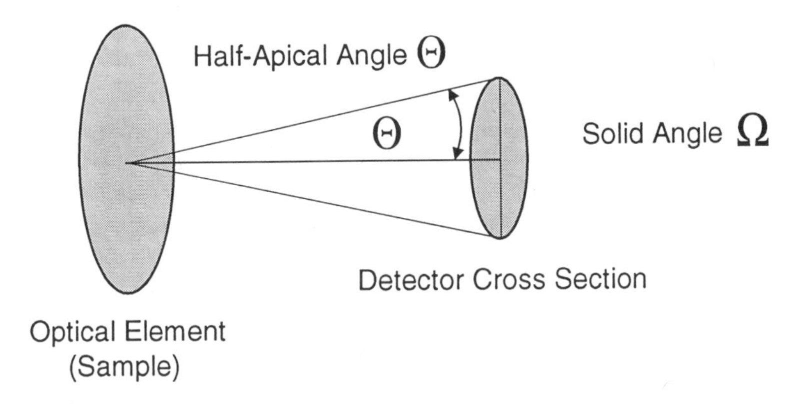

Figure 3. The solid angle and cross section used for calculating energy throughput. The sample may be envisioned as a collection of point sources, one of which is illustrated in the figure. Each point source is the apex of a cone defined by the detector. Ω is the solid angle defined by the cone; thus if A is the area of the sample, the energy throughput $E = \Omega A$.

A similar diagram could illustrate the etendue from the detector "viewpoint," where each point on the detector is a collection point of radiation from a cone containing the entire sample. Note that for an efficient optical system, the etendue should be the same for both the detector and the sample. Matching the detector etendue to the sample etendue is recommended for optimum *S/N* [9]. Any remaining unfilled detector just contributes to additional noise, whereas a larger sample wastes information because the energy passing through the sample does not impinge on the detector. Note that field stops are sometimes added for eliminating unwanted energy from regions other than the sample and that the sample is usually smaller than the full image of the source.

The etendue is rarely used as a specification for either infrared spectrometers or microscope objectives. Therefore, a brief discussion involving the more common methods to cite optical performance of spectrometers (the *f*-number) and of microscopes (the numerical aperture, N.A.) will be included here, and then we will return to the etendue, our preferred method for comparing optical preformance.

The N.A. is defined as the product of the refractive index n of the medium outside the lens and the sine of half the apical angle, which is in a plane dividing the cone of light collected by the objective, or

$$\text{N.A.} = n \sin \theta$$

The higher the numerical aperture, the more radiant flux is collected. The highest possible (theoretical) value for the N.A. in air is 1, since the refractive index of air is approximately 1, and because θ cannot exceed 90°, using conventional optics. Microscope objectives are currently available with an N.A. as high as 0.95 [10].

The *f*-number is often defined as the focal length of the lens (or optical element) divided by its diameter [6, p. 277] (although this relationship does not really hold when an object is very close to the lens; i.e., it holds only for small angles). For example, a system with a focal length of 35 mm and a diameter of 7 mm has an *f*-number of approximately 5 (i.e., *f*/5). Just as the N.A. cannot exceed 1.0 in air, correspondingly, the *f*-number of the lens cannot be lower than *f*/0.5. In other words, a lens with *f*/0.4 is unknown.

To determine if the infrared spectrometer is properly coupled to the microscope, the N.A. of the microscope needs to be compared to the *f*-number of the spectrometer. The relationship between these two optical variables is

$$\text{N.A.} = \frac{1}{2(f\text{-number})}$$

Ideally, the N.A. values of both the spectrometer and the microscope are identical to each other. As an example, suppose that the N.A. of an objective is 0.867 and the *f*-number of the infrared spectrometer is *f*/5. The N.A. of the spectrometer is then calculated to be 0.10, which leaves considerable room for improvement.

We prefer to compare the etendue of both spectrometer and microscope, which generally requires considerable more effort to calculate. To approximate the etendue, the area of the optical element A as well as the solid angle Ω are required, since $E = \Omega A$, as stated above. For example, suppose that the microscope objective has a focal length of 3 mm and an N.A. of 0.867. θ is found to be 60.1° from the definition of the numerical aperture. The radius is found to be 2.60 mm (see Figure 4a) using the focal length and the sine function, resulting in a calculated area of 21.24 mm^2. Calculating Ω from $2\pi(1 - \cos\theta)$ [8, p. 333; 10] gives 3.15 sr. Thus an etendue of 66.93 sr · mm^2 is calculated for the microscope objective. [Note that if the solid angle is estimated as $\Omega = \pi\,(\text{N.A.})^2$, which is applicable only for small values of θ, an unacceptable value of 2.36 sr is calculated.]

In comparison, if the spectrometer has an f-number of f/5 and an optical diameter of 7 mm, A is calculated to be 38.48 mm^2. The solid angle Ω is calculated to be 0.031 sr, and the etendue of the spectrometer is calculated to be 1.21 sr · mm^2. This match between the etendue of the spectrometer and that of the microscope could be better.

A common error in creating a representative diagram for the calculation above is to visualize the surface of the lens as a plane, and the angle between a line representing the lens radius and a line representing the focal length f as 90°, as we show in Figure 4b. The error is to envision the focal length as one

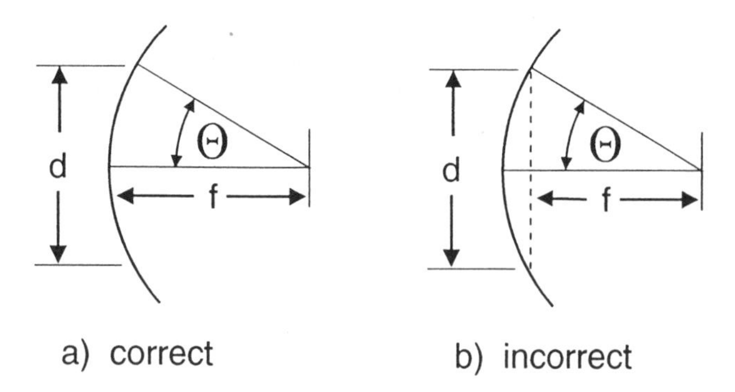

Figure 4. Apical angle determination. (a) The correct diagram shows one side of the "right triangle" as an arc of a sphere whose center is the focal point of the lens. The length of both sides of the triangle have magnitude f, and no right angle truly exists. θ is calculated to be about 60° from N.A. = sin θ. (b) The incorrect diagram shows f as one side of a right triangle and angle θ erroneously calculated to be about 40° using tan θ.

side of a right triangle and to use tan θ (rather than sin θ) to calculate the angle θ. In reality, one side of the ''right triangle'' is an arc of a sphere whose center is the focal point of the lens. The length of both sides of the ''triangle'' have the magnitude *f*, and no right angle truly exists, as shown correctly in Figure 4a. The reason for this misconception is that one fails to realize that the Abbe sine correction must be employed [11] to correct for coma and spherical aberrations. In this particular example, the angle would erroneously be calculated to be 40.9° rather than the correct value of 60.1°—a sizable error!

3.3.4. Least Resolvable Separation/Spatial Resolution

The resolving power of a microscope depends not so much on the magnifying power as on the N.A. of the objective lens. Back in 1950, Grey listed the numerical aperture as number one on his list of the most desired improvements in microscopes [12]. Now, decades later, numerical aperture values of objectives have improved considerably and a variety of different objectives are available to vary the spatial resolution (*not* spectral resolution).

Of significance in microscopy is that the N.A. may be used to determine the least resolvable separation (LRS) of tiny specks or defects in a sample, by employing the equation [11]

$$\mathrm{LRS} = \frac{\lambda}{2\mathrm{N.A.}}$$

For wavelengths of 2500 cm^{-1} (4 μm) and an N.A. of 0.87, the LRS is approximately 2 μm. Fraser [13] has indicated that a large numerical aperture is important to increase the energy collected from a given tiny sample and to reduce the diffraction pattern. Since the wavelengths of impinging radiation are longer for infrared rays (used to collect the spectrum) compared to the visible rays used to position the sample, multiple defect specks which may be visibly seen as separate by the observer, may not be separable in the infrared mode of infrared microscopes.

3.4. Spectral Collection Procedures

Typically, when we collect IR data, we compare the energy from the source, (both with and without sample) for each wavelength, realizing that the aperture must be the same for the sample and the background. We must therefore be concerned with the overall stability, especially thermal stability, of the spectrometer over the entire background and sample collection time periods. We must not forget the fact that the instrument and source must be held at a constant temperature ± 0.01°C (or better).

The possibility of heating the sample is greater in infrared microscopy because the infrared beam has been condensed. Particularly in *quantitative* infrared microscopic analysis, one should consider the sample heating effect. Miller [14] suggested that in a dispersive system about 1 W per 100 mm^2 is focused on the sample. Using an ΩEOMEGA TL-F-105 test strip, we have measured a temperature of between 40.6 and 43.3°C in a conventional FT-IR spectrometer sample compartment, which is about 20°C above room temperature. For a sample with a 10-μm-diameter cross section with 1 W of power impinged upon it, a considerably higher temperature would be realized if all the energy were absorbed.

Data collection procedures involve decisions as to variables, including the apodization, spectral range, and level of zero-filling. Generally, we use the Beer–Norton apodiziation, a mirror velocity of 0.2 cm/s, a spectral range of 4000 to 400 cm^{-1}, one level of zero-filling, and an aperture consistent with the size of the sample. Typical numbers of scans for transmission, ATR-IR, and RAIRS are 150, 200, and 250, respectively. An MCT (type B) detector is used. The spectrometer is purged with nitrogen gas from a liquid nitrogen supply to remove the carbon dioxide and water in the air. The sample region is not purged, so we do observe weak carbon dioxide and water bands.

Note that the microscope work described below in Section 4 involved the use of an objective with an N.A. of 0.65, a number that applies to both the viewing microscope and to the IR microscope because we are using mirrors, not refractive optics. We use 32× magnification, a resolution of 4 cm^{-1}, and a typical collection time of 10 min to obtain spectra of defects less than 100 μm in diameter. The etendue for typical collections, using 100 μm as the cross section of the sample, is estimated by multiplying the area (0.0078 mm^2) by the solid angle (1.508 sr) and found to be 0.0118 sr · mm^2.

3.5. Infrared Ultramicrospectroscopy

The so-called *diffraction limit,* defined for an optical element for which the aberrations are so small (i.e., negligible) that the image of a point is not larger than the central spot of the Airy disk, is alleged to be the greatest hindrance to achieving good-quality infrared microspectrometric results of samples smaller than about 10 μm.

Messerschmidt has discussed a technique known as *infrared ultramicrospectroscopy,* for which he has reported infrared analysis of particles which have a cross section of only a few microns [15]. In this technique of visible ultramicroscopy, pinholes as small as 30 μm are placed above the sample and *very close to the sample surface,* so that the light does not have a chance to spread out into a diffraction pattern. For infrared ultramicroscopy, Messerschmidt has used pinholes in the size range 1 to 50 μm, which are readily available. He can actually look through these holes and scan across the sample until the specimen or speck of interest is aligned. He then switches to the IR detect mode and collects the spectrum. Thus for pinholes the size of 1 to 10 μm, the spectroscopist can collect spectra below the infrared diffraction limit.

An alternative method for conquering the diffraction limit has been developed by Pam Martoglio and John Reffner of Spectra-Tech [16], who have obtained infrared results using a 6-μm × 6-μm aperture with the National Synchrotron Light Source. They feel confident that they can use a 1-μm × 1-μm aperture, and possibly even go below this value.

4. INDUSTRIAL EXAMPLES

4.1. Diffusion Studies

A block sample of a polymer measuring about 2 cm × 1 cm × 1 cm thick was prepared by standard techniques [17] for diffusion studies. Half the fused sample was pure polymer, and the other half was the polymer with a carbonyl-containing additive. A diffusion profile of the additive in this fused sample was to be determined by infrared analysis. The time for the experiment was varied to achieve a measurable distribution and the diffusion rate may be calculated from this time and profile.

This block was microtomed to obtain a sample with approximate dimensions of 5 mm × 26 mm × 62 ± 11 μm thick (measured with a *Laserule* from Pratt and Whitney). The sample was positioned in a sample holder consisting of two flat metal plates with slits on the order of 300 μm wide and about 3.2 cm long. Approximately 75 IR spectra were collected at about 10-μm intervals across the sample (as viewed through the slit in the metal sample holder) using the IRμs scanning microscope. Redundant aperturing was required. Each spectral "view" was directly adjacent to the next view or viewing area. (Other options were overlapping the spectral viewing area or increasing the intervals so that gaps or spaces between each spectral viewing area existed.) Each sampling area was 10 μm × 250 μm, with the smaller distance parallel to the diffusion direction. The entire IR spectrum was collected for each position.

To determine the diffusion profile of the additive, a specific absorption peak was chosen that is present only in the additive and a specific peak was chosen that is present only in the base polymer. In this example the carbonyl absorption band from the additive is strong and occurs at a wavelength where no absorption due to the polymer obscures it. It should be noted that the substrate absorption peak may vary slightly due to causes including variations in sample thickness, although ideally, the height of the substrate peak is constant. Variations in this peak intensity were used to scale the additive peak. The height of the additive peak compared to the height of the base polymer peak may be plotted against

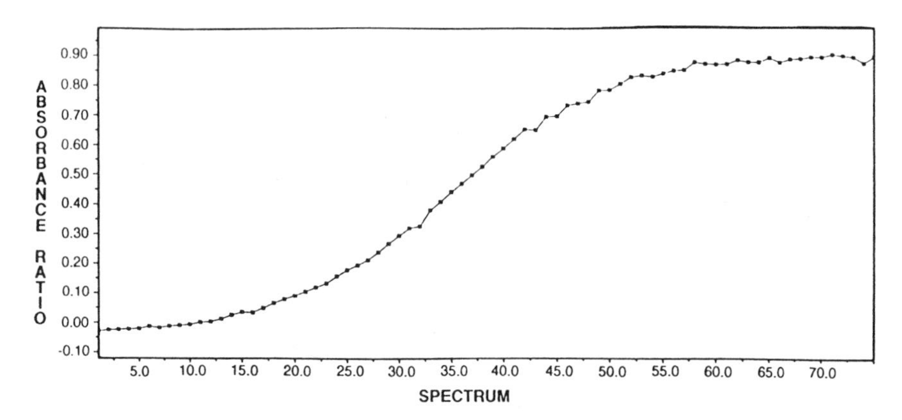

Figure 5. Diffusion of material into a substrate, plotting the ratio of carbonyl band to base polymer band versus spectral number. The contour of the plot reveals the diffusion profile.

distance, and thus the resulting variation in additive absorbance is due to the concentration gradient across the sample.

In Figure 5 we plot the peak ratio versus spectrum number (which is easily related to distance) as we scan across the microtomed sample. This plot shows the relative change in carbonyl intensity with position in the fused sample. Spectra with the lower numbers are farther from the position where the two polymers were fused. The carbonyl absorption peak decreases with distance from the point where the two blocks were fused, giving the extent of the additive diffusion into the substrate.

A three-dimensional array of wavenumbers versus absorbance versus distance (spectral number) from another series of collections from the same sample is shown in Figure 6. Seventy-five spectra, over 10-μm × 250-μm intervals, are partially superimposed, ending at a portion of the sample with additive. The magnitude of the additive peak begins to increase at about the ninth spectrum and is essentially constant beginning with the seventieth spectrum. Minor variations in the additive peak height are probably due to spectral noise. Obviously, a variety of plots may be obtained with different additives, fusing times, different temperatures, and different base polymers.

4.2. Defects in Lithographic Plates

A small speck, called a "hickey," measuring about 150 μm × 30 μm and about 25 μm thick and visible to the naked eye, was present on a lithographic plate. A photomicrograph is shown in Figure 7. We wished to identify the source of the speck, since it caused visible changes in the printed impression. Possible

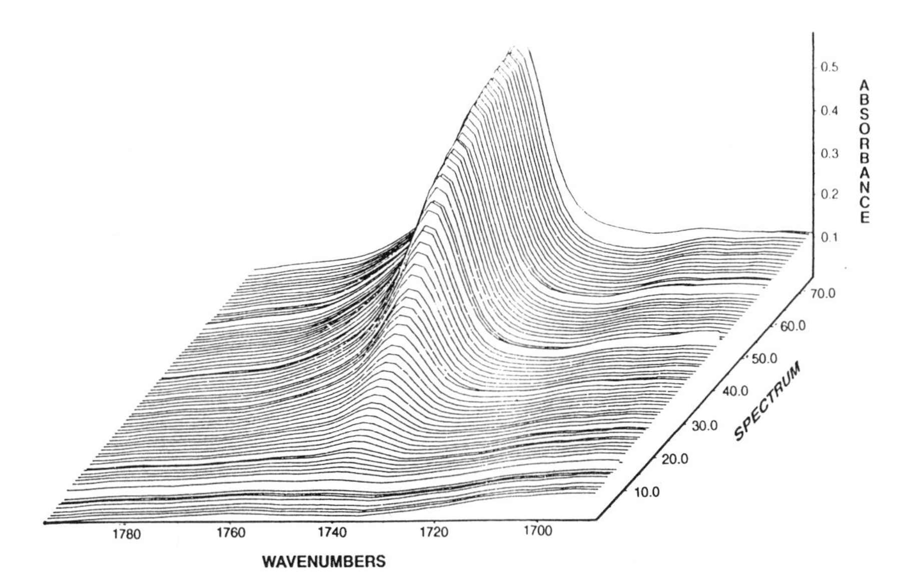

Figure 6. Three-dimensional array of spectra for diffusion studies. Seventy-five scans, over 10-μm intervals, visually indicate the extent of diffusion of an additive into a pure polymer after fusion of the two materials.

sources for the contamination that were suggested included the ink, gum, lithographic coating, and paper components.

For reasons of convenience and speed (the customer was pressing for a response), we opted for ATR-IR analysis. Two hundred ATR-IR scans utilizing the ZnSe crystal gave the spectrum in Figure 8. The speck was quickly identified as consisting primarily of a rubberlike material similar to that used in the rubber roller in the printing press. Replacement of the rollers alleviated the problem.

Hickies as small as 25 μm across could be examined. Thicknesses under about 0.2 μm would be more difficult to analyze due to spectral interferences from the lithographic plate coating or other substrate.

4.3. Embedded Defects in Extruded Polymeric Materials

Although we have not included a specific example, we frequently analyze embedded defects in extruded polymeric materials. These defects are usually caused by overheating the extruder, by hot regions in the extruder, or by insufficient cleaning of the extruder between production runs. Packaging material or environmental dust may also be the source of contamination. Most of these are analyzed by transmission IR after removing them from the matrix, using either a small aperture or the diamond cell.

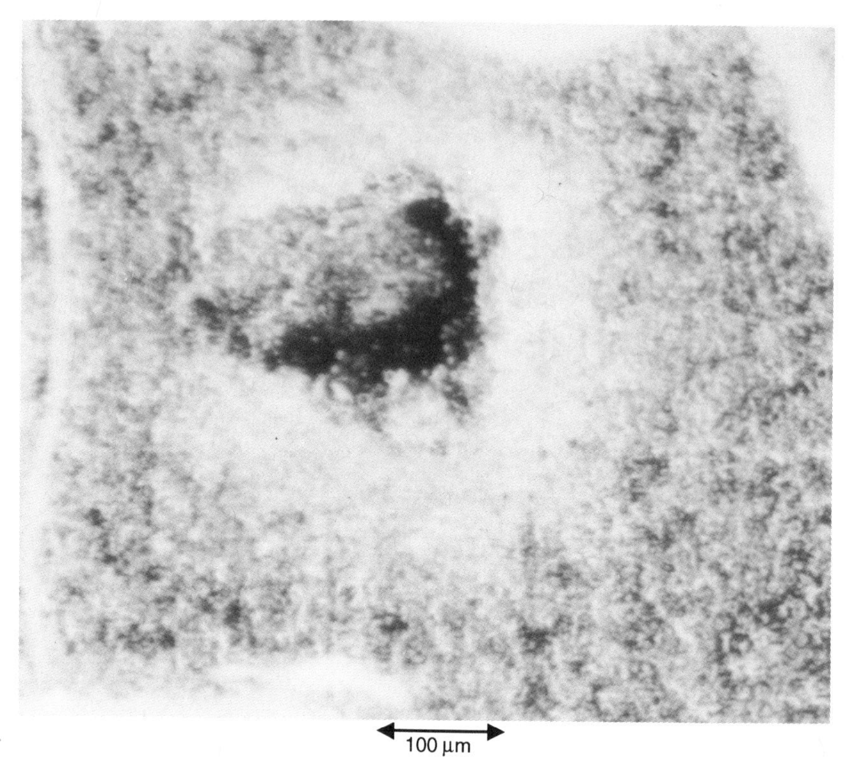

Figure 7. Photomicrograph of a lithographic plate "hickey." Easily discernible due to the light background, the speck measures about 150 µm × 30 µm.

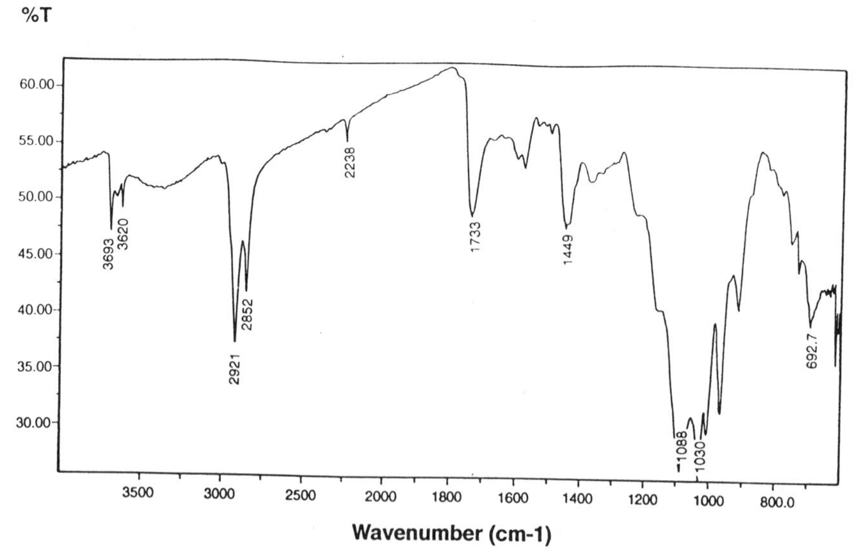

Figure 8. ATR-IR spectrum of lithographic plate "hickey." Bands suggest components including a synthetic rubber-containing material.

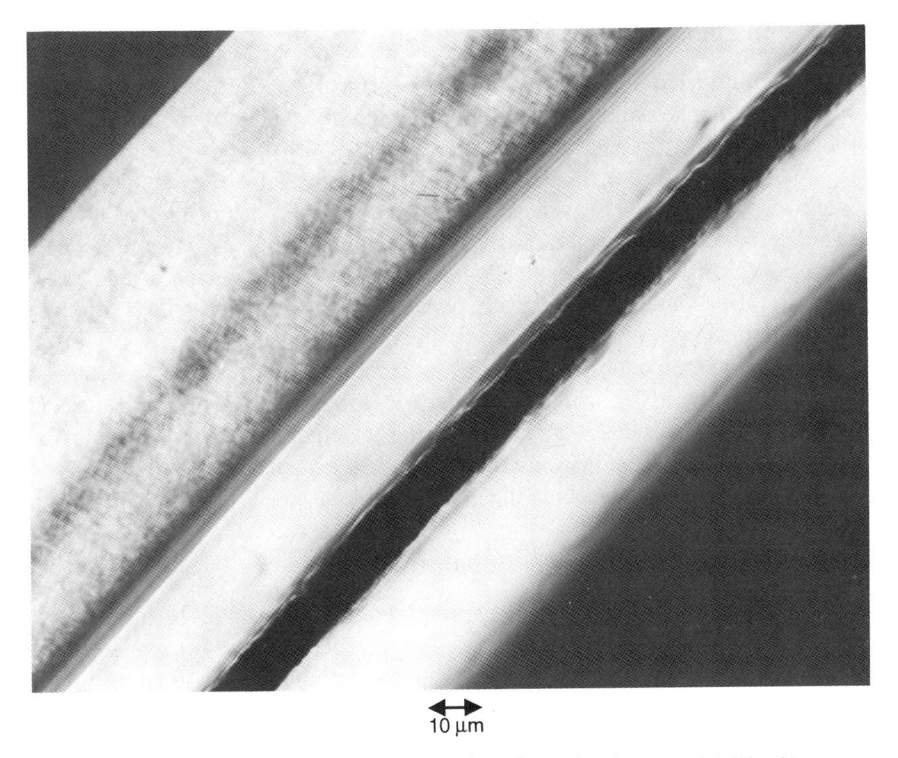

Figure 9. Photomicrograph of the cross section of a packaging material. The film cross section is about 100 μm wide.

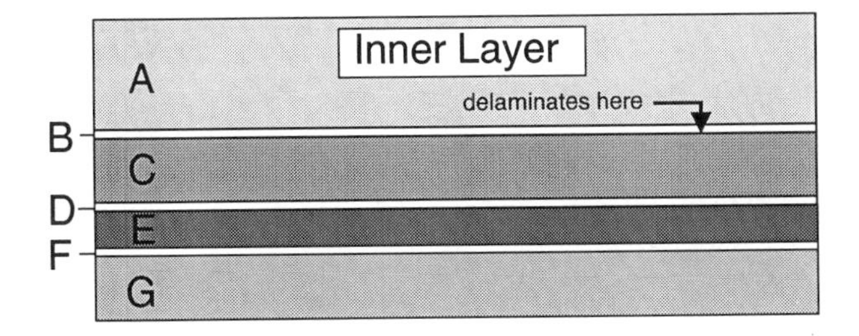

Figure 10. Cross-sectional diagram of a packaging material. All the components were identified using a combination of ATR-IR and transmission techniques.

4.4. Foil Laminates

A packaging material was microtomed cryogenically and examined by optical microscopy. A photomicrograph of a cross section having a thickness of about 100 μm is shown in Figure 9, which suggests seven layers, as illustrated in Figure 10 (not to scale). Both ATR and transmission infrared techniques were employed in the analysis using a 32× objective. The functionality of layer A was determined by transmission IR after delamination at the layer B/layer C interface. Layer B, which remained attached to layer A, was then examined by ATR-IR. The aluminum layer E dissolved in about a minute with a strongly basic solution made by dissolving one pellet of KOH in a few drops of deionized water. The exposed adhesive prime layers D and F were rinsed with deionized water and analyzed by using the IRμs ATR objective. Finally, transmission spectra of layers C and G were obtained after the aluminum layer was removed. Note that layers A, C, and G were thick enough to examine normal to the surface of the cut so that components in the other layers caused no spectral interference.

4.5. Smudge on a Metallic Substrate

An example of a problem solved by employing RAIRS involved the identification of a smudge on a metal surface. The advantage of the IRμs microscope in this analysis was the ability to move rapidly from one region to another until a sample of the proper thickness could be located. Once the critical spot was

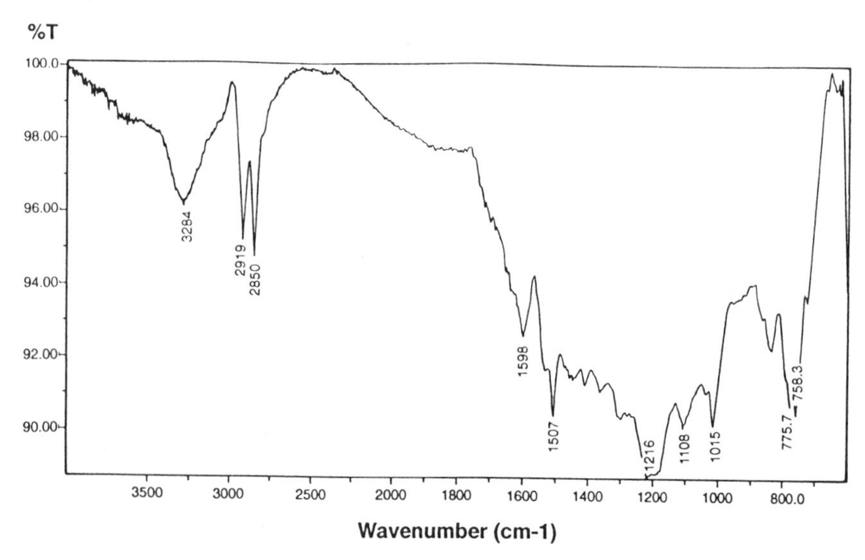

Figure 11. IR spectrum of contaminant on a metallic substrate. RAIRS allowed for easy optimization of the IR spectrum, and thus identification of the smudge.

located, an excellent spectrum was obtained, as illustrated in Figure 11. Note that the range of percent transmittance is under 10%, yet the spectrum is quite acceptable. The ability for rapid relocation of the position of the analysis area to achieve the proper spectrum saved hours of effort.

4.6. Compositional Contour Mapping of an Interpenetrating Network

A rapid infrared technique has been developed to monitor the composition of a film from a production line, which consisted of an interpenetrating network built by using a porous polymeric substrate material and a second polymer applied to the first, which filled the voids in the substrate. The desire was to determine how uniform the material was that was being produced. Relating the infrared results to those obtained from hydrolysis of the polymeric network is aided by a compositional contour map, which permits a more accurate determination of

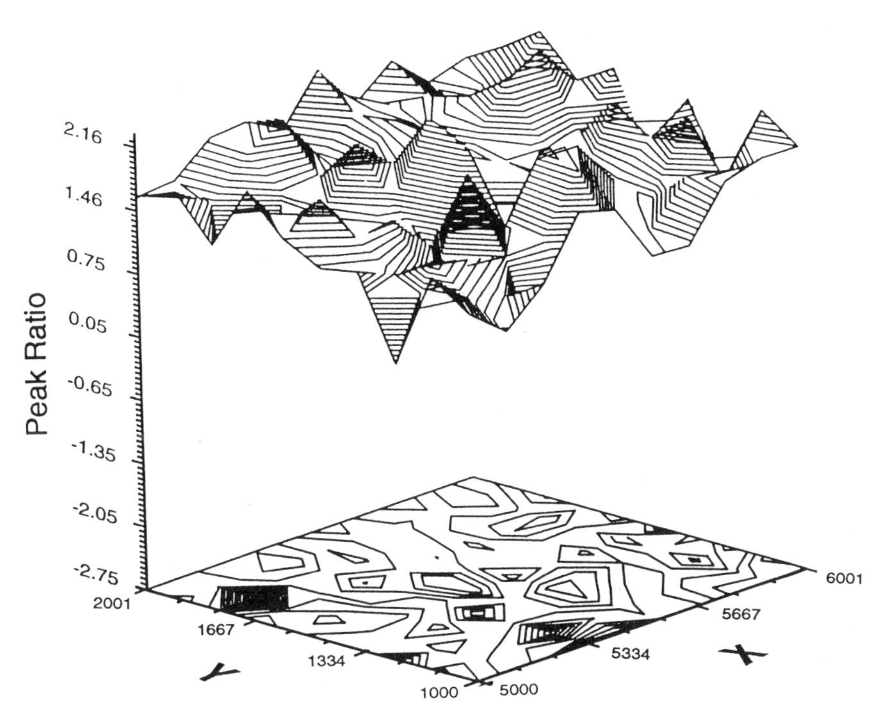

Figure 12. Molecular mapping of an interpenetrating network. A three-dimensional map of the peak ratios of characteristic peaks of each component is shown in the upper portion of the diagram. A two-dimensional projection of the map is shown as well in the lower portion of the diagram. The *X* and *Y* values are in microns using arbitrary initial values.

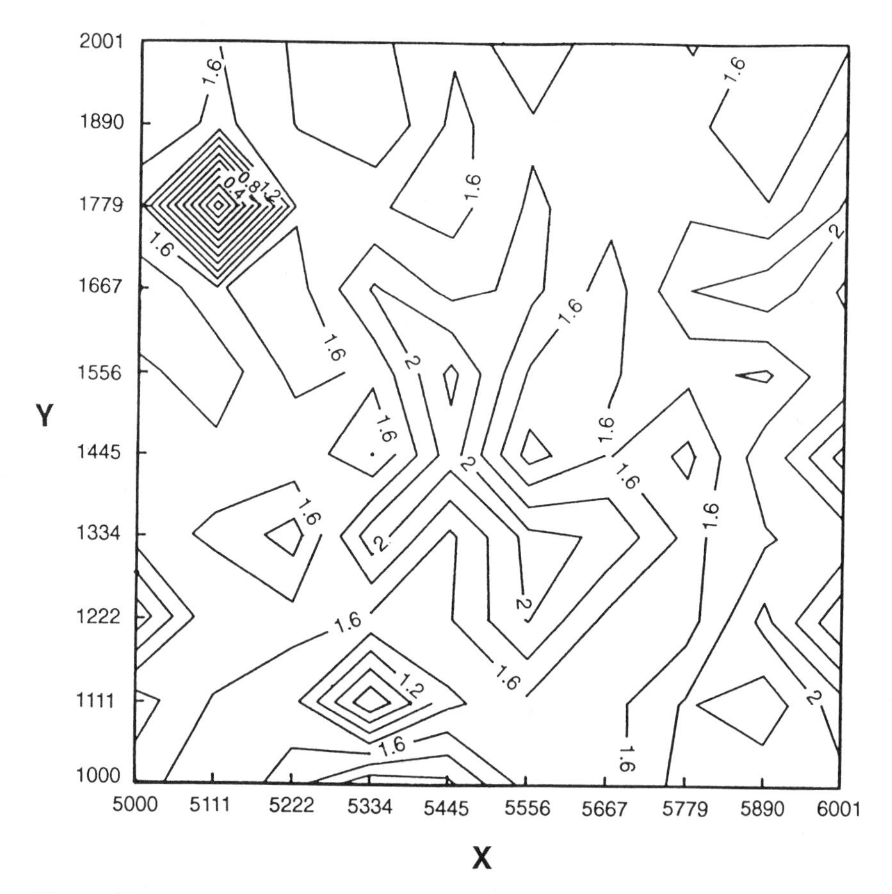

Figure 13. Two-dimensional molecular relief map. Contours of constant peak ratios, in intervals of 0.4 unit, are shown as connected lines and are to be compared with the three-dimensional view of Figure 12.

the actual distribution of materials present in the film. Figure 12 illustrates a three-dimensional map of the peak ratio (calculated by employing peaks specific to each component) as a function of the position across the film surface area. An alternative two-dimensional contour map is illustrated in Figure 13.

5. FINAL REMARKS

Some samples we receive for defect analysis have a cross section less than 0.1 μm. For such samples a gap of nearly two orders of magnitude exists between our capabilities and our needs. Reffner has reported analyses approaching the 1-μm range, by using a more intense source. Could this methodology extend our capabilities to the submicron range? Or could our analytical range be extended to samples considerably smaller than a few microns by possibly employing an array detector and the appropriate algorithm? The wavelengths each have their own signature, and the diffraction pattern phenomenon is well understood mathematically. In addition, improvements in detectors, such as the AIGaAs quantum well detector [18] and improvements in the range of the A/D converter, are anticipated to extend our capabilities in the future.

The authors welcome comments and questions regarding the material presented in this chapter and seek the perspectives of others in the field. Please direct communications to R.W.D.

ACKNOWLEDGMENT

The authors wish to thank Rebecca J. Duerst for assistance with the figures.

REFERENCES

1. Compton, S. V., and P. Stout, *Am. Lab.*, May 1983, p. 36N.
2. Needham, G. H., *The Practical Use of the Microscope Including Photomicrography*, Charles C. Thomas, Springfield, IL, 1958.
3. Coates, V. J., A. Offner, and E. H. Siegler, Jr., *J. Opt. Soc. Am.*, *43*(11): 984 (1953).
4. Griffiths, P. R., and J. A. de Haseth, *Fourier Transform Infrared Spectrometry*, Wiley, New York, 1986.
5. Anderson, R. J., and P. R. Griffiths, *Anal. Chem.*, *47*:2339 (1975).
6. Strobel, H. A., *Chemical Instrumentation: A Systematic Approach*, 2nd ed., Addison-Wesley, Reading, MA, 1973.
7. *Opt. Radiation News*, No. 59, Spring 1993, p. 7.
8. Chantry, G. W., *Long-Wave Optics: The Science and Technology of Infrared and Near-Millimetre Waves*, Vol. 1, Principles, Academic Press, New York, 1984.
9. Jones, R. Clark, *Appl. Opt.*, 1:607 (1962).
10. Oriel, *Light Sources, Monochromators, Detection Systems*, Vol. II, Oriel Corp., Stratford, CT, 1989, p. 21.
11. Kingslake, R., *Optical System Design*, Academic Press, New York, 1983, p. 17.
12. Grey, D. S., *J. Opt. Soc. Am.*, *40*(5): 283 (1950).
13. Fraser, R.D.B., *Disc. Faraday Soc.*, No. 9, 378 (1950).
14. Miller, R. G. J., and B. C. Stace, eds., *Laboratory Methods in Infrared Spectroscopy*, 2nd ed., Heyden, New York, 1972, p. 13.
15. Messerschmidt, R. G., Photometric considerations in the design and use of infrared microscope accessories, *The Design, Sample Handling, and Applications of Infrared Microscopes*, ASTM STP 949, P. B. Roush, ed., American Society for Testing and Materials, Philadelphia, 1987, p 21 f.

16. Martoglio, P., and J. Reffner, Probing the ultimate limits in microspectroscopy, *FACSS XX*, Oct. 19, 1993.
17. Klein, J., and B. J. Briscoe, *Proc. R. Soc. London A*, *365*:53 (1979).
18. *Far-Infrared Imaging Radiometer*, Technical Support Package, Goddard Space Flight Center, NASA Tech. Briefs, GSC-13467.

6

Infrared Microspectroscopy of Forensic Paint Evidence

Scott G. Ryland Florida Department of Law Enforcement, Orlando, Florida

1. INTRODUCTION

Forensic: It means for use in courts of law or open for debate. Most chapters in this book deal with topics open for scientific debate. But forensic science matters are typically intended for open debate in courts of law and as such usually involve the examination of physical evidence encountered in criminal investigations, with subsequent presentation of the results in a courtroom setting. This chapter deals with one very small area of the vast field of forensic science, the examination of paint evidence by infrared microspectroscopy.

1.1. Nature of the Forensic Paint Examination Problem

Paint may be defined as a pigmented liquid composition that is converted to an opaque solid protective film after application as a thin layer. But since not all protective films are pigmented, it is more proper to speak of ''coatings'' as opposed to paints. The two terms will be used interchangeably for the purpose of this chapter, keeping in mind the subtle yet important distinction. The forensic coatings examiner is responsible for the recognition, collection, and examination of this type of physical evidence, as well as for the interpretation of its evidential significance. Paint evidence occurs as transfers in a variety of crimes, including homicides, vehicular hit-and-runs, sexual assaults, and burglaries. Unlike the specimens examined routinely by the industrial paint chemist, the samples are typically in-service coatings plagued by a combination of sample preparation problems. First, they are cured and any solvent previously present has already evaporated. Furthermore, they are often multilayered from previous applications or from use as a coating system and are usually small (on the order of 1 mm^2 or less). In many instances they are found as trace smears on the surface of substrates, such as fabric, metal, or even other painted objects. Other times they are encountered as minute intact multiple-layered ''chips'' or fragments fused to surfaces or buried in a mound of microscopic debris recovered from a garment. They are usually weathered and are seldom found in a pristine condition. The thickness of layers often varies from one fragment to another, and surface contaminants can often be found entrapped between successive layers. Occasionally, specimens may be encountered in the form of a partially used can with the remaining paint still in the liquid state; however, that is not the norm. The partially cured inhomogeneous drippings adhering to the lip of the container may be the only material with which to work. As can be imagined, the analytical challenges are formidable in all of these situations and further complexed by the legal requirement of retaining a portion of the sample, whenever possible, for examination by the opposing legal counsel.

The questions typically asked of the forensic coatings examiner are of two types. First, can any suggestion be given as to what type of surface might be the source of the minute flecks of recovered paint? This often aids in an ongoing investigation where a search for the source of the transferred paint will ensue. Is it from an architectural application or from a motor vehicle? Is it an interior paint or an exterior paint? Could it be road line paint or is it more typical of a maintenance paint, such as that used on a bridge or on a tool? More frequently, the problem comes down to listing the years, makes, and models of vehicles which have paint like that of the fragment recovered, and accordingly, could be potential donors of the paint in question. For that endeavor to be fruitful, it is important to recognize whether the paint is an original finish or a refinish.

Second, given a recovered paint transfer (questioned sample) and a sample of paint from an object suspected of being the source of the transferred paint (standard sample), the forensic coatings examiner is asked to compare the two and reach a conclusion as to whether or not the questioned sample could have come from the suspected object. Furthermore, is the paint on this object unusual in any way? How common is it? How limited is the group of potential donor sources?

Finally, in questions of fraud or noncompliance with specifications, the forensic coatings examiner will sometimes be asked to determine whether or not certain components are in a coatings formulation. Although this question is more frequently asked of forensic laboratories in the private sector, it does occasionally arise in government crime laboratories.

1.2. Types of Materials Encountered

Coatings from all types of end-use applications may be encountered in the forensic laboratory. By far the most common is *automotive paint*, occurring as transfers in vehicular homicides or any type of crime where a motor vehicle is employed. The transfers may be deposited on the clothing of pedestrians, victim's bicycles, other vehicles, or structures. *Architectural paint* is also routinely found transferred to tools used for breaking and entering or on the clothing of a person gaining access to a structure through an area of flaking paint. *Maintenance paint* includes a wide range of applications, such as tool paint, bridge paint, and paint used on maintenance equipment. The paint from tools used for breaking and entering or as weapons may be transferred to structural substrates or wound tracts, while bridge paint may be deposited on vehicles fleeing the scene of a crime. *Marine* and *aircraft paints* are encountered in collision investigations as well as in the investigation of drug-smuggling incidents, where the "dropoff" boat or plane leaves transfers on other objects which are later used to provide evidence that the boat or plane in question was indeed at some time at the dropoff location.

Occasionally, other types of related materials are so closely associated with the transferred paint that they too are transferred. These include such materials as automotive trim tapes consisting of colored plastic backing with a contact adhesive, plastics from vehicle grills or accessory items, and building materials such as caulks and sealants, wallboard, masonry, pressed fiberboard, and so on. The polymeric materials, such as plastics, trim tape backings, adhesives, and sealants may be analyzed in a similar manner to that of the coatings. Since they do not possess the increased number of discriminating characteristics afforded by multiple-layer paint fragments, they require a more rigorous attempt to compare all components. Plasticizer extraction and analysis [by infrared microspectroscopy and gas chromatography–mass spectrometry (GC-MS)] along with detailed polarized light microscopy (PLM) examination of the material are typically employed at a minimum.

1.3. Basics of Paint Chemistry

As pointed out by Thornton [1], a reasonably comprehensive understanding of paint chemistry is essential in order to derive the most information from paint evidence. Rodgers et al. [2] further stress the importance of component classifications, rather than just comparisons, in establishing the credibility of the forensic coatings examiner. For these reasons, as well as understanding the terminology to be used later in the classification of coatings as to their end-use types, it will behoove us to take a brief look at some fundamentals of paint chemistry.

Paint is comprised of both a pigment portion and a vehicle portion. The pigment portion contains transparent extender pigment and opaque coloring pigment, while the vehicle portion is comprised of solvent, resin (or ''binder''), and additives. As noted previously, the solvent has evaporated in a cured coating. A variety of additives may or may not remain in cured coatings, including driers, plasticizers, thickeners, mildewcides, and antifouling agents. Thornton provides a concise list of definitions for basic coatings technology terms, as well as a compilation of typical materials comprising the various portions of a paint [1]. More detailed discussions of the chemistry and use of resins, pigments, and additives can be found in the *Federation Series on Coatings Technology* [3] or other basic texts on coatings [4–7]. For the purpose of this chapter, it is important to concentrate only on the primary components that will come into play when a paint film is analyzed by infrared microspectroscopy.

The binders found in coatings can cure by several different mechanisms. *Oxidizing vehicles* incorporate drying oils in the resin to permit air drying to a nonconvertible enamel film. The double bonds of these drying oils react with oxygen to form peroxides, which decompose to free radicals and thus provide cross-linking of the base polymer. *Polymerizing vehicles* are composed of synthetic resins that undergo a continuation of the polymerization process prompted by heat or a catalyst also forming a nonconvertible enamel film. *Solvent evaporation vehicles,* typically referred to as lacquers, cure by solvent evaporation alone and undergo coalescence on a molecular scale. They form convertible films that can be redispersed in their original solvent. *Coagulating vehicles* cure primarily by coalescence on a macromolecular scale and may be placed into one of two groups, waterborne emulsions or organic solvent–borne dispersions. The waterborne emulsions are commonly recognized as latex paints. The organic solvent–borne dispersions used in the automotive industry fall into two subcategories, nonaqueous dispersion lacquers (NAD lacquers) and nonaqueous dispersion enamels (NAD enamels).

These modes of cure hold a practical importance for the forensic coatings examiner, for if recognized by some means, they can often be related to specific end uses. For example, latexes are waterborne emulsions and are typical of architectural applications. Nonaqueous dispersion lacquers are a cross between an organic solvent–borne dispersion (coagulating vehicle) and a solvent evaporation vehicle and are characteristic of certain General Motors automobiles. When oxidizing vehicles are encountered in automotive finish coat paints, they are indicative of a refinish and not an original finish. Each end-use classification provides meaningful information to the investigator or the court evaluating the significance of the evidence.

Just as the modes of cure are often important in recognizing end-use applications for a questioned paint sample, so is the chemistry of the resins (binders) in the vehicle. A detailed discussion of the functional groups present in the various polymer classes is beyond the scope of this book; however, recognition of the types of polymers most frequently used as binders is essential for understanding classifications discussed later in this chapter. They are listed in Table 1 along with some brief highlights for each type. Take note that the main difference between an alkyd and a polyester binder is the presence of drying oils in the alkyd resin, thus providing its air-dry capability. Also realize that modified

Table 1. Typical Coating Binders

Alkyds
Cure primarily by oxidation of drying oils
Short, medium, long classes, depending on amount of drying oil
Degree of unsaturation in drying oils used affects drying time
Frequently modified (styrene, vinyl toluene, acrylic, etc.)
Frequently cross-linked with an amine (melamine, urea)
Polyesters
Similar to alkyds but no drying oils for oxidation
Typically two-package systems
Acrylics
Polyesters with polymerization at the vinyl site
Used in lacquers, baking enamels, and latexes
Frequently modified (amines, alkyds, polyurethanes, styrene, epoxies, vinyl acetate, etc.)
Urethanes
Frequently modified (acrylics and alkyds)
One-package (moisture cure) or two-package (catalyst) systems
Limited pot life but "tougher" than acrylics
Epoxies
Ether linkage, consequently strong bonding
Frequently modified (alkyds, acrylics, amines, polyurethanes)
Vinyls
Substituted ethylene derivatives (chloride, dichloride, acetate, alcohol, benzene)
Frequently used in latexes, especially copolymerized with acrylics
Used as a modifier for other resin types
Cellulosics
Substituted cellulose derivatives (nitro, acetate, acetate butyrate, ethyl, methyl, etc.)
Typically, lacquer systems when used as base resin
Used as modifiers in other resin systems
Silicones
Polymer of silicone oil
High heat stability
Often used as modifier for other resin systems

(mixed) resin systems can exist either as copolymers or as mere mixtures of the polymers. This point is important in understanding the variations found in automotive refinishes.

Although the pigment portion of the paint (including both extender and coloring pigments) is not as important as the resin in predicting end use of the material, it does comprise the other half of a cured paint film and is quite important in the comparison of two samples. The binders in two paints may be the same while their pigments differ. A list of commonly used extender pigments can be found in Table 2. Note that there can be different crystalline forms of the same chemical material. The choice of extender(s) is governed by cost, along with the manufacturer's desire to adjust physical properties of the paint, such as bulk, viscosity, gloss, weatherability, and abrasion resistance. Coloring pigments add even more variation in the paint formulation. They impart the all-important color and opacity to the film and are the focus of a never-ending quest by the manufacturer to achieve new colors while maintaining lightfastness and

Table 2. Common Extender Pigments

Silicates	
Quartz	Silicon dioxide
Diatomaceous earth	Silicon dioxide
Synthetic silica	Silicon dioxide
Kaolin (China clay)	Aluminum silicate
Bentonite	Aluminum silicate
Talc	Magnesium silicate
Asbestine	Magnesium calcium silicate
Mica (muscovite)	Potassium aluminum silicate
Mica (phlogopite)	Potassium magnesium aluminum silicate
Wollastonite	Calcium silicate
Synthetic calcium silicate	Calcium silicate
Sulfates	
Barytes	Barium sulfate
Blanc fixe	Barium sulfate
Gypsum	Calcium sulfate
Precipitated calcium sulfate	Calcium sulfate
Calcium sulfate anhydrite	Calcium sulfate
Carbonates	
Precipitated calcium carbonate	Calcium carbonate
Calcite	Calcium carbonate
Limestone	Calcium carbonate
Aragonite	Calcium carbonate
Dolomite	Calcium magnesium carbonate

weatherability. Table 3 lists the major classes of both inorganic and organic coloring pigments, with attention drawn to those used most frequently in automotive paints. Of course, it is important not to overlook the most ubiquitous coloring pigments, the white pigments. Even though they may not vary from one paint to another in their chemical composition, they can differ in the quantity that is present. Titanium dioxide is found in most paint films as a white tinting pigment and exists in two common forms, rutile and anatase. The rutile form is more opaque and is more commonly used. The anatase form is usually used as a modifier, owing to its lower opacity and tendency to chalk. Zinc oxide is also employed to some extent, primarily for specific properties other than hiding power. It protects organic binders from photodegradation and buffers acids, thus prolonging the life of exposed films [8]. Antimony trioxide is much less common, being used chiefly for its fire-retardant properties [8]. Metal or pearlescent flake can also be added to finish paints to impart a flamboyant appearance. Aluminum metal flake is the most common but one may also encounter bronze metal flake, consisting of various alloys of copper and zinc each of which impart a slightly different color to the flake. Pearlescent flake typically consists of fine mica flakes coated with an extremely thin layer of titanium dioxide or iron oxide, producing a rainbow of pastel colors arising from the interference phenomena of light.

Table 3. Common Coloring Pigments

Inorganics	
Chromates[a]	(yellows, oranges)
Ferrocyanides[a]	(blues, maroons)
Mixed chromates and ferrocyanides[a]	(greens)
Sulfides[a]	(yellows, oranges, reds)
Oxides	
Iron[a]	(large variety of colors)
Lead[a]	
Copper	
Chromium[a]	
Silicate/phosphates	(blues, violets)
Metallic flake[a]	(aluminum, copper/zinc)
Organics	
Azos[a]	(yellows, oranges, reds)
Phthalocyanines[a]	(blues, greens)
Vats and anthraquinones	(all colors)
Quinacridones[a]	(reds, violets)
Carbon and lampblack[a]	(blacks)

[a]Used extensively in automotive finishes.

Of the additives used in paints, plasticizers are typically the only ones present in cured films possessing high enough concentrations to be detected easily by infrared spectroscopy. They may be internal or external, which ultimately determines whether they must, respectively, be analyzed in situ or may be extracted and analyzed independent of the primary binder resin. Thornton provides a list of those types commonly used in paints and points out that only a few of them are used routinely [1]. Consequently, only a limited improvement in discrimination potential is realized when paint plasticizers are isolated and compared. Some other additives, especially those used in latex paints, do have the potential for selective solvent extraction and subsequent infrared analysis. They are, however, present in low concentrations and are quite difficult to isolate from a single layer of paint contained within a multiple-layer paint fragment measuring less than 1 mm^2.

1.4. Application of Infrared Microspectroscopy to the Forensic Examination of Paints

As can be seen, there are a variety of materials in both the binder and the pigment portions of paint which have the potential for being detected, compared, and even identified using infrared spectroscopy. The literature validates this, citing use of the technique for identifying components from binders to inorganic pigments [9–12]. Forensic infrared spectroscopic analysis of automotive paints focusing on component classifications, rather than just comparisons, was advanced significantly by a series of papers by Rodgers et al. [2,13,14]. The authors employed a diamond cell and beam condenser mounted in the sample compartment of a dispersive spectrometer to afford the small sample capability required for the analysis of individual layers in a minute paint fragment. The effort not only improved the quality of comparisons, but also led directly to the ability to estimate the sample's uniqueness [2].

The 1983 introduction of the first commercially available infrared microscope coupled to a Fourier transform infrared spectrophotometer (FTIR) has brought about an amazing revolution in the examination of forensic trace evidence. Previously, nonroutine experiments, such as those accomplished with the diamond cell technique used by Rodgers et al., in 1976, could now be performed easily. Sample preparation and mounting for single-layer paint analysis was made much simpler and consequently faster. A reduction in the required sample area from 750 μm^2 to 100 μm^2 (or less) further extended the capability of the technique. Samples of individual layers from a multiple-layer paint fragment measuring less than 1 mm^2 can be acquired and analyzed in minutes for both binder and pigment components, still leaving sufficient sample remaining for other complementary analytical techniques. And in cases where the sample size is extremely limited, the technique has the further advantage of being nondestructive.

Individual layer analysis by infrared microspectroscopy permits the forensic coatings examiner to recognize uncommon finish systems and interpret their significance appropriately to courts of law. In doing so, he or she demonstrates a knowledge of the components in the sample, thereby improving credibility. Furthermore, the examiner is no longer challenged by the comparison of two samples having varying layer thicknesses, as depicted in example A of Figure 1. Bulk analysis and comparison of the intact chips ground and dispersed in potassium bromide would result in uninterpretable differences owing to the thickness difference, and thus component ratio differences, of the first and second layers. In addition, the analyst can easily maximize discrimination potential in a comparison of two samples having layer structures similar to that depicted in example B of Figure 1. Had a bulk analysis of the intact chips been performed, subtle differentiating characteristics that might be present in layer 4, the thin original lacquer coat, would probably be overshadowed by the absorptions contributed by the top three thicker refinish layers.

For the task of providing investigative lead information, a simple analysis of the topcoat binder present in an automotive paint smear found on a hit-and-run victim's clothing may indicate that the suspect vehicle is of foreign manufacture. Collection of the infrared absorption spectrum of the topcoat binder can also reveal whether or not a recovered paint fragment is an original finish and worthy of a make/model search. In the case where several similar minute fragments are recovered from the debris adhering to the clothing of a homicide victim, classification of the binders along with a knowledge of binder end-use applications may suggest the source of the paint, and consequently, the previous location of the victim. Analyses may even provide corroborative evidence that the deceased was the victim of a vehicular hit-and-run and not a premeditated murder.

Thus infrared microspectroscopy has truly found its niche in the analysis of paint and coatings evidence. As we discuss later, it should not stand alone as the only analytical technique used. But thanks to its speed, sensitivity, and informative nature, it is most certainly one of the primary techniques in the arsenal of today's forensic paint examiner.

2. METHODS

As noted by Humecki [15], "as technology improves, the limiting factor in our ability to analyze microscopic specimens will not depend on instrumentation but on our ability to properly prepare small specimens for analysis," and the past 10 years have seen improvements in sample preparation techniques for coatings. There is still much to be said for a good stereomicroscope, a scalpel, forceps, and a steady hand, but mini-diamond compression cells, microtoming, and im-

EXAMPLE A

QUESTIONED SAMPLE

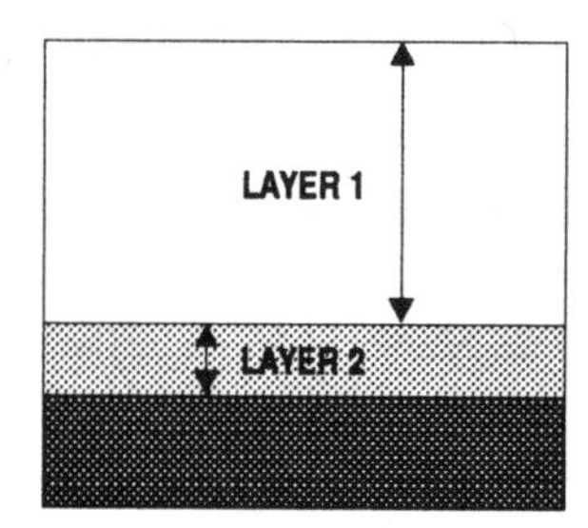

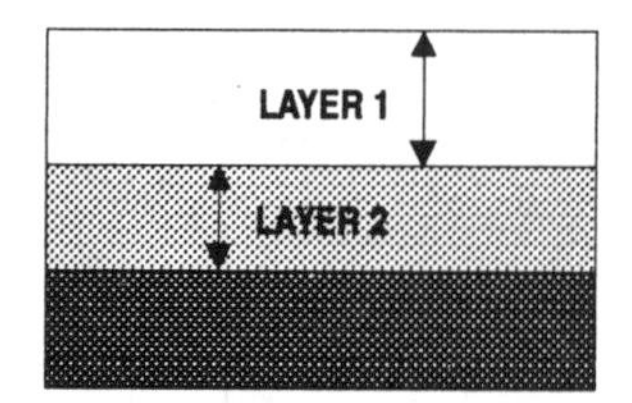

EXAMPLE B

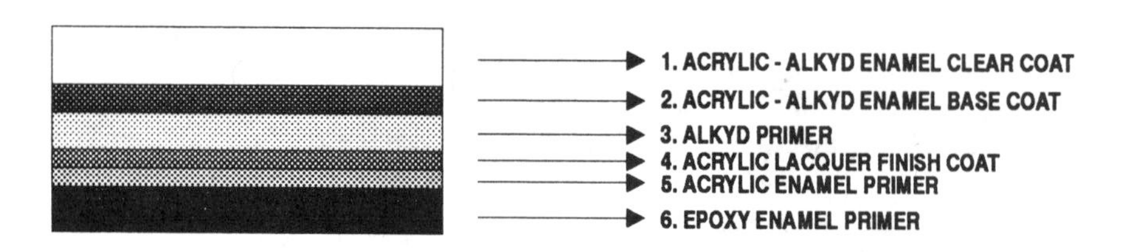

Figure 1. *Example A*: Cross-sectional representation of layer thickness variations that may be found in the comparison of a paint fragment from a known sample (standard) with one from a questioned sample. The respective layers correspond to one another if analyzed individually, yet if the fragments were analyzed intact one could expect to see differences due to layer thickness variations. *Example B*: Cross-sectional representation of a six-layer automotive refinish where the detailed binder characteristics of the layer 4 original top coat would be masked by the binder constituents of the top three refinish layers had the sample been analyzed as an intact chip.

proved microscope reflectance capabilities have helped to ease the difficult task of presenting the sample to the infrared microspectrometer. Of course, these techniques are to no avail if the forensic coatings examiner cannot find, isolate, and morphologically characterize the questioned evidentiary material in the first place. For this reason, let us first take a look at the recovery of paint trace evidence, stereomicroscopic characterization of the material based on its morphology, and some basic solubility and microchemical tests that complement the information subsequently acquired by infrared spectrometry. As the name suggests, infrared microspectroscopy is truly a meld of the two previously distinct fields of microscopy and infrared spectroscopy.

2.1. Stereomicroscopy

As discussed previously, forensic coatings evidence can occur in a variety of forms and may be deposited on a variety of substrates. Automotive paint transfers are encountered as smears on pedestrian's or cyclist's clothing, as smears on painted surfaces such as other vehicles or bicycles, or as smears on plastic substrates such as bicycle seats, handlebar grips, or vehicle grill parts. Each of these situations presents potential contamination of the transferred material by the substrate. Architectural paint transfers are typically found as smears on tools used for gaining entry to a structure. Although the receiving substrate is often unpainted metal, complications typically arise as a result of the smearing and mixture of multiple layers in the transferred paint. All of these situations require the use of a good stereomicroscope, not only to discover the transfer, but to physically pick or peel off the paint for further comparison and analysis. The tools are familiar to every microscopist; high-quality fine-pointed tweezers, a fine-tipped metal probe, and a sharp number 11 scalpel blade affixed to an appropriate handle. Bright episcopic light provided by a quartz-halogen light source equipped with dual fiber optic light guides is important for visual clarity and color perception. The ability to adjust the light source for oblique lighting sometimes aids in sample characterization. The minimal sample-size requirements afforded by today's infrared microspectrometers make physical separation of a sufficient amount of uncontaminated sample commonplace.

Automotive and architectural paint transfers may also be found as tiny multiple-layer fragments adhering to the surface of clothing or fused to substrates. In that this type of transfer provides evidence with more characteristics for comparison than single-layered smears and which more closely resembles the color and gloss of the paint on its source of origin, a thorough stereomicroscopic search for multiple-layer paint chips adhering to the surface of a tool or weapon or buried in the debris recovered from clothing is of great value. Clothing debris may be collected by suspending garments from a rack over large sheets of white "butcher" paper and carefully scraping their surfaces in a downward motion with the blade of a large spatula. This serves to dislodge the particulate surface debris and deposit it onto the surface of the white paper, where it can subsequently be concentrated to the center and ''packaged'' by cutting off the periphery of the paper and making a large pharmaceutical fold with that remaining. Of course, torn or abraded areas of garments should be searched under the magnification of a stereomicroscope prior to scraping so that paint fragments fused to the yarns or entangled in the frayed fabric can be recovered and directly associated with the damaged area, thus substantiating a forceful rather than a passive transfer. Some examiners prefer to use sticky tape for ''patting down'' clothing and subsequently recovering the trace evidence from the tape adhesive. If such an approach is used in lieu of scraping, great care must be exercised in avoiding contamination of the minute paint sample by the resins comprising the tape's contact adhesive. External surfaces of the specimen must be carved away and discarded while constantly cleaning the cutting blade with an appropriate solvent so as to avoid recontaminating the newly exposed specimen surface with residual adhesive adhering to the blade. In microsampling, a little bit of polystyrene or acrylic resin goes a long, long way in distorting spectral characteristics.

When liquid paint samples are submitted to the laboratory as standards for comparison, they are usually applied to a clean glass microscope slide and permitted to dry and cure. These specimens may then be examined like those mentioned previously. If both the questioned and standard samples are in a liquid form, there is another portion of the paint to be analyzed. Organic solvents are usually compared by head-space gas chromatography, and external plasticizers may be extracted by selective solvents and either cast onto salt plates for infrared analysis and/or subjected to gas chromatographic–mass spectrometric (GC-MS) analysis. Additional sample may then be examined as mentioned previously.

Architectural paint fragments are often recognized microscopically by their particulate texture and/or rough surfaces. They frequently consist of several layers differing in an apparently random sequence of colors. Intact paint chips viewed in cross section should always be oriented such that the layer interfaces are parallel to the axis of the oblique episcopic illumination. This helps to avoid overlooking thin layers due to ''shadowing'' by neighboring layers. The latexes are usually quite soft and flexible, yet tear under tension. The glossy alkyd or urethane oil-based finishes are difficult to recognize as solely architectural, maintenance, or automotive. Rough-surface topology may give some indication as to a nonautomotive application, but this may be difficult to observe on minute fragments. Substrate materials such as wood fibers or traces of masonry material may also be adhering to the underside of the bottom paint layer, further indicating an architectural application.

Fragments of automotive coating systems can usually be recognized by a characteristic sequence of finish coats and primers [16,17]. Examples of original finish systems may be found in the first four diagrams of Figure 2. Older systems employed an opaque pigmented topcoat followed by one to three primers, usu-

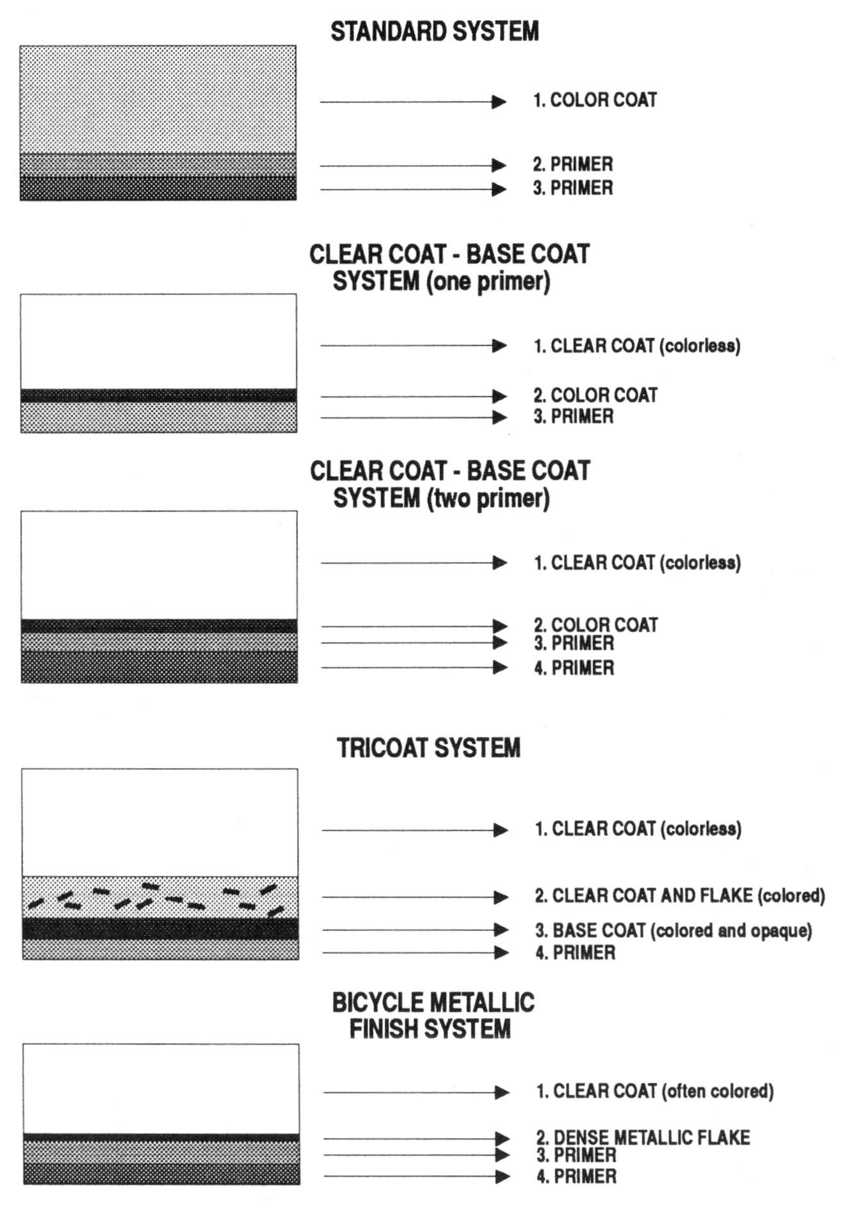

Figure 2. Cross-sectional representations of four common original-finish automotive coating systems and a bicycle original metallic finish. In the case of metallic automotive finishes, the decorative metal flake is usually found dispersed throughout the opaque color coat in the standard system as well as in the one- and two-primer clear coat/base coat systems. This is not the case in the automotive tricoat system or in the typical bicycle finish system.

ally light or medium gray, black, taupe (medium red-brown), or dark red-brown in color. The three-primer system is usually found on foreign-manufactured vehicles. The underside of the primer often has a rather rough symmetrical topology, resembling the surface of the skin of an orange (''orange peel''). The top coat could be either nonmetallic or metallic, the latter having aluminum flake dispersed throughout the finish coat. Newer systems employ a colorless transparent ''clear coat'' as the top coat, followed by a thinner opaque-colored coat, or ''base coat.'' The clear coat imparts the mechanical attributes to the film as well as protecting the pigments, especially the organic pigments, from atmospheric exposure and ultraviolet light degradation [18]. Approximately 96% of car finishes and 55% of truck finishes used on 1993 domestic manufacturer vehicles employ base coat/clear coat technology [19]. Approximately 81% of 1993 foreign-make vehicles also use base coats/clear coats [19]. The metallic types have the aluminum flake dispersed throughout the base coat. The lightly colored metallics were the first to be converted during the phase-in of this automotive coating technology, which started in about 1978. Their aluminum flake tended to oxidize rapidly, promoted by the deep penetration of ultraviolet radiation and subsequent degradation of the binder system. Microscopic recognition of a clear-coated paint system suggests an automotive application.

The late 1980s saw the introduction of a new automotive coating system, the tricoat. As can be seen in the fourth diagram of Figure 2, it employs a three-layer finish coat followed by primer(s). The colorless clear top coat is followed by a colored layer, as in the standard base coat/clear coat system; however, it is *transparent*, with either aluminum or pearlescent flake dispersed throughout. The full opacity and color is provided by the opaque color-coordinated third layer. The finish system provides a very deep wet look.

Refinish systems exhibit microscopic characteristics similar to those noted for the original finishes and are usually accompanied by the presence of a repeat sequence of the layers, that is, multiple finish coats. Sometimes the original top-coat color is changed and sometimes it is not, depending on the whim of the customer. Refinish clear-coat systems are offered to match closely the color and gloss of the original finish. A refinish primer/sealer may or may not be present between the new top coat and the original top coat. In addition, body panel repair materials may also be seen in the layer structure. Spot putty is somewhat particulate when contrasted to the primers and usually has rather vivid colors of blues, greens, and reds, with randomly dispersed large grains of colorless extender grains. Body filler, used to smooth out deeper panel imperfections, is usually a pale tan or pink with large extender grains and somewhat resembles tapioca pudding under low-power magnification. Occasionally, sanding striae may be seen when carving through layer structures, indicating preparation of existing coatings for refinish. When in doubt, infrared microspectroscopy enables the examiner to recognize repainted finish coats based on binder type.

Special attention has been paid to automotive and architectural paints since they account for the overwhelming majority of the kinds of paint encountered as physical evidence. In that vein, one more type of finish should be described. Bicycle paints can also be of the nonmetallic or metallic type. The nonmetallics are often difficult to distinguish from automotive paints unless they employ an unusually colored primer. The metallics, however, typically exhibit the layer structure depicted in the last diagram of Figure 2. The transparent clear coat is colored if the bike is not silver and is followed by a thin, opaque, dense aluminum flake film, imparting a flamboyant appearance to the finish. Although this type of layer structure may be encountered on a very few custom automotive refinishes, it can be considered quite indicative of a metallic bicycle paint.

When characterizing the morphology of the paint sample under the stereomicroscope, a few basic solvent and microchemical tests are helpful in classifying binders and comparing specimens. The tests can be performed simply by taking a small peel of a layer with a number 11 scalpel blade, placing the peel on a glass microscope slide, applying the solvent or test reagent, and observing its reaction under the stereomicroscope. Figure 3 presents a flowchart using acetone, xylene, chloroform, and diphenylamine test reagent[1] to permit classification of an automotive coating as an enamel, a dispersion lacquer, or a solution lacquer of either the acrylic- or nitrocellulose-based type. All but the xylene test for dispersion lacquers are equally useful for other coatings end-use types. Other classification schemes and tests exist; however, this basic scheme provides quick binder classification information to augment infrared spectroscopic interpretation [20–23].

2.2. Sample Preparation for Infrared Microspectroscopy

Most infrared microspectrometers afford the ability to record the spectra of paint samples in either the transmittance or the reflectance mode. The reflectance mode has the lure of limited sample preparation but suffers several drawbacks. First, and foremost, the surface to be analyzed must be visible. This means that internal layers in intact paint fragments must be imaged and analyzed in cross section, requiring the use of small apertures (on the order of 15 μm $\times$ 100 μm for automotive primers and base coats) which produce even more signal-to-noise problems than normally encountered with the energy-limited technique. Comparison of paint layer spectra requires a high degree of reproducibility, even in the unresolved shoulders riding on the major absorption bands. Signal-to-noise problems seriously affect this precision. Furthermore, refractive index changes and differences in the infrared absorption coefficient for different layers result

[1]Diphenylamine test reagent consists of 1.5 g of diphenylamine dissolved in 100 mL of concentrated sulfuric acid, which is then slowly added to 50 mL of glacial acetic acid.

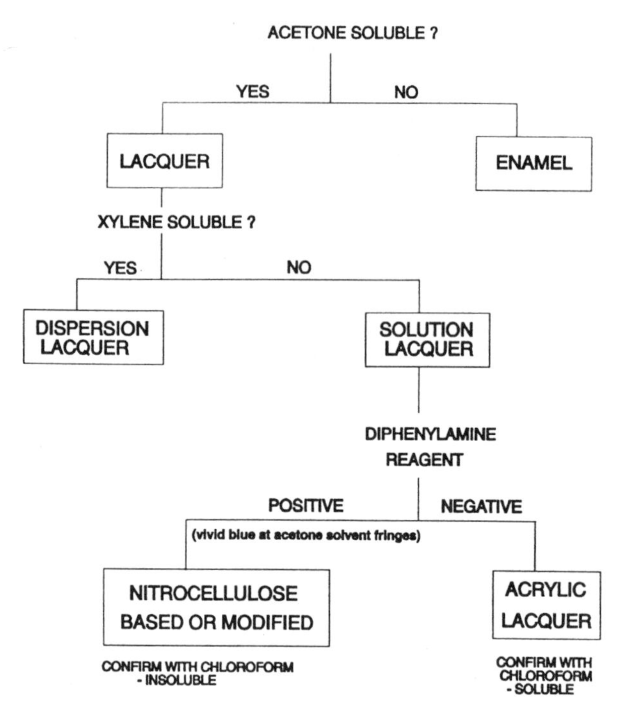

Figure 3. Basic solvent and microchemical flowchart to aid in the classification of automotive paint binder types.

in a significant distortion of the reflectance spectrum. Even when a Kramers–Kronig algorithm is used to correct for these differences, absorption band shifts of up to 30 cm^{-1} can result when compared to the absorption maxima determined by transmission analysis [24]. This can be devastating when using transmission data, as are most of the spectra available in printed references, for classification criteria. Additionally, reproducibility problems resulting from inhomogeneity in the distribution of large-grained extender agglomerates in low-gloss architectural paints and variations in the distribution and/or orientation of aluminum flake in "metallic" vehicle paints occur frequently. The same problem is encountered in the examination of paint samples by scanning electron microscopy/energy-dispersive x-ray spectrometry (SEM-EDX), where the problems start at 30-μm^2 raster areas [25] and have been experienced even at 100-μm^2 raster areas on low-gloss ("flat") structural paints. Wilkinson et al. [26] report that use of reflected light in visible microspectroscopy is also fraught with unexplainable variations resulting from small misadjustments in the instrument or inconsistencies in preparation of the surface. This author has experienced the same problems with that technique as well as with reflected infrared microspectrometry.

The transmission mode avoids all of these problems at the price of a little more sample manipulation. It is a price that is well worth paying. Larger-aperture areas can be used if the layers are sampled individually and compressed to thin films, thus reducing extender inhomogeneity effects. Microtomed cross sections may be prepared if the specimen is just too small to handle safely without the risk of losing it. The more robust signal acquired in the transmittance mode throughput of the Cassegrainian objective reduces signal-to-noise limitations. Furthermore, surface imperfections and topography problems are no longer relevant. Therefore, in most instances, transmitted infrared microspectroscopy is the method of choice.

There are several approaches to preparing multiple-layer paint fragments for such analysis. The most straightforward, and the one that this author still prefers, requires a small sample to be taken from each layer using a sharp number 11 scalpel blade while holding the paint chip in place with a pair of high-quality tweezers, such as those marketed by scanning electron microscopy accessory vendors. The scalpel balde must be held at a very low angle relative to the benchtop so as to provide thin peels. Given the small amount of sample required, peeling the layers while observing the specimen under a stereomicroscope and resting your hands on a benchtop is not extremely difficult with some practice. One may take repeated thin peels, working down through the layer structure of the paint chip, remembering to carve *toward* the point at which one is holding the fragment in place with the tips of the tweezers. If one carves away from that point, the paint chip will have a tendency to fragment. Clean samples, uncontaminated by neighboring layers, can then be harvested from the multiple peels. Once a sample is acquired, it can be placed on a glass microscope slide and compressed with the flat beveled surface of the scalpel blade while observing it under the stereomicroscope, employing a rocking and smearing motion. This significantly increases the working area of the sample. The thin film (weighing approximately 1 to 2 μg) may then be picked up, using the scalpel blade much like a spatula, and simply placed on the selected window material for subsequent placement on the infrared microscope stage. Several prepared compressed peels can be placed on the window material at one time and quickly positioned in the infrared beam using appropriate episcopic and transmitted white light illumination in combination with a low-power observation objective, as are available on most infrared microscopes. An alternative to this approach is to use a small hand-held compression press with 1-mm dies or a pressure diamond cell. Sample placement and recovery are often difficult with hand-held presses, and it is not as easy to gauge the appropriate thickness without frequently removing and observing the sample. Both high- and medium-pressure diamond cells are currently commercially available, with the medium-pressure cells being of an appropriate size to fit onto most infrared microscope stages while enabling the objective to be focused on the specimen plane.[2] The medium-pressure cells often supply insufficient pressure to obtain a thin enough film when preparing the high-bake automotive enamels for analysis. They are, however, extremely valuable when dealing with elastomeric samples that are difficult to slice into thin peels and tend to roll up into balls when compressed and smeared [15], such as the urethane finishes used on many modern vehicle header panels and air dams. Pressure can be held on the specimen during analysis so that the requisite short pathlength may be maintained. Furthermore, the thinner diamonds used in the minicells permit complete ratio cancellation of the absorption of the diamonds in the 2000-cm^{-1} region, while the thicker diamonds used in the high-pressure cell do not [27]. Although this is not a particularly characteristic absorption area for coatings, the flat 100% line with at least some detector sensitivity remaining for monitoring any sample absorption is desirable.

The selection of an appropriate infrared window support material should take several factors into consideration. The windows should be readily available and hopefully, relatively inexpensive. It would be advantageous for them to be nonhygroscopic, so that one could keep them out on the workbench while preparing and placing sample peels on their surface. A 1-mm thickness, as opposed to a 2-mm thickness, would be desirable in an attempt to reduce refractive index effects to the infrared beam. This is not as important if corrective infrared objectives, such as Spectra-Tech's Reflachromat series of objectives, are used. They should be transparent to infrared radiation down to 450 cm^{-1} so that one can take full advantage of the wideband mercury–cadmium–telluride (MCT) detectors. Furthermore, they should not be brittle and fracture under pressure or powder with abrasion for those special cases when the sample is more of a trace powder than a thin film and the analyst must press and smear the specimen directly onto the window. This wish list presents some significant problems for the traditional window materials, such as sodium chloride (590-cm^{-1} transmission cutoff), and potassium bromide (highly hygroscopic). Barium fluoride suffers a transmission cutoff at approximately 800 cm^{-1}. Zinc selenide meets the criteria fairly well but is rather expensive. Spectra-Tech offers 1-mm-thick silver chloride windows [for its Qwik-Cell Part No. 0026-071, package of six windows] which cost about $15 apiece and meet the other criteria quite admirably. Although they are made of a somewhat soft and malleable material that mars easily, surface scratches and slight hazing due to surface imperfections have not proven to be any problem. This disadvantage can sometimes be an advantage, in that the material readily accepts embedded powdered specimens and holds them in place, forming a minipellet of sorts. This permits spectra to be obtained

---

[2]High Pressure Diamond Optics, Inc., 7400 North Oracle Road, Suite 372, Tucson, AZ 85704, and Spectra-Tech, Inc., 652 Glenbrook Road, Stamford, CT 06906.

on very limited samples of highly extended paints. The windows suffer one other disadvantage. Prolonged exposure to light will cause them to turn purple gradually, resulting in reduced transparency. To minimize this problem, they should be stored in a dark place when not in use. Considering advantages versus disadvantages, the 1-mm-thick silver chloride windows are the support material of choice.

Another approach for preparing multiple-layer paint fragments for transmitted infrared microspectroscopy involves the creation of a thin cross section of the intact layer structure followed by placement of the cross-section peel on a window support material and subsequent aperturing and analysis of each individual layer. Although it sounds relatively simple at first blush, the task is often somewhat difficult as well as time consuming. It is the method of choice, however, if the sample is too small to permit manual sampling of individual layers, if a detailed profile of chemical variations across the depth of a layer is desired, or if the analyst is not adept at handling small specimens under the stereomicroscope. While observing the sample under the stereomicroscope, one may be able to hold the paint chip on its top and bottom with a pair of sharp tweezers and then manually take a very thin peel of the cross section with a very sharp scalpel blade. The resulting thin ribbon can then be compressed on a clean glass microscope slide with the scalpel blade, producing a thin cross section with slightly increased sampling area. If this approach will not work due to limited sample size or layer texture restrains, the more elaborate and time-consuming method of using a rotary microtome to prepare the sample after it is embedded in an appropriate mounting medium may be undertaken [26,28]. An embedding medium as simple as white paraffin may prove to be suitable [29]. Michele Derrick presents an excellent discussion of embedding media and paint microtoming in Chapter 8 of this volume.

When performing transmitted light infrared microspectroscopy on multiple-layer paint cross sections, the analyst will be daunted by the same homogeneity problems described previously for the reflectance technique. Furthermore, automotive paint primers are often on the order of only 15 to 20 μm thick, and as such, require small apertures to mask the area of interest. This can result in diffraction effects causing a significant portion of the light collected by the objective to come from outside the apertured area of interest—in other words, neighboring layers [30]. The problem can be reduced by using an infrared microscope employing dual apertures, one below the specimen plane and another above the specimen plane. In such a system, the operator can control the area and location of the infrared beam striking the specimen with the source (lower) aperture and then, by adjusting the image (higher) aperture, collect the transmitted light only from an area approximately 75% of that illuminated by the source aperture. Of course, the problems introduced due to diffraction and homogeneity effects do not exist if such small sampling areas can be avoided by choosing a method of sample preparation other than cross sectioning.

2.3. Instrumentation and Sample Analysis

There are currently a variety of Fourier transform infrared spectrometers (FTIRs) on the market, and each offers either its own infrared microscope accessory or compatability with a second-party infrared microscope accessory. Since forensic coatings specimens can typically be compressed to yield a relatively large working area (on the order of 100 to 200 μm^2), the spectrometers are not strained for sensitivity. Keep in mind that apertured areas greater than 50 μm^2 not only provide better signal but also tend to avoid extender inhomogeneity problems and diffraction effects. With a high-quality infrared microscope accessory, suitable transmittance spectra can be obtained by coadding 16 to 64 scans at a resolution of 4 cm^{-1}. The spectra presented in this chapter were all collected under such conditions, taking approximately 45 s for the maximum collection time.

Most infrared microscope accessories on the market today offer easy conversion from the transmission mode to the reflectance mode. Side-angle reflectance is offered on many models to accommodate in situ analysis on larger samples. Kramers–Kronig correction programs are offered to correct for reflectance spectra anomalies. On-axis all-reflective optics have improved signal-to-noise ratios and have reduced both visible and infrared light aberrations. Kohler-style illumination has further improved the intensity of the signal. Lower- and high-power infrared objectives (15× and 32 to 36×) are available offering the analyst a choice between the trade-offs of higher magnification and increased numerical aperture versus lower magnification and longer working distances. Both narrowband and wideband small-target MCT detectors are available and are often supplied with a prealigned condenser objective to permit easy interchange. The improved signal, visual image clarity, and application versatility attained over the last nine years have made the infrared microspectrometer an easy analytical tool to use in performing previously nonroutine microanalysis.

A Digilab FTS-40 with a UMA 300 infrared microscope accessory was used for collection of the spectra presented in this chapter. The microscope is equipped with a 4-power magnification viewing objective and a 36-power magnification Cassegrainian all-reflecting infrared objective. A wideband MCT detector was employed, having a 250-μm^2 target area. The microscope optics are of the off-axis type, with neither a substage condenser objective nor a source aperture. Both the FTIR spectrometer and microscope were purged with compressed air supplied by an air compressor. The purge gas was dried and carbon-dioxide scrubbed by an in-line Balston 75-60 filter, reducing moisture to a −100 dew point and carbon dioxide content to less than 1 part per million. A rectangular image aperture was used employing a 100- to 200-μm^2 aperture area.

Spectra were collected from 4000 to 460 cm^{-1} coadding 16 to 64 scans at a resolution of 4 cm^{-1}.

The samples were thin peels that were compressed on a glass microscope slide using either the beveled edge of a number 11 scalpel blade or a roller bearing, as found on Spectra-Tech's roller knife (Part No. 0036-521). The compressed peels were placed on the surface of a silver chloride window (13 mm diameter, 1 mm thickness), placed on the microscope stage, and analyzed in the transmission mode. Background spectra were collected on an unused area of the silver chloride window using the same aperture configuration as that used for the collection of the sample spectrum, so as to minimize any observed diffraction effects in the resultant ratioed spectrum. Spectra were displayed as percent transmittance, in that many of the printed reference texts and spectrum collections are in that format.

The choice of detectors is a difficult one. The narrowband MCT detector offers the advantage of significantly improved sensitivity over the wideband MCT; however, it suffers a constrained spectral range with a cutoff at 700 cm^{-1}. The wideband MCT detector, on the other hand, offers an extended cutoff at 450 cm^{-1} while commanding a significantly improved sensitivity when compared to nonquantum detectors. As will become apparent later, the spectral range from 700 to 450 cm^{-1} is invaluable in recognizing the various inorganic extenders used in coatings. Since the samples encountered in routine casework may typically be flattened out to larger areas, the trade of sensitivity for increased spectral range is a wise one. If smaller samples are encountered or cross-section analysis of thin layers becomes important, a modular narrowband MCT detector affording quick and easy interchangeability can be employed.

2.4. Spectrum Evaluation

A majority of the information necessary for classification of binders or recognition of extenders can be found in region 2000 to 450 cm^{-1} of the spectrum. There are some exceptions, such as the recognition of an alkyd versus a polyester resin (where the same acid or anhydride was used in the polymerization) based on the intensity of the methyl stretching bands between 2700 and 3000 cm^{-1}, the corroboration of the presence of clay or talc by the water of hydration bands between 3600 and 3700 cm^{-1}, or the recognition of iron ferrocyanide (iron blue) pigment at 2100 cm^{-1}. In an effort to maximize the subtle details of band shapes and absorption maxima, the spectra presented will encompass only the range 2000 to 450 cm^{-1}. Of course, one should always examine the complete spectral range from 4000 to 450 cm^{-1}, especially when performing comparisons.

As noted by Katon et al. [27], one of the most troubling aspects of infrared microspectroscopy is specimen thickness. The analyst wants a specimen thin enough to avoid band broadening at the absorption maxima, yet thick enough to avoid noise artifacts and to permit observation of the finer details in a spectrum. Most coatings have polyester functional groups present in either their base resin or in plasticizers added to their binder. A basic rule of thumb is to make the sample thin enough so that the cabonyl absorption at 1730 cm^{-1} exhibits an absorption of 50 to 90% transmittance units; then display the spectrum with the carbonyl band full scale. If the sample is too thick, it may simply be removed from the supporting window and compressed further. It is also useful to bevel cut a peel from the specimen layer slightly larger than is necessary, compress it, and then aperture down so that different thickness areas of the peel may be sampled by the infrared beam.

3. RESULTS AND DISCUSSION

As mentioned in Section 1, paint examinations by infrared spectroscopy can involve both classifications and comparisons. FTIR microspectroscopy is well suited to both these endeavors on extremely small samples and provides the forensic coatings examiner with a wealth of information for significance interpretation. First, let us take a look at the technique as it relates to paint *classifications.*

Infrared microspectroscopy is one of the best techniques available for coatings binder classification [1,31]. It is quite reproducible over long periods of time and therefore ideal for building data bases. It provides for the recognition of basic functional groups characteristic of certain generic types of resins as well as for the reproducible detection of minor characteristics indicative of certain binder types within a generic resin class. This leads to detailed binder classifications of individual paint layers in minute multilayered in-service coating specimens, permitting the analyst to better assess the significance of the trace evidence. In those instances when the examiner has trouble reproducing a spectrum from an unusually thick paint layer, the technique pays off tenfold by quickly proving that the thick layer actually consists of two coats of paint having the same well-matched color, yet different binders. Simply taking a sample from the top of the layer and comparing it with one taken from the bottom of the layer solves the problem. Furthermore, as mentioned previously, one gets a look at both binder and pigment simultaneously without destroying the sample. If a sufficient amount of extender is present (usually greater than 10 wt %), different crystalline forms of some extenders may even be recognized.

On the other hand, higher concentrations of pigments may present problems. Sometimes, such as with the low-gloss architectural finishes, the extender pigment is present in such high concentrations that it masks the binder absorptions, thus making binder classification difficult if not impossible. Other times, the concentration of organic coloring pigments is such that numerous sharp absorption bands appear "scattered" throughout the spectrum, overlaid on top of the binder absorptions. This may cause concern that the characteristic absorption bands noted for a specific binder type are in fact originating from the organic pigment and not the binder. The problem can usually be sorted out by looking for confirmatory absorption bands, or if necessary, by analyzing the sample by a complementary technique, such as pyrolysis infrared spectrometry, pyrolysis gas chromatography, or pyrolysis mass spectrometry.

Another shortcoming of infrared spectroscopy lies in its inability to permit classification of an automotive acrylic lacquer as being of the dispersion or solution type (see the modes of cure in Section 1.3). This is valuable information, for as will be seen later, the dispersion acrylic lacquers were used only as original finishes on General Motors cars. The most obvious difference between the dispersion and solution acrylic lacquer binders following curing is the absence of cellulose acetate butyrate (used as a flow controller) in the dispersion type [32]. Unfortunately, cellulose acetate butyrate has a rather nondescript infrared absorption spectrum which tends to be buried under the acrylic resin absorption spectrum. The two types can, however, be recognized by the simple solvent tests cited in Section 2.1 as well as by pyrolysis gas chromatography.

Not withstanding these drawbacks, coatings binder classification by FTIR microspectroscopy provides the forensic paint examiner with overall chemical information regarding individual paint layers in a minute paint fragment more quickly than any other available technique.

3.1. Automotive Coatings Binder Classification

Since automotive paints are the most frequent end-use type of paint encountered in crime laboratories in the United States, it would be prudent to concentrate on them initially. Due to their diversity, they offer a good introduction to a broad range of paint binders used in other end-use applications. In surveying the types of binders utilized, one must consider original finishes used on both domestic and foreign manufactured vehicles as well as those coatings used as refinishes. Manufacturer product information, reference collections, and FTIR microspectroscopy of in-service finishes provide the forensic coatings examiner with a profile of the types of binders in use both currently and previously.

3.1.1. Binder Types

Table 4 lists the binders currently in use as original finish coats on domestically manufactured vehicles. The acrylic lacquers (solvent evaporation vehicles) stand out as being used solely on General Motors vehicles for almost 40 years. The nonaqueous dispersion form (coagulating vehicle) has not been used in the automotive refinish business. Thus, recognition of a *dispersion* acrylic lacquer finish coat is indicative of an original finish on a General Motors automobile.

Table 4. Current Original-Finish-Coat Binders on Domestic Automobiles

Acrylic lacquers	
Solution:	GM cars since mid-1950s (slow conversion to dispersions during 1970s)
Dispersion:	GM cars: 1970–early 1990s
Acrylic–melamine enamels	
Solution:	All domestic manufacturers: cars and trucks
Dispersion (NAD):	All domestic manufacturers: cars and trucks
Waterborne:	GM plants in California since mid-1970s
Polyester–melamine enamels	Ford–Lincoln–Mercury: sporadic to routine use since 1983
Polyurethane enamels	All domestic manufacturers: flexible fascia

You will remember that they cannot be recognized solely by infrared spectroscopy, but can be characterized as such by solvent tests and pyrolysis gas chromatography. Other than the polyurethane enamel finishes used on the flexible fascia of domestically produced vehicles, the other binders are all melamine cross-linked (polymerizing vehicles). Substantial melamine cross-linking is atypical of refinishes. Thus, *if substantial melamine copolymerization can be recognized, it assures the examiner that he or she is dealing with either an original finish or a factory refinish.* Factory refinishes occurring after the initial high-bake portion of the assembly line typically use the same types of binders as mentioned previously, except that the cross-linking is catalyst driven as opposed to heat driven [32]. They may be recognized by the appearance of multiple melamine cross-linked finish-coat layers of similar color in the paint fragment layer structure. It is also worthwhile to note that the only *water-based* melamine cross-linked finish coat binders used are those on General Motors vehicles manufactured in their two California plants. The rationale finds its seed in emission control standards, and the practicality of recognition is obvious to the forensic paint examiner. This author currently knows of no way to recognize such finishes using infrared spectroscopy; however, Thornton suggests a way to characterize the finish based on microchemical tests [20]. Of course, as with any manufactured product, time brings about change. Table 5 lists the major changes in finish-coat binder types for General Motors, Ford, Chrysler, and American Motors Corporation over the past 40 years. As will become apparent later, *it is worthy of note that alkyd–melamine enamels have not been used on domestically manufactured automobiles since 1964.*

Table 6 lists the major changes in original finish coat binder types for domestically produced trucks. Again, *it is worthy of note that alkyd-melamine*

Table 5. Original-Finish-Coat Binder History for Domestic Automobiles

General Motors		Ford/Chrysler/AMC	
1954–1958	Conversion to acrylic lacquers	Pre-1964	Alkyd–melamine enamels
1969	Conversion to NAD acrylic lacquers	1964–1966	Conversion to acrylic–melamine solution enamels
1974	Waterborne acrylic–melamine enamels in California plants	1966–1971	Acrylic–melamine solution enamels in full use
1975–1976	Initial conversions from NAD acrylic lacquers to acrylic–melamine enamels	1974	Conversion to acrylic–melamine NAD enamels
1982–1985	Base coat/clear coat systems on metallics and nonmetallics	1978–1979	Base coat/clear coat systems introduced on metallics
1992	Phase out of most all NAD acrylic lacquers complete	1985–1986	Base coat/clear coat systems introduced on nonmetallics

Table 6. Original-Finish-Coat Binder History for Domestic Trucks

General Motors		Ford/Chrysler/AMC	
Pre–1956	Alkyd enamels	Pre–1956	Alkyd enamels
1956–1969	Alkyd–melamine enamels	1956–1972	Alkyd–melamine enamels
1969–present	Acrylic–melamine enamels	1972–1974	Metallics: acrylic–melamine solution enamels Nonmetallics: alkyd–melamine enamels
		1974–present	Acrylic–melamine enamels

enamels have not been used since 1974, approximately 20 years ago. Furthermore, domestic truck manufacturers have not used acrylic lacquers or polyester–melamine enamels over the past 40 years. The picture is similar for vehicles of foreign manufacture but a bit less clearly defined. Table 7 lists original finish coat binders used between the late 1970s and mid-1980s, based on

Table 7. Original-Finish-Coat Binders on Foreign-Manufactured Vehicles

Nonmetallics	Metallics
Acrylic–melamine enamels	
Used sporadically by some Asian manufacturers (Nissan/Honda/Hyundai/Subaru)	Used by most Asian manufacturers in base and clear coats (Acura/Honda/Isiuzu/Mazda/Nissan/Mitsubishi/Subaru/Suzki/Toyota)
Seldom used by European manufacturers (one 1980 BMW and one 1980 Fiat)	Used in most European clear coats and some base coats
Polyester–melamine enamels	
Used by some Asian manufacturers in non-clear-coated post-1986 models Acura/Isuzu/Nissan/Toyota) *Note*: also used by Ford–Lincoln–Mercury since mid-1980s, but all are clear coated	Not encountered with Asian manufacturers
Not encountered with European manufacturers	Used in most European base coats and some clear coats (Audi/BMW/Fiat/SAAB/VW/Volvo) *Note*: Also used by Ford–Lincoln–Mercury sporadically since 1983
Alkyd–melamine enamels	
Used by many Asian manufacturers (Acura/Honda/Isuzu/Mazda/Nissan/Mitsubishi/Subaru/Toyota)	Not encountered with Asian manufacturers
Used by most European manufacturers (Audi/BMW/Fiat/Mercedes-Benz/MG/Porsche/SAAB/Triumph/VW/Volvo)	Encountered only on some Mercedes-Benz with acrylic–melamine clear coat
Acrylic lacquers	
Pre-1987 Jaguars	Pre-1987 Jaguars

the FTIR microspectroscopic analysis of a collection of samples in the author's laboratory obtained from approximately 180 vehicles awaiting repair in local body shops [33]. The trends are consistent with those reported by Beckwith, who listed the finish-coat binders in use on both domestic and foreign vehicles in 1973 [34]. *The most notable point is the continued use of alkyd–melamine enamels on foreign nonmetallic original finish coats.* The European manufactured vehicles were still using this type of binder into the late 1980s, while the Asian manufacturers were changing to either polyester–melamine enamels or acrylic–melamine enamels on their nonmetallics in that same time frame. Both the European and Asian manufacturers have converted to either acrylic–melamine or polyester–melamine enamel technology on their metallic finishes, with the exceptions of Mercedes-Benz, which was using an acrylic–melamine clear coat over an alkyd–melamine base coat on 1984 to 1986 vehicles. It should also be noted that polyester–melamine binders were used on some Asian-manufactured nonmetallics since 1986; however, they were not clear coated. This is in contrast to the Ford–Lincoln–Mercury line, which has sporadically used polyester–melamines as domestic original finishes since 1983, but has routinely clear-coated them. Jaguar was the final holdout for acrylic lacquer technology, but the literature indicates that they changed to acrylic enamels in 1987.

Table 8 lists the types of binders typically encountered in today's finish-coat repaints. Just as with the original finishes, refinish lacquers (solvent evaporation vehicles) are easier to paint with and easier to repair. They do, however, require numerous coats to build up acceptable film thickness and must be buffed out to produce the high luster for which they are noted. Furthermore, they lack chemical resistance and do not weather as well as the enamels. Alkyd enamels

Table 8. Automotive Repaint Finish-Coat Binders

Nitrocellulose lacquers
Usually acrylic modified
Acrylic lacquers
All solution type, no dispersion type
Alkyd enamels
Acrylic modifiers available
Urethane "hardeners" available
Acrylic–alkyd enamels
Urethane "hardeners" available
Acrylic–urethane enamels
Two-package systems
Will not air-dry
New-generation "oxithane type" available
Flexible urethane enamels

(oxidizing vehicles) were the mainstay of the vehicle refinish business for many years. They offered improved chemical resistance but did not afford the high-gloss luster of the lacquers, especially with environmental exposure. In an effort to overcome this drawback, slightly higher-priced one-package copolymerized acrylic–alkyd binder systems were introduced in the early 1970s [18]. They offered improved gloss retention while retaining the air-dry enamel characteristics of the alkyd enamels. The air-dry capability of the resin is quite important, not only to the body shop painter who does not have high-temperature drying rooms, but also to the forensic paint examiner. At the risk of being redundant, it should again be noted that these resins do not incorporate substantial amounts of melamine cross-linkers as do the original finishes. This would require either high-temperature baking to drive the polymerization (so high that it would melt the plastic and rubber on the vehicle) or expensive chemical catalysts to drive the reaction at ambient temperatures. Therefore, *the absence of melamine absorption bands in the infrared spectrum of a nonflexible vehicle enamel indicates that it is a repaint.*

In a continuing effort to find refinishes that performed like the original finishes, a urethane additive was introduced that would improve gloss and mar resistance when added to the original product. The major refinish manufacturers referred to the additive as a "hardener," and it was made available for both alkyd and acrylic–alkyd resin bases. In the quest for even more durable and high-gloss finishes, the major refinish paint suppliers began marketing two-package copolymerized acrylic–urethane binder systems around the late 1970s (polymerizing vehicles). They are quite a bit more expensive than the acrylic–alkyd refinishes and are much more difficult to apply, although they do offer a very durable high-gloss finish which is especially enhanced when applied as a clear coat/base coat system. In that they do not contain an alkyd component, they will not air-dry. The two-package product must be mixed and then used prior to exceeding the pot life of the resin. There are two generations of this binder type, with the newer being earmarked by the 1987 introduction of DuPont's "polyoxithane"-type resin [35,36].

Finally, as newer automobile styles brought on increased use of flexible body panels, such as air dams, header panels, and front and rear bumper shrouds, more flexible refinishes had to be developed to function like the polyester–urethane enamel original finishes. To meet this demand, "flex additives" are offered by the refinish manufacturer. They are essentially a separate package that can be added to the basic paint, thus imparting elasticity through a high content of urethane polymer in the film.

Vehicle paint primers tend to be somewhat similar in both the original finish and the refinish systems. The primary difference, again, is the absence of melamine cross-linking in the refinish primers. Acrylic-, epoxy-, and alkyd-based binders are prevalent in the original finishes with many being melamine cross-linked. All sorts of resin combinations can be found, from epoxy-modified alkyd–melamines to acrylic-modified epoxy enamels. Binder classifications follow the same lines as those for the finish coats; that is, the major resins are denoted as the base binder and the minor resins are denoted as modifiers. The refinish primer binders tend to be alkyd or epoxy based and may have acrylic or urethane modification. Acrylic or nitrocellulose–acrylic lacquers (solvent evaporation vehicles) may also be encountered, unlike the primers used in most original finishes.

3.1.2. Binder Classifications by Infrared Microspectroscopy

The major infrared absorption group frequency regions used in automotive paint binder classification are depicted in Figure 4. In interpreting a spectrum, two approaches may be used. First, one can approach the problem initially as a functional group recognition task followed by a detailed comparison of major absorption band frequencies with tables of such bands for each binder type. These tables were formerly developed by Rodgers et al. in 1976 [13]. The tables presented in this work are an attempt to bring that information up to date for the binders currently in use in both original and repaint finish coats. Accordingly, the spectrum is initially examined for the presence of any original finish melamine or urea cross-linking, as evidenced by absorption bands at 1550 and 815 cm^{-1} for melamine and 1650, 1540, and 770 cm^{-1} for urea. Although extremely uncommon in automotive binders, urea is mentioned as an amine cross-linker since it is similar to melamine cross-linking and is found as a component in some bicycle finishes and tool paints. If the layer may be a primer, the 1510 cm^{-1} region should then be checked for the presence of a sharp epoxy band and if found, confirmed by the presence of an 830 cm^{-1} band resulting from the out-of-plane bending of adjacent hydrogens in the *para*-substituted aromatic ring of bisphenol A. Next, a look at the 1525- to 1530-cm^{-1} region should indicate whether or not any urethane modification is present. Finally, the 1300 to 1100 cm^{-1} region should be examined to indicate whether the *most intense absorption band* indicates an alkyd-based resin (1260 to 1280 cm^{-1}), a polyester-based resin (1235 to 1245 cm^{-1}), or an acrylic-based resin (1150 to 1180 cm^{-1}). The microchemical tests illustrated in Figure 3 should be used to alert oneself to an acrylic or nitrocellulose lacquer. Once these determinations have been made, the analyst can proceed to the most likely binder types listed in Tables 9 and 10 and find the best fit of characteristic absorption frequencies. One should be careful to rely on the broader binder absorption bands and not be confused by the sharp absorption bands that sometimes arise from organic pigments. These bands are usually quite numerous and tend to ride on top of the characteristic binder bands. As always, the spectrum should then be compared to a representative reference spectra of the binder type suspected.

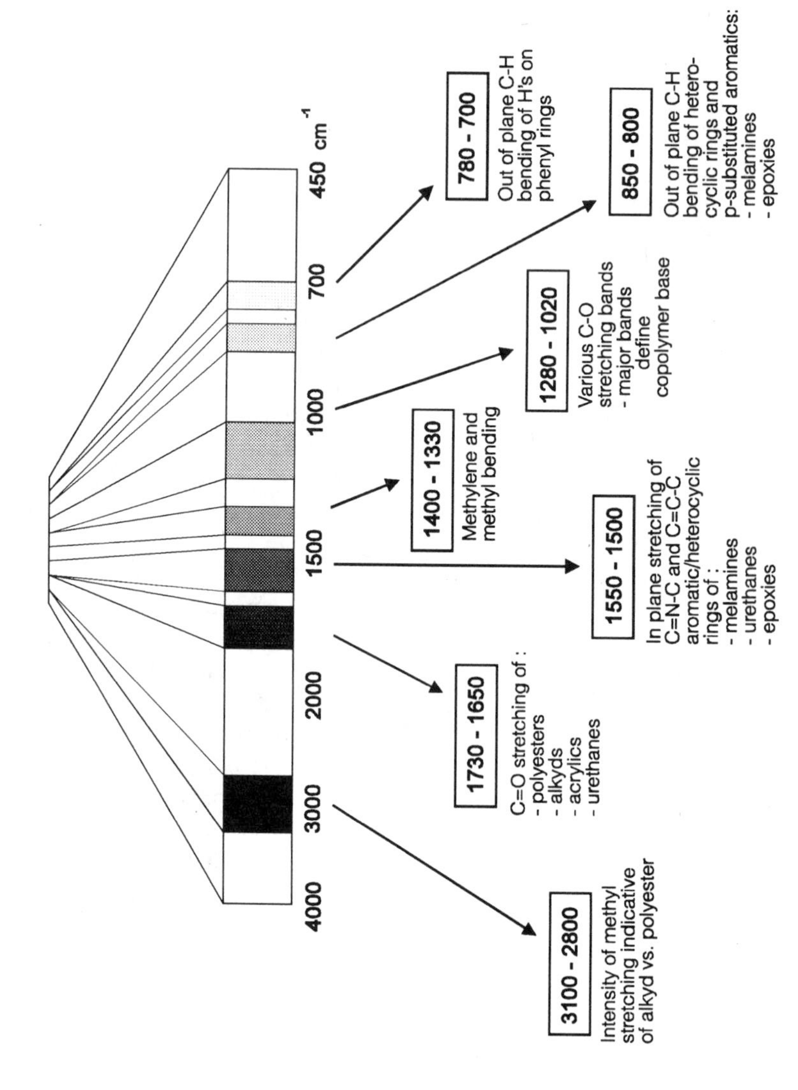

Figure 4. Mid-infrared spectrum showing the major functional group frequency regions employed in automotive paint binder classifications.

Table 9. Major Absorption Bands for Original Automotive Finishes

	Acrylic lacquers	
1150 cm^{-1} (dominant)	705 cm^{-1}	(weaker)
1270 cm^{-1}/1240 cm^{-1} (doublet)	750 cm^{-1}	(stronger)
1070 cm^{-1}	840 cm^{-1}	
	Acrylic–melamine enamels	
1550 cm^{-1} (usually dominant (C—N))	815 cm^{-1}	(ring C—N)
1480cm^{-1}	1090 cm^{-1}	
1370 cm^{-1}	760 cm^{-1}	(weaker)
1170 cm^{-1}	700 cm^{-1}	(stronger)
1240/1270 cm^{-1} (sometimes)		
	Polyester–melamine enamels	
1550 cm^{-1} (C—N)	815 cm^{-1}	(ring C—N)
1240 cm^{-1}	730 cm^{-1}	(strong)
1300 cm^{-1}	705 cm^{-1}	
	750 cm^{-1}	
	Alkyd–melamine enamels	
1550 $cm^{-1}cm^{-1}$ (C—N)	815 cm^{-1}	(ring C—N)
1270 cm^{-1}	740 cm^{-1}	(stronger)
1120 cm^{-1}	710–705 cm^{-1}	(weaker)
1070 cm^{-1}		
	Alkyd–urea enamels	
1540 cm^{-1}	770 cm^{-1}	
1650 cm^{-1}	*Note*: 815 cm^{-1} absent	
1270 cm^{-1}	740 cm^{-1}	(stronger)
1120 cm^{-1}	710–705 cm^{-1}	(weaker)
1070 cm^{-1}		

Second, one can use a flowchart such as that presented in Figure 5. Although requiring less knowledge of the chemistry of the binders and their characteristic functional group absorptions, flowcharts cannot predict all possible pitfalls, and a minor turn down the wrong path can be disastrous. Remember to use major bands when making decisions. Although this flowchart has been tested by several different examiners, final reference to band assignment tables and reference spectra is a must. If a collection of known vehicle paint spectra is not

Table 10. Major Absorption Bands for Automotive Refinishes

Acrylic lacquers		
See table 9		
Nitrocellulose lacquers		
1650 cm^{-1}	840 cm^{-1}	
1280 cm^{-1}	750 cm^{-1}	
(Often have acrylic lacquer modification)		
Alkyd enamels		
1270 cm^{-1} (dominant)	710–705 cm^{-1}	(weaker)
1120 cm^{-1}	740 cm^{-1}	(stronger)
1070 cm^{-1}	1450 cm^{-1}	
Acrylic–Alkyd enamels		
1260–1270 cm^{-1} 1150–1180 cm^{-1} 1130 cm^{-1} 1070 cm^{-1}	Absorption bands differ in stepwise intensity as compared to alkyd	
Urethane modification 1530–1520 cm^{-1} (C—N) No 815 cm^{-1} 1730 band is often complexed by 1690 and 1640 cm^{-1} stair step		
Acrylic–urethane enamels		
1530–1520 cm^{-1} (C—N) 1240 cm^{-1} 1150–1180 cm^{-1} 1070 cm^{-1}	Often accompanied by 1690- and 1640-cm^{-1} stair step	
Newer-generation "oxithane" type 1460 cm^{-1} band is almost as strong as 1730 cm^{-1} and is accompanied by strong 1690 cm^{-1} (looks like 1730/1690 cm^{-1} doublet)		

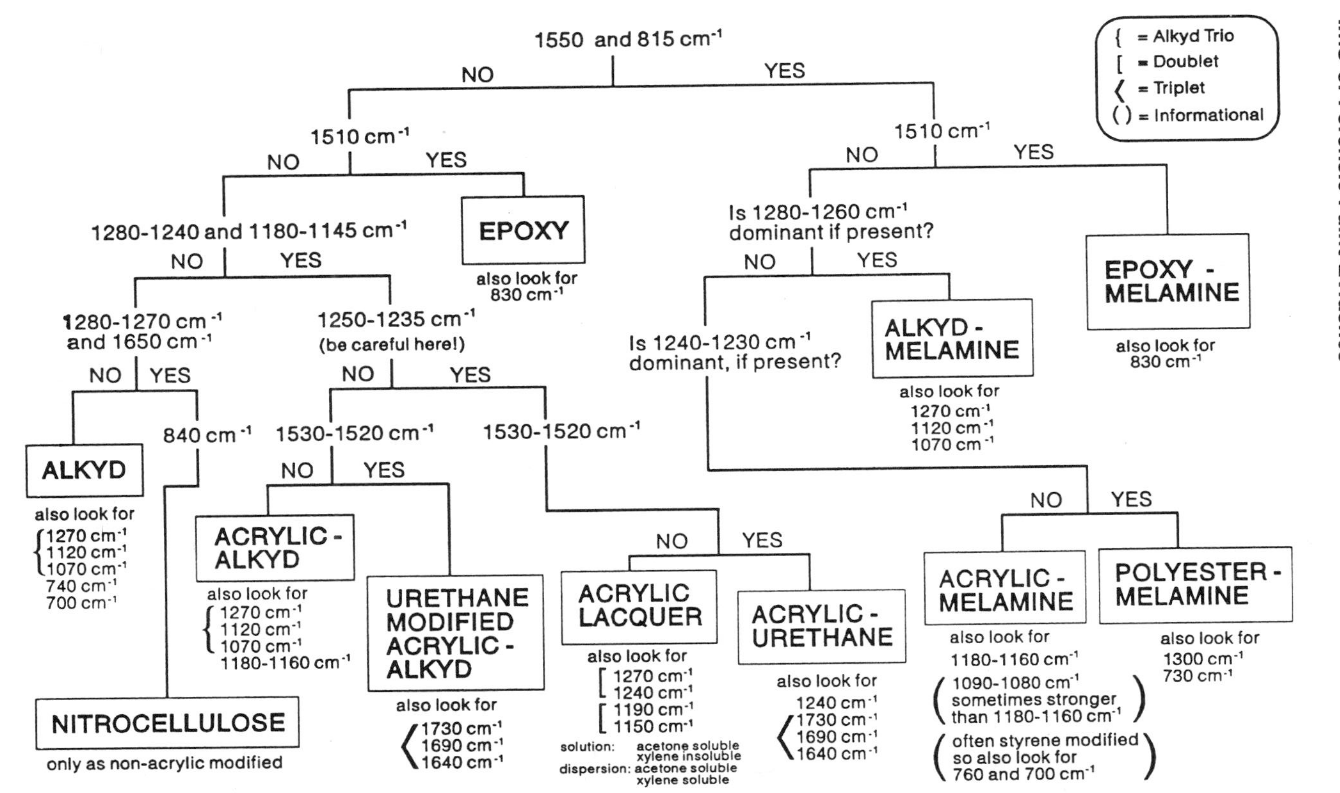

Figure 5. Automotive paint binder infrared classification flowchart for use as an aid in the recognition of various original-finish and repaint binders currently employed. (Some revisions from the latest version published in [50].)

available, a good starting place is *An Infrared Spectroscopy Atlas for the Coatings Industry* [10].

Figures 6 through 13 contain representative spectra of the various types of automotive finish-coat paint binders listed. Several key points are worthy of note. Two types of lacquers are presented in Figure 6. As noted in Section 3.1, the acrylic lacquers are used as both original and repaint finishes, while the nitrocellulose lacquers are encountered only in repaints (other than perhaps Rolls-Royce). Figure 6a depicts the typical characteristics of an acrylic lacquer with the pronounced "double doublet" arising from a large concentration of methyl methacrylate. The 1150-cm^{-1} band is dominant and accompanied by the less intense 1240- and 1270-cm^{-1} doublet. Of course, as noted in Section 2.1 and Figure 3, if it is a dispersion lacquer it will be both acetone and xylene soluble, whereas if it is a solution lacquer, it will be xylene insoluble. Figure 6b depicts the typical characteristics of a nitrocellulose lacquer. Although nitrocellulose alone has no carbonyl band at 1730 cm^{-1} and is characterized by its intense 1650-, 1280-, and 840-cm^{-1} N=O stretching bands, it is usually plasticized with a phthalate ester or modified with an acrylic polymer giving rise to a 1730 cm^{-1} band. If the flowchart alone is used for classification, this might lead one to an acrylic lacquer classification; accordingly, the diphenylamine microchemical test (Section 2.1 and Figure 3) should always be used on a lacquer to alert oneself of the presence of nitrocellulose.

Figures 7 and 8 contain spectra of the three major original finish enamels currently in use. Their major characteristic absorption bands are listed in Table 9. All three types display the 1550- and 815-cm^{-1} melamine absorption bands characteristic of an original finish. The acrylic–melamine enamel in Figure 7a depicts the strong acrylic absorption in the 1160 to 1180 cm^{-1} region, along with the presence of styrene absorptions at 760 and 700 cm^{-1}. This is not always the case, for variation in the formulation can result in the 1090-cm^{-1} band being more intense than that of the 1160 to 1180 cm^{-1} band, as seen in Figure 7b. Furthermore, the styrene bands at 760 and 700 cm^{-1} may occasionally be inverted in intensity, with the 700-cm^{-1} band being less intense than the 760-cm^{-1} band, or even absent. Of course, there are always exceptions. Figure 8 shows the spectra of an alkyd–melamine enamel and a polyester–melamine enamel. The automotive alkyd–melamine enamels are typically based on an *ortho*-phthalic acid or anhydride, while the polyester–melamine enamels are typically based on an isophthalic (*meta*-substituted) acid. Their spectra behave accordingly. The alkyd–melamine enamels display the typical "alkyd trio" at 1270, 1120, and 1070 cm^{-1}. The polyester–melamine enamels display absorption bands typical of an isophthalic alkyd (refer to Ref. 10 for typical spectra), with the major C=O stretch at 1230 to 1240 cm^{-1} accompanied by a characteristic band at 1300 cm^{-1} and a sharp intense band at 730 cm^{-1}. Hence, once melamine cross-linking is recognized and an acrylic-based copolymer is ruled out, the

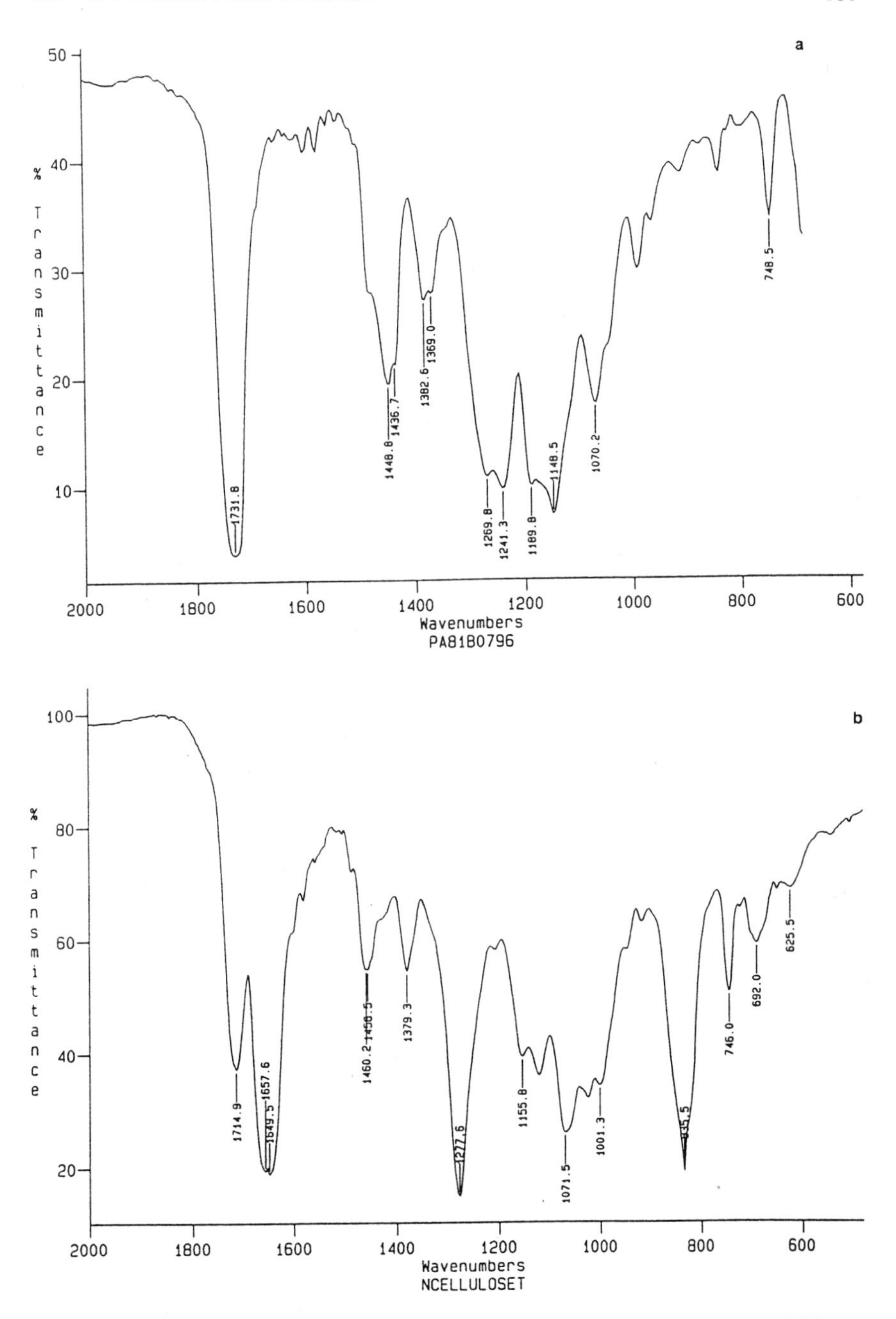

Figure 6. Infrared spectrum of (a) automotive acrylic lacquer finish coat and (b) automotive nitrocellulose lacquer finish coat, depicting major absorption bands.

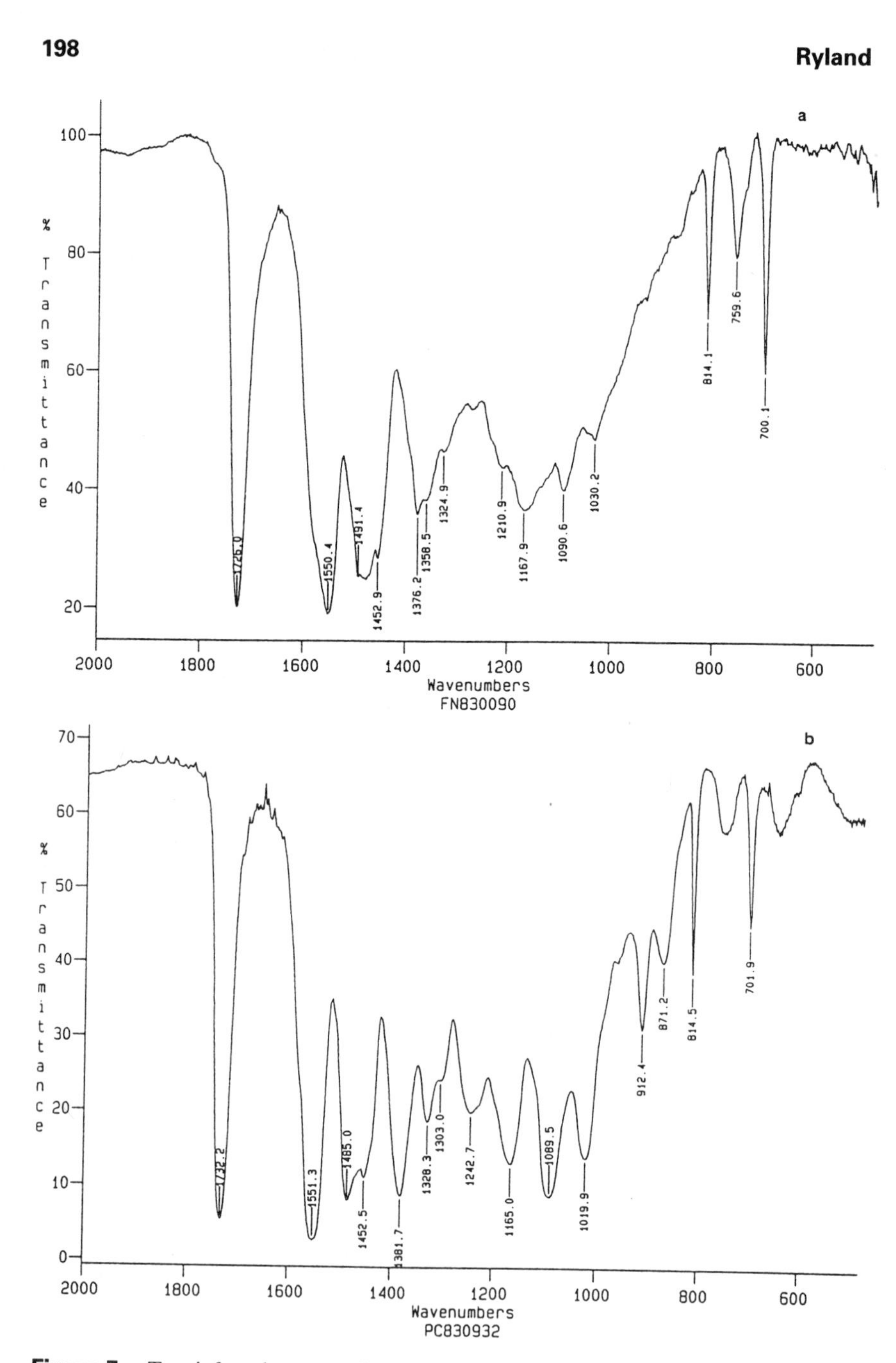

Figure 7. Two infrared spectra of automotive acrylic–melamine enamel original-finish coats depicting the major absorption bands as well as the variations that may occur in the absorption band intensities in the 1000- to 1300-cm^{-1} region of the spectrum.

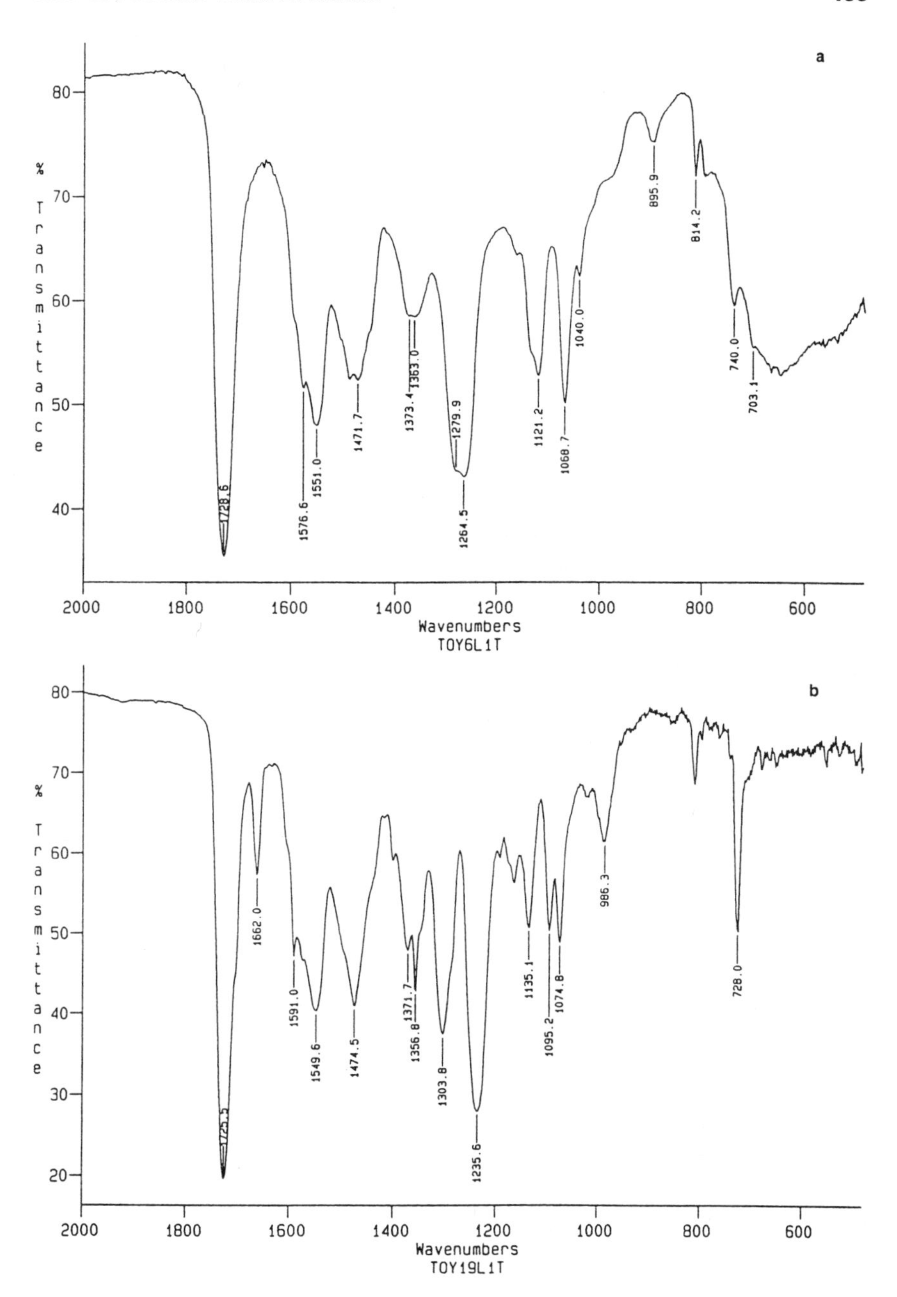

Figure 8. Infrared spectrum of (a) automotive alkyd–melamine enamel original-finish coat and (b) automotive polyester–melamine enamel original-finish coat, depicting major absorption bands.

original-finish enamel choice comes down to an alkyd or a polyester, which are easily differentiated by the presence of a 1270- to 1280-cm^{-1} band versus the 1230- to 1240-cm^{-1} band, respectively.

Figure 9 contains the spectrum of a typical urea cross-linked alkyd as may be found in bicycle finishes, tool, and other maintenance paints. When looking initially for melamine cross-linking at 1550 cm^{-1}, note that the band is slightly shifted to 1540 cm^{-1}. Melamine cross-linking may be ruled out due to the absence of the triazine ring C—H bending band at 815 cm^{-1}. Additional examination of the spectrum reveals the presence of a pronounced 1650-cm^{-1} band (urea carbonyl stretch) riding on the side of the 1730 cm^{-1} alkyd carbonyl stretching band. Urea cross-linking can then be corroborated by the presence of a 770-cm^{-1} band.

Now that the original finishes have been covered, let us move on to the refinishes. The major characteristic absorption bands are listed in Table 10, while the refinish products examined are listed in Table 11. The acrylic and nitrocellulose lacquers are no different from those depicted in Figure 6. As noted previously, dispersion acrylic lacquers are not used in the automotive refinish industry, only solution lacquers. Figure 10 contains the spectra of the two types of air-drying alkyds used, an alkyd enamel and an acrylic–alkyd enamel. As

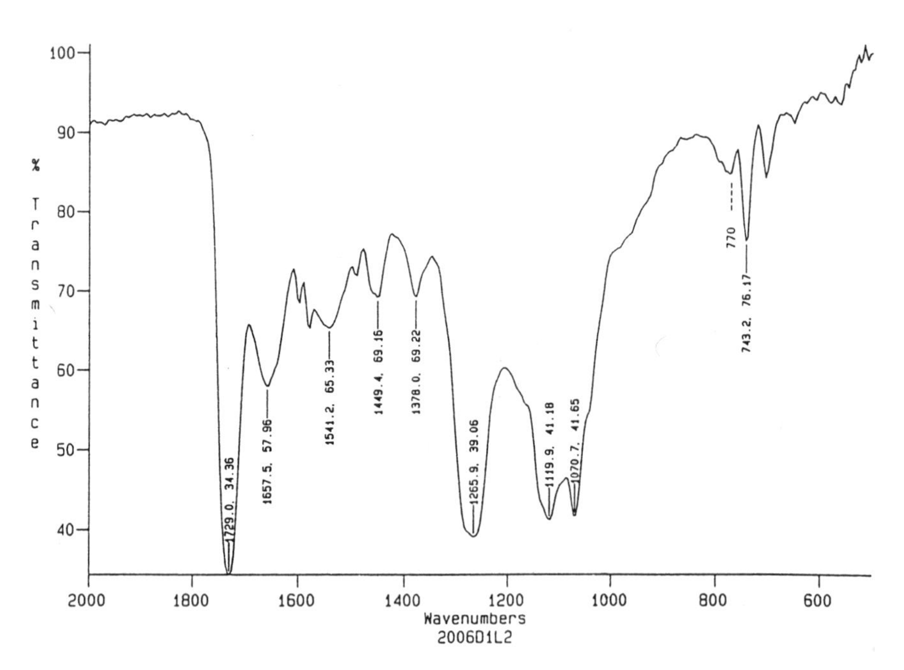

Figure 9. Infrared spectrum of an alkyd–urea enamel original bicycle finish coat, depicting its major absorption bands.

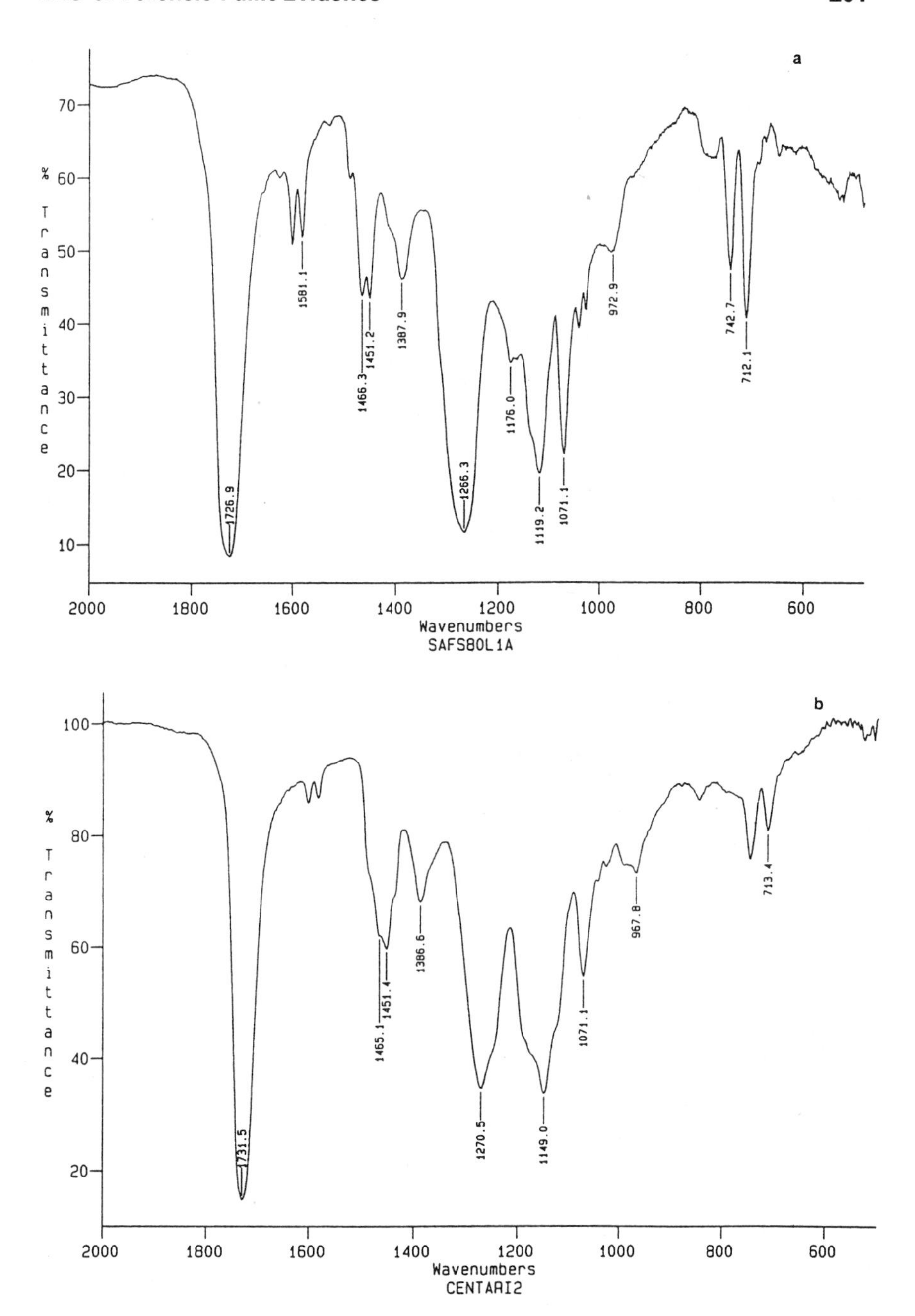

Figure 10. Infrared spectrum of (a) automotive alkyd enamel refinish top coat and (b) automotive acrylic–alkyd enamel refinish top coat, depicting major absorption bands.

can be seen in Figure 10a, the alkyd enamel displays the typical 1270-, 1120-, and 1070-cm^{-1} alkyd trio of an *ortho*-phthalic alkyd. Although theoretically possible, isophthalic alkyd enamels have not been encountered in refinishes. There is an absence of any melamine cross-linking at 1550 and 815 cm^{-1} and the typical styrene pattern at 760 and 700 cm^{-1} has given way to the alkyd pattern of 740 cm^{-1} and 710 to 705 cm^{-1}, with the 740 cm^{-1} band typically being more intense than the 710- to 705-cm^{-1} band. Various manufacturers' alkyd enamel spectra look very similar to one another, with the primary differences being in the C=H stretch area around 2700 to 3100 cm^{-1} and the C=C stretch region around 1580 to 1650 cm^{-1}. This is a result of the variety of different types of drying oils that may be used, each having different degrees and types of unsaturation. The acrylic–alkyd (a copolymerized formulation) shown in Figure 10b retains the basic alkyd characteristics, but is complexed by impingement of the acrylic absorptions around 1150 to 1180 cm^{-1}. The band shapes in this area can change with the maxima shifting, depending on the various manufacturers' acrylic formulation. There is an absence of melamine bands and the major C=O stretching band falls between 1260 and 1280 cm^{-1}. The formulations may even be styrene modified, resulting in the 760- and intense 700-cm^{-1} pattern noted in the acrylic–melamine enamels.

As mentioned previously, urethane additives are offered for both the alkyd enamels and the acrylic–alkyd enamels. They are usually called "hardeners" by refinish manufacturers. Their addition to the normal refinish paint at the body shop prior to spraying the automobile will result in a coatings spectrum with a very small band around 1530 cm^{-1} in the case of alkyd enamels and a more intense band around 1520 to 1530 cm^{-1} in the case of the acrylic–alkyd enamels. The intensity of this C=N stretch is dependent on the quantity of urethane modification imparted to the resin. There may be some concern for confusion of this urethane band with the 1550 cm^{-1} band of melamine; however, as noted in Table 10, the absence of the 815 cm^{-1} absorption band of melamine avoids the misclassification. Depending on the type of urethane modification employed by the refinish paint manufacturer, a series of "stair steps" may be noted on the side of the 1730 cm^{-1} carbonyl band of the acrylic–alkyd enamel, as noted in Table 10 and depicted in Figure 11. These are probably due to the carbonyl stretches of the particular urethane(s) used. It is not surprising to expect a shift of the carbonyl absorption from that of the alkyd's polyester since the urethanes employed in automotive topcoats are usually aliphatic or alicyclic [18], not aromatic as is the phthalate of the alkyd's polyester. The 1640 cm^{-1} stair-step band is close to that of the urea cross-linked alkyd (1650 cm^{-1}); however, no confusion should result since the alkyd–urea resin's spectrum is not accompanied by the 1690 cm^{-1} stair-step shoulder.

In summary, there are four types of alkyd resins (air drying) that may be encountered in automotive refinishes: alkyd enamels, urethane-modified alkyd

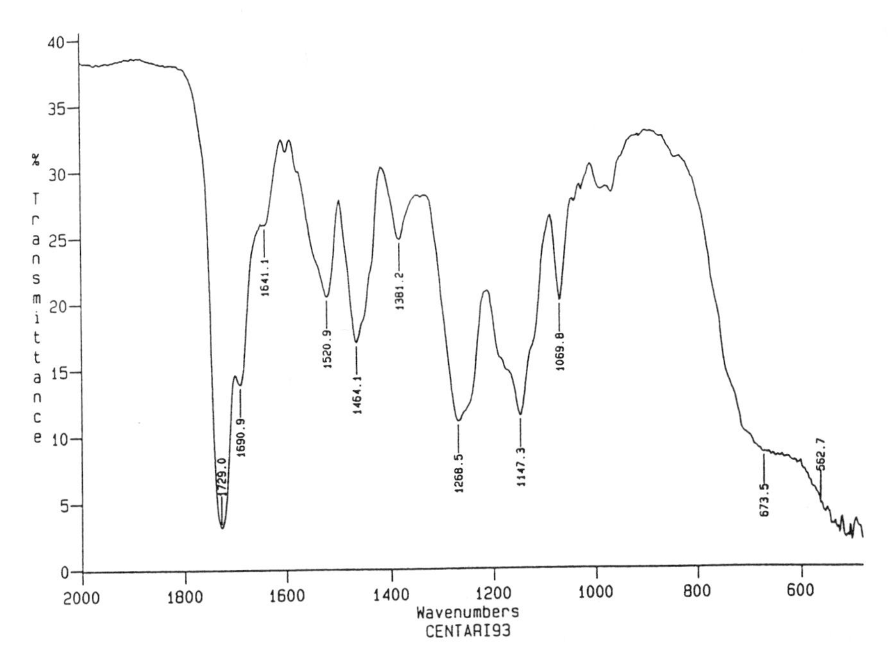

Figure 11. Infrared spectrum of an automotive urethane-modified acrylic–alkyd enamel refinish top coat, depicting its major absorption bands along with the characteristic stair-step bands along the side of the 1730-cm^{-1} carbonyl band. These stair-step absorption bands may or may not be present with urethane modification; however, the 1530- to 1520- cm^{-1} C—N stretch will always be present.

enamels, acrylic–alkyd enamels, and urethane-modified acrylic–alkyd enamels. There are also acrylic modification packages available for the low-cost alkyd enamels; however, they are usually present in such low concentrations that they are difficult to recognize due to the slight absorption of the alkyds in the 1180-cm^{-1} region.

Figure 12 contains the spectra of the older and newer generations of acrylic–urethane enamels used in automotive refinishes. The older generation, as can be seen in Figure 12a, is very similar to the urethane-modified acrylic–alkyds, with the most notable difference being the shift of the alkyd 1270- to 1280-cm^{-1} absorption band to 1235 to 1250 cm^{-1}. Just as in the urethane-modified acrylic–alkyds discussed previously, there is a pronounced 1520- to 1530-cm^{-1} band from the urethane component and a strong 1150- to 1180-cm^{-1} band from the acrylic components. The spectrum may or may not have the characteristic 1730-, 1690-, 1640-cm^{-1} urethane stair step. The difference between the urethane-modified acrylic–alkyd enamels and the acrylic–urethane enamels is very subtle, yet reproducible for all of the major manufacturers' products listed in Table 11

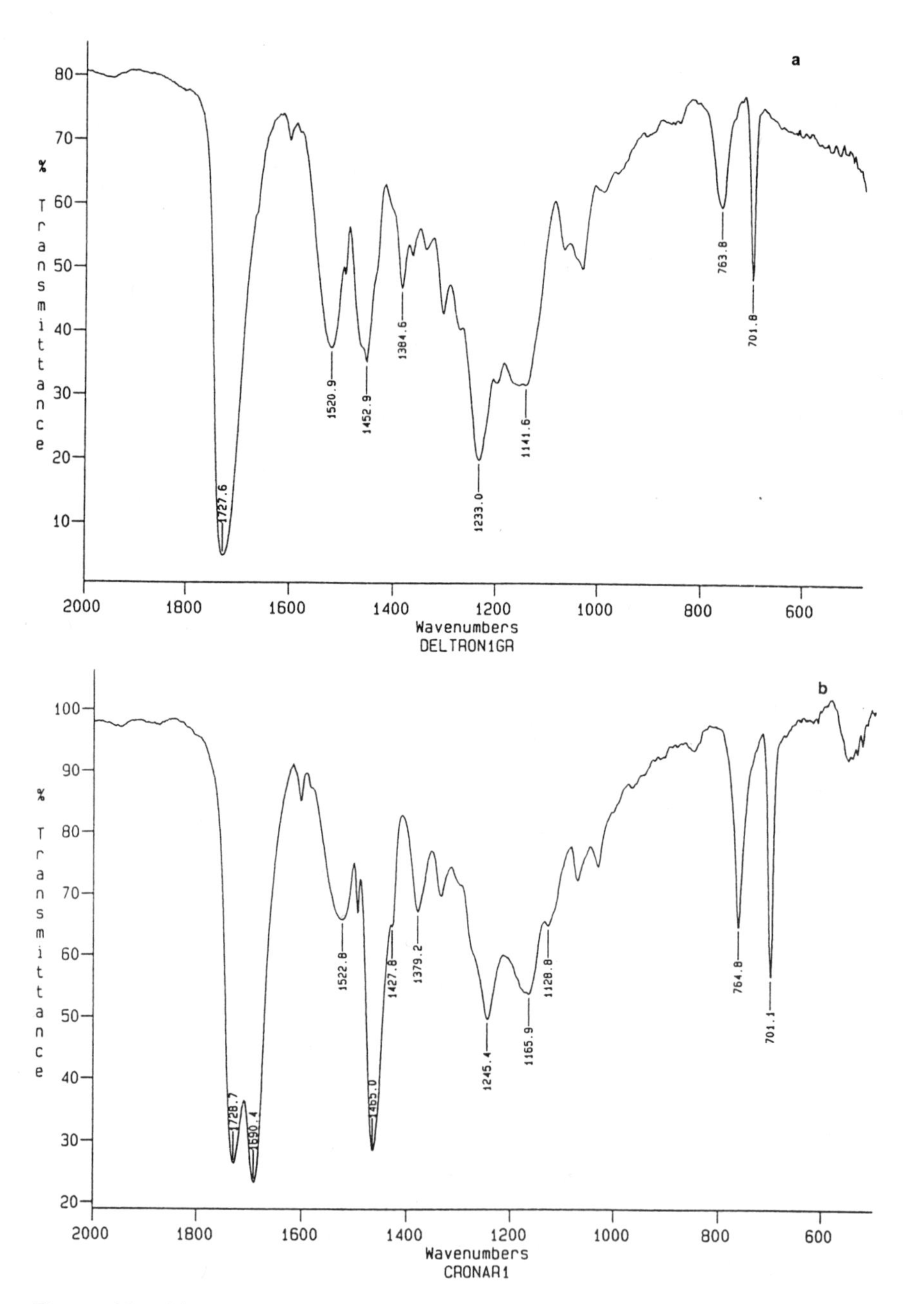

Figure 12. (a) Infrared spectrum of an older-generation automotive acrylic–urethane enamel refinish top coat, depicting its major absorption bands. Note the shift of the alkyd's C—O stretching band from 1270 to 1280 cm^{-1} to 1235 to 1250 cm^{-1}. (b) Infrared spectrum of a newer-generation automotive acrylic–urethane enamel refinish top coat, depicting its major absorption bands.

Table 11. Automotive Repaint Finish-Coat Binder Products

Acrylic Lacquers	
PPG/Ditzler	Duracryl
Dupont	Lucite
Inmont/R-M	Alpha-Cryl
Sherwin Williams	Acrylic lacquer
Numerous others	
Alkyd enamels	
PPG/Ditzler	Ditzco
Dupont	Dulux (Dulux Plus)[a]
Inmont/R-M	Enamel
Sherwin Williams	Kem Transport (Fast-Cat)[a]
Numerous others	Many with both acrylic and urethane modification packages available
Acrylic–alkyd enamels	
PPG/Ditzler	Delstar (Delthane)[a]
Dupont	Centari (urethane hardener)[a]
Inmont/R-M/BASF	Super-max (Star-Rock)[a]
	Miracryl II (urethane hardener)[a]
Sherwin Williams	Acrylyd (Polasol Plus)[a]
American Lacquer and Solvents	Syn-A-Cryl
Others	
Acrylic-urethane enamels	
PPG/Ditzler	Deltron
Dupont	Imron
	Chromabase clear coat/base coat
Inmont/R-M	Miracryl II Clearcoat
Sherwin Williams	Sunfire 421
American Lacquer and Solvents	Amerflint II
Others	
Newer-generation acrylic-urethane enamels	
PPG	NCT clear coat (for use with Deltron base coat)
	Concept clear coat/base coat
DuPont	Cronar clear coat/base coat
Inmont/RM/BASF	Diamont clear coat/base coat
ACME	Probase clear coat/base coat

[a]Urethane ''hardener'' available.

except DuPont's Imron. That product is described by the manufacturer as a ''polyester urethane.'' Its infrared spectrum is shown in Figure 13. The infrared absorption characteristics meet the criteria for a urethane-modified acrylic–alkyd. Therefore, whenever a coating is tentatively classified as a urethane-modified acrylic–alkyd, be sure to compare it directly to the spectrum of DuPont's Imron

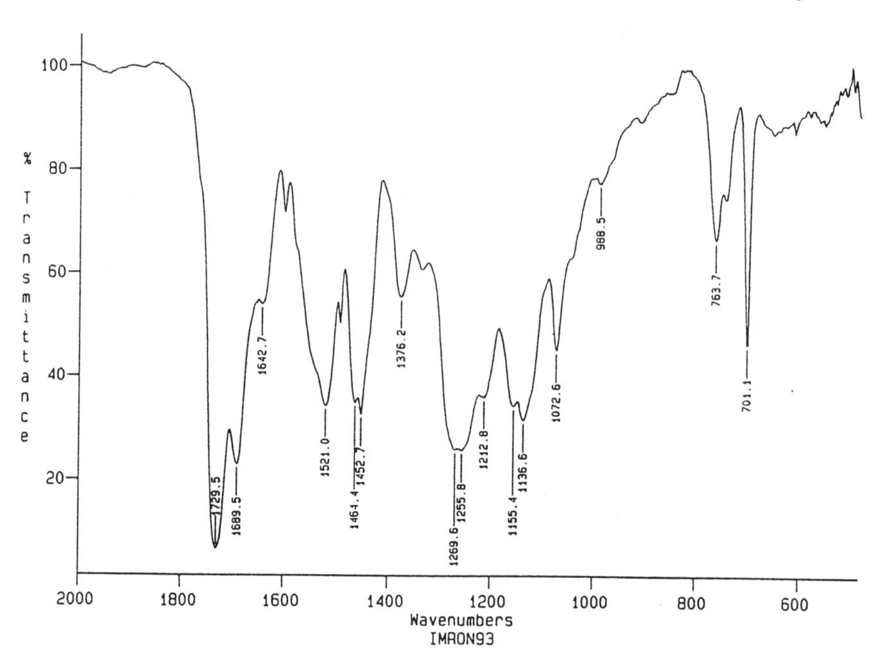

Figure 13. Infrared spectrum of DuPont's Imron automotive acrylic–urethane enamel refinish top coat. It is the only finish coat examined that defied the classification scheme, classifying as an acrylic–alkyd as opposed to an acrylic–urethane.

to avoid any misclassifications. As can be seen in Figure 12b, the new generation of acrylic–urethane enamels (post-1987) can easily be recognized by a strong doublet in the carbonyl stretch area, as opposed to a stair step. As noted in Table 10, the doublet is comprised of the 1730-cm^{-1} band and an intense 1690-cm^{-1} band. In addition, the 1460-cm^{-1} methyl and methylene deformation band is quite intense compared to the first-generation acrylic–urethane spectra. The effect of the flex additives on the infrared spectra of refinishes has not been characterized fully at this point. The problem becomes somewhat moot in practice, for it is quite obvious that they are highly elastomeric when attempting to obtain a thin peel under the stereomicroscope.

Finally, a word about automotive primers. As noted previously, acrylics, alkyds, and epoxies are typically encountered. The epoxies are the most prevalent, even in the refinish primers. When alkyds or acrylics are used, they are often epoxy modified. In addition, the epoxies may be modified with alkyd or acrylic resins. Rodgers provides typical spectra for these modifications along with band assignments [13]. The spectrum of a typical epoxy primary is shown in Figure 14. The epoxy resin can be recognized by the presence of a sharp 1510-cm^{-1} absorption band (C=C stretch of the *para*-disubstituted aromatic in

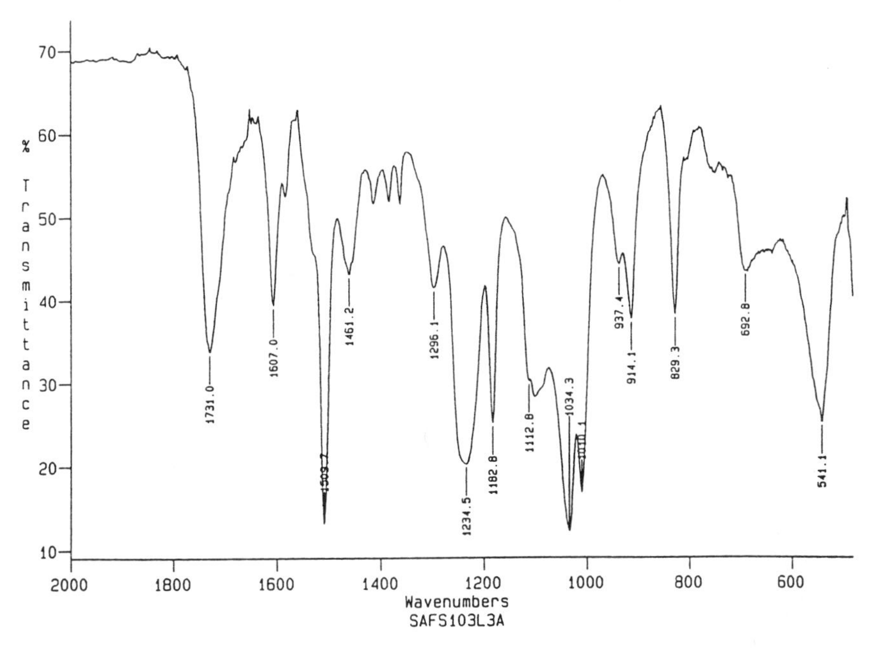

Figure 14. Infrared spectrum of an automotive epoxy enamel primer with clay extender, depicting its major absorption bands.

bisphenol A) along with a pronounced 830-cm^{-1} band (out-of-plane bending of adjacent hydrogens of the *para*-disubstituted aromatic in bisphenol A). The characteristic epoxy 1240- and 1180-cm^{-1} C═O stretch bands should not be confused with those of an acrylic–urethane finish coat, since current automotive finish coats do not utilize heavy epoxy resin modification. This region of the spectrum is important, however, for it permits recognition of alkyd or acrylic modification to the epoxy resin. If there is alkyd modification, there will be a distinct shoulder at approximately 1280 cm^{-1}, and if there is acrylic modification, there will be a marked shoulder or yet another broad absorption band at approximately 1150 to 1180 cm^{-1}. The addition of the acrylic or alkyd polyesters will also introduce additional absorption bands in the 760 to 700 cm^{-1} region of the spectrum. The alkyd and alkyd–melamine-based primers look very much like their finish-coat counterparts. The acrylic enamel primers tend to produce a rather nondistinct spectrum with a broad major absorption band (other than the 1730-cm^{-1} carbonyl stretch) falling around 1150 to 1180 cm^{-1}.

One of the weaknesses of infrared microspectroscopy becomes readily apparent when observing the clay extender absorptions present in the 1000-cm^{-1} area of the Figure 14 spectrum. Primers contain substantially higher extender pigment loads than finish coats, and the broadband absorptions resulting from these pigments often obliterate useful binder classification information. If the paint layer is soluble in a volatile solvent, such as the lacquers (see Figure 3), a small-scale extraction can be performed in the well of a ceramic spot plate to separate the binder and additives from the interfering extender. The paint-layer peel is dissolved in several drops of chloroform or acetone and then left to stand a few moments to permit the extender pigment grains to settle. The supernatant is slowly drawn off with a micropipette or syringe and then cast onto the silver chloride window one drop at a time. The resulting film may then be analyzed by transmitted infrared microspectrometry. If the paint sample is too small to handle by this method, the specimen may be placed directly on the window support material and ''extracted'' in place simply by placing a small amount of an appropriate solvent directly on it. The solvent will extract some of the binder and additives and carry them to the perimeter of the solvent evaporation ring. This area may then be apertured and analyzed. Although this method is quite useful for small-scale extractions of both binders and plasticizers, the extract is not quite as homogeneous as in the preceding method. Sadly, this approach for avoiding extender interference will not be of any aid in the case of insoluble binders, as are the overwhelming majority of the intractable finishes encountered in automotive primers as well as low-gloss architectural finishes.

On the other hand, these sometimes troublesome absorption bands provide classification information about the inorganic pigment, often turning the weakness into a strength. The ratio of extender pigment to resin in automotive primers is usually such that reliable classification can be reached for both components. If you will recall, Table 2 listed a majority of the extenders used in the paint industry. Table 12 lists the characteristic absorption bands for the most commonly encountered extenders, while Figure 15 provides a flowchart to assist in their initial recognition. Always remember to use flowcharts with caution and confirm all classifications with comparisons to reference spectra, such as published in *An Infrared Spectroscopy Atlas for the Coatings Industry* [10]. This is especially true for the extenders, for there will often be more than one present in any given paint layer.

Extender pigments tend to have broad absorption bands in the fingerprint region of the infrared spectrum. This severely hampers their reliable identification. Furthermore, several extenders may have absorptions falling in the same area, such as clay, talc, mica, and calcium silicate or silica, barium sulfate, and calcium sulfate. Complex mixtures of silica with talc and clay further complicate the problem of extender identification. The extender pigment absorption bands that fall below 700 cm^{-1} can often be of great value in sorting out these identification problems, as can be seen in Table 12. For this reason, employing a wideband MCT detector is a very wise choice. Although these binder classification interferences caused by inorganic extenders are somewhat of a problem with automotive paint primers, they become even more so in the following end-use type of paint to be discussed: architectural coatings.

Table 12. Major Absorption Bands for Common Extender Pigments[a]

Barium sulfate	Calcium carbonate
1120 cm^{-1} and 1080 cm^{-1} [B]	1440 cm^{-1} [B]
1180 cm^{-1}	875 cm^{-1}
980 cm^{-1}	715 cm^{-1} (marked in calcite)
610 cm^{-1} and 640 cm^{-1}	
Clay	Talc
1040 cm^{-1} and 1010 cm^{-1} [B]	1020 cm^{-1} [B]
1105 cm^{-1}	670 cm^{-1}
910 cm^{-1}	460 cm^{-1}
540 cm^{-1}	450 cm^{-1}
470 cm^{-1}	
Silica (crystalline)	Silica (hydrolyzed)
1100 cm^{-1} [B]	1100 cm^{-1} [B]
800 cm^{-1} and 780 cm^{-1}	470 cm^{-1}
700 cm^{-1}	
520 cm^{-1} and 470 cm^{-1}	
Silica (diatomaceous)	Calcium sulfate
1100 cm^{-1} [B]	1120 cm^{-1} [B]
800 cm^{-1}	1160 cm^{-1}
480 cm^{-1}	675 cm^{-1} and 600 cm^{-1}
	612 cm^{-1} (not in gypsum)
Mica	Calcium silicate
1030 cm^{-1} [B]	1020 cm^{-1} and 1090 cm^{-1} [B]
540 cm^{-1}	920 cm^{-1} (multiple bands) [B]
470 cm^{-1}	470 cm^{-1}
	570 cm^{-1}
	650 cm^{-1} and 680 cm^{-1}

[a][B], broad.

3.2. Architectural Coatings Binder Classification

Architectural coatings are the second most frequently encountered type of paint in the typical forensic laboratory concentrating on the examination of physical evidence occurring in criminal investigations. One should remember that their binders are often quite different from those employed in the automotive industry, and consequently, the forensic coatings examiner should not blindly follow the same classification guidelines put for in Section 3.1. There are some similarities, but care must be taken to explore all avenues of binder variation. The examiner should fully utilize the power of visual microscopy discussed in Section 2.1 to assess the morphology of the specimen and reach a preliminary decision as to

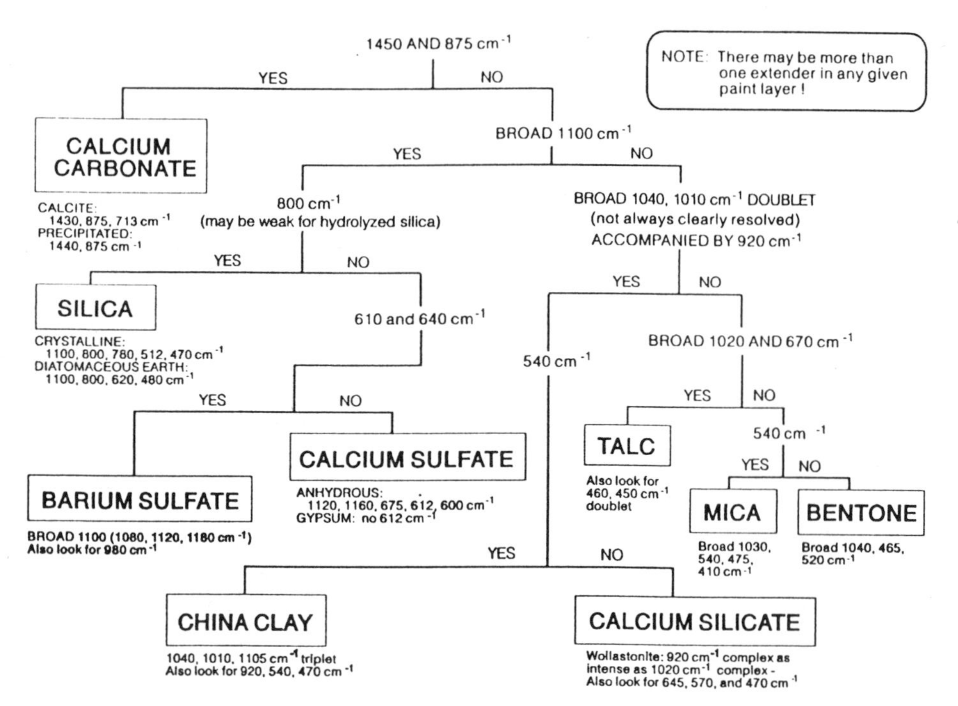

Figure 15. Paint extender pigment infrared classification flowchart for use as an aid in recognition of the various inorganic fillers currently employed.

the end-use application of the paint. This may very well prevent the analyst from going down the wrong path in binder classification. Initially, let us review what general types of binders are typically used.

3.2.1. Binder Types

Architectural finishes may be broken down into two groups, waterborne coatings and organic solvent–borne coatings. In addition, each group has applications for both interior and exterior finishes. First, let us take a look at the waterborne coatings, commonly known as latexes (coagulating vehicles).

Table 13 lists the resins commonly used in architectural latex paints. The polyvinyl acetate–acrylic binders command the greatest share of the market for interior flat and eggshell gloss paints. The acrylic binders are more durable but more expensive and consequently more commonly found in interior semigloss and gloss latexes. The styrene–butadiene binders were popular in the mid-1970s but gave way to the polyvinyl acetate–acrylic (PVA acrylic) resins. They are the lowest cost of the three major binders in common use today. There is also

Table 13. Common Architectural Finish Binders

Latex systems
Polyvinyl acetate–acrylics
Acrylics
Styrene–butadienes
Polyvinyl acetate–ethylenes
Organic solvent–borne systems
Nitrocellulose and acrylic lacquers
Acrylic enamels
Alkyd enamels
Urethane enamels

some limited use of a polyvinyl acetate–polyethylene copolymer resin as a latex binder, but it is typically higher priced than the PVA–acrylic resin and consequently, not that popular.

The exterior latex market finds the more durable acrylic resins as the major binder in use. They offer ease of cleanup, yet admirable durability. The available resins permit formulation of paints with surface reflectivity ranging from flat to gloss. There are other binder types in use as well as a variety of modifications of the acrylic-based resin, but the overwhelming majority of the exterior latexes sold are acrylics.

The second group of architectural finishes is the organic solvent–borne resins. They are also listed in Table 13. Nitrocellulose and acrylic lacquers (solvent evaporation vehicles) may be found as trim or accessory coatings, especially on metal objects associated with the structure. They are often packaged as spray paints, and acrylic enamels may also be found in this form. The alkyd resins, however, hold the major portion of the market in this group. They are available as both interior and exterior enamels, but their major use is for exterior trim. They provide a highly durable gloss or semigloss air-dry finish (oxidizing vehicle), their only drawback being that of cleanup ease. In interior applications, they are ideal for high-humidity environments (such as bathrooms) or heavy-wear areas (such as kitchens or industrial interiors). They also provide excellent adhesion to metal surfaces (such as handrails) when applied over an appropriate primer. Polyurethanes hold a minor share of the market, being used primarily for metal trim or stained wood trim coatings. They are typically of two types. The first is a drying-oil modified type, which cures by oxidation in much the same way as the alkyd resins do. The second type cures by exposure to water, typically in the form of humidity (polymerizing vehicle). Both types are one-package systems, unlike the urethanes and acrylic–urethanes used in the automotive paint industry. They are employed for the same reasons as they are used in the automotive paint industry, flexibility and mar resistance.

3.2.2. Binder Classifications by Infrared Microspectroscopy

Table 14 list the characteristic absorption bands for the various binders discussed. The infrared spectra of the acrylic lacquers, nitrocellulose lacquers, and urethane enamels are quite similar to their counterparts in the automotive paints section. Many of the alkyd enamel spectra encountered will also be very similar to the spectra of the automotive alkyd enamels. There is one critical variation, however. There is much more diversity in the architectural finish industry, with many more manufacturers producing a variety of paints in both large and small quantities. Alkyds, for instance, can be found in two major types. They may be

Table 14. Major Absorption Bands for Architectural Finishes

Latex systems	
Polyvinyl acetate–acrylic	
1735 cm^{-1}	1135–1020 cm^{-1} (characteristic shape)
1370 cm^{-1} (stronger than 1435 cm^{-1})	
1240 cm^{-1} (dominant)	945 cm^{-1}
1175 cm^{-1}	605 cm^{-1}
Acrylic	
1730 cm^{-1}	
1370 cm^{-1} (weaker than 1435 cm^{-1})	
1240 cm^{-1}	
1170–1150 cm^{-1} (dominant)	
Styrene–butadiene	
760 cm^{-1} (very strong)	1730 cm^{-1} (possible plasticizer)
700 cm^{-1} (very strong)	
1450 cm^{-1}	
1495 cm^{-1}	
1600 cm^{-1}	
Oil-based systems	
Alkyd enamels	
1730 cm^{-1}	745 cm^{-1} (stronger)
1270 cm^{-1}	710–705 cm^{-1} (weaker)
1120 cm^{-1}	
1070 cm^{-1}	
Urethane enamels	
1730 cm^{-1}	
1530 cm^{-1}	
1240 cm^{-1}	

either *ortho*-phthalic *or* isophthalic (*meta*-substituted)-based resins and may employ several types of modifications, including styrene, vinyl toluene, rosin, urethane, and even silicone. Typically, the resins used for house paints are not modified, and if they are, it is usually with styrene or vinyl toluene. The infrared spectrum will look like either an *ortho*-phthalic alkyd (resembling the automotive alkyd enamels) or an isophthalic alkyd (resembling the automotive polyester enamels), with any modification appearing with the appropriate styrene absorption bands at 760 and 700 cm^{-1}, or the vinyl toluene absorption bands at 700, 780, and 815 cm^{-1}. One can often confirm that the paint is an alkyd enamel and not a polyester enamel by examining the intensity of the methyl and methylene stretch absorptions in the 2800- to 3100-cm^{-1} region of the infrared spectrum. Due to the large quantity of such functional groups in the drying oils of the alkyd resins, the methyl and methylene stretch absorptions are quite intense compared to those in the typical unmodified polyester resin. Excellent spectra illustrating these points can be found in *An Infrared Spectroscopy Atlas for the Coatings Industry* [10].

Turning our attention back to the waterborne coatings (latexes), one can see that the characteristic infrared absorption bands for the PVA–acrylic binders and the acrylic binders listed in Table 14 appear very similar at first glance. In recognizing these paints, the analyst should remember to use his or her microanalytical skills and note that these resins tend to be somewhat soft and flexible in texture and tear easily when examined microscopically. Infrared spectra for unpigmented PVA–acrylic and acrylic binders can be found in Figure 16. As can be seen, they are easily differentiated, with the most intense fingerprint region absorption band for the PVA–acrylic binder falling around 1240 cm^{-1}, while that of the acrylic binder appears at around 1150 to 1180 cm^{-1}. You will also note that there is an inversion in the relative intensities of the 1435- and 1370-cm^{-1} methylene and methyl bending absorption bands, with the 1370-cm^{-1} methyl band being more intense for the PVA–acrylic resins. Two further characteristics that are quite useful in recognizing the PVA–acrylic resins are the 945- and 605-cm^{-1} absorption bands. Any concern for confusion with the alkyd enamel resins can be further allayed by noting the absence of aromatic or olefinic methyl stretching above 3000 cm^{-1}, unlike the phthalate and unsaturated drying-oil-based alkyds. So the three are easily differentiated given the absence of heavy pigment loading. But that is not always real life!

The Figure 17 alkyd architectural primer illustrates the complications presented by the incorporation of two common extender pigments and one common coloring pigment in a low-gloss (flat) architectural coating. Calcium carbonate absorbs strongly in the 1440-cm^{-1} region, talc absorbs strongly in the 1030-cm^{-1} region, and titanium dioxide adsorbs strongly in the 450- to 700-cm^{-1} region, all interfering with important binder classification areas. The problem can be compounded even more by the addition of clay, which will then obliterate

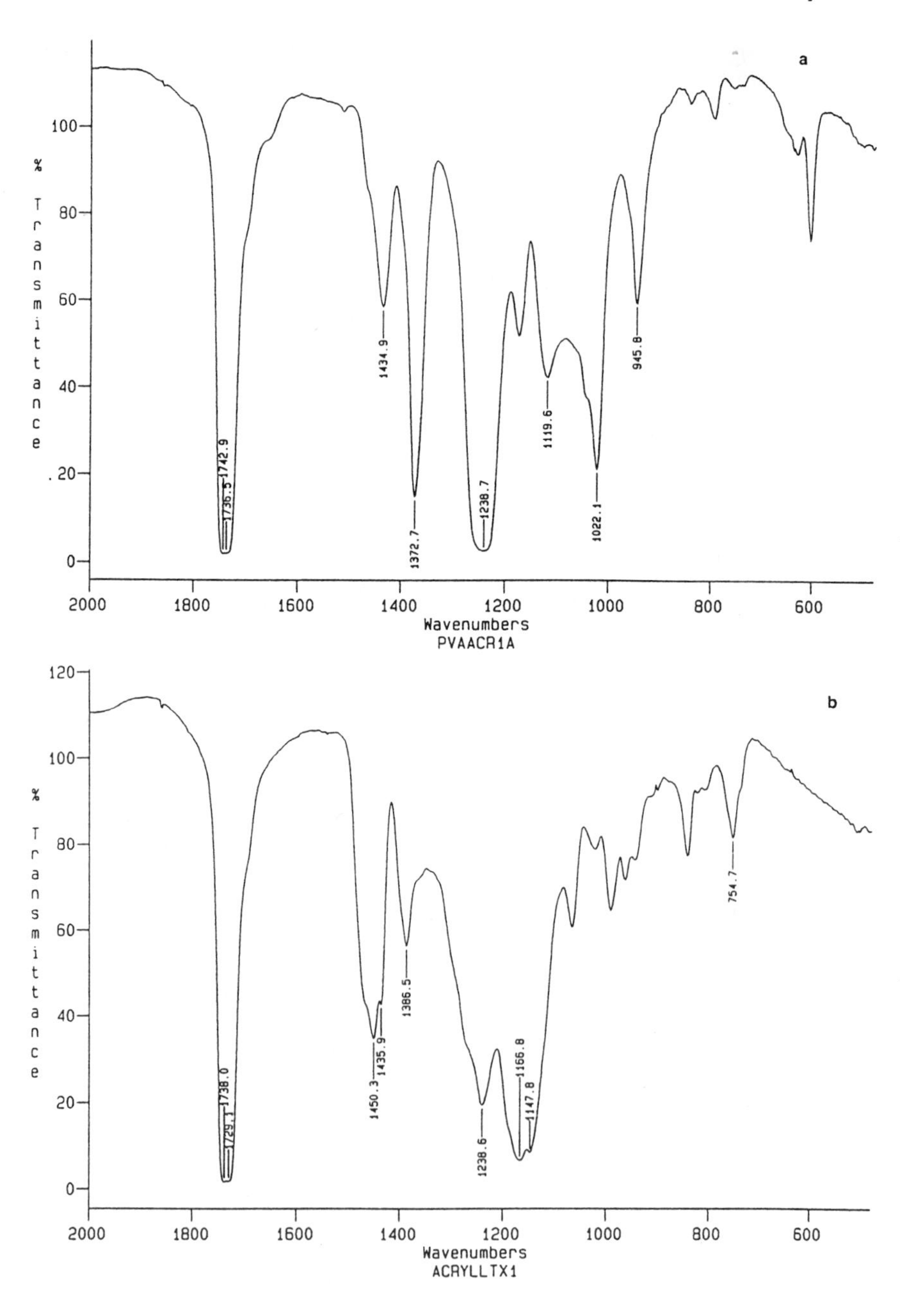

Figure 16. Infrared spectrum of (a) unpigmented PVA–acrylic latex architectural top coat and (b) unpigmented acrylic latex architectural top coat, depicting major absorption bands.

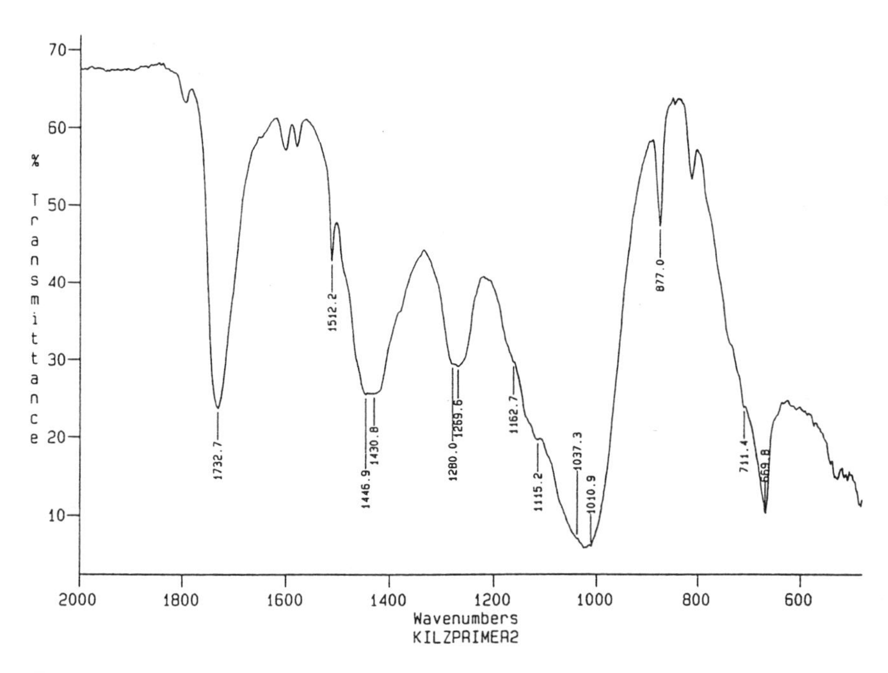

Figure 17. Infrared spectrum of an alkyd architectural primer with heavy loading of calcium carbonate and talc extenders as well as titanium dioxide coloring pigment.

the acrylic recognition area of the mid-1100-cm^{-1} region. All that remains unobstructed are the regions 1240 to 1280 cm^{-1} and 3000 cm^{-1}. The polyvinyl acetate 945-cm^{-1} band may be recognized riding on the broad 1030-cm^{-1} talc band *if* there is no interference from a barium sulfate extender. Consequently, low-gloss architectural finish coats or primers may present such overwhelming problems in binder classification that conventional infrared microspectroscopy cannot lead to a conclusive opinion as to the binder type. Alternative sample preparation methods, such as pyrolysis infrared microspectroscopy, or complementary analytical techniques, such as pyrolysis gas chromatography or pyrolysis mass spectrometry, may then be needed to affect a classification.

The styrene–butadiene binders are easily recognized by their characteristic absorption bands. A problem arises when they are so heavily filled with extender pigment that their resin bands become very minor components of the infrared spectrum. In that they are the third most common resin employed in architectural flat latexes, they may initially be recognized by the absence of absorption bands characteristic of the two other major binder types. Again, alternative sample preparation methods or complementary analytical techniques may also be required for confirmation of the classification.

The polyvinyl acetate–polyethylene binders are not very common, but their characteristics do deserve mention. With lower polyethylene compositions, they will appear very similar to the PVA–acrylic binder spectra (Figure 16A) except for the absence of the 1150- to 1180-cm^{-1} acrylic absorption band. With higher ethylene compositions, the 1435-cm^{-1} absorption band of the PVA–acrylic resin will give way to a 1475-cm^{-1} band (methylene environment change) while the 1120- and 605-cm^{-1} absorption bands will decrease in intensity. This is accompanied by a marked increase in the intensity of the methyl/methylene C—H stretching bands around 2800 to 2950 cm^{-1}, resulting from the long polyethylene polymer chains.

3.3. Binder Classifications of Other End-Use Types of Coatings

As discussed in Section 1.2, the forensic coatings examiner will not only encounter automotive and architectural paints as evidence, but will be called upon to examine and compare maintenance paints, marine paints, and aircraft paints. Fortunately, most of these end-use applications utilize finishes that have binders similar to those already discussed.

Maintenance finishes may generally be considered as durable working coatings designed to maximize product life. They are found on hand tools, power tools, appliances, road signs, metal bridges, and industrial equipment and components. Their finish coat binders will typically be epoxies, alkyds, modified or cross-linked alkyds, or acrylic–urethanes. The recognition of these binder types by infrared microspectroscopy has been addressed in previous sections. There is one binder type that has not been discussed and may be encountered in maintenance finishes designed specifically for heat resistance. Silicone or silicone-modified binders are typically recognized by their strong broad absorptions at 1020 and 1100 cm^{-1} resulting from the Si—O—Si stretching vibrations. These are accompanied by strong sharper absorption bands at 800 and 1260 cm^{-1} resulting from S—C stretching and methyl bending of the S—CH_3 bond, respectively. Fortunately, the paints encountered are usually higher-gloss finishes, permitting assignment of the 1020- and 1100-cm^{-1} bands to the resin and not a clay or talc extender pigment.

The typical marine finish involved in casework originates from a pleasure craft and utilizes an alkyd, modified alkyd, or acrylic–urethane binder. The alkyds are not well suited for immersion service, however, and epoxy, vinyl, or chlorinated rubber resins are usually found in these applications. When used, polyurethanes are found as wood varnishes or cosmetic coatings on upper-deck structures. Again, the characteristic infrared absorption bands for most of these binders have been described previously. The chlorinated rubber resins are encountered so infrequently that they will not be discussed. The vinyl resins are typically copolymers of polyvinyl chloride and may be recognized by the characteristic C—Cl infrared absorption bands at 620, 640, and 690 cm^{-1}, assuming that a wideband MCT detector is being employed. Although polyvinyl chloride alone does not have a 1730-cm^{-1} carbonyl absorption, its modifiers or plasticizers often do. A quick chloroform wash of the paint peel followed by evaporation of the solvent and subsequent infrared microspectroscopic analysis should remove most of the spectral interference originating from the plasticizer. The characteristic polyvinyl chloride 1250-, 1335-, 960-, and 1430/1440-cm^{-1} doublet absorption bands should then be apparent, along with a marked reduction in the intensity of the 1730-cm^{-1} carbonyl band. If the resin is a copolymer, the sharp 1430/1440-cm^{-1} doublet, along with the 620-, 640-, and 690-cm^{-1} C—Cl stretching absorptions are the easiest way initially to recognize the polyvinyl chloride component.

The aircraft coatings commonly encountered originate from smaller propeller-driven airplanes. Their finish coat binders include acrylic lacquers, alkyd enamels, modified alkyd enamels, acrylic–urethanes, urethanes, and occasionally epoxies, with their primers typically being epoxies or alkyds. The characteristic infrared absorption bands for all these binders have been listed in the previous sections.

3.4. Significance of Binder Classifications

Provided with the tools of microspectroscopy, a knowledge of the various characteristic infrared absorption bands associated with commonly encountered binders and a general appreciation of the end-use applications of coatings binders, the forensic paint analyst can begin to improve his or her interpretation of the significance of paint trace evidence. The fruits of this labor can be harvested routinely from individual layers of multiple-layer paint fragments as small as 0.5 mm^2.

By combining the information acquired from careful stereomicroscopic and microsolubility/microchemical examinations with that obtained from infrared microspectrometry analysis, potential end-use applications of paint fragments recovered from victim's or suspect's clothing debris can be predicted. A polyvinyl acetate–acrylic binder is typical of an interior architectural latex coating, not an automotive paint. An acrylic dispersion lacquer is characteristic of a car manufactured by General Motors between 1970 and 1992. A high-gloss alkyd–urea enamel finish coat recovered from a hit-and-run victim's clothing is consistent with originating from a bicycle, not an automobile. Several fragments of a moderate-gloss epoxy enamel finish coat with a rough-surface topology recovered from the clothing or wound tract of a beating victim is indicative of a maintenance paint source, quite possibly originating from a tool used as a weapon.

The technique can also prove invaluable in the examination of recovered automotive paint fragments. Occasionally, in the microscopic examination of a multiple-layer chip having both original finish and refinish layers present, it becomes very difficult to determine whether a particular layer is a primer or a finish coat. There is much more variation in the binder formulations of finish coats than in primers, so the examiner may want to know which layers to concentrate on for further comparisons using more discriminating techniques, such as pyrolysis gas chromatography or pyrolysis mass spectrometry. The recognition of any substantial epoxy modification in the paint layer immediately alerts the analyst that he or she is probably dealing with a primer and not a finish coat.

In the case where paint fragments are recovered from the debris adhering to the clothing of a hit-and-run victim with no suspect vehicle in custody, it is desirable to attempt to predict the make/model of the vehicle from which they originated as an aid to investigators searching for the perpetrator. If the paint is a refinish, one does not want to attempt to match it with existing reference standards of original finishes. A refinish paint matching an original-finish color can be applied to any make/model vehicle, not just the vehicle for which it was originally intended. To suggest potential make/models based on such a comparison may very well mislead investigators. Usually, binder classification based on the information acquired from the infrared spectrum of the finish coat can quickly indicate whether the paint in question is an original finish or a refinish. Acrylic dispersion lacquers, acrylic–melamine enamels, alkyd–melamine enamels, and polyester–melamine enamels are all original finishes; while alkyd enamels, acrylic–alkyd enamels, and acrylic–urethane enamels are refinishes. Nitrocellulose lacquers are most probably refinishes. Acrylic solution lacquers and flexible polyurethane enamels may be used for both original finishes and refinishes. Furthermore, if one studies the information acquired and listed in Section 3.1, it will become obvious that an unweathered automotive alkyd–melamine enamel finish coat indicates that it probably originated from a foreign-manufactured vehicle. This type of binder was last used approximately 20 years ago as a domestic original finish coat on Ford, Chrysler, and American Motor trucks. Additionally, as mentioned previously, a nonmetallic polyester–melamine finish coat without a clear coat is indicative of some newer (post-1986) foreign automobiles manufactured in Japan.

Finally, binder classifications can provide the forensic coatings examiner with the means to evaluate how common or uncommon a particular finish may be in our environment. Names are very important in this endeavor. It is not extremely difficult to acquire frequency of occurrence data without classifying samples as long as the sample population is limited. For example, one could take a flower, carefully note all of its features, and then simply conduct a detailed survey of every plant within a two-city-block area to determine how many times a flower having these particular features was encountered. But is this frequency of occurrence representative of that for the entire city or the entire state? Without naming the flower, the task of collecting the data to answer such a question is so time intensive that it becomes impractical. But if a systematic nomenclature is developed for flowers in the state, one can then compare their data with that of others and better answer the later question; and so it is with paint binders. If the nomenclature used by the paint industry can be associated with the infrared absorption characteristics of the various binders, the paint industry's sales experience can be incorporated with the forensic paint analyst's practical experience, producing significance assessments unconstrained by sample populations. For example, styrene–butadiene latex paints are relatively uncommon and are currently marketed only as "low-end" interior finishes. Alkyd enamel automotive refinishes are quite common and are typical of low-cost repaints. The newer-generation acrylic–urethane refinish enamels are designed for professional application, they are expensive, and they have been on the market only since mid-1987. They are much more uncommon. An acrylic–melamine enamel clear coat over an alkyd–melamine base coat is quite uncommon (see Table 7). Each of these observations can help improve the investigator's and the court's understanding of how significant it is that a questioned paint sample matches that of a given source, and that is one of the functions of the forensic paint examiner when he or she enters the courtroom.

3.5. Pigment Classifications by Infrared Microspectroscopy

The recognition and classification of extender pigments was covered in Section 3.1. Some inorganic coloring pigments will contribute absorption bands to the midinfrared region of the spectrum [12]. These may or may not permit a tentative identification of the coloring pigment present, depending on how much interference is encountered from the binder and extender pigment absorptions. The bands are usually quite broad and require additional examination to effect an identification, such as that afforded by polarized light microscopy, energy-dispersive x-ray spectrometry (EDX), or x-ray diffraction spectrometry (XRD). They are, however, quite useful as complementary data to that obtained by EDX for identifying the anion form (e.g., carbonate, sulfate, etc.). If warranted, interfering binder may be removed either by solvent extraction in the case of lacquers or low-temperature oxygen plasma ashing in the case of intractable enamels.

Organic pigments, on the other hand, are rich in absorption bands in the fingerprint region of the infrared spectrum. They are usually present in relatively low concentrations when compared to that of the binder and extender pigment and are consequently difficult to identify routinely. Occasionally, such as in high-gloss coatings, they may be present in sufficient concentrations to permit classification of at least the generic type of pigment by carefully examining the sharp absorption bands riding on top of the binder absorption bands (see Figure 25). Again, referral to a spectral reference collection, such as *An Infrared Spectroscopy Atlas for the Coatings Industry*, is quite valuable [10].

3.6. Coatings Comparisons Using Infrared Microspectroscopy

As noted in Section 1, the second purpose of a forensic paint examination is to compare a questioned paint to a representative sample taken from a known source (the "standard") to determine whether or not the questioned paint could have originated from the source represented by the standard. With that new motive comes new challenges.

Reproducibility of minute spectral detail now becomes of paramount importance. Reliable discrimination between sources of very similar paints is a must if the technique is to be of value. If the precision of the technique is poor, variations in samples actually originating from the same source may be interpreted as significant differences indicative of origin from different sources, ultimately resulting in *false exclusions.* On the other hand, variations in samples actually originating from different sources may be of the same magnitude as the variation observed within multiple runs of the same sample, ultimately resulting in *false inclusions.* Fortunately, infrared microspectroscopy does not routinely fall prey to either of these situations. The technique has proven to be quite reproducible in most instances. But several challenges to attaining this reproducibility do exist and, as with any microsampling technique, one must be cautious if attempting to evaluate results without the benefit of multiple runs.

The first challenge is that of inhomogeneity. Although usually not a problem for the automotive coatings or the medium- to higher-gloss architectural finishes and maintenance paints, the situation can occur when collecting spectra on low-gloss coatings using microscope aperture sizes less than 100 μm^2. This is typically a result of the larger extender grains expanding in area when compressed during sample preparation. As noted previously, Gardiner reported on variations due to this effect when analyzing architectural paint samples in bulk by scanning electron microscopy in conjunction with energy-dispersive x-ray spectrometry (SEM-EDX). He reports that reproducible elemental peak ratios could not be obtained when rastering the electron beam under an area of 30 μm^2 [25]. Depending on the paint sample, this inhomogeneity can easily occur on analysis areas of up to 100 μm^2 following sample compression for infrared microspectrometry. In some instances, the infrared microscope's aperture can be reduced sufficiently to avoid these crushed extender sites and permit the absorption spectrum of the binder to predominate. Of course, both the questioned and standard samples must be analyzed in a similar manner.

Inhomogeneity can also become a problem in the comparison of partially used or improperly mixed architectural coatings. Occasionally, the questioned material will be deposited as a spill or splash from a can of unmixed paint or a spray from an unshaken can of spray paint. The standard will be submitted to the laboratory as a partially consumed can of paint. Quite obviously, there will be homogeneity problems in the sample questioned, which must be assessed followed by the systematic preparation of a series of samples from the remaining liquid paint in the standard designed to cover the full range of possible variation. Comparison of the results will usually permit a conclusion to be reached concerning the question of common origin; however, even if a correspondence is found, there will be an increased possibility for alternative sources of the sample questioned, given the range of variation in the specimen.

The second challenge to ideal reproducibility finds its roots in the method of sample preparation. The most popular way of reducing a sample's thickness for transmission analysis is to compress it, as discussed in Section 2.2. This may introduce orientation to the polymer molecules in the specimen, resulting in an apparent change in the crystallinity of the resin [37]. The induced changes will result in intensity or band-shape variations in those infrared absorption areas sensitive to changes in crystallinity, such as the methylene deformation bands around 1450 cm^{-1} [38]. Again, multiple sample preparation and analysis will help the analyst evaluate the degree of variation expected for a particular specimen under the various pressures employed.

The third challenge to precision comes with the reality that the forensic paint examiner, like his or her colleagues examining textile fibers, metals, glass, and so on, cannot typically choose the condition or the size of the sample questioned. One must work with whatever the trace evidence transfer provides and glean all possible information from it, all the time being fully aware of the limitations placed on the results, conclusions, and significance. Often, coatings transfers occur as smears on a polymeric substrate. The transfer process is accompanied by heat resulting from friction between the depositing and receiving surfaces. This may very well produce a transferred paint smear that is mixed intimately with the substrate, be it another coating or synthetic fibers. Although FTIR provides the distinct advantage of being so reproducible in wavelength calibration (the Connes advantage) that spectral subtractions are possible, accomplishing this feat on realistic samples with such precision as to permit reliable comparisons of minute spectral detail is seldom possible. That is not to say that spectral subtraction is useless. This approach can certainly be of value in classifying the binder or extender pigment of substrate-contaminated smears and may provide some degree of comparative capability to help the examiner decide on which standards might be potential donors. Actual comparisons, however, will usually require the preparation of mixtures of the two materials in an attempt to determine whether or not all of the spectral details can actually be reproduced.

Notwithstanding these challenges, infrared microspectroscopy is one of the most valuable tools of the forensic paint examiner. It provides for highly reproducible nondestructive individual layer comparisons with a very limited time investment. Most often, both the binder and pigment portions of the paint are represented in the data obtained providing for quick eliminations without the use of multiple instrumental techniques. It is quite discriminating for most binder types, especially those employing copolymer modification. The technique is also sensitive to changes in organic pigment composition, a capability not shared by several other instrumental techniques normally used in the analytical scheme. But even if all of these attributes are capitalized upon and the challenges of reproducibility are conquered, infrared microspectroscopy still has some limitations and should not be the only instrumental technique used for a full comparison of a questioned and standard sample unconstrained by sample size and condition.

Infrared spectroscopy is not classically recognized as a technique providing sensitive detection of minor components. Constituents present in concentrations less than 5 wt % will typically go unnoticed. This situation can exist for some binder copolymers or internal plasticizers, especially if the minor components are similar in chemical structure to the major polymeric species. A typical example can be seen in Figure 18, where the infrared spectra of two different original-finish acrylic–melamine enamel automotive top coats manufactured in the same red nonmetallic color are presented. Although quite similar to one another, differences can be seen in the absorption band shapes in the region 1100 to 1000 cm^{-1}. The pyrolysis gas chromatograms of the same two finishes can be seen in Figure 19 and leave no doubt as to the differentiation of the two paints. Another less straightforward example can be found in Figures 20 and 21, where the infrared spectra of two different original-finish acrylic–melamine enamel automotive top coats manufactured in the same green metallic color for Ford Motor Corporation are very difficult to differentiate by infrared microspectroscopy, yet are easily discriminated by pyrolysis gas chromatography. The data in Figures 20a and 21a were obtained from the nonaqueous dispersion formulation manufactured by PPG, while those in Figures 20b and 21b were obtained from the same type of finish manufactured by Cook Paint and Varnish Company.

The problem of poor sensitivity for minor components is not limited to the organic portion of the coating. As mentioned previously, broadband inorganic pigments tend to be somewhat nondescript, especially in the 2000- to 700-cm^{-1} region of the infrared spectrum. If they are present in low concentrations, they will be masked by the major inorganic extenders or organic binders. Hydrolyzed silica is quite prone to this, with its major absorption band at 1110 cm^{-1} being hidden by barium sulfate, calcium sulfate, or even clay. Of course, there may be other minor additives in the paint that may also go undetected, such as the drier metal complexes used in drying oil resins (e.g., alkyds) and rheology modifiers used in latexes.

Furthermore, there are some types of binders whose infrared spectra change very little from one formulation to another. Both the alkyd enamels and the acrylic lacquers are prone to this problem. Quite often, the polyester portion of an alkyd resin is based on the same precursors, with the primary variation being

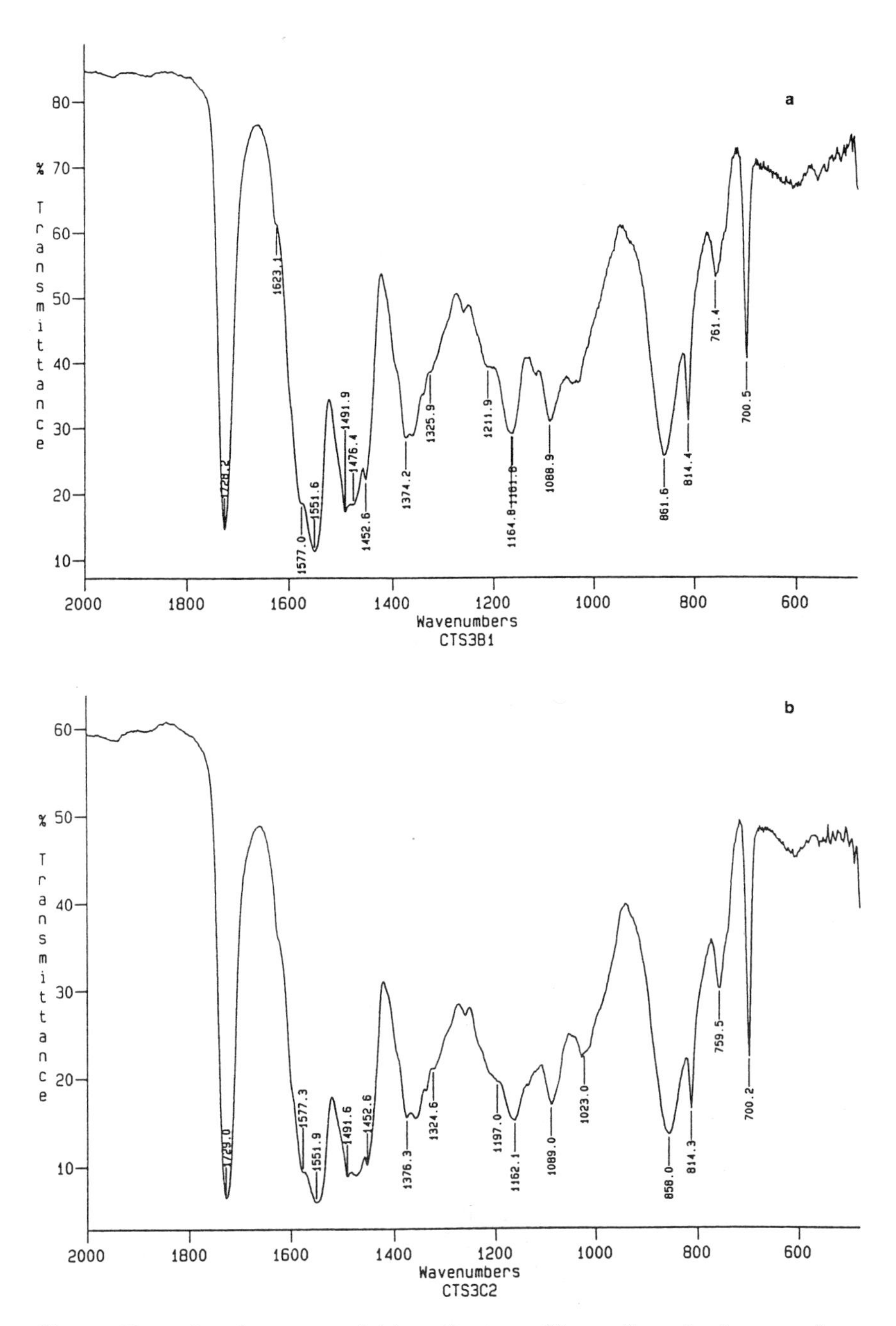

Figure 18. Infrared spectrum of (a) a red nonmetallic acrylic–melamine enamel automotive original finish top coat and (b) the same color acrylic–melamine enamel automotive original finish top coat supplied by a different paint manufacturer for use on the same make/model vehicle.

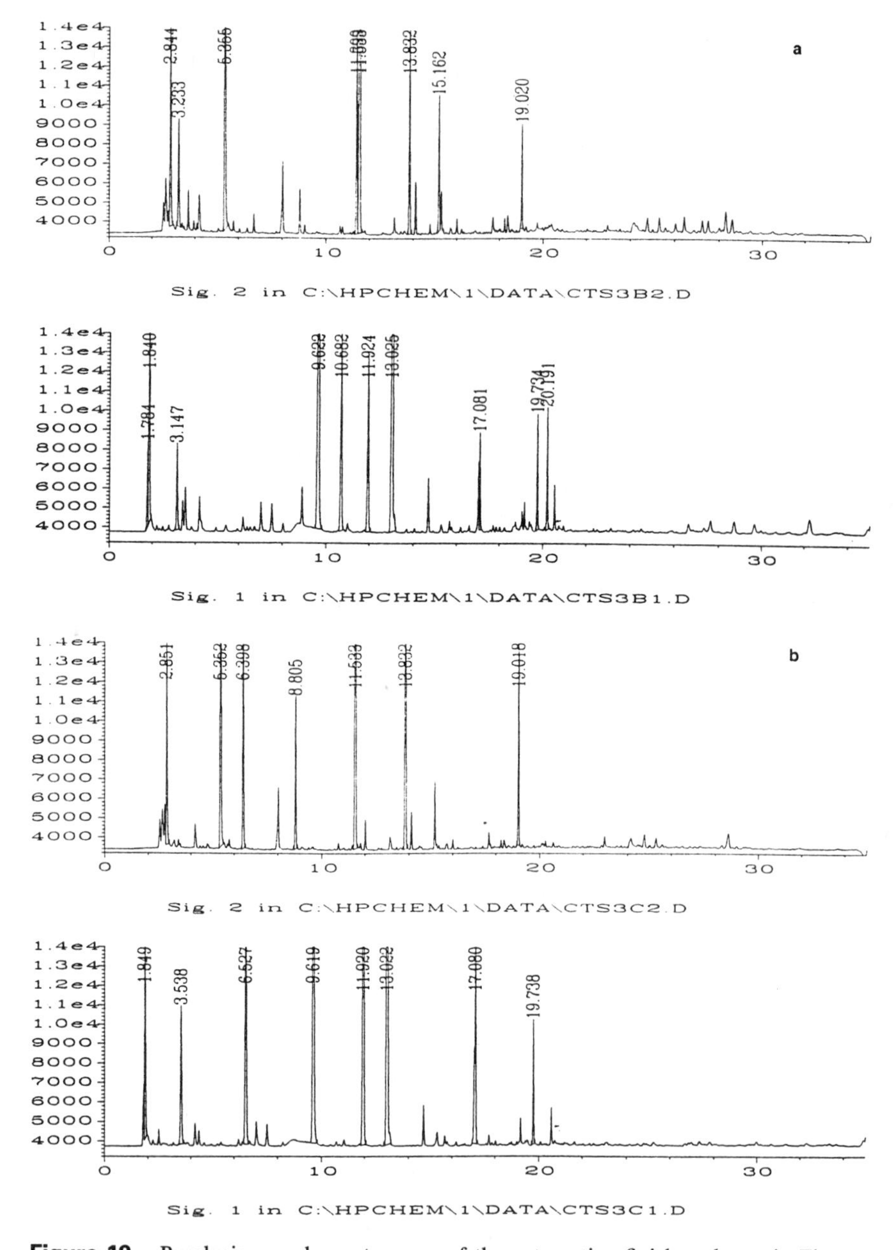

Figure 19. Pyrolysis gas chromatograms of the automotive finishes shown in Figure 18 (a) and (b) respectively. The pyrograms were obtained on a Hewlett-Packard 5890 gas chromatograph employing dual fused-silica capillary columns with flame ionization detectors and a Chemical Data Systems coil probe pyrolyzer employing a final pyrolysis temperature of 750°C with the heating-rate ramp turned off and a pyrolysis time of 20 s. Signal 1 depicts the pyrogram from the 25-m high-polarity FFAP column, while signal 2 depicts the pyrogram from the 30-m low-polarity methyl silicone column.

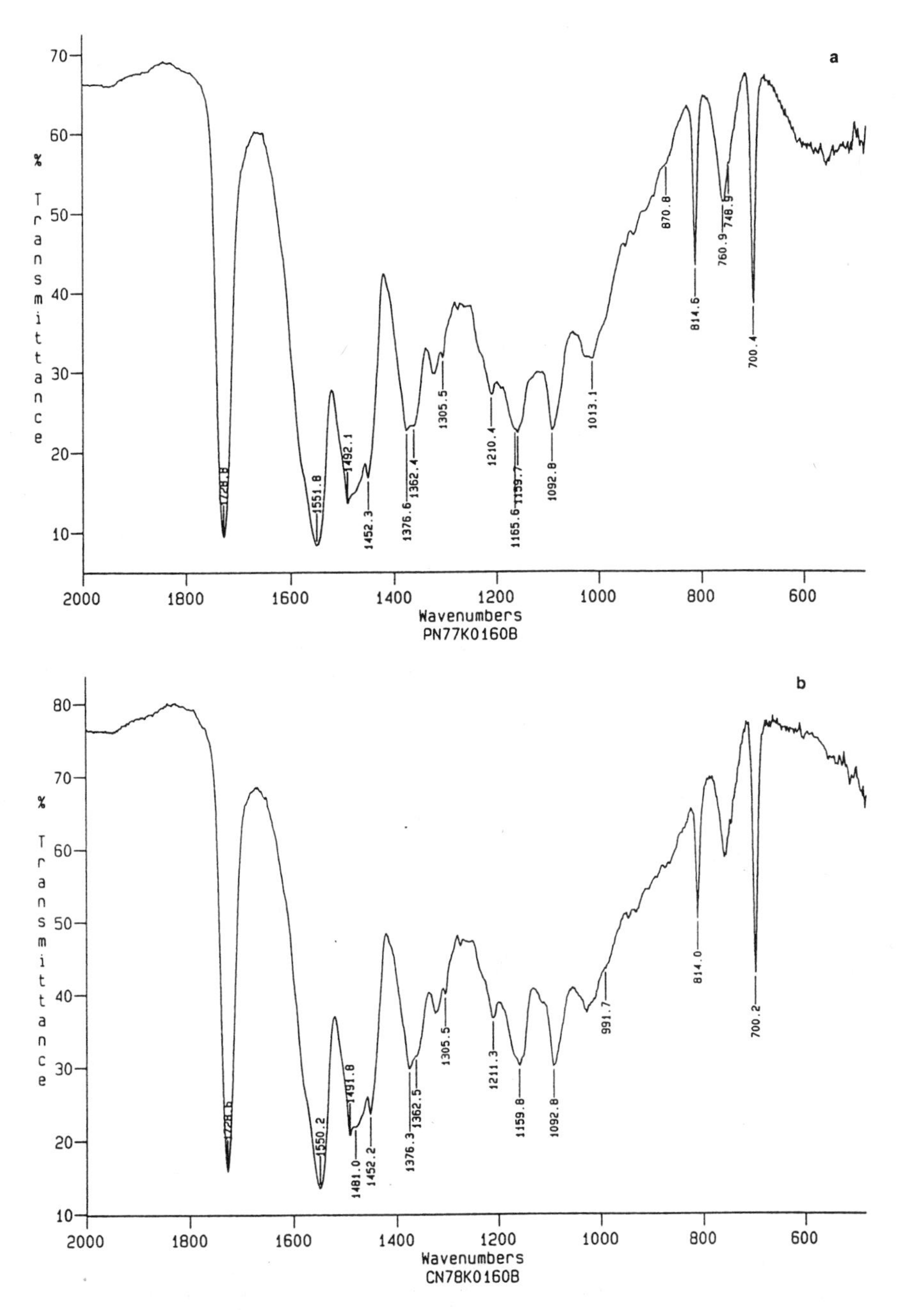

Figure 20. Infrared spectrum of (a) a green metallic acrylic–melamine enamel automotive original top coat manufactured by PPG for use on Ford Motor Corporation vehicles in 1977 and (b) the same green metallic acrylic–melamine enamel automotive original top coat manufactured by Cook Paint and Varnish Company for use on Ford Motor Corporation vehicles in 1978.

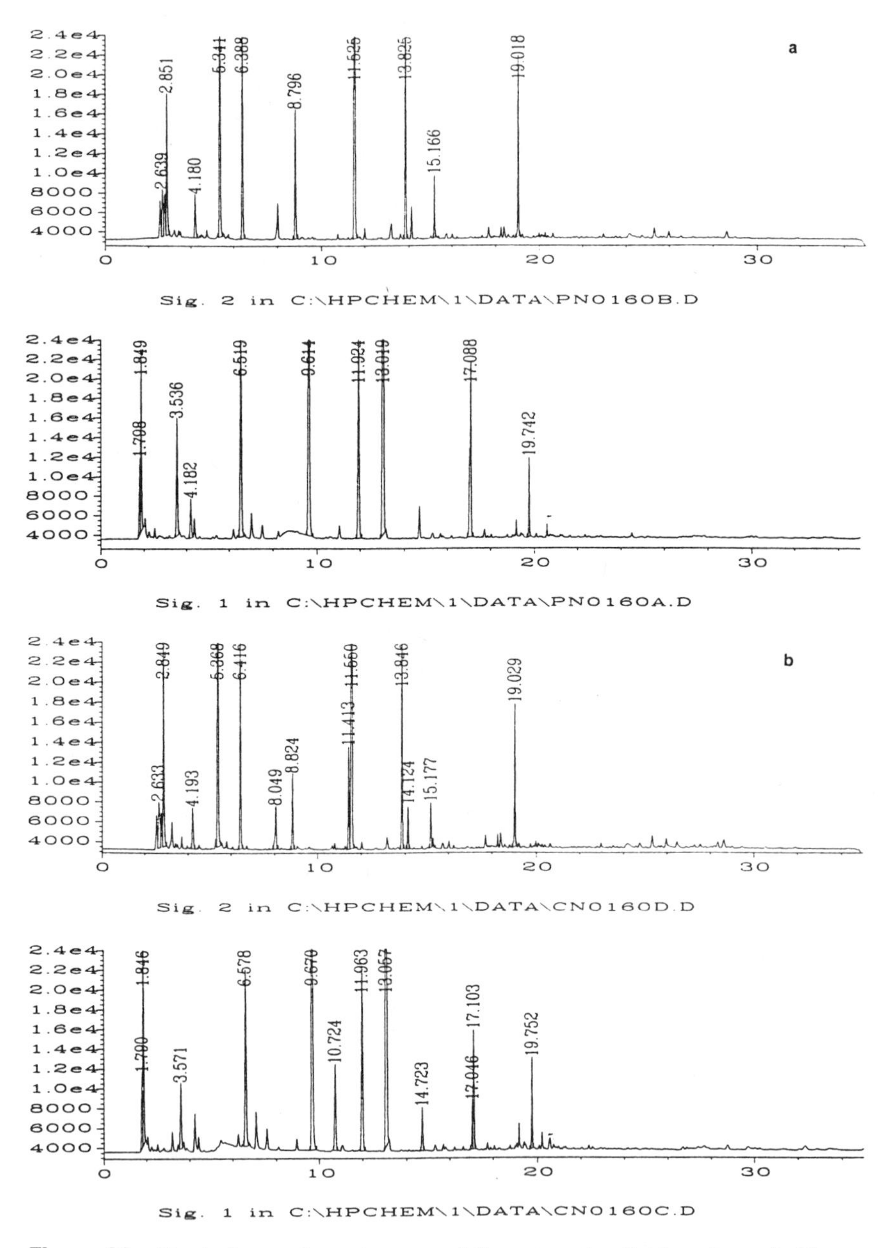

Figure 21. Pyrolysis gas chromatograms of the automotive finishes shown in Figure 20. The pyrograms were obtained on a Hewlett-Packard 5890 gas chromatograph employing dual fused-silica capillary columns with flame ionization detectors and a Chemical Data Systems coil probe pyrolyzer employing a final pyrolysis temperature of 750°C with the heating-rate ramp turned off and a pyrolysis time of 20 s. Signal 1 depicts the pyrogram from the 25-m high-polarity FFAP column, while signal 2 depicts the pyrogram from the 30-m low-polarity methyl silicone column.

found in the type or quantity of drying oils employed. The drying oils, being composed of straight-chain saturated and unsaturated hydrocarbons, all have similar infrared spectra. If similar, yet different types of oil (e.g., fish oil and linseed oil) are used in one alkyd resin formulation compared to another, there will be little to no perceivable change in the infrared spectrum. The binders used in acrylic lacquers for the automotive industry typically employ polymethyl methacrylate as the primary polymer, with a variety of other acrylates and acrylate derivatives being combined to modify the resin's physical properties. A polyester external plasticizer may also be used. If one examines the infrared spectra of various acrylates and methacrylates, it becomes apparent that there is not a large difference between many of them. In mixtures, the distinctive infrared absorption bands of these minor modifying acrylates or polyester plasticizers tend to be masked by those of the polymethyl methacrylate. It then becomes quite difficult to conclude that the very minor differences observed between spectra are indeed significant differences indicating different formulations. Consequently, discrimination power between these binders suffers. An example can be found in Figures 22 through 24. Figure 22a and b are the spectra obtained from two automotive acrylic dispersion lacquers of the same color, the first manufactured by PPG and the second by Inmont. Small yet reproducible differences can be noted in the absorption band intensities around 1370 and 1020 cm^{-1}. Although these differences are sufficient to discriminate between the two finishes, the pyrolysis gas chromatograms of the respective coatings presented in Figure 23a and b leave no doubt of the differentiation of the two. Figure 24a contains the infrared spectrum of the Dupont formulation for the same colored paint and is quite another matter. It is very similar to the infrared spectrum of the PPG formulation in Figure 22a and makes discrimination of the two paints tenuous, if not impossible. The pyrolysis gas chromatogram of the DuPont finish appears in Figure 24b. Careful comparison of the pyrograms in Figures 23a and 24b reveal small, yet reproducible differences in the intensity of the pyrolysis fragmentation products in the 13.0-, 19.0-, 21.5-, and 27.7-min retention time regions, to mention a few. A reproducible difference in the ratio of the 2.4-min peak to the 11.3-min peak can also be noted. Although this differentiation is not straightforward and requires several runs to establish reproducibility, it does afford reliable discrimination of the two resins, which cannot be established by infrared microspectroscopy alone.

The presence of high concentrations of extender pigments can cause problems in binder comparisons just as they do in binder classifications. If the resins are quite similar, as often occurs in the PVA–acrylic and acrylic latex architectural paints, the extender pigments may mask the areas crucial for binder differentiation. Additionally, not all extender pigments having similar chemical compositions, yet different crystalline forms can be differentiated by infrared spectroscopy. This is especially true when the extenders are present as a mixture

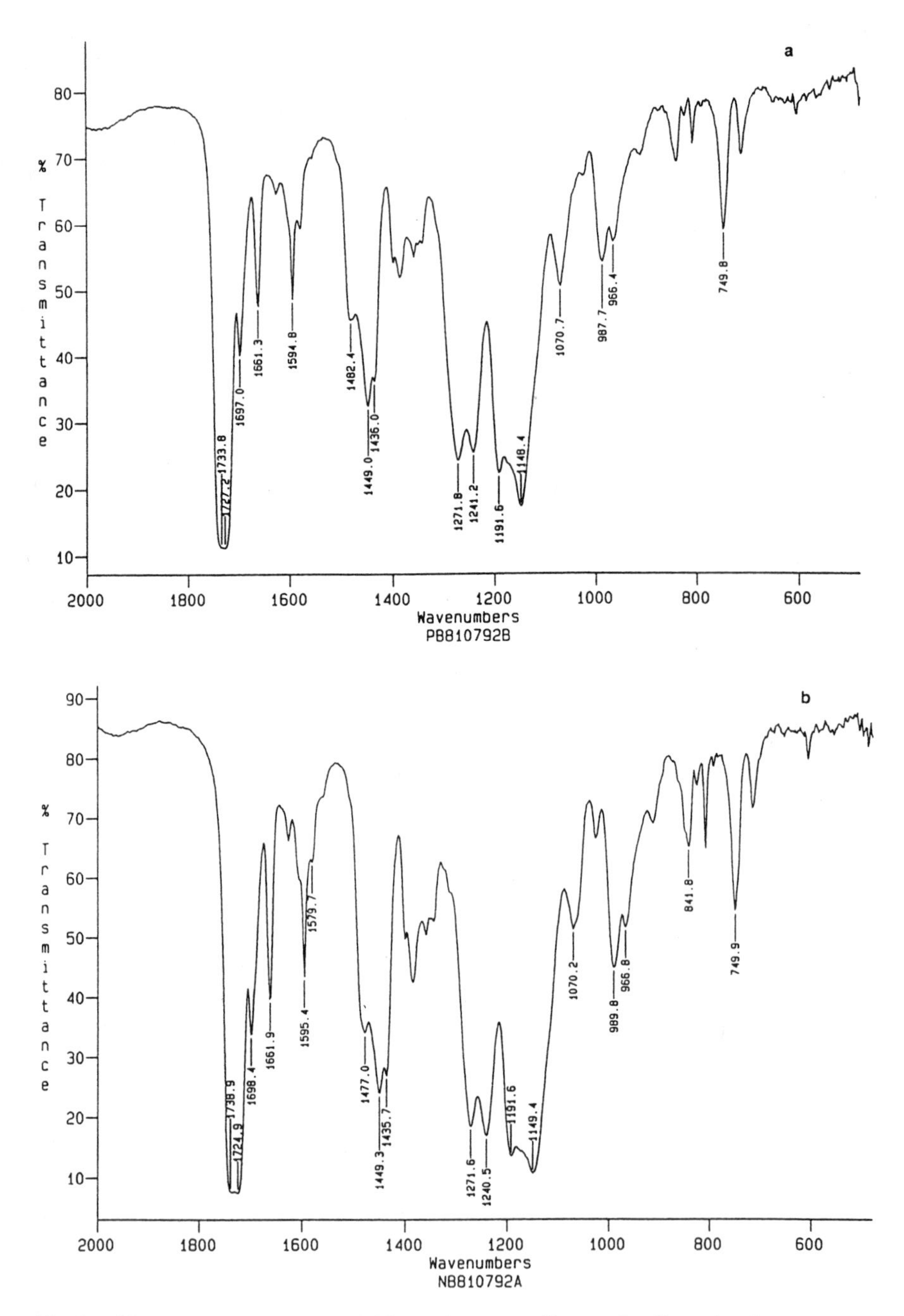

Figure 22. Infrared spectrum of (a) a red nonmetallic acrylic dispersion lacquer automotive original top coat manufactured by PPG for use on General Motors Corporation vehicles in 1981 and (b) the same color red nonmetallic acrylic dispersion lacquer automotive original top coat manufactured by Inmont Corporation for use on General Motors Corporation vehicles in 1981.

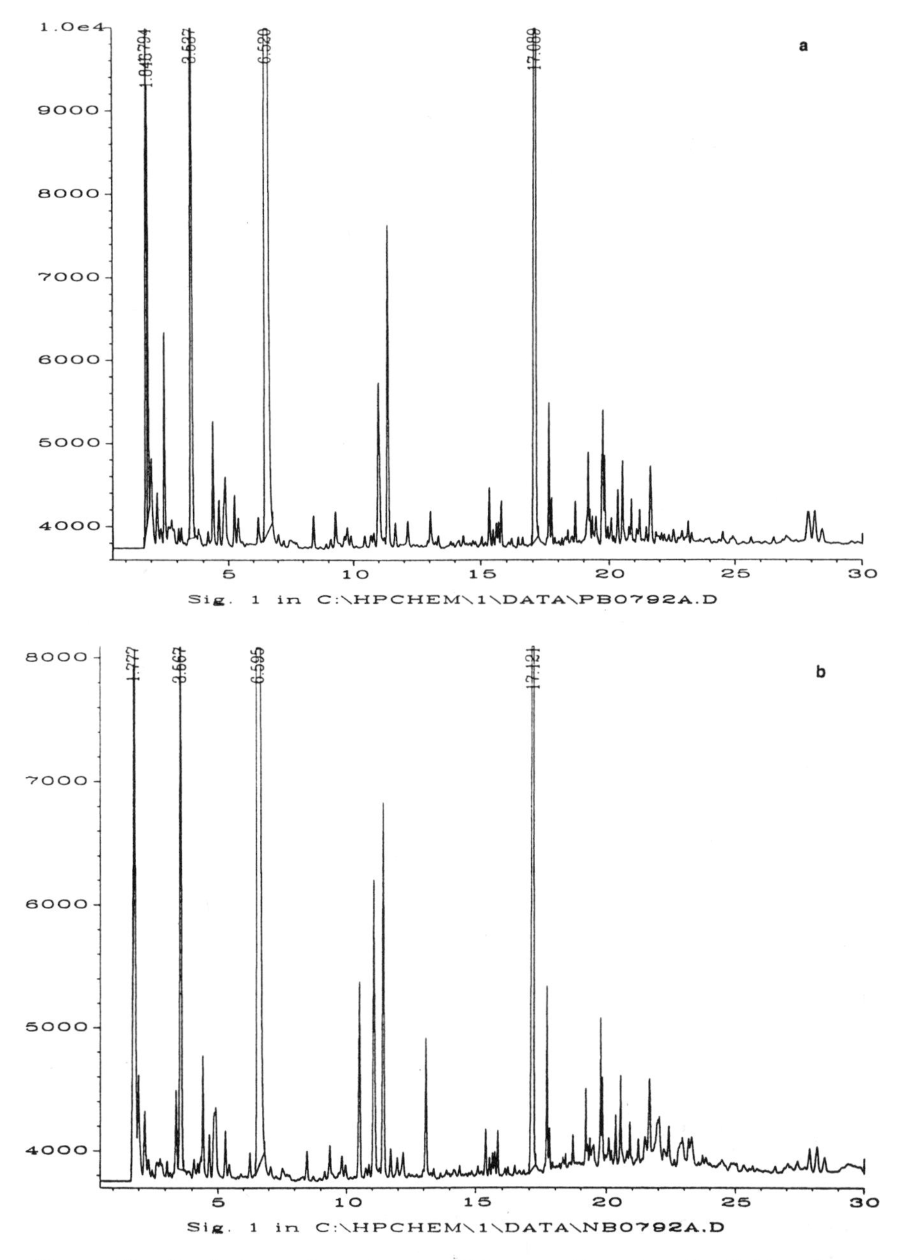

Figure 23. Pyrolysis gas chromatograms of the respective automotive finishes shown in Figure 22. The pyrograms were obtained on a Hewlett-Packard 5890 gas chromatograph employing a high-polarity 25-m FFAP fused-silica capillary column with a flame ionization detector and a Chemical Data Systems coil probe pyrolyzer employing a final pyrolysis temperature of 750°C with the heating-rate ramp turned off and a pyrolysis time of 20 s.

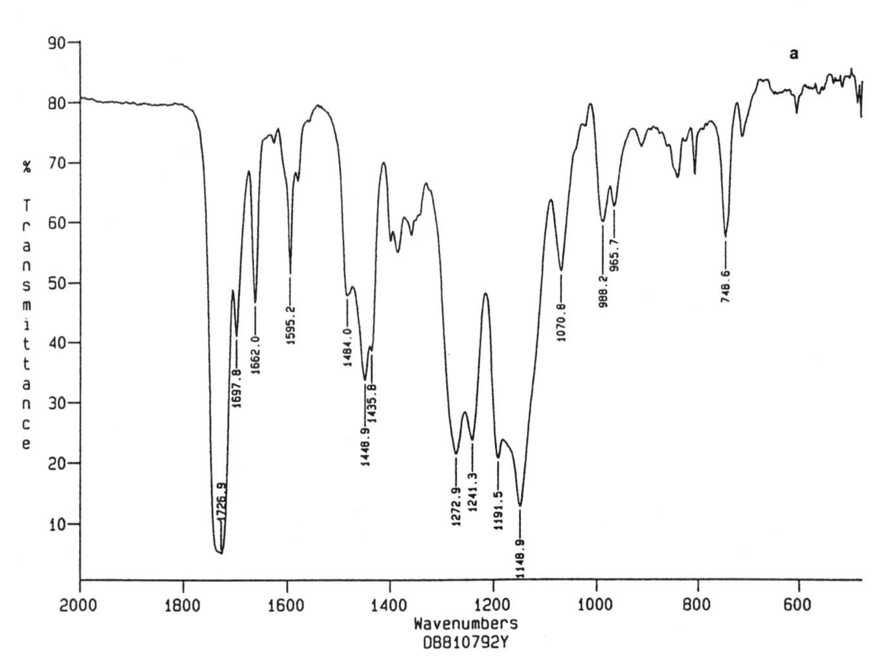

Figure 24. (a) Infrared spectrum of the same color red nonmetallic acrylic dispersion lacquer automotive original top coat as depicted in Figures 22 and 23 but manufactured by DuPont for use on General Motors Corporation vehicles in 1981. (b) Pyrolysis gas chromatogram of the DuPont formulation of the red nonmetallic acrylic dispersion lacquer automotive original top coat. The pyrogram was obtained on a Hewlett-Packard 5890 gas chromatograph employing a high-polarity 25-m FFAP fused-silica capillary column with a flame ionization detector and a Chemical Data Systems coil probe pyrolyzer employing a final pyrolysis temperature of 750°C with the heating-rate ramp turned off and a pyrolysis time of 20 s.

of two or three varieties, with one being present in a relatively low concentration compared to the others.

Thus infrared microspectroscopy should be viewed as a valuable tool in the battery of complementary analytical techniques available to the forensic coatings analyst. Burke et al. reported the results of a study comparing the discrimination capabilities of infrared spectroscopy, pyrolysis gas chromatography (PGC), and pyrolysis mass spectrometry (PMS) when used to analyze approximately 60 unpigmented paint binders representing a broad range of chemical types [31]. They confirmed that infrared spectroscopy is admirably suited for classifying paint resins. Although they found that infrared spectroscopy was not as well suited for discrimination as were the other two techniques, it was pointed out that each technique yields different information about the resins, with infrared

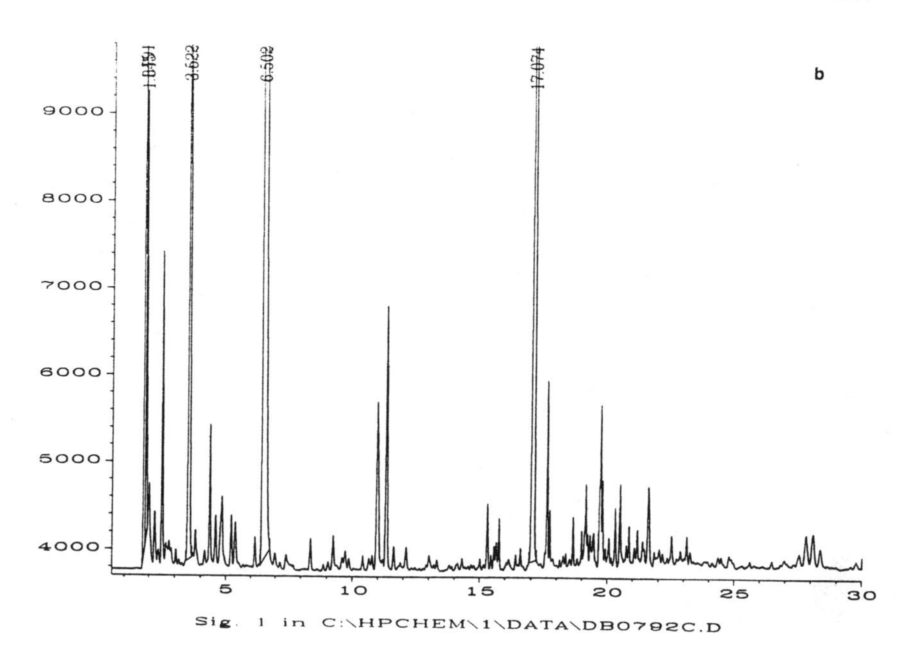

spectroscopy occasionally providing classification of binder modifications when the other techniques could not. Fukuda used infrared spectroscopy and pyrolysis gas chromatography to attempt to discriminate between a series of white top coats and clear coats found as original finishes on Japanese automobiles [39]. He, too, found infrared spectroscopy to be useful, yet inferior in discrimination potential. Both of these studies, however, did not address the positive influence of pigments on the discrimination potential of infrared spectroscopy. Burke et al. point out that coloring pigments and extenders will probably hamper the ability to classify paint binders [31]. But at the same time, this disadvantage for classification ability may in some instances turn out to be an advantage for discrimination. Neither PGC nor PMS reflect variations in the inorganic pigment-to-binder ratio, as does infrared spectroscopy. Infrared spectroscopy is also quite sensitive to different organic pigments. Both of these points would appear to improve the technique's ability to differentiate between similar paints. May and Porter published a study of 31 household gloss paints (most probably alkyd enamels) in 1975 which supports this thought [40]. The 11 white, 10 red, and 10 green samples were prepared by 10 separate manufacturers to meet a British Standard color specification. The authors reported greater discrimination by infrared spectroscopy than by PGC in both the red and green groups, although they were using the less discriminating packed column versus capillary column pyrolysis gas chromatography. Discrimination capability for infrared spectroscopy was double that for PGC in the green paint group. It must be kept in mind, of course, that these results are based on a particular end-use type of resin and may not be justifiable for other application types, such as the original automotive finish coats discussed previously.

In summary, infrared microspectroscopy is a fast nondestructive technique having the capability for individual layer comparisons on casework-sized samples. It can provide information about both the binder and the pigment portions of a cured paint specimen, as well as their relationship to one another. It possesses the ability to discriminate between a great number of paints having similar colors and similar binders. As such, it should be an integral part of any forensic paint comparison, with its results used in conjunction with those of complementary analytical techniques to reach a conclusion as to potential source.

3.7. Complementary Techniques

The most basic technique, which is complementary to any type of trace evidence examination, is of course that of microscopy. The mere size of the forensic sample usually requires its use to some extent; however, the power of microscopy can be extended beyond that afforded by just stereomicroscopy alone. Both bright-field and polarized-light microscopy can be invaluable in the recognition of various pigments or extenders based on their morphology and birefringent characteristics. Full characterization often requires that the pigment be separated or at least dissociated from the binder. In the case of architectural latex and low-gloss alkyd enamels, this can usually be accomplished by placing a small sample of the paint into a drop of two separate Cargille refractive index oils (typically, 1.53 and 1.60 nDs) each on a separate glass microscope slide, fragmenting and dispersing the specimens with a drawn glass rod, and subsequently compressing them under glass coverslips with a rotating motion. The task is much more difficult and often impossible in the case of automotive paints. In these situations, thin peels of a paint layer may be placed on a microscope slide, covered with a small drop of mounting medium or refractive index oil and compressed between a coverslip and the slide, thus permitting the analyst to at least characterize the size and distribution of the pigment grains, extender grains, and decorative metal flake. Recognition of these morphological characteristics becomes increasingly important to paint discrimination when the number of layers in the specimen is limited.

As mentioned and demonstrated in Section 3.6, pyrolysis gas chromatography and pyrolysis mass spectrometry can provide a detailed ‘‘signature’’ of the binder along with its additives and will typically provide improved discrimination in comparison situations. That is not to imply that they are useless in effecting binder classifications, for they are the saving grace when heavy extender pigment loads obscure the binder characteristics of an infrared spectrum.

They essentially perform an in situ extraction of the organic portion from the inorganic portion of the paint and consequently avoid pigment interference. Acrylic latex, PVA–acrylic latex, and alkyd enamel binders may then be identified easily in low-gloss (flat) architectural finishes at the same time that detailed reproducible comparisons are being undertaken. Furthermore, PGC enables the analyst to classify automotive acrylic lacquers as dispersion or solution types, thus corroborating microsolubility tests and facilitating recognition of many General Motors original automotive finishes.

The complications encountered in comparing and classifying inorganic coloring and extender pigments by infrared microspectroscopy have been mentioned previously. To combat these shortfalls, it is prudent to employ some sort of elemental analysis technique in the overall analytical scheme. Situations do occur where finishes are indistinguishable by their infrared spectra yet differ in their elemental constituents. Such an example is shown in Figure 25, which contains the infrared spectra of two red nonmetallic polyester–melamine enamel automotive finish coats manufactured by PPG in 1988 and 1989. They are essentially indistinguishable given the normal precision afforded by the sample preparation method described in Section 2.3. The Figure 25a formulation was reportedly used on 1988 and 1989 Mazda MX6s produced by Mazda North America, while the Figure 25b formulation was reportedly used on 1989 Ford Probes supplied by Mazda North America [45]. The numerous sharp absorption bands riding on top of the broader binder absorption bands originate from red organic pigment. The colors are slightly different from one another if larger full-thickness samples are compared; however, discrimination is tenuous in the case of smaller thin peels. Furthermore, the samples are indistinguishable by dual-capillary-column pyrolysis gas chromatography. The energy-dispersive x-ray spectra of the respective finishes is presented in Figure 26 and indicates an obvious and reproducible difference in the titanium level in the two paints.

The forensic paint examiner will typically rely on either scanning electron microscopy in conjunction with energy-dispersive x-ray spectrometry (SEM-EDX), energy-dispersive x-ray fluorescence spectrometry (XRF), x-ray diffractometry (XRD), or some sort of optical emission spectrometry, such as direct-current (dc) arc-emission spectrography (DCES) or inductively coupled plasma emission spectrometry (ICP), to address this need. Occasionally, a combination of techniques might be used depending on the type of paint, such as using XRF in conjunction with SEM-EDX for the detection of low-level metallic driers in alkyd binders. Some of the advantages and disadvantages of each are discussed in detail by Thornton [1].

SEM-EDX is probably the most versatile of the techniques for the analysis of minute multilayered specimens. It offers rapid detection of most elemental constituents over a broad dynamic range, permits elemental profiles to be collected for each individual layer and is nondestructive. By rastering the electron

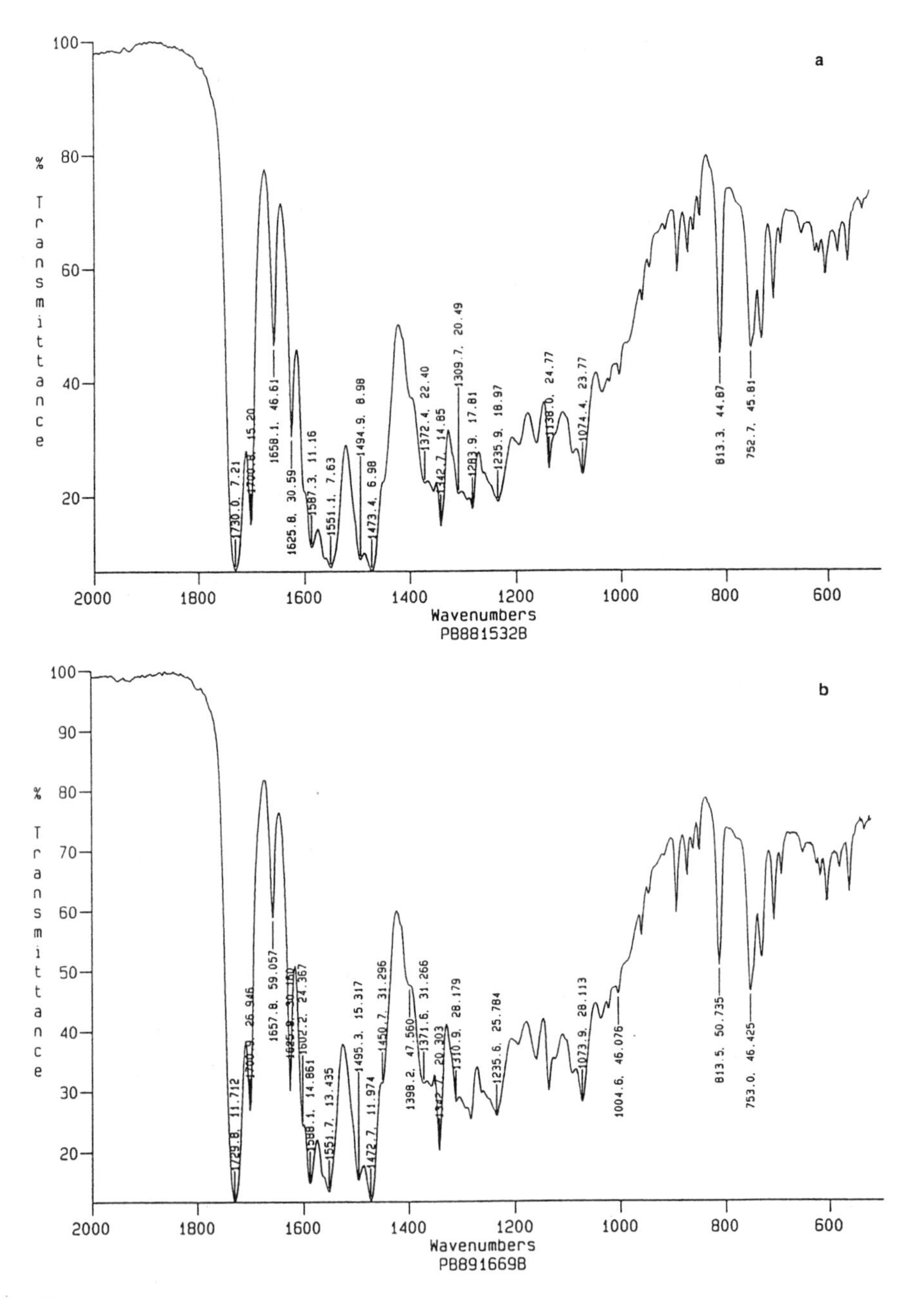

Figure 25. (a) Infrared spectrum of a red nonmetallic polyester–melamine enamel automotive finish coat manufactured by PPG for use on 1988 and 1989 Mazda MX6s produced by Mazda North America (CTS Automotive Paint Collection number PB881532). (b) Infrared spectrum of a red nonmetallic polyester–melamine enamel automotive finish coat manufactured by PPG for use on 1989 Ford Probes supplied by Mazda North America (CTS Automotive Paint Collection number PB891669).

20-JAN-94 09:07:03 EDAX READY
RATE= 0CPS TIME=1552LSEC
FS= 13661CNT PRST= OFF
A =PB88-1532 L1/TOP/NO PRIMER

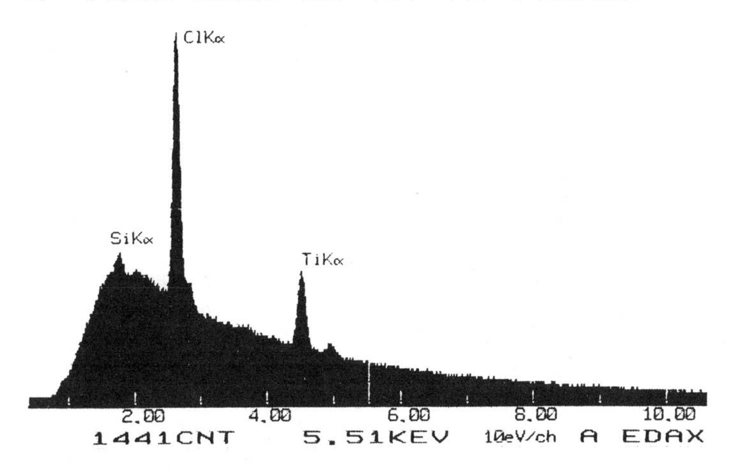

20-JAN-94 09:09:00 EDAX READY
RATE= 0CPS TIME= 795LSEC
FS= 11779CNT PRST= OFF
B =PB89-1669 L1/TOP/NO PRIMER

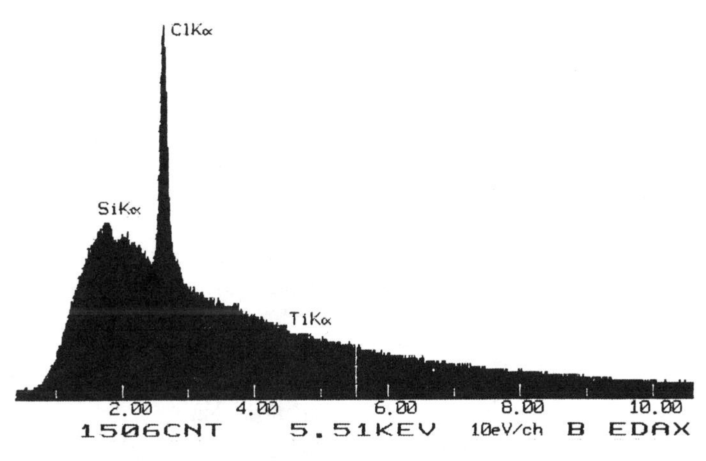

Figure 26. Energy-dispersive x-ray spectra of the two respective finishes whose infrared spectra appear in Figure 25. The spectra were obtained on a Leica-Cambridge S360 scanning electron microscope with an EDAX PV9800-ECON 4 energy-dispersive x-ray spectrometer employing a 20-kV beam potential, 23-mm working distance, 0° stage tilt, and beryllium-windowed EDX detector. The specimens consisted of carbon-coated single-layer paint chips mounted on aluminum specimen stubs using carbon adhesive disks.

beam over a large area of each exposed layer, the resulting elemental spectra can be used for layer-by-layer comparisons of questioned and standard samples since the data reflect the type of extenders and pigments present as well as their relative quantities. Elemental profiles can also be used to deduce which specific extenders might be present when used in conjunction with the data collected via infrared microspectroscopy. Also, occasionally, the x-ray spectrum will indicate the presence of an additional extender not noted in the infrared spectrum, such as the appearance of a pronounced potassium peak in the spectrum of a typical aluminum silicate, suggesting the presence of mica. If there are a limited number of layers present in a paint fragment, providing few discriminating characteristics, the intact chip may be subjected to low-temperature oxygen plasma ashing to ablate the binder resin from the extender and pigment particles. The high spatial resolution of SEM-EDX then permits individually exposed grains to be analyzed for their morphology and elemental constituents [41]. In that EDX analysis alone cannot identify the crystalline form, consideration of the data obtained from infrared microspectroscopy often confirms the extender identification. The two techniques truly work hand in hand. Furthermore, most SEM-EDX systems permit x-ray dot mapping to be carried out. This approach can visually depict the spatial distribution of various elements throughout a paint layer cross section and can be useful in corroborating the presence of pigment settling originally detected either microscopically or by variations in the infrared spectra of several thin peels taken from various depths within a single layer.

The final technique which should be mentioned is that of ultraviolet/visible light microspectrophotometry (UV-VIS). The technique's application to forensic paint examination is relatively new, with the bulk of the initial research work being done by the British Home Office Central Research Establishment. In cases where extremely limited sample is present, comparison by UV-VIS microspectrophotometry aids in the recovery of discrimination potential lost due to poor visual color perception of minute or thin specimens. Although no work has been published indicating an increase in discrimination afforded by this technique following use of the previously mentioned analytical techniques, UV-VIS microspectrophotometry has been shown to offer promise for the compilation of data bases targeted for vehicle finish make/model identifications [42–44]. When used in conjunction with similar data bases constructed from infrared microspectroscopy data, a powerful team should emerge.

3.8. Reference Collections and Databases

One of the benefits stemming from the reproducibility of infrared microspectroscopy is its use for reliable computerized databases. An in-house spectral library built from a collection of specimens representing the end-use application of interest provides the ideal database. Concerns over sample preparation, instrumental, and search algorithm variations are then put to rest. Unfortunately, collecting the number of specimens required and performing the analyses is not always practical or possible. For this reason, prepared data bases or spectral atlases may be the only alternative. Both approaches can quickly solve classification problems and provide indications of potential end-use applications.

Although prepared reference sample collections of architectural finishes do not exist, with some diligence the analyst can collect samples of known binder and extender types from friends, family, and associates and ultimately build quite a nice collection. A spectral data base constructed from these samples may aid in classifications of binders and extenders. Unfortunately, there are too many architectural paint manufacturers to permit collection of all formulations for the purpose of identifying the manufacturer of a questioned paint sample. This variety can be of benefit to the forensic coatings examiner, however, when considering the evidential significance of two ''matching'' architectural paints. The more variation there is in the sample population, the less likely the chance for random overlap. This translates to fewer potential sources for the same kind of paint and therefore increases the likelihood that the paint sample questioned originated from the source represented by the standard sample.

In the case of automotive paints, the opportunity to build a database is not quite so bleak. Collaborative Testing Services has marketed a collection of original automotive finish coats applied to domestically manufactured vehicles from 1974 to 1989 [45]. Starting in 1975, the samples included the various chemical formulations used by different paint manufacturers to produce the same color paint for a given vehicle manufacturer. The collection came in two parts, book and chemical. The book collection consisted of one of the paint manufacturer's formulations for each color used, each sample being checked for accurate color rendition against the vehicle manufacturer's color standards. The chemical collection consisted of the various formulations used by other paint manufacturers for each book collection color. In the late 1980s, domestically manufactured foreign vehicle paints were included in the collection. Sadly, Collaborative Testing Services discontinued manufacturing any form of the Automotive Paint Collection in 1991. The only other broad-based domestic automotive paint collection currently in existence is maintained by the Federal Bureau of Investigation (FBI) Laboratory in Washington, D.C., and is referred to as the National Automotive Paint File (NAPF). Access to this collection is granted only to law enforcement agencies, due to funding restraints.

Reference collections of polymers used in paints as well as in materials encountered along with paint evidence are also available. The ResinKit[3] and Scientific Polymer Products, Inc.'s Polymer Collection[4] are excellent starting places. Scientific Polymer Products, Inc., also offers a Plasticizer Kit, which

---

[3]The ResinKit Company, 1112 River Street, Woonsocket, RI 02895.

[4]Scientific Polymer Products, Inc., 6265 Dean Parkway, Ontario, NY 14519.

provides samples of a broad range of the various types of plasticizers in use. To augment these collections, samples of actual products can usually be obtained by requesting them directly from the various manufacturers.

Considering that these reference sample collections are starting points in building a comprehensive database of infrared spectra, most examiners will also choose to acquire prepared databases marketed by various suppliers. Spectral atlases are offered by several companies, including Aldrich and Sadtler. A comprehensive coatings infrared spectral atlas of binders, extenders, coloring pigments, plasticizers, solvents, and additives is published by the Federation of Societies for Coatings Technology and should be near at hand for every forensic paint examiner [10]. Digital spectral libraries are also available from these same sources, with *An Infrared Spectroscopy Atlas for the Coatings Industry* collection [10] currently being available only from Nicolet Analytical Instrument Corporation and Spectra-Tech, Inc. Personal computer (PC)–based operating systems are now becoming commonplace for the various FTIR spectrometer manufacturers, however, which creates new avenues for sharing and acquiring digital libraries previously constrained to a particular FTIR vendor. Microsoft Corporation's Windows software supports several vendors' infrared manipulation and search programs, which usually have import–export facilities to convert acquired spectra to a format compatible with the chosen digital library and search software.

Access to a digital library for automotive finish coats is another matter. Tillman, formerly of the Georgia Bureau of Investigation, has developed a digital infrared spectral database currently containing approximately 2800 of the 3700 paint samples contained in the Collaborative Testing Services' (CTS) Automotive Paint Collection and evaluated its success for determining the make and models of vehicles on which the paint was reportedly used [46]. The database is available through most infrared spectrometer vendors, although care should be taken as to which version of the database one acquires. Initial versions, containing approximately 2000 entries, contained spectra obtained from only the chemical file samples with none of the book collection samples being represented. The book collection samples are very important, for they were typically manufactured by the major paint supplier for a given automotive paint color. Even though his spectra were acquired on a Nicolet 20SXB spectrometer with a first-generation Spectra-Tech Spectrascope, a direct comparison of database spectra with those acquired from corresponding CTS samples analyzed on the author's Digilab FTS-40 FTIR spectrometer with a Digilab UMA-300A infrared microscope accessory revealed very good correspondence of even minute spectral detail. Unfortunately, even after selecting "hits" based on color and metallic–nonmetallic characteristics, this author has not experienced quite the success reported by Tillman and Bartick [46], especially with the more common blue and green colors. Perhaps this may be attributed to variations in the different infrared spectrometer manufacturer's search software. Tillman and Bartick also stress the importance of preparing a thin enough sample to avoid excessive absorbance of the 1730-cm^{-1} carbonyl band as well as collecting a sufficient number of scans to minimize noise effects. Baseline correction is a necessity, especially if Euclidean distance or absolute-value search algorithms are being used. They tend to see the area under a drifting baseline as a real artifact of the material analyzed. Another problem occurs in the decision of how to handle clear coat/base coat specimens. Several of the database spectra representing clear-coated nonmetallic finishes were apparently acquired from the clear coat and not the color coat. This is evidenced by the absence of any titanium dioxide absorbance in a "white" paint and coloring pigment in "red" paint. Infrared spectra acquired from the clear coats of several of these CTS paint samples correspond to the library spectra. The variation in clear-coat spectra from one paint manufacturer to another is significantly less than that in the color coats. Furthermore, the variation in clear-coat spectra from one color to another for a given paint manufacturer is often negligible. This can hamper discrimination power if colored base coats are not included in the database. Preparing the existing database was a monumental undertaking, however, and its utility has certainly been demonstrated. Of course, one should always treat the results of a database search with caution, keeping in mind that it does not contain many of the domestic finishes used (900 out of 3700 in the CTS collection alone) as well as foreign finishes (which account for approximately 35% of the vehicles on the road). What will the future bring, now that the CTS collection has been discontinued?

Perhaps time will see the development of a nationally accessible digital infrared spectral database constructed from samples in the FBI Laboratory's National Automotive Paint File. The Bureau continues to expend quite a bit of effort in maintaining their sample collection, and vehicle manufacturers are more prone to cooperate with one agency rather than numerous ones. Only time will tell. Until then, complete queries including chemical characteristics will probably have to include sample searches conducted by the FBI Laboratory.

4. CONCLUSIONS AND SUMMARY

The value of infrared microspectroscopy to the forensic coatings examiner has been discussed and demonstrated. The technique is one of the most popular instrumental methods currently in use for coatings examinations in most crime laboratories throughout the world. Its success is due to many attributes, including reproducibility, speed, sensitivity such that individual layer analysis on minute multiple-layer paint fragments is an easy task, simultaneous access to binder and pigment information in many instances, amenability to binder and pigment classifications, ability to search digital libraries, and its nondestructive nature.

The conclusions drawn from the use of infrared microspectroscopy answer many of the questions proposed in Section 1.1 addressing the nature of the forensic paint examination problem. When used in conjunction with complementary analytical techniques, it can provide a strong foundation for the conclusion that a questioned paint sample could have come from, or probably came from, a particular source. Studies have been reported to help the forensic paint examiner decide when to use instrumental analyses in addition to microscopic examination and to indicate what significance those additional techniques add to the paint evidence [47–49]. But infrared microspectroscopy stands alone as the only instrumental method used routinely no matter how complex the layer structure of a paint specimen might be. It has earned this esteemed stature because it permits a more complete description of each paint layer, going beyond that of just color. It demonstrates that the analyst has a reasonably comprehensive understanding of paint chemistry, as called for by Thornton [1], and goes a long way in establishing the credibility of the criminalist as a forensic coatings examiner, as pined for by Rodgers et al. [2]. For the forensic paint examiner, it is a technique that is here to stay.

ACKNOWLEDGMENTS

My thanks to Marianne Hildreth and Jan Taylor for carefully reviewing this manuscript and spending hours convincing me to change this or add that. As forensic paint examiners, they, Tammy Jergovich, and Lynn Henson continue to share casework and background experiences with me and have played no small part in the collection of the information presented herein. A special note of appreciation also goes to Warren Tillman, formerly of the Georgia Bureau of Investigation, for his hours of work in developing the initial version of the automotive paint binder infrared classification flow chart and incorporating many of the recommendations given. The current version includes some additional fine tuning. I would also like to thank Jim Corby, of the FBI Laboratory, for his years of support and willingness to share information acquired from automotive paint manufacturers. Through combined efforts, we all benefit.

REFERENCES

1. Thornton, J. I. (1982). Forensic paint examinations, in *Forensic Science Handbook* (R. Saferstein, ed.), Prentice Hall, Englewood Cliffs, N.J., pp. 529–571.
2. Rodgers, P. G., et al. (1976). The classification of automotive paint by diamond window infrared spectrophotometry, Part II: Automotive topcoats and undercoats, *Canadian Society of Forensic Science Journal*, *9*, pp. 49–68.

3. Units 3–22, *Federation Series on Coatings Technology*, Federation of Societies for Coatings Technology, Philadelphia.
4. Morgans, W. M. (1984). *Outlines of Paint Technology*, Vols. 1 and 2, Charles Griffin & Company, Ltd., London.
5. Martens, C. R. (1981). *Waterborne Coatings: Emulsions and Water-Soluble Paints*, Van Nostrand Reinhold Co., New York.
6. Oil and Colour Chemists' Association (1976). *Introduction to Paint Technology*, 4th ed., Watford Printers Ltd., Watford, Herts, England.
7. Crown, D. A. (1968). *The Forensic Examination of Paints and Pigments*, Charles C Thomas, Springfield, Ill.
8. Madson, W. H. (1967). In *White Hiding and Extender Pigments* (W. R. Fuller, ed.), Federation Series on Coatings Technology, Federation of Societies for Paint Technology, Philadelphia.
9. Nielsen, H. K. R. (1984). Forensic analysis of coatings, *Journal of Coatings Technology*, *56*(718), pp. 21–32.
10. Infrared Spectroscopy Committee, Chicago Society for Coatings Technology (1980). *An Infrared Spectroscopy Atlas for the Coatings Industry*, Federation of Societies for Paint Technology, Philadelphia.
11. Mathias, L. J. (1985). Analyzing coatings with FT spectroscopy, *Modern Paint and Coatings*, Nov., pp. 38–50.
12. Harkins, T. R., et al. (1959). Identification of pigments in paint products by infrared spectrophotometry, *Analytical Chemistry*, *31*(4), pp. 541–545.
13. Rodgers, P. G., et al. (1976). The classification of automotive paint by diamond window infrared spectrophotometry, Part I: Binders and pigments, *Canadian Society of Forensic Science Journal*, *9*, pp. 1–14.
14. Rodgers, P. G., et al. (1976). The classification of automotive paint by diamond window infrared spectrophotometry, Part III: Case histories, *Canadian Society of Forensic Science Journal*, *9*, pp. 103–111.
15. Humecki, H. J. (1987). Specimen preparation for microinfrared analysis, in *The Design, Sample Handling, and Applications of Infrared Microscopes* (P. B. Roush, ed.), ASTM, Philadelphia, pp. 39–48.
16. Deaken, D. (1975). Automotive body primers: their application in vehicle identification, *Journal of Forensic Sciences*, *20*(2), pp. 283–287.
17. Dabdoub, G., and Severin, P. (1989). The identification of domestic and foreign automobile manufacturers through body primer characterization, *Journal of Forensic Sciences*, *34*(6), pp. 1395–1404.
18. McBane, B. N. (1987). In *Automotive Coatings, Series II* (D. Brezinski and T. J. Miranda, eds.), Federation of Societies for Coatings Technology, Philadelphia.
19. Refinish Marketing Group (1993). *DuPont Refinish Color Book*, Domestic and Import Editions. E.I. duPont de Nemours & Co., Wilmington, Del.
20. Thornton, J. I., et al. (1983). Solubility characterization of automotive paints, *Journal of Forensic Sciences*, *28*(4), pp. 1004–1007.
21. Klug, F., et al. (1959). A microchemical procedure for paint chip comparison, *Journal of Forensic Sciences*, *4*(1), pp. 91–96.
22. Linde, H. G., and Stone, R. P. (1979). Application of the LeRosen test to paint analysis, *Journal of Forensic Sciences*, *24*(3), pp. 650–655.

23. Home, J. M., et al. (1983). The discrimination of small fragments of household gloss paint using chemical tests, *Journal of the Forensic Science Society, 23*(1), pp. 43–47.
24. McEwen, D. J., and Cheever, G. D. (1993). Infrared microscopic analysis of multiple layers of automotive paints, *Journal of Coatings Technology, 65*(819), pp. 35–41.
25. Gardiner, L. R. (1981). The homogeneity of modern household paints using the scanning electron microscope-energy dispersive x-ray analyses (SEM-EDXA), personal communication, Home Office Central Research Establishment, Aldermaston, Reading, Berkshire, England.
26. Wilkinson, J. M., et al. (1988). The examination of paint as thin sections using visible micro-spectrophotometry and Fourier transform infrared microscopy, *Forensic Science International, 38*, pp. 43–52.
27. Katon, J. E., et al. (1987). Instrumental and sampling factors in infrared microspectroscopy, in *The Design, Sample Handling, and Applications of Infrared Microscopes* (P. B. Roush, ed.), ASTM, Philadelphia, pp. 49–63.
28. Cartwright, et al. (1977). A microtome technique for sectioning multilayer paint samples for microanalysis, *Canadian Society of Forensic Science Journal, 10*(1), pp. 7–12.
29. Carl, R. (1992). The analysis of paint by stereomicroscopy and FTIR microscopy with an emphasis on binder classification and end-use characterization, *SAFS/SWAFS Combined Spring Seminar Workshop* (S. Ryland and R. Carl, eds.), Apr. 8–9, Shreveport, La.
30. Messerschmidt, R. G. (1987). Photometric considerations in the design and use of infrared microscope accessories, in *The Design, Sample Handling, and Applications of Infrared Microscopes* (P. B. Roush, ed.), ASTM, Philadelphia, pp. 39–48.
31. Burke, P., et al. (1985). A comparison of pyrolysis mass spectrometry, pyrolysis gas chromatography and infra-red spectroscopy for the analysis of paint resins, *Forensic Science International, 28*(3/4), pp. 201–219.
32. Hayes, (1984). Automotive finishing at general motors, *Proceedings of the International Symposium on the Analysis and Identification of Polymers*, Quantico, Va., U.S. Department of Justice, Federal Bureau of Investigation, Washington, D.C., pp. 4–5.
33. Jergovich, T., et al. (1988). Foreign automotive paints: are they really foreign? *1988 Fall Seminar of the Southern Association of Forensic Scientists*, Clearwater Beach, Fla.
34. Beckwith, N. P. (1973). Automotive refinishing: a world-wide review, *Paint and Varnish Production*, Apr., pp. 15–20.
35. DuPont Sales Division (1987). Cronar: the most advanced refinish system ever developed, *DuPont Refinisher News*, Mar./Apr., pp. 6–10.
36. DuPont Sales Division (1987). Cronar: a closer look, *DuPont Refinisher News*, May/June, pp. 5–7.
37. Bartick, E. G. (1987). Considerations for fiber sampling with infrared microspectroscopy, in *The Design, Sample Handling, and Applications of Infrared Microscopes* (P. B. Roush, ed.), ASTM, Philadelphia, pp. 64–73.
38. Daniels, W. W., and Kitson, R. E. (1958). Infrared spectroscopy of polyethylene terephthalate, *Journal of Polymer Science, 32*, pp. 161–170.

39. Fukuda, K. (1985). The pyrolysis gas chromatographic examination of Japanese car paint flakes, *Forensic Science International*, *29*(3/4), pp. 227–236.
40. May, R. W., and Porter, J. (1975). An evaluation of common methods of paint analysis, *Journal of Forensic Sciences*, *15*(2), pp. 137–146.
41. Brown, R. (1993). Light and electron microscopy of inorganic paint constituents, Proceedings of Scanning '93, *Scanning*, *15*, Suppl. III, p. III-35.
42. Cousins, D. R. (1989). The use of microspectrophotometry in the examination of paints, *Forensic Science Review*, *1*(2), pp. 141–162.
43. Cousins, D. R., et al. (1984). The variation in the colour of paint on individual vehicles, *Forensic Science International*, *24*, pp. 197–208.
44. Nowicki, J., and Patten, R. (1986). Examination of U.S. automotive paints: I. make and model determination of hit-and-run vehicles by reflectance microspectrophotometry, *Journal of Forensic Sciences*, *31*(2), pp. 464–470.
45. Collaborative Testing Services (1989). *Reference Collection of Automotive Paints*, Collaborative Testing Services, Herndon, Va.
46. Tillman, W. L., and Bartick, E. G. (1990). The evaluation of an infrared spectroscopic automobile paint data base, presented at the *12th International Meeting of Forensic Scientists*, Adelaide, Australia, Oct. 24–30.
47. Tippet, C. F., et al. (1968). The evidential value of the comparison of paint flakes from sources other than vehicles. *Journal of the Forensic Science Society*, *8*(2/3), pp. 61–65.
48. Gothard, J. A. (1976). Evaluation of automobile paint flakes as evidence, *Journal of Forensic Sciences*, *21*(3), pp. 636–641.
49. Ryland, S. G., and Kopec, R. J. (1979). The evidential value of automobile paint chips, *Journal of Forensic Sciences*, *24*(1), pp. 140–147.
50. Tillman, W. (in press). Paint identification systems, *Proceedings of the International Symposium on the Forensic Aspects of Trace Evidence*, U.S. Government Printing Office, Washington, D.C.

7

Forensic Examination of Synthetic Textile Fibers by Microscopic Infrared Spectrometry

Mary W. Tungol Hairs and Fibers Unit, FBI Laboratory, Washington, D.C.

Edward G. Bartick Forensic Science Research and Training Center, FBI Academy, Quantico, Virginia

Akbar Montaser The George Washington University, Washington, D.C.

1. INTRODUCTION

1.1. Fibers as Trace Evidence

The most important evidence in most criminal cases is not the smoking revolver or the blood-stained knife, but rather, small, often microscopic-sized bits of materials referred to as trace evidence. One or more types of trace evidence have been reported in 80% of criminal cases in a recent study [1]. Had evidence recognition and handling procedures been optimized, there would probably be one or more types of trace evidence found in nearly every case. Trace evidence can provide (1) important associations between people, places, and various objects, (2) a description of the occupation or environment of the principals in a case, and occasionally (3), unequivocal association between people, places, and things. Trace evidence can be divided into two major categories: fibrous substances and particulate matter. Fibrous substances may be further subdivided into human hair, animal hair, and fibers. The fiber category includes synthetic, vegetable, and mineral fibers. This chapter is concerned with synthetic fibers.

1.2. Traditional Methods for Forensic Fiber Examinations

Forensic fiber examinations typically are conducted in two phases: fiber identification and fiber comparison. Occasionally, the examination may involve identification alone, when the analyst is requested to provide information regarding the possible origin of a sample. Analytical approaches for the forensic analysis of textile fibers have been reviewed recently [2–5]. Of the many techniques discussed for the identification and comparison of fiber evidence, visible-light microscopy is used most frequently in the forensic science laboratory. Physical characteristics of the fibers, such as color, diameter, and cross-sectional shape, are easily observed with transmitted light. Optical properties such as birefringence and refractive index are quickly determined by polarized-light microscopy (PLM). The fibers' generic class (i.e., polyester, nylon, etc.) can usually be identified by examination of the optical properties alone. Microscopic examination may be supplemented by solubility testing, UV-visible microspectrophotometric analysis, examination of fluorescent properties, and a myriad of other techniques. A representative protocol for forensic fiber examinations is shown in Figure 1.

1.3. Role of Infrared Analysis in Forensic Fiber Examinations

Aside from the traditional methods, and among the analytical techniques reported for forensic examinations of fibers, infrared (IR) spectroscopy is the best established technique for identifying polymeric fibers. Infrared analysis provides more specific chemical information about polymer composition than optical microscopy, thus increasing the evidential value of the fiber match. (The term *match* is used in the sense that one fiber is the counterpart of the other in all the properties studied.) In addition, the American Society for Testing and Materials (ASTM) lists IR spectroscopy as the preferred method of analysis for identifying synthetic fibers, stating "Where the data are consistent and the spectra obtained and interpreted by an experienced spectroscopist, the infrared procedure has no known bias" [6].

Many investigators [7–15] have addressed IR spectroscopy as a technique for the identification of single textile fibers. Unfortunately, the traditional IR techniques, such as cast films, KBr pellets, and diamond cells coupled with beam condensers, exhibit three disadvantages for single-fiber analysis: sample preparation techniques are time consuming, sample preparation procedures are either destructive or require a large amount of sample, and long analysis times are necessary to obtain good-quality spectra. Consequently, IR analysis was not incorporated as a routine part of forensic fiber examinations in most laboratories until the introduction of Fourier transform IR instruments and microscope attachments in the mid-1980s.

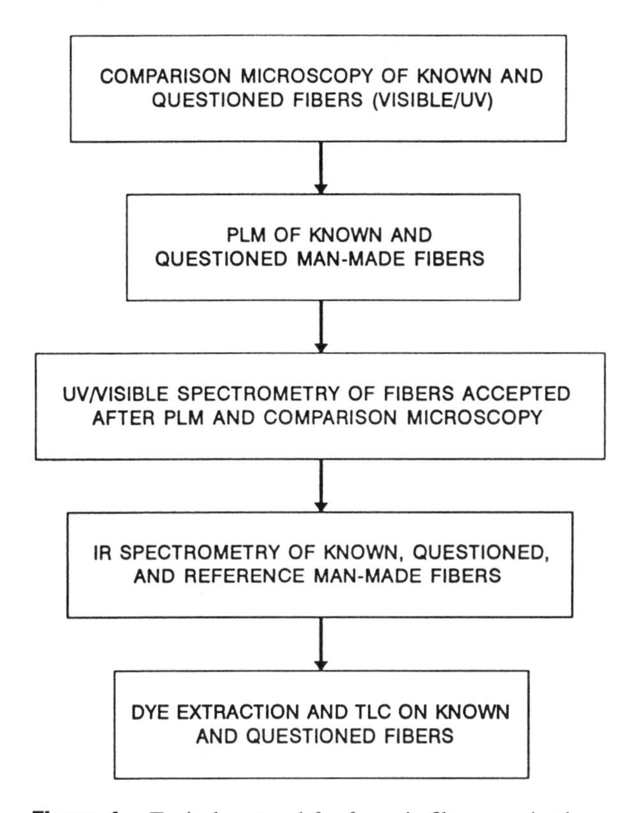

Figure 1. Typical protocol for forensic fiber examinations.

1.4. Microscopic FT-IR for Single-Fiber Analysis

Since the introduction of Fourier transform (FT) techniques coupled with microscope accessories in 1983, IR analysis of single fibers has expanded significantly. Rapid scan rates, multiplexing capabilities, and minimal sample preparation have greatly reduced the time and effort required to obtain IR spectra of single fibers. In addition, the technique is essentially nondestructive because the fiber (or a small portion of it) need only be flattened. These considerations have made IR analysis of single fibers feasible on a routine basis in the high-throughput setting of the forensic laboratory and have led to a renewed interest in the application of IR analysis to forensic examinations of fibers.

2. EXPERIMENTAL CONSIDERATIONS

2.1. Microscope Features

Many IR spectrometers and microscopes are commercially available. Most microscopes will fit on available spectrometers to permit fiber analysis. Before purchasing an IR microscope, its features must be examined carefully to ensure that the desired results are obtained. Radiation intensity in the IR region is relatively low and fiber samples are thin. Thus a high-throughput microscope should be used to enhance the signal-to-noise ratio (*S/N*) of the spectra in fiber analysis. The essential features of a microscope for fiber analysis are discussed below.

2.1.1. Reflecting Optics

The most important feature of an IR microscope is its optical configuration. The Cassegrainian optical design is used for the objective on most microscopes, and also for on-axis condensers. The standard magnification is 15× for the objective, while objectives of up to 36× are available. Higher magnification does not enhance energy throughput but does allow improved viewing, which is useful for examining fibers with diameters less than 20 μm. This capability is gaining more importance due to the increasing emphasis on microdenier fiber technology.

The radiation throughput of the Cassegrainian optics is controlled by the numerical aperture (N.A.):

$$\text{N.A.} = n \sin \frac{\text{AA}}{2} \tag{1}$$

where n is the refractive index of the medium above the specimen and AA, the angular aperture, is the maximum angle of light rays that pass through the optical element. As N.A. is increased, light collection is enhanced, and improved IR spectra are obtained. The numerical aperture of most 15× objectives available ranges from approximately 0.25 to 0.60. If the IR beam enters at the bottom of the microscope, the objective is located after the specimen in the beam path. For this configuration it is important to have a large N.A., to allow collection of transmitted light through irregularly shaped fibers. A condenser with a large N.A. is advantageous if the IR beam enters at the top of the microscope. Large N.A. must be used because smaller-diameter fibers are pushing the limitations of the instrument performance.

Infrared window materials, used to support or compress samples during analysis, defocus the beam due to refraction and cause spherical aberration. Because the specimen is no longer in the focal plane of the beam, less IR radiation reaches the fiber specimen, and the *S/N* ratio is reduced. Higher-quality spectra are produced when microscopes allow correction for spherical aberration (i.e., the sample and condenser are brought into sharp focus).

The minimum distance between two points that can still be distinguished in the observed image is controlled by the spacial resolving power of a microscope. This power is increased with N.A. In visible-light microscopy, the resolution is the fineness of detail observable with the microscope. Spatial definition in IR microscopy is analogous to spatial resolution in visual microscopy. Spatial definition is the ability to define an area for analysis in order to obtain spectral purity. Because N.A. is fixed for a given objective, the positioning of apertures must be changed to define the analysis area.

2.1.2. Apertures

The area of the sample to be analyzed in IR microscopy is isolated from the remainder of the field of view with one or two rectangular or circular apertures placed at focal planes. In general, rectangular apertures are better suited for masking fiber samples due to the fiber shape.

Apertures not only define an area of the sample for measurement, but they block unwanted radiation originating outside the fiber specimen. If stray light is allowed to reach the detector, absorption intensity is reduced. Further, if unwanted radiation passes through materials outside the defined area, additional absorption bands may appear in the spectrum. Sample isolation in the IR region is not all that simple. Diffraction results when light strikes a high-contrast edge. Diffraction effects are greater in the infrared range than in the visible range. The result of diffraction is light intensity (i.e., diffracted light), which appears in the region which visually appears to be blocked by the aperture edge. Diffraction effects in microscopic IR have been described in detail by Messerschmidt [16].

Two situations can occur when light diffraction occurs. First, if only the fiber edges are isolated by an aperture, the diffracted radiation will reach the detector through the surrounding air and absorption band intensities are reduced. This is illustrated clearly by the data in Figure 2A and B for a large polystyrene film (Figure 2A) and a 54-μm-wide strip of the same film. Note that only one aperture was placed after the sample in the beam path for the IR spectrum in Figure 2B. Bands that were totally absorbing in Figure 2A show a loss of intensity in Figure 2B. Figure 2C shows the spectrum of the same polystyrene strip obtained with one aperture located before the sample in the beam path. A loss of photometric accuracy is apparent under these conditions, particularly at longer wavelengths (lower frequency), where diffraction effects are greater. Any quantitative spectral measurements acquired under these conditions will result in significant errors. Note that the qualitative results are accurate.

The second diffraction problem occurs when the area defined by an aperture

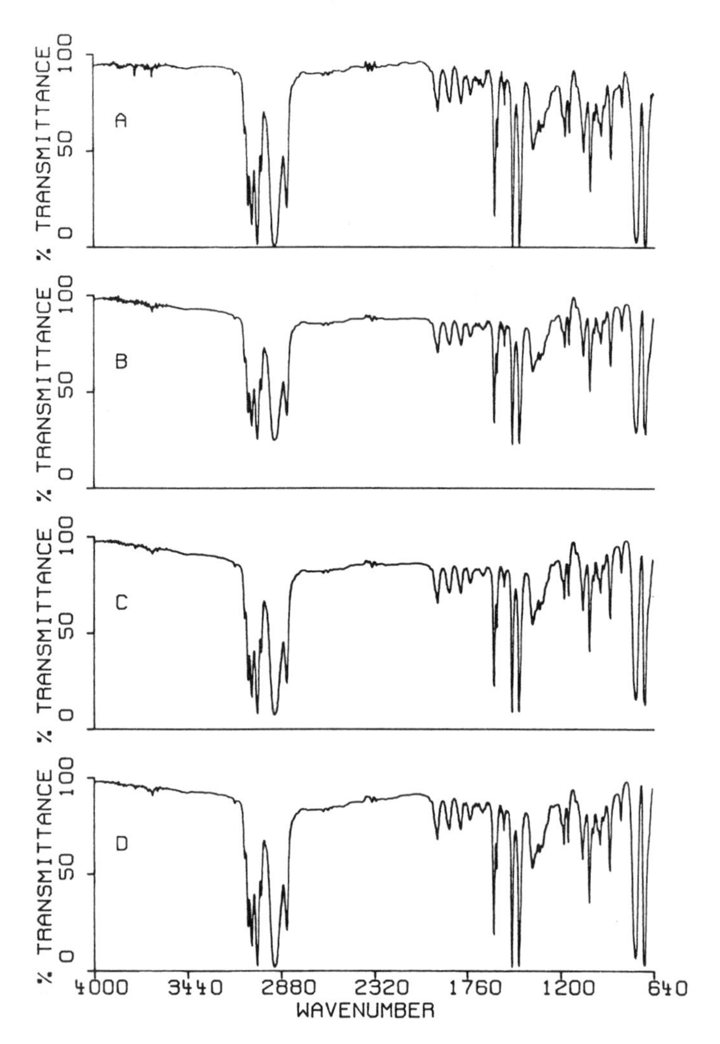

Figure 2. Infrared spectra of polystyrene film with different sample width and aperture conditions: (A) wide reference film; (B) 54-μm-wide ribbon, single aperture placed after the sample; (C) 54-μm-wide ribbon, single aperture placed before the sample; (D) 54-μm-wide ribbon, two apertures. (From Ref. 17.)

is a portion of a larger sample and diffracted light passes through the surrounding material. The spectral features of the surrounding material will appear in the spectrum of the area visually selected for examination. This problem occurs when analyzing one component of a bicomponent fiber. Bicomponent fibers are discussed further in Section 4.4. If the diffracted light passes through the surrounding sample, qualitative identification of most spectra can still be made provided that the surrounding material has been identified.

Significant improvement, however, can be made by using two image-plane apertures placed both before and after the sample in the beam path [18]. Figure 2D shows the spectrum of the same 54-μm-wide polystyrene strip discussed earlier, obtained using two apertures. The photometric accuracy is significantly improved. If one has to use only one aperture, placing it before the sample in the beam path produces better accuracy. With the aperture located before the sample, only the area of interest on the sample is illuminated by the source image [19]. Because diffraction spreads the light in an infinite pattern, even with two apertures, diffracted radiation has been observed 40 μm from a sample edge [19]. Although at this distance the diffracted radiation is of low intensity, multiple fibers should be spaced to allow collection of sample and background spectra at least 50 μm away from other fiber samples.

2.1.3. Detectors

The overall requirements of the laboratory usually dictate detector type. The most frequently used detector for IR microscopes is the mercury–cadmium–telluride (MCT) detector. The narrow-range MCT offers the greatest sensitivity but limits the lower end of the frequency range to approximately 750 cm^{-1}. Wide-range MCT detectors are four to eight times less sensitive than the narrow-range detectors but provide a lower-frequency limit of approximately 450 cm^{-1}. Because very small fiber samples are being analyzed, a narrow-range detector is usually used. However, the additional information obtainable in the lower-frequency region may warrant sacrificing sensitivity for range. A narrow-range detector will permit identification of all fiber types; however, the additional information gained with a wide-range detector may be useful. For samples other than fibers, such as those containing inorganic materials, the wide-range detector may be desirable.

2.2. Sample Preparation Methods

There are many methods for preparing and mounting fiber samples for analysis by microscopic IR spectroscopy. Choice of a method depends on the individual analyst, the particular requirements of the laboratory, and whether or not subsequent analyses may be required. Several methods of sample preparation are discussed below along with their advantages and disadvantages. To prevent band intensity changes in the spectrum, fibers must be oriented in the same direction.

2.2.1. Unflattened or As-Is Fibers

Unflattened fibers may be mounted across an aperture or placed on an IR window and analyzed "as is." The aperture may simply be a hole in a metal disk with the fiber mounted on double-sided adhesive tape. Adhesive paper sample holders that have small apertures are commercially available. The cross-sectional shape of the sample is preserved and there is no alteration of the fine structure of the fiber or the spectrum.

This method suffers from two major disadvantages. First, fibers greater than 50 μm in diameter are generally too thick for acquiring good-quality spectral data [20]. Although polymer films thicker than 5 μm are typically too thick to yield good spectra, good-quality spectra can be obtained from fibers up to 50 μm in diameter [20]. Fibers with circular cross sections have a variable pathlength and act like a lens because the beam passes through the sides and the center. This can be likened to the lens cell described by Hirschfeld [21]. Under these conditions the ratios of less intense bands to stronger bands are greater than those in spectra obtained with a uniform pathlength. Qualitative interpretation is still readily accomplished. Second, circular and irregular cross-sectional shapes tend to diffract and scatter incident IR radiation as described in Section 2.1. The spectral baselines are often irregular due to scattered radiation. In general, the use of unflattened fibers is good for obtaining initial spectra of weaker crystalline bands. The fiber should subsequently be flattened by some means to obtain a good overall spectrum. Note that crystallinity changes may occur during fiber flattening and the spectra may be slightly distorted.

2.2.2. Flattened Fibers

Flattening the fibers prior to analysis offers several advantages. Deviations from Beer's law are reduced because samples have more uniform thickness. The reduction in pathlength is also important for fibers possessing high absorptivity. Additionally, flattening increases the surface area of the sample available for analysis, thereby enhancing *S/N* while reducing diffraction effects at the fiber edges.

Fibers may be flattened with a roller knife or between dies in a laboratory press. A pair of dies should be reserved for this purpose, because certain fibers may produce indentations in the die faces. Nylon is the only common generic class of polymer fiber for which simple rolling frequently does not produce a sufficiently thin sample. The hardness and high absorptivity of nylon fibers make flattening with a roller insufficient for most samples. The thick samples result in overabsorption of the amide I and II bands, a problem that makes computer searching and quantitative analysis more difficult. Thinner samples may be produced by pressing in a hydraulic die press at 15,000 lb. Often, however, the die is indented before the sample is sufficiently thin to produce high-quality spectra. Slicing a thin section off the edge of the flattened fiber with a scalpel and repressing in the press will produce yet a thinner sample, but the procedure is more time consuming than is ideally desirable. Sufficiently thin nylon fiber samples may be prepared by flattening the fiber with a roller, slicing off a narrow section along the edge of the fiber, and flattening this section further with the roller. This procedure can be repeated until high-quality spectra are obtained.

Flattening does, however, destroy the physical shape of the fiber. In addition, the application of pressure during rolling can change the crystalline composition of the polymer. A change in crystallinity can result from shear forces as the polymer is forced to flow (J. W. Brasch, J.B. Labs, Columbus, Ohio, 1989). Although no major spectral differences have been observed as a result of applying pressure to the fiber samples, minor changes in peak frequencies, intensities, and shapes do occur for certain fibers. The data in Figure 3 (left side) show variation in the C—O—C stretching absorptions near 1250 and 1100 cm^{-1} in polyethylene terephthalate (PET) for the same sample preparation procedure. Because all fibers are oriented in the same direction, these differences are not due to differences in orientation, as reported for the case of clumped fibers [20]. Peak ratio inversions also are observed (Figure 3 right side) for the absorption peaks near 1409 cm^{-1} (phenyl, in plane) and 1339 cm^{-1} (CH_2 deformation perpendicular to the plane of CH_2). Peak intensities are known to change, depending on the degree of crystallinity of PET [22]. The intensities of the absorption bands near 1339, 973, and 874 cm^{-1} increase with percent crystallinity, thus causing the variation in intensity ratios. For certain nylon 6,6 fibers

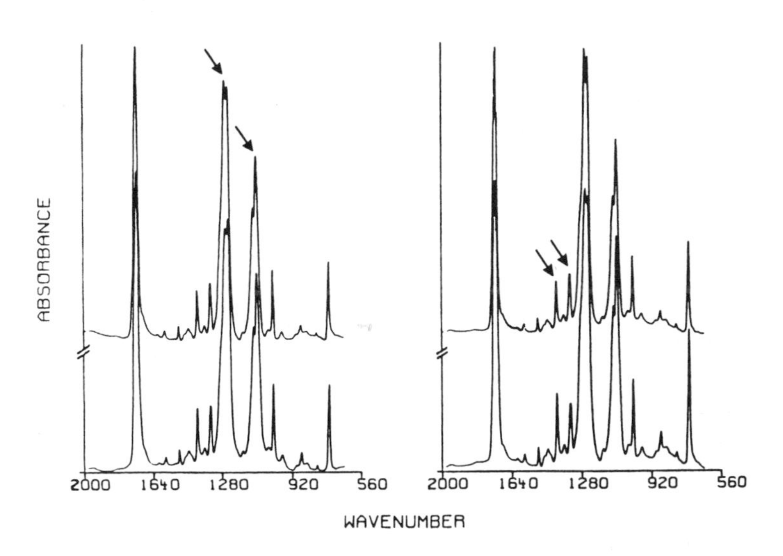

Figure 3. Infrared spectra of PET fibers from two different sample preparations. Note (left side) differences in absorptions near 1250 and 1100 cm^{-1} and (right side) 1339, 973, and 874 cm^{-1} due to differing degrees of crystallinity, resulting during sample preparation. (From Ref. 25.)

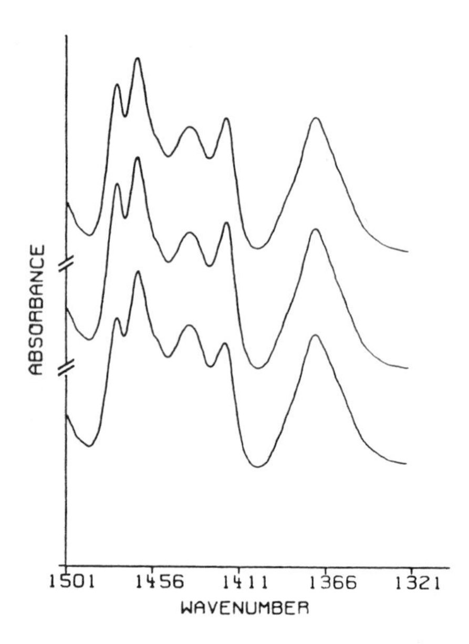

Figure 4. Infrared spectra of three different sample preparations of nylon 6,6 fibers. Differences in intensities of absorption bands are due to differing degrees of crystallinity resulting during sample preparation. (From Ref. 25.)

(Figure 4), variation in peak intensities also are observed in the region from 1400 to 1500 cm^{-1}. These variations are due to crystalline effects similar to those noted in PET fibers [23].

Flattening the fibers on a hard surface with a roller also effectively produces a thin film and may result in interference fringes in the IR spectra. These fringes do not interfere with qualitative interpretation of the spectra, but they can adversely affect quantitative analysis and computer searching of large databases. Interference fringes in the IR spectra may be greatly reduced when the surface of the roller is slightly roughened with silicon carbide sandpaper. Interference fringes may still be observed occasionally, however, especially for acrylic and modacrylic fibers. Further reduction of interference fringes can be accomplished by flattening the fiber on the ground-glass portion of a microscope slide.

2.2.3. Fibers on or Between IR Windows

Certain fiber samples are either too small or too brittle to place across an aperture in a sample holder. Additionally, many forensic fiber samples are too small to sacrifice a 2-mm length for permanent mounting. In these cases, a small length of the fiber (easily less than 100 μm) may be removed with a scalpel, flattened, and placed on a KBr window for analysis. The sample spectrum is ratioed against a background spectrum of the KBr window acquired adjacent to the sample.

Alternatively, the flattened fiber may be placed between two KBr disks and compressed with a compression cell. The sample spectrum is ratioed against a background spectrum of a KBr crystal placed adjacent to the fiber between the KBr windows. Use of a KBr crystal eliminates the two KBr–air interfaces between the windows and reduces interference fringes in the background spectrum. Sandwiching the fiber between two KBr windows greatly reduces interference fringes in most spectra, due to the refractive index match between KBr and most synthetic fibers. Note that the passage of the IR radiation through KBr (or other IR transparent material) results in a shift in the focal point or spherical aberration due to refraction. Ideally, the microscope design will allow adjustment of the objective and/or condenser to compensate for spherical aberration (Section 2.1.).

2.2.4. Fibers in a Moderate-Pressure Diamond Cell

The use of a diamond anvil cell involves sandwiching a sample between two diamond windows. Elastomeric fibers such as spandex and rubber (and any other type of fiber) are easily analyzed in a moderate-pressure or miniature diamond cell (MDC). An MDC may be purchased as such or may be effected by the use of diamond windows in a compression cell designed for 13-mm KBr windows.

With an MDC, large interference fringes are produced if the background spectrum is acquired adjacent to the sample. The problem has several solutions. The first approach entails ratioing the sample spectrum against a background spectrum of the two diamond faces placed in direct contact. This approach, used with beam condensers [24], works well for that application because the aperture size is constant and only one background spectrum needs to be acquired. With a microscope, however, the aperture size varies from sample to sample, and the background spectrum must be recorded for each sample spectrum. This process scratches the diamond faces and is time consuming as well. A second approach involves ratioing the spectrum of the sample in the diamond cell against an air background followed by subtraction of a diamond reference spectrum to obtain the sample spectrum [25]. With this method, it is necessary to adjust the focus before acquiring the background spectrum. This method is convenient and protects the diamond faces from scratches. Third, a small KBr crystal, placed next to the fiber prior to flattening in the MDC, may be used as a background. Finally, the diamond windows may be separated prior to analysis and a spectrum acquired of the flattened fiber adhering to only one of the windows [26]. The background spectrum is then acquired of the diamond window adjacent to the fiber. Obviously, this single-anvil technique will not work for elastomeric fibers.

The pressure applied to samples in the MDC is greater than that used in rolling. The effect of mechanical pressure (shear forces) on the spectrum of a

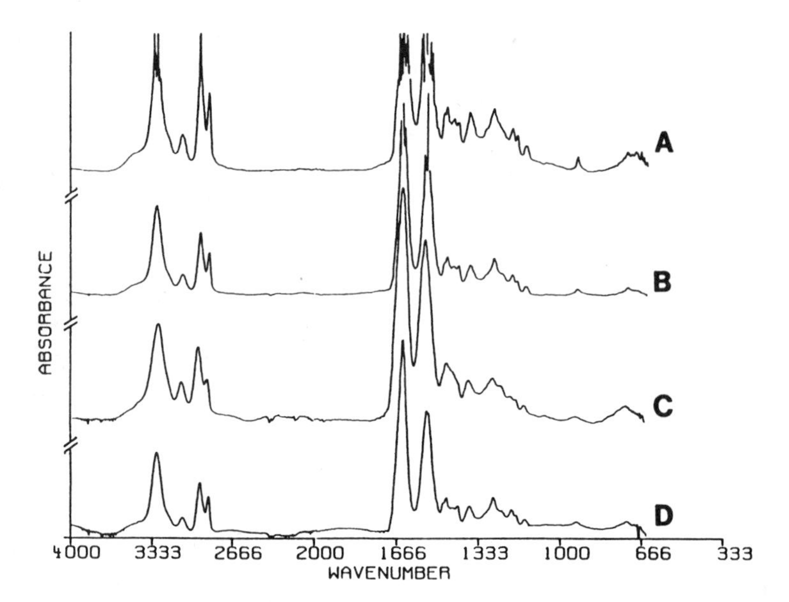

Figure 5. Infrared spectra of a nylon 6,6 fiber in a miniature diamond cell (A) before force was applied, (B) with force applied, (C) with maximum force applied, and (D) after force was released. (From Ref. 25.)

nylon 6,6 fiber mounted in an MDC is shown in Figure 5. As the sample is compressed and flattened, spectral quality is enhanced, but the peak sharpness is diminished to an extent that the spectrum of nylon 6,6 becomes difficult to distinguish from that of nylon 6 (Figure 6). Separating the diamond windows after compression does not necessarily restore the absorption bands to their original sharpness. Thus one must acquire two spectra: one with the fiber only slightly compressed in order to preserve the fine structure of the spectrum, and a second with the fiber sufficiently flattened to obtain a good overall spectrum.

3. INFRARED SPECTRAL DATABASES

3.1. Initial Databases

Infrared spectral databases are valuable tools in identifying a textile fiber. Several approaches to the compilation of these fiber databases have been taken. Early efforts were based both on text information [27] and on collections of IR spectra identified by manufacturer and trade name (Laser Precision Analytical, Irvine, California). The text information database includes physical characteristics, optical properties, and polymer composition as determined by IR spec-

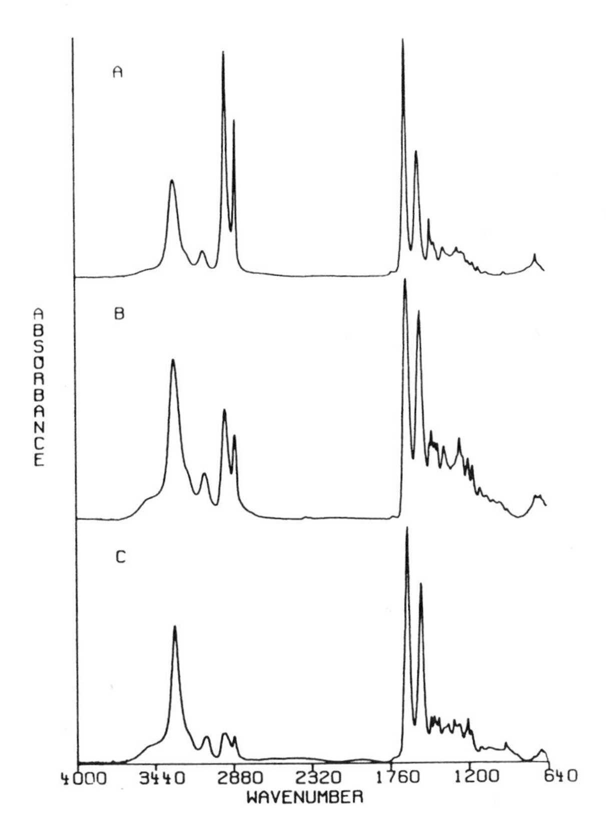

Figure 6. Infrared spectra of three types of nylon fibers: (A) nylon 12; (B) nylon 6; (C) nylon 4. (From Ref. 28.)

troscopy. This approach is useful for compiling statistical data and narrowing down possible sources for a particular fiber; however, the absence of the actual IR spectra makes it difficult to distinguish fibers exhibiting nearly similar IR spectra. On the other hand, databases containing only large numbers of IR spectra of fibers, identified by manufacturer and trade name, are inadequate when the polymer type is quite common.

3.2. Polymer Composition Database

A small spectral library of fibers, identified originally by polymer composition [25], has been published in electronic form [28]. Samples targeted for inclusion in the library were those representative of each generic class of fiber and of each subclass described in the forensic literature. Because the contents of this library include most common fibers in today's textile market, the contents are described in detail below to provide an overview of synthetic fiber types.

3.2.1. Library Contents

The spectral library currently contains 19 generic classes of fibers [28]. The definitions by the U.S. Federal Trade Commission (FTC) for common generic

Table 1. Definitions by the Federal Trade Commission for Certain Common Fiber Classes

Generic name	Definition of fiber-forming substance[a]
Acetate	Cellulose acetate; triacetate where not less than 92% of the hydroxyl groups are acetylated
Acrylic	At least 85% acrylonitrile units
Aramid	Polyamide in which at least 85% of the amide linkages are attached directly to two aromatic rings
Azlon	Regenerated naturally occurring proteins
Modacrylic	Less than 85% but at least 35% acrylonitrile units
Nylon	Polyamide in which less than 85% of the amide linkages are attached directly to two aromatic rings
Olefin	At least 85% ethylene, propylene, or other olefin units
Polyester	At least 85% an ester of a substituted aromatic carboxylic acid, including but not restricted to substituted terephthalate units, and *para*-substituted hydroxybenzoate units
Rayon	Regenerated cellulose, and regenerated cellulose in which substituents have replaced less than 15% of the hydrogens of the hydroxyl groups
Saran	At least 80% vinylidene chloride units
Spandex	Elastomer of at least 85% segmented polyurethane
Vinal	At least 50% vinyl alcohol units and at least 85% total vinyl alcohol and acetal units
Vinyon	At least 85% vinyl chloride units

[a]All fibers are defined as being manufactured and all percentages are by weight. Except for acetate, azlon, and rayon, the fiber-forming substance is described as a long-chain synthetic polymer of the composition noted.

fiber classes are listed in Table 1. Spectra of major chemical subclasses within each generic class have been included. A complete listing of the library contents is given in Table 2. While acrylic, modacrylic, and polyester fiber subclasses had previously been identified [8,10,29,30], the use of IR analysis was infrequent due to the difficulties cited earlier. The emphasis is on synthetic fibers, but a small group of natural fibers was included because they are frequently encountered.

3.2.2. Origin of Fiber Subclasses

Fiber subclasses arise in two fashions. The subclass can result from the use of different monomers that fit the FTC definition for the generic class. Nylon fibers, for example, fall into two major categories, both of which use different monomers to yield various subclasses. Nylons based on polyamides of amino acids may vary according to the length of the carbon chain in the amino acid, resulting in nylon 4, nylon 6, nylon 11, and so on. Similarly, nylons (nylon 6,6, nylon 6,10, nylon 6,11, etc.) produced from the reaction of diamines with diacids may vary according to the length of the carbon chains in the diamine and diacid. Other variations may also be seen, as in DuPont's Qiana, which has cyclohexane rings incorporated into the diamide monomer. Different nylon subclasses may be distinguished by careful examination of their IR spectra (Figure 6). Other fiber classes in this category are aramids and polyolefins.

Fiber subclasses can also arise when fibers are synthesized as copolymers; the major comonomer dictates the generic class, while the minor comonomers are varied by the manufacturers for particular fiber types. For example, acrylics are composed of at least 85% acrylonitrile. Minor comonomers include varying combinations of vinyl acetate, methyl acrylate, methyl methacrylate, and methyl vinyl pyridine, which constitute up to 15% of the fiber copolymer. Smalldon classified acrylics into eight types [8]. More recently, Grieve has extended the number of types to 20 by the inclusion of residual solvent and minor termonomer bands [31]. Spectra of several common acrylic subclasses are shown in Figure 7. Other fiber classes in this category are modacrylic, saran, spandex, vinal, and vinyon.

Other generic classes, such as polyesters and fluorocarbons, contain both types of subclasses. Polyester fibers, for example, may be formulated from different ester monomers, such as ethylene terephthalate or cyclohexylene dimethylene terephthalate. Polyester fibers may also be formulated as copolymers such as poly(terephthalic acid:*p*-hydroxybenzoic acid:ethylene glycol). Fluorocarbons may be formulated either from different single monomers such as tetrafluoroethylene or vinylidenefluoride, or as copolymers such as poly(ethylene: chlorotrifluoroethylene).

Table 2. Fibers in the IR Spectral Database

Generic class	Subgeneric class[a]
Acetate	Acetate
	Triacetate
Acrylic	Polyacrylonitrile
	Poly(AN:MA)
	Poly(AN:MA:MVP)
	Poly(AN:VA)
	Poly(AN:VA:MVP)
	Poly(AN:MMA)
Aramid	Kevlar
	Nomex
Azlon	Alginate
	Casein
	Ground nut
Fluorocarbon	PTFE
	PVDF
	Poly(ethylene:TFE)
	Poly(ethylene:CTFE)
Glass	
Modacrylic	Poly(AN:VDC:MAA)
	Poly(AN:VDC)
	Poly(AN:VC)
Natural	Cotton
	Silk
	Wool
Nylon	Nylon 4
	Nylon 6
	Nylon 6,6
	Nylon 6,10
	Nylon 6,11
	Nylon 6,12
	Nylon 11
	Nylon 12
	Qiana
PBI	Two types[b]

Table 2. *Continued*

Generic class	Subgeneric class[a]
Polycarbonate	
Polyester	PET PCDT Poly(TA:PHBA:EG) PHEB PBT
Polyolefin	PE PP
Rayon	
Saran	Three types[b]
Spandex	Two types[b]
Sulfar	PPS
Vinal	
Vinyon	Poly(VC:VA)

Source: Reproduced in part from Ref. 28.
[a]AN, acrylonitrile; CTFE, chlorotrifluoroethylene; EG, ethylene glycol; MA, methyl acrylate; MAA, methyl acrylamide; MMA, methyl methacrylate; MVP, methyl vinyl pyridine; PBT, polybutylene terephthalate; PCDT, poly(cyclohexyl dimethylene terephthalate); PE, polyethylene; PET, polyethylene terephthalate; PHBA, *p*-hydroxybenzoic acid; PHEB, polyhydroxyethoxybenzoate; PP, polypropylene; PPS, polyphenylene sulfide; PTFE, polytetrafluoroethylene; PVDF, polyvinylidene fluoride; TA, terephthalic acid; TFE, tetrafluoroethylene; VA, vinyl acetate; VC, vinyl chloride; VDC, vinylidene chloride.
[b]Variation of polymer composition is currently unknown.

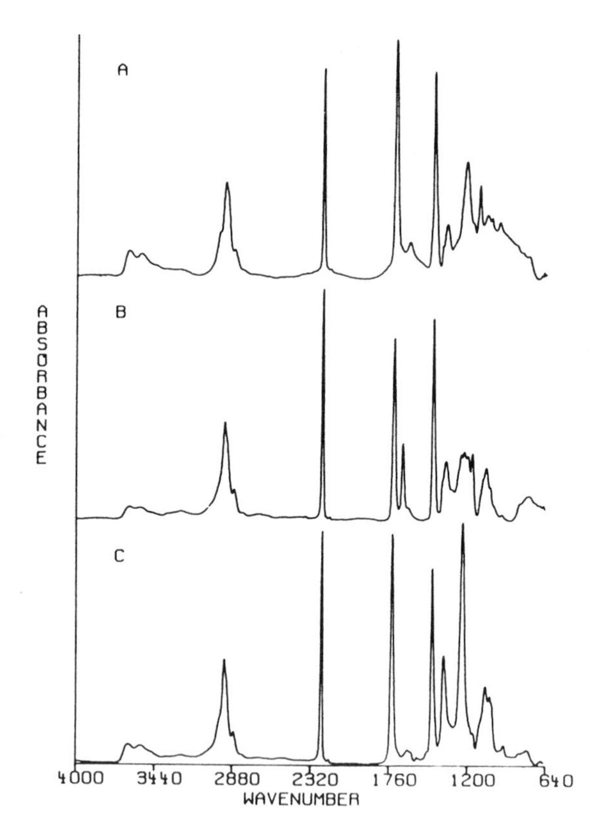

Figure 7. Infrared spectra of three types of acrylic fibers: (A) poly(acrylonitrile:methyl methacrylate); (B) poly(acrylonitrile:methyl acrylate); (C) poly(acrylonitrile:vinyl acetate). (From Ref. 28.)

3.3. Combined Spectra and Text Database

Infrared spectra and text information have been combined to form a new fiber database [32]. The initial database contains IR spectra of fibers from the Collaborative Testing Service, Inc. (CTS) (Herndon, Virginia) collection. Included with the IR spectra are manufacturer information on chemical and physical data. Physical data, such as cross-sectional shape and diameter, can be used for a presearch to filter out only those files that contain parameters of interest. The spectral search can then be limited to this subset. When the entire library is searched by IR spectra alone, many fibers with the same chemical composition are often listed as the top hits. By first selecting fiber files with the correct physical characteristics, the chances are greatly improved that the IR spectra search will result in the correct manufacturer as the first hit.

4. VALUE OF IR ANALYSIS IN FIBER CASEWORK

Infrared spectra have been obtained of fibers encountered in 92 fiber matches made during routine casework at the FBI Laboratory [33]. No fiber matches made by optical microscopy were shown to be different by their IR spectra; however, additional compositional information was obtained which enhanced the evidential value of the match. Polymer identification by microscopic IR analysis can enhance forensic fiber examinations in the four ways discussed below.

4.1. Identification of Generic Class

The fiber generic class is unequivocally identified by microscopic IR analysis. Although this information is obtainable by PLM in most cases, the results are subjective and require an experienced microscopist. If the fibers are opaque or of a type previously unencountered by the examiner, identification by PLM is difficult, if not impossible. The IR spectrum, on the other hand, provides an objective, absolute identification of the generic class. With the aid of reference spectra in a database, an examiner can quickly and easily identify the polymer composition of the fiber. Opacity of heavily dyed fibers presents no problems. A few minor peaks may appear in the spectrum due to the high concentration of the dye or dye modification of the polymer, but polymer identification is never hindered. Should new types of fibers be encountered in a case, they may at least be characterized, if not identified, by microscopical IR. For example, when efforts to identify a new brown fiber by optical microscopy failed, the fiber's IR spectrum was found to be quite similar to that of a polybenzimidazole (PBI) in a fiber spectral library [33]. The fiber was later confirmed to be a new variation of PBI. Had the PBI fiber not been included in the fiber spectral library, a similar polymer could probably have been found in the commercial IR polymer libraries. With rapidly changing fiber technology, similar situations are certain to become more commonplace. A fiber examiner without IR microscopy will certainly be at a disadvantage when identifying new fibers.

4.2. Identification of Fiber Subclass

In addition to the generic class, microscopic IR analysis can identify the specific polymer composition or fiber subclass, thus enhancing the evidential value of a fiber match. In short, fiber subclass provides another point of comparison in a forensic fiber examination.

Identification of subclass also permits the compilation of statistics. Two such studies have been reported [33,34], the most thorough being that of the Home Office in the United Kingdom. The statistical information may be used to assess the relative frequency of occurrence of different fiber types within a generic class. For example, the most frequently encountered type of polyester in both studies was the subclass PET. This statistical finding greatly enhances the evidential value of a match between two polyester fibers belonging to a subclass other than PET. Similarly, most nylon fibers examined belonged to either the nylon 6 or nylon 6,6 subclasses. The acrylic fibers were, for the most part, copolymer fibers consisting of either poly(acrylonitrile:methyl acrylate) (AN:MA), poly(acrylonitrile:methyl methacrylate) (AN:MMA), or poly(acrylonitrile:vinyl acetate) (AN:VA). As discussed above for polyester fibers, a fiber match involving nylon or acrylic fibers belonging to a uncommon subclass would greatly enhance the evidential value of the match. Note that casework statistics probably differ from fiber production statistics due to three factors: (1) the degree fibers are shed, (2) whether or not the fibers are readily transferred from the donor to recipient item, and (3) the degree to which the fiber will persist on the recipient item.

In certain cases, identification of fiber subclass may lead to the identification of the fiber's manufacturer. For example, dyed acrylic homopolymer (AN) fibers are produced only by Mann Industries. Such information is usually not needed in casework; however, production statistics by a manufacturer can be used to determine the commonality of the fiber type. End uses (such as garments or carpets) of a fiber from a sole manufacturer also may be important when looking for possible donor sources of a questioned fiber. Thus the significance of a fiber match is increased when the fiber's manufacturer is identified.

4.3. Characterization of Contaminants

Fibers submitted for microscopical IR analysis are frequently contaminated with a wide variety of materials that will contribute additional bands to the IR spectrum. If the fibers have been subjected to a PLM examination, mounting media such as Permount may be encountered. Blood or other body fluids may be present on fibers recovered from a body. The presence of a contaminant can, in certain instances, increase the evidential value of a fiber match when the contaminant is shown to be present on both the questioned and known fibers. Thus, when additional bands are noted in the spectrum, it is desirable to identify their source prior to washing the fibers.

In one reported case [33], IR spectra (Figure 8) were obtained of questioned (Q) fibers from a body and known (K) fibers from a suspect's automobile trunk. Fibers Q and K were both identified as polypropylene; however, fiber K contained additional absorption peaks near 1730, 1651, 1020, 702, and 672 cm^{-1}. Spectra shown in Figure 9 were obtained after aperture size and locations on the fiber were changed. An increase in aperture size resulted in the appearance of the additional bands in fiber Q, whereas a decrease in aperture size resulted in a reduction in intensity of the additional bands in fiber K. The fluctuation

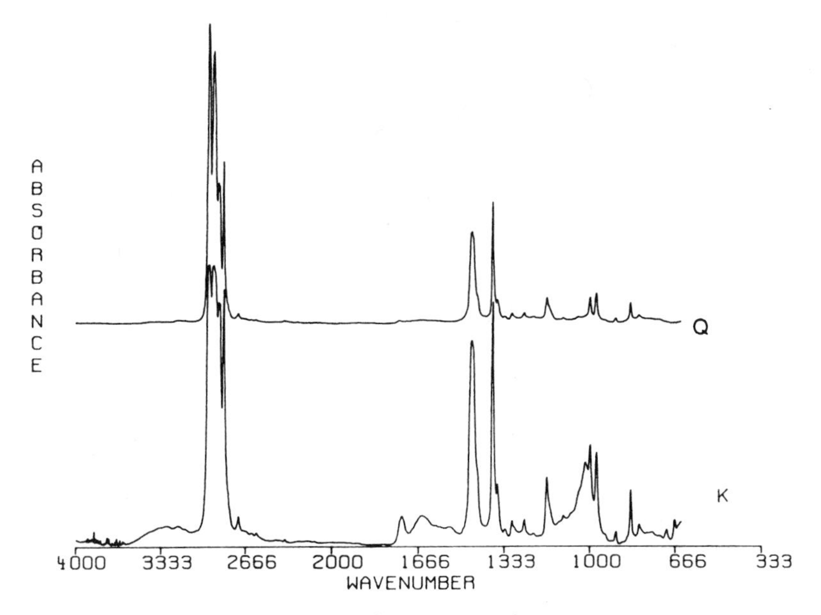

Figure 8. Infrared spectra of polypropylene fibers Q and K. Note additional absorption bands near 1730, 1651, 1020, 702, and 672 cm^{-1} in the spectrum of fiber K. Aperture sizes: Q = 60 × 92 μm; K = 116 × 188 μm. (From Ref. 33.)

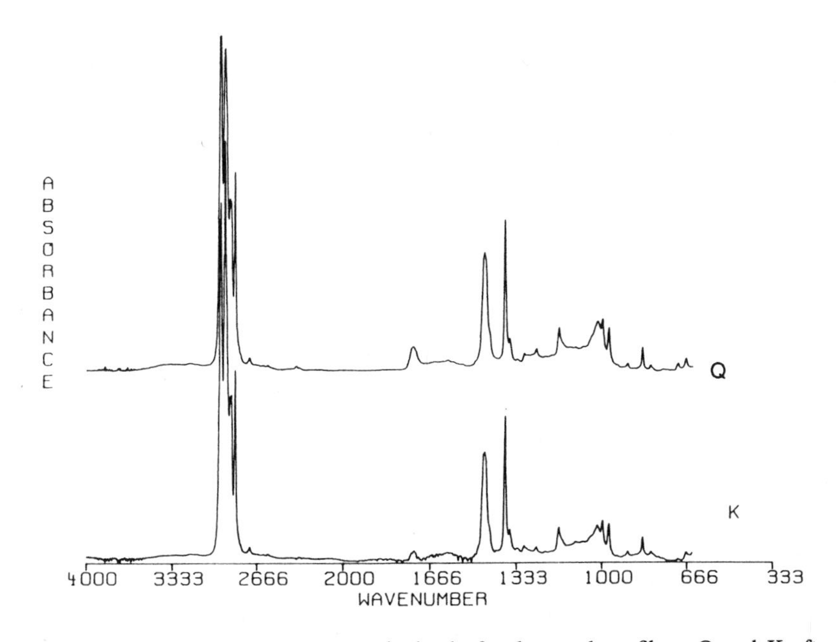

Figure 9. Second set of IR spectra obtained of polypropylene fibers Q and K after location of analysis on fiber and aperture size were changed. Aperture sizes: Q = 100 by 240 μm; K = 24 by 48 μm. (From Ref. 33.)

with aperture size and location revealed that the source of the additional bands was associated with the presence of a contaminant rather than a difference in composition between fibers Q and K. Obviously, the increase in aperture size resulted in the inclusion of the contaminant in the sampling field, whereas the reverse process excluded most of the contaminant from the sampling field.

An MDC was used to obtain a spectrum (Figure 10) of a black particle found adhering to one of the fiber samples. This spectrum was compared with the difference spectrum obtained by subtracting Q number 1 from Q number 2. From the spectral similarities in Figure 10, the black particle was concluded to consist of the same material that produced the spectral differences between fibers Q and K shown in Figures 8 and 9. The spectrum of the black particle is similar to a polyester resin (Figure 11), with additional absorption bands at 1020 and 672 cm^{-1}. These peaks may be accounted for by the presence of talc (Figure 12). In short, the ''contaminant'' was probably a glue used in the construction of the automobile trunk liner from which the fibers originated. The identification of the glue on both Q and K fibers greatly increased the evidential value of the fiber match.

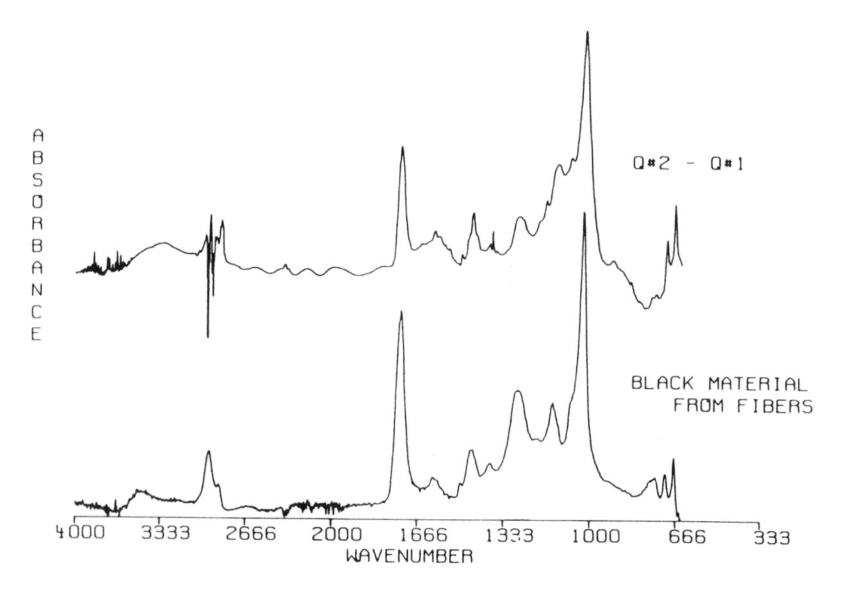

Figure 10. Difference spectrum, obtained by subtracting the first spectrum of Q from the second, and the spectrum of a black particle found adhering to one of the fibers. (From Ref. 33.)

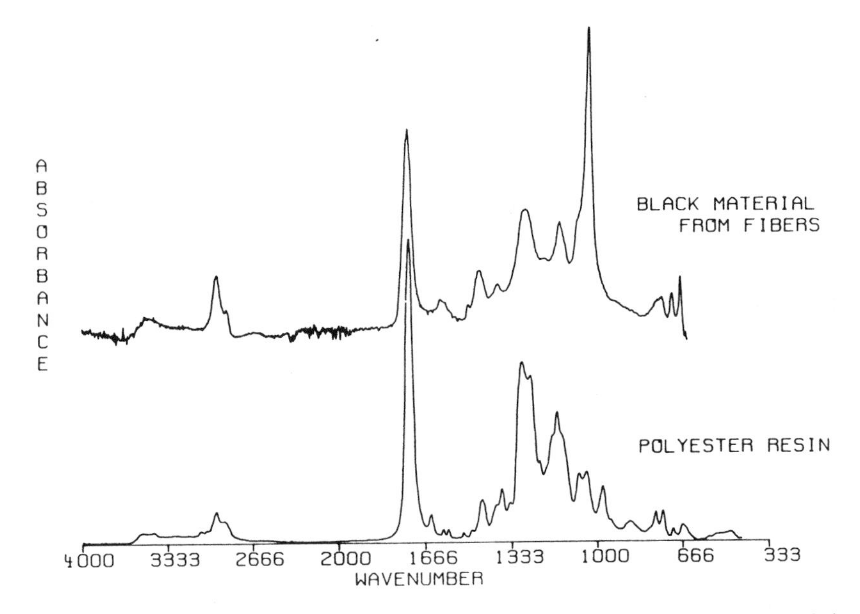

Figure 11. Infrared spectrum of the black material found adhering to a fiber and the IR spectrum of an unsaturated polyester resin. Peaks near 1020 and 672 cm^{-1} are not accounted for by the resin. (From Ref. 33.)

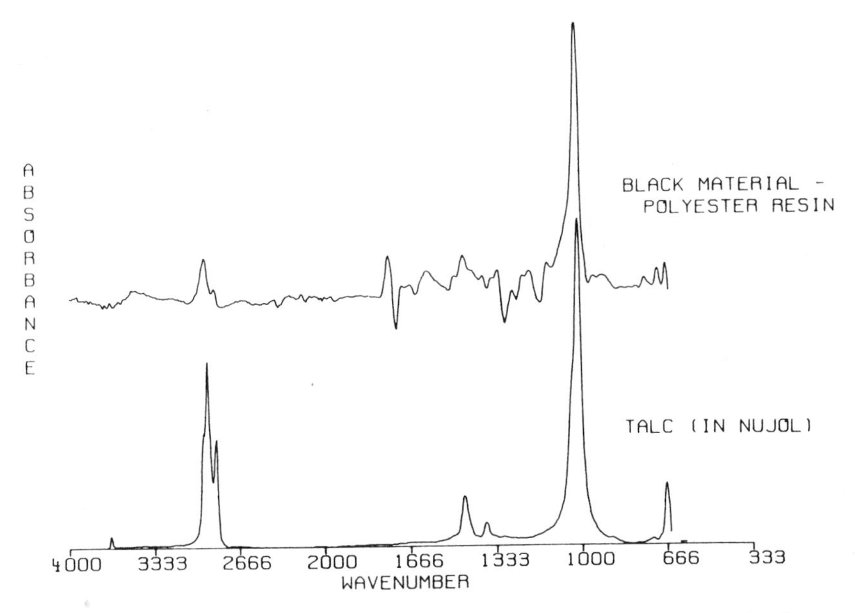

Figure 12. Difference spectrum, obtained by subtracting the spectrum of polyester resin from the spectrum of the black material, and the spectrum of talc. The talc spectrum contains C—H absorption bands that originate from Nujol. (From Ref. 33.)

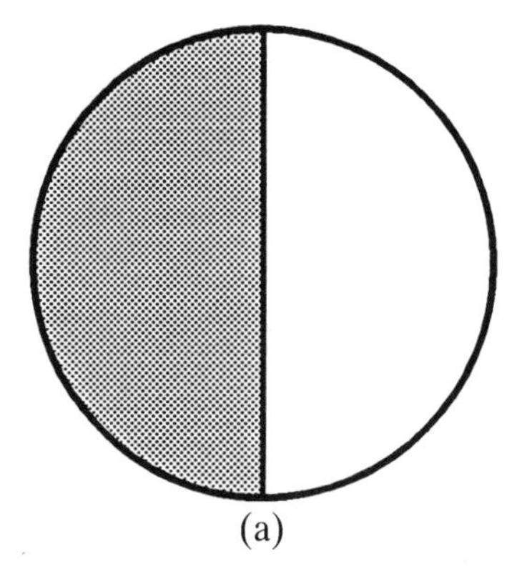

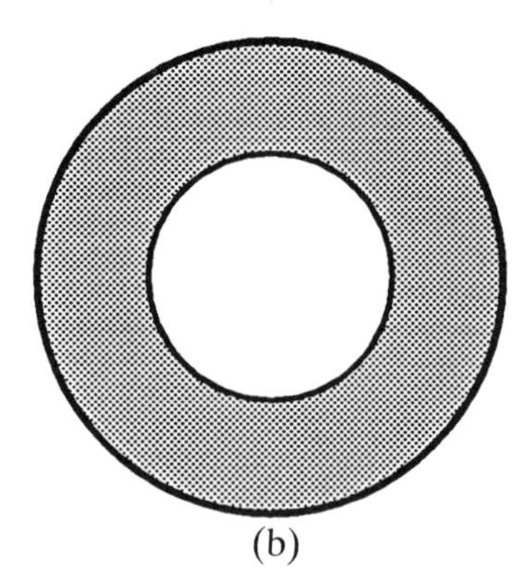

Figure 13. Schematic diagram of the cross section of (a) a side by side, and (b) a sheath-and-core bicomponent fiber.

4.4. Analysis of Bicomponent Fibers

Certain fibers are manufactured as bicomponent fibers having two separate polymer types continuous along the fiber length and permanently joined at the interface. Two predominant types are the "side-by-side" and "sheath-and-core" configurations illustrated in Figure 13. Identification of the two generic fiber types by PLM for a "side-by-side" configuration is a straightforward matter, as is the microscopical IR analysis of each side [35]. With these fibers, as noted previously, the diffraction problem adversely affects spectral purity and thus must be taken into consideration when identifying spectra of the two components. Identification of the two components of a "sheath-and-core" arrangement by PLM, however, is not an easy task. Cross-sectioning the fiber is usually required. With IR microscopy, the fiber is flattened and IR spectra (Figure 14) are obtained near the edge of the fiber (sheath) and close to the center of the fiber (predominately core material). The core spectrum will include weaker features of the sheath material. The spectrum recorded from the edge of the flattened fiber identifies the sheath composition as polypropylene, while the core is easily identifiable as PET, although weak polypropylene bands are apparent, as expected. The spectrum of the sheath could easily be subtracted from that of the core to remove the polypropylene bands.

5. PEAK AREA RATIO ANALYSIS OF ACRYLIC COPOLYMER FIBERS

Copolymer subgeneric classes may be subdivided further on the basis of comonomer ratios provided that adequate variations exist between different fibers

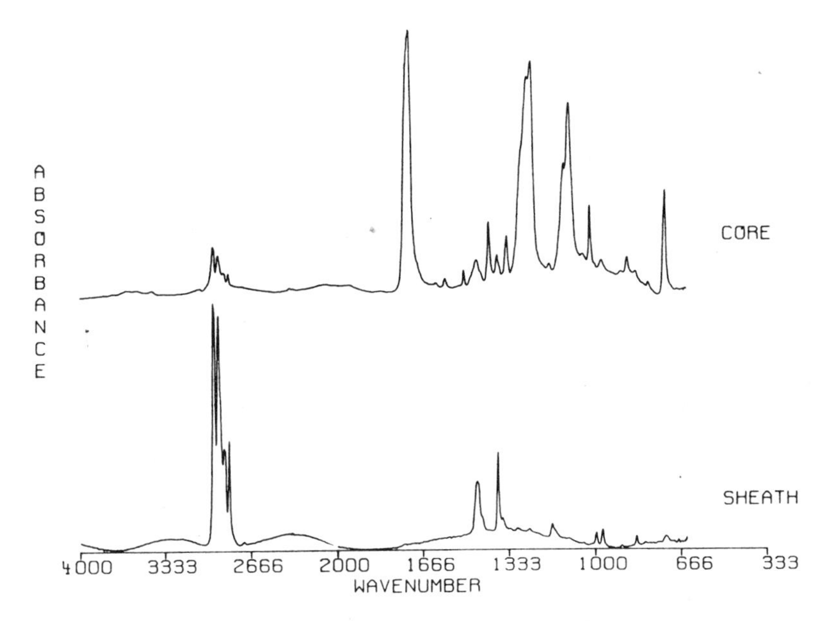

Figure 14. Infrared spectra of the sheath and core of a flattened bicomponent fiber. The spectrum of the polypropylene sheath was obtained by analyzing a location near the edge of the fiber, while the spectrum of the PET core was obtained by analyzing the center of the fiber. Note the weak polypropylene absorption bands present in the PET spectrum. (From Ref. 33.)

on the commercial market. Several studies have been conducted on single acrylic fibers to ascertain the capability of microscopical IR to yield valid quantitative information [36,37], one of which has focused on forensic applications [37]. As stated previously, an acrylic fiber is a manufactured fiber in which the fiber-forming substance is any long-chain polymer composed of at least 85 wt % of acrylonitrile (AN) units [38]. Most acrylic fibers are composed of copolymers of AN with an ester such as methyl acrylate (MA) or vinyl acetate (VA). A third component, such as vinylbenzenesulfonic acid or a vinyl pyridine derivative, may also be present, to enhance dyeability.

5.1. Calculation of Peak Ratios

The absorption peaks used to monitor the relative comonomer ratio are the nitrile (C≡N stretch) and the carbonyl (C=O stretch) absorption bands near 2240 and 1730 cm^{-1}, respectively. The nitrile band results from the AN comonomer only, while the carbonyl band originates from the MA comonomer (and to a small extent from any other minor carbonyl-containing comonomers or additives that may be present). As shown in Figure 7, both absorption bands are strong and relatively free from overlapping bands. Both peak heights and peak areas have been used to monitor comonomer ratios [36,37]. With modern computing capabilities, peak areas are easily calculated and should be used.

The precision of two methods of determining the area under the absorption bands have been compared [37]. The first method entailed selecting the absorption minima on both sides of the absorption band, constructing a baseline between the two points, and integrating the peak area above the baseline. This method has one drawback when the absorption band is located on an interference fringe or sloping baseline. In this situation, no minima exist, and points for baseline construction must be estimated. No such limitation existed when a fixed frequency range was selected to encompass approximately 95% of each of the two absorption bands, excluding the band wings (edges). For this second method, baselines were constructed between the points where the spectrum intercepted these frequency ranges, and peak areas were integrated above the baselines. The same frequency ranges were used for all spectra of a given copolymer type. Better precision was obtained with the fixed-frequency-range method [37].

5.2. Effect of Aperture on Peak Area Ratios

The effect of aperture width has been examined [37] by calculating $A_{C=O}/A_{C\equiv N}$ for a series of spectra acquired of a single location on an acrylic fiber approximately 30 μm in diameter. The fiber measured 60 μm in width after flattening. Because many IR microscopes only have one aperture, data for both single and double apertures are shown in Figure 15. The single aperture used was placed before the sample in the IR beam path (upper aperture).

No significant variation in $A_{C=O}/A_{C\equiv N}$ was noted for either the single- or double-aperture spectra until the aperture width approached 10 μm. At this point, the decrease in the signal-to-noise ratio caused $A_{C=O}/A_{C\equiv N}$ to vary slightly. $A_{C=O}/A_{C\equiv N}$ was consistently lower in the single-aperture spectra compared to the double-aperture spectra. This is a result of deviation from Beer's law due to diffraction effects as the diameter of the aperture approaches the wavelength of the IR radiation [16]. At the lower frequency (longer wavelength), the carbonyl region suffers more from diffraction effects than does the nitrile region, which appears at a higher frequency (shorter wavelength). Thus, ideally, both upper and lower apertures should be used and aperture widths kept as large as possible, while excluding the fiber edges.

5.3. Effect of Homogeneity on Peak Area Ratios

The effect of homogeneity on comonomer ratios was examined for AN:MA fibers [37]. For each of five AN:MA fiber samples, spectra were acquired 10 times at a single location and 10 times at adjacent segments (120 μm long)

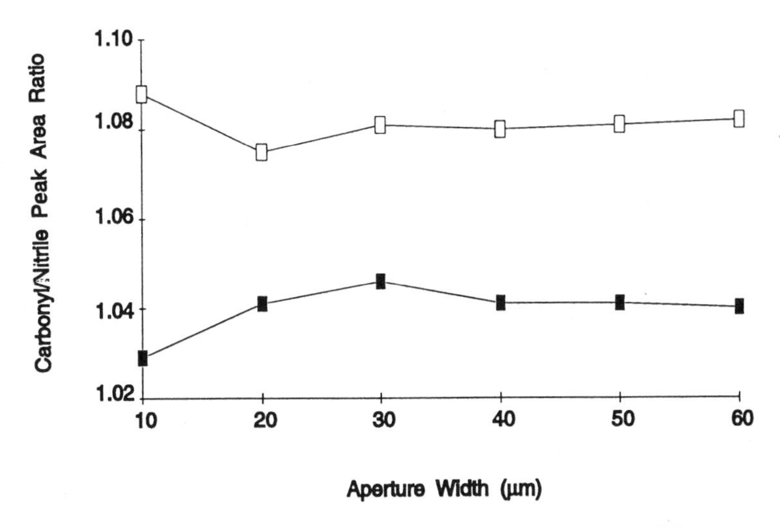

Figure 15. $A_{C=O}/A_{C≡N}$ for a 60-μm-wide flattened acrylic fiber using different aperture widths with two apertures (upper) and one aperture placed before the sample in the beam path (lower).

along the length of a single fiber. Spectra were also acquired of 10 different fibers for each AN:MA fiber sample. Peak areas were integrated between 2266 and 2220 cm^{-1} for the nitrile band and between 1755 and 1702 cm^{-1} for the carbonyl band.

The values of $A_{C=O}/A_{C≡N}$ for the five AN:MA fiber samples are presented in Table 3. The technique is very precise for repeated analysis of a single location. Relative standard deviations (RSDs) ranging from 0.19 to 0.32% were measured. Samples analyzed from adjacent locations along the length of a fiber yielded higher RSDs (0.54 to 7.04%) indicating a certain degree of inhomogeneity in the fibers. In general, the percent RSD increased further when different specimens of the same fiber type were tested, demonstrating yet more inhomogeneity between different fibers. The RSD for sample 81A0105 (7.04%), when adjacent locations were analyzed, was high in comparison to results for other fibers examined similarly (0.54 to 1.5%). The reason for this, other than greater than expected inhomogeneity within the particular fiber, was not known. Note that the two Badische A201 fibers, which were believed to differ only in the amount of delusterant (they share the same manufacturing code) yielded very similar results when 10 fibers were averaged. This clearly supports the applicability of the method.

Overall, $A_{C=O}/A_{C≡N}$ was significantly higher for the Badische fibers than for the DuPont fibers (1.7 versus 1.1). This observation supports the use of peak area ratios for differentiating copolymer fibers within the same subclass. Quan-

Table 3. Average (First Number in Each Block) and Relative Standard Deviation of $A_{C=O}/A_{C\equiv N}$ Measured for AN:MA Copolymer Fibers

Fiber type	One location (10 times)	Along length (10 locations)	10 fibers
DuPont Orlon 42 CTS 81A0012	1.066 0.19%	1.083 1.3%	1.103 6.1%
DuPont Orlon TR01 CTS 81A0031	1.258 0.32%	1.068 1.0%	1.138 2.0%
Badische A201 CTS 81A0105	1.734 0.23%	1.717 7.04%	1.688 5.5%
Badische A201 CTS 81A0106	1.721 0.29%	1.658 1.5%	1.681 5.3%
Badische A302 CTS 81F0111	1.540 0.26%	1.672 0.54%	1.656 4.6%

Source: Reproduced in part from Ref. 37.

titative results appear to be reproducible to within about 5% in most cases. Fibers from more manufacturers must be examined before any conclusions can be drawn about identifying the fiber manufacturer by $A_{C=O}/A_{C\equiv N}$.

Because certain IR microscopes have only a single aperture, the experiment was repeated with sample 81A0012 using only the upper aperture. As shown in Table 4, the use of a single aperture does not appear to affect precision significantly, but $A_{C=O}/A_{C\equiv N}$ is consistently lower. This is due to diffraction effects, as discussed in Section 2.1.

Table 4. Average (First Number in Each Block) and Relative Standard Deviation of $A_{C=O}/A_{C\equiv N}$ Measured for DuPont Orlon 42 (CTS 81A0012) Using One and Two Apertures

	Both apertures	Upper aperture only
One location (10 times)	1.066 0.19%	1.041 0.29%
Along length (10 locations)	1.083 1.3%	1.068 1.5%
10 fibers	1.103 6.1%	1.082 3.3%

Source: Reproduced from Ref. 37.

5.4. Effect of Physical Characteristics on Peak Area Ratios

The ratio $A_{C=O}/A_{C\equiv N}$ was determined for 13 types of Monsanto Acrilan fibers [37]. These poly(acrylonitrile:vinyl acetate) (AN:VA) fibers varied in manufacturer code and physical characteristics, such as diameter and amount of delusterant. The nitrile and carbonyl bands were integrated from 2260 to 2230 cm^{-1} and 1755 to 1715 cm^{-1}, respectively. The ratio $A_{C=O}/A_{C\equiv N}$ ranged from 1.225 to 1.318 (Figure 16). Relative standard deviations ranged between 0.87 and 7.8%. Little variation in $A_{C=O}/A_{C\equiv N}$ was present within the manufacturer trade name. Differences were present, however, both within and between fiber types. The minor variations are possibly due to minor carbonyl-containing additives, termonomers, or surface treatments on the polymer fibers. Lower precision was reported as the fiber diameter was reduced (Figure 17). Fibers with smaller diameters had shorter pathlengths and required narrower aperture widths; consequently, S/N was reduced.

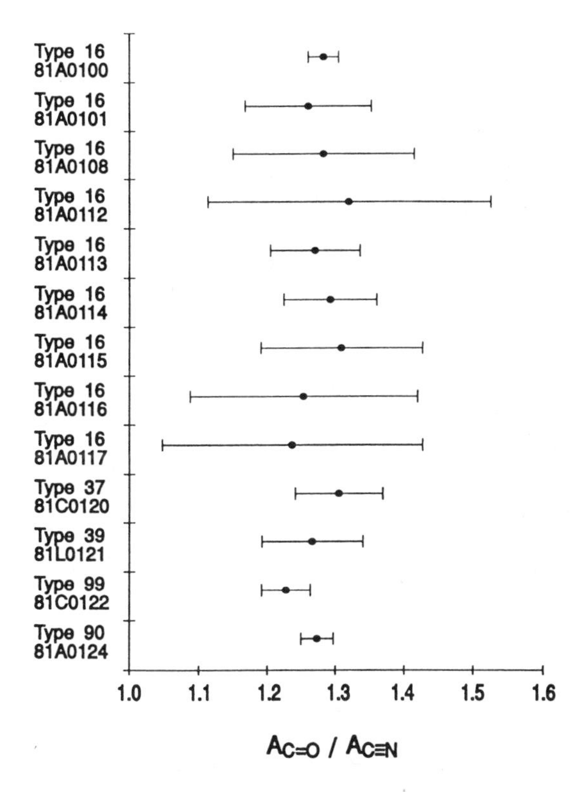

Figure 16. $A_{C=O}/A_{C\equiv N}$ with 95% confidence limits for 13 types of Monsanto Acrilan AN:VA fibers (average of 10 analyses). (From Ref. 37.)

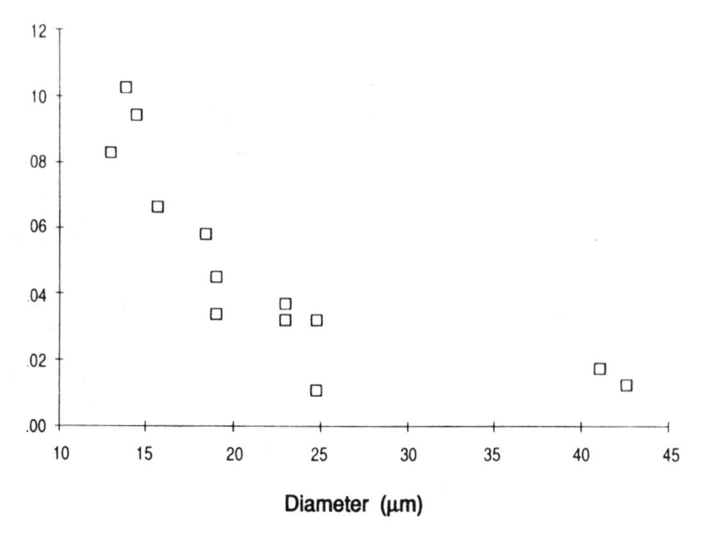

Figure 17. Relative standard deviation (RSD) versus average fiber diameter for 13 Monsanto Acrilan fibers (average of 10 analyses). (From Ref. 37.)

5.5. Variations of Comonomer Ratios in Acrylic Fibers Encountered in Casework

To examine variation in $A_{C=O}/A_{C\equiv N}$ between fibers on the commercial market, spectra of acrylic fibers recorded in case studies [33] were examined for AN:MA, AN:VA, and AN:MMA fibers (Figure 18a–c) [37]. The greatest range (1.131 to 2.034) for $A_{C=O}/A_{C\equiv N}$ was noted for AN:MA fibers. The AN:MA fibers appear to fall into three distinct groups according to $A_{C=O}/A_{C\equiv N}$. For AN:VA and AN:MMA, the ranges were 1.224 to 1.400 and 1.797 to 2.074, respectively. While ideally, more spectra would have been acquired to obtain these average values, there are obvious differences in $A_{C=O}/A_{C\equiv N}$ values within acrylic fiber types on the market. Thus difference in $A_{C=O}/A_{C\equiv N}$ can be used to further differentiate between acrylic fibers belonging to the same subgeneric class and the potential for classification within the subgeneric class exists based on this peak area ratio.

6. DICHROISM STUDIES OF SINGLE FIBERS

Certain generic classes of fibers consist primarily of only one or two polymer compositions. For example, as noted in Section 4.2, the vast majority of polyester fibers are composed of polyethylene terephthalate, and most nylon fibers are either nylon 6 or nylon 6,6. Identification of chemical composition does not provide much discriminative information unless the fiber happens to have an unusual composition. Because these fiber types are so common, the undyed fibers are usually considered to have little or no evidential value unless they have a particularly unusual cross-sectional shape.

In addition to chemical composition, fibrous polymers may be characterized by molecular chain length, degree of molecular orientation, and degree of crystallinity. While molecular chain length could potentially be useful for characterizing and comparing fibers, size-exclusion chromatography is the standard method and typically requires more sample than is available for destruction in forensic situations. Molecular orientation of both amorphous and crystalline regions, however, is easily examined by dichroism using polarized IR spectroscopy.

Dichroism is a much more powerful technique than is birefringence used to examine fibers by PLM. Birefringence measures average values over all the components and phases present in a sample. Dichroism, on the other hand, is specific to particulr absorption bands and can be applied to different groups on the same molecule, different segments in a copolymer, and different components of a polymer blend.

The dichroic ratio (R) is defined according to

$$R = \frac{A_{\parallel}}{A_{\perp}} \tag{2}$$

where $A_{\parallel}$ and $A_{\perp}$ are the absorbances obtained with parallel and perpendicular polarized radiation. Actually, R represents the ratio for the integral absorbance, although $A_{\max}$ may be sufficient in certain cases. For R greater or less than 1, the band is said to exhibit π or σ dichroism, respectively. Ideally, the true dichroic ratio (DR) should be determined according to

$$\mathrm{DR} = \frac{A_{\parallel} - A_{\perp}}{A_{\parallel} + A_{\perp}} \tag{3}$$

For DR = 1, the dipole moment vector of the bond and the fiber axis are parallel. When DR = -1, the dipole moment vector and fiber axis are perpendicular. The fiber is assumed to have radial symmetry, valid for fibers having circular cross sections.

Molecular orientation is developed in fibers during the manufacturing process to impart desired properties to the fibers. For example, melt spinning and high-speed spinning are two common fiber manufacturing processes. In melt spinning, the molten polymer solidifies almost immediately after extrusion through spinneret orifices. Filaments are then drawn or stretched to develop orientation and crystallinity. In high-speed spinning, the molten polymer is extruded as in melt spinning, but the filament windup speed is so high relative to the extrusion speed that orientation and crystallinity are developed during elongational flow on the spin line. The net result of these processes is the formation of a fiber fine structure characterized by a high degree of crystallinity and orientation for both the crystallites and the polymer chain segments in the amorphous domains.

6.1. Feasiblity of Dichroism Measurements on Single Fibers

The capability of microscopic IR spectroscopy to yield valid polarization data have been demonstrated for single fibers [39–42]. To arrive at a dichroic ratio, two spectra from a nonideal sample must be quantified. To preserve the secondary structure of the fiber, the fiber must not be flattened; therefore, stray radiation presents a problem. Errors are also produced by the high numerical aperture of the IR microscope objectives (i.e., the radiation striking the sample is highly convergent). Thus one must assume that dichroic ratios are only an estimate of the true value. However, if a high degree of precision and a suitable spread of dichroic ratios are obtainable, dichroic ratios should be another feasible point of comparison for fibers previously matched by the standard means.

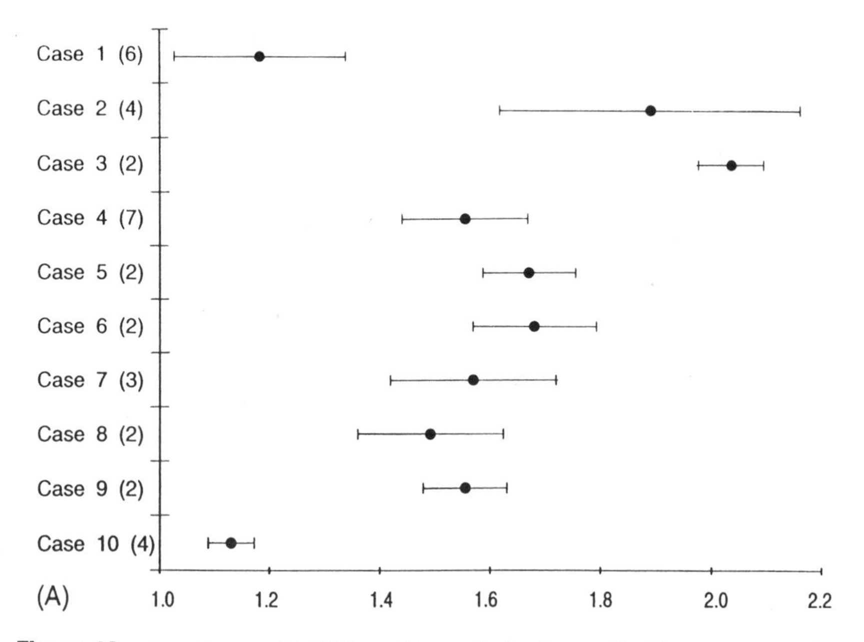

Figure 18. $A_{C=O}/A_{C\equiv N}$ with 95% confidence limits for acrylic fibers encountered as part of routine casework. The number of results averaged is shown in parentheses. (A) AN:MA fibers; (B) AN:VA fibers; (C) AN:MMA fibers.

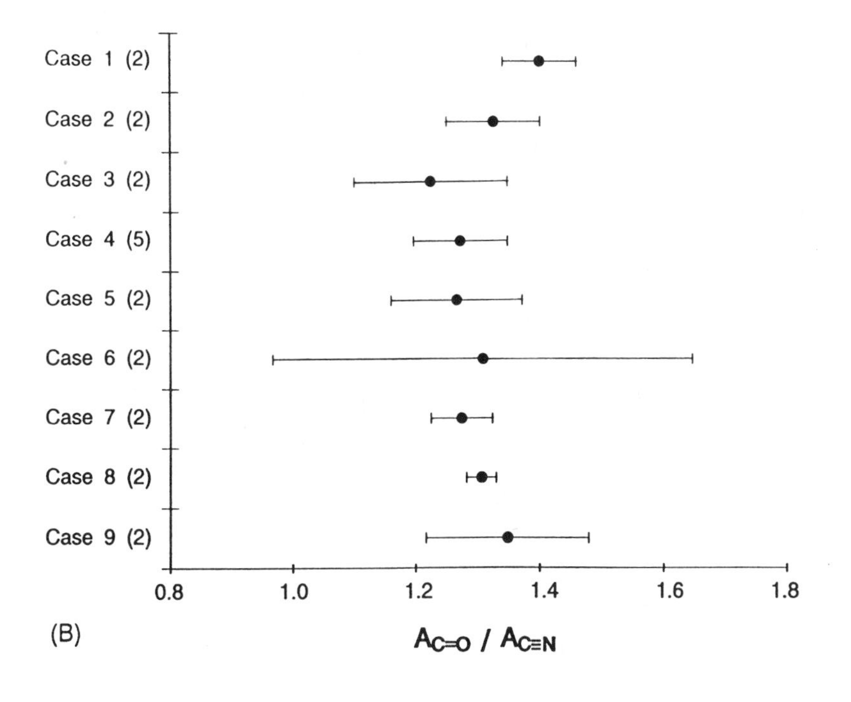
Case 1 (2)
Case 2 (2)
Case 3 (2)
Case 4 (5)
Case 5 (2)
Case 6 (2)
Case 7 (2)
Case 8 (2)
Case 9 (2)
0.8
1.0
1.2
1.4
1.6
1.8
(B)
A C=O / A C≡N

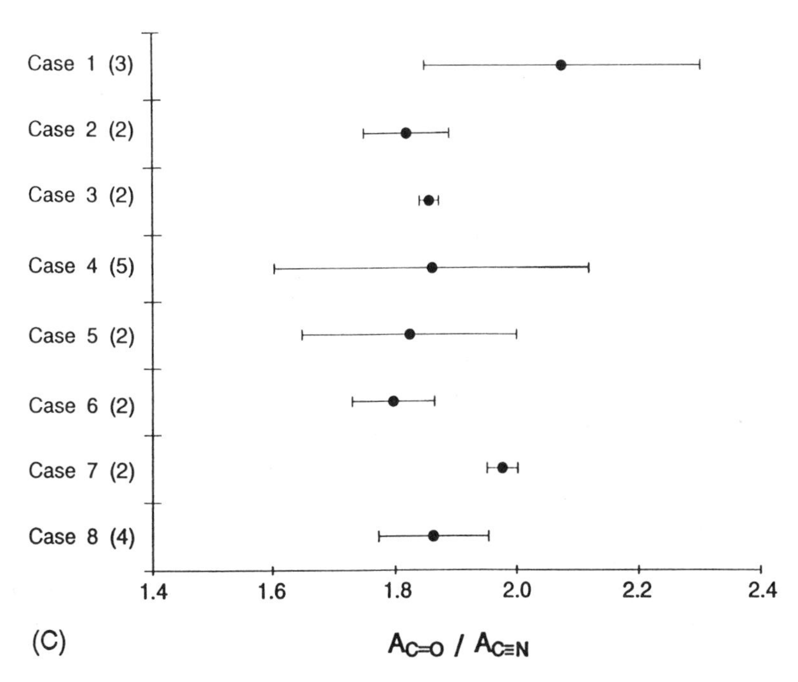
Case 1 (3)
Case 2 (2)
Case 3 (2)
Case 4 (5)
Case 5 (2)
Case 6 (2)
Case 7 (2)
Case 8 (4)
1.4
1.6
1.8
2.0
2.2
2.4
(C)
A C=O / A C≡N

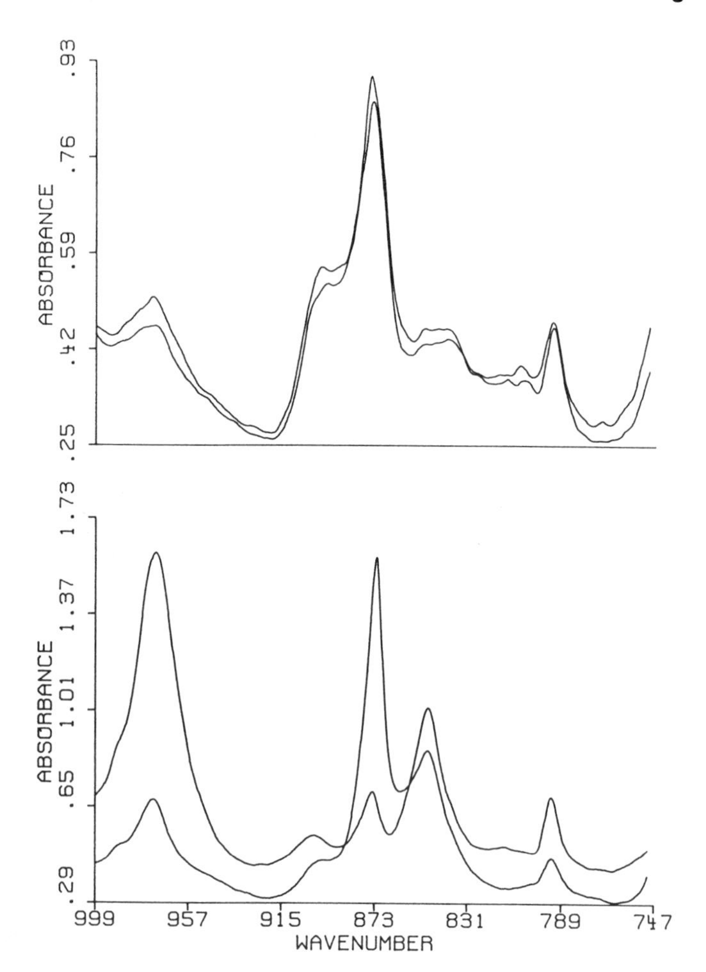

Figure 19. Polarized IR spectra of an undrawn polyethylene terephthalate (PET) fiber (upper) and a drawn PET fiber (lower). Note the similarity between the parallel and perpendicular polarized spectra in the undrawn fiber and the differences in the drawn fiber. (From Ref. 17.)

A striking example of differences in dichroism among polyester fibers is shown in Figure 19. The two polarized spectra on the top were obtained from a polyester fiber (produced for melt bonding) that was not drawn during the manufacuring process. The parallel and perpendicular polarized spectra are very similar because little orientation was developed in the fiber. The two polarized spectra on the bottom were obtained from a polyester fiber drawn during the manufacturing process. Here molecular orientation was produced in the fiber and the parallel and perpendicular polarized spectra show considerable differences in band intensity.

6.2. Dichroism in Polyester Fibers

A more detailed forensic study of dichroism in PET fibers has recently been reported [43]. Parallel and perpendicular polarized IR spectra were acquired of 11 different types of PET fibers. Ten fibers of each type were analyzed. Peak areas were calculated using both the minima-to-minima and fixed-frequency-range methods described in Section 5.1. The dichroic ratio (R) and true dichroic ratio (DR) were determined for the 1579-, 1505-, 973-, and 872-cm^{-1} absorption bands.

Varying degrees of precision were obtained with the different parameters chosen. For the 1505- and 872-cm^{-1} bands, the minima-to-minima baseline yielded the best precision, with RSDs ranging from 1.2 to 5.9% and 2.3 to 6.8%, respectively. The fixed-frequency-range method produced better precision for the 1579- and 973-cm^{-1} bands, with RSDs ranging from 1.0% to 5.2% and 1.0% to 5.7%, respectively. These differences are probably due to interferences by overlapping bands. Additionally, DR yielded better precision than did R. The best range of values was reported for the 872-cm^{-1} band (Figure 20). Although there is overlap between some fibers, certain other fibers can be differentiated. To provide better differentiation between fibers, simultaneous comparison of multiple bands may prove useful (Figure 21). Principal components analysis [44] may be helpful for further differentiation of various fibers.

7. FUTURE DIRECTIONS

Applications of microscopic infrared to forensic fiber analysis are just beginning. The advent of new techniques is rapidly opening up many new avenues of research. Three areas in molecular spectroscopy that show great promise for forensic fiber analysis are microscopic internal reflection spectroscopy, IR photoacoustic spectroscopy, and Raman spectroscopy.

7.1. Microscopic Internal Reflection Spectroscopy

Internal reflection spectroscopy (IRS), also referred to as attenuated total reflectance (ATR), has been applied extensively to routine-sized samples for opaque objects such as polymers [45]. The recent development of an ATR microscope objective has made microscale analysis possible by this method [46,47]. Microscopic ATR offers three advantages for fiber analysis. First, ATR is a surface technique; that is, the process of fiber thinning for transmission is eliminated and fiber spectra may be obtained quickly and easily. Second, because the depth

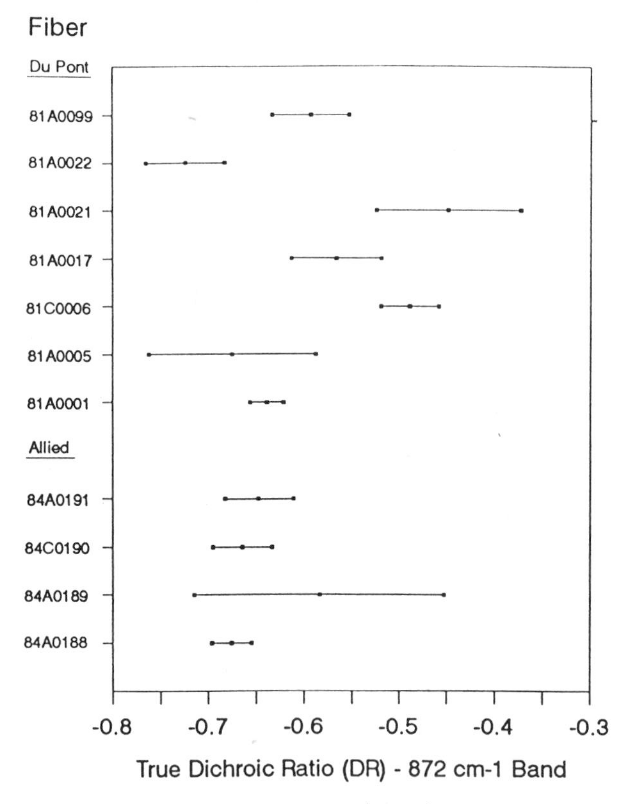

Figure 20. True dichroic ratios (DR) for the 872-cm^{-1} band of 11 types of PET fiber. (Peak areas were integrated between minima.)

of penetration of the IR beam into the sample increases with wavelength, absorption bands at lower frequencies are more intense than those in the transmission spectra. This enhances the fingerprint region of the spectrum and allows minor spectral features to be more easily distinguished. Third, because only the outer few micrometers of the fiber is analyzed, analysis of surface finishes on fibers is feasible.

7.2. Photoacoustic Spectroscopy

Photoacoustic spectroscopy (PAS) has recently been reported as a single-fiber technique [48]. The photoacoustic signal is generated when IR radiation absorbed by the sample is converted into heat within a sample. The diffusion of the heat from the sample surface into the adjacent atmosphere causes thermal

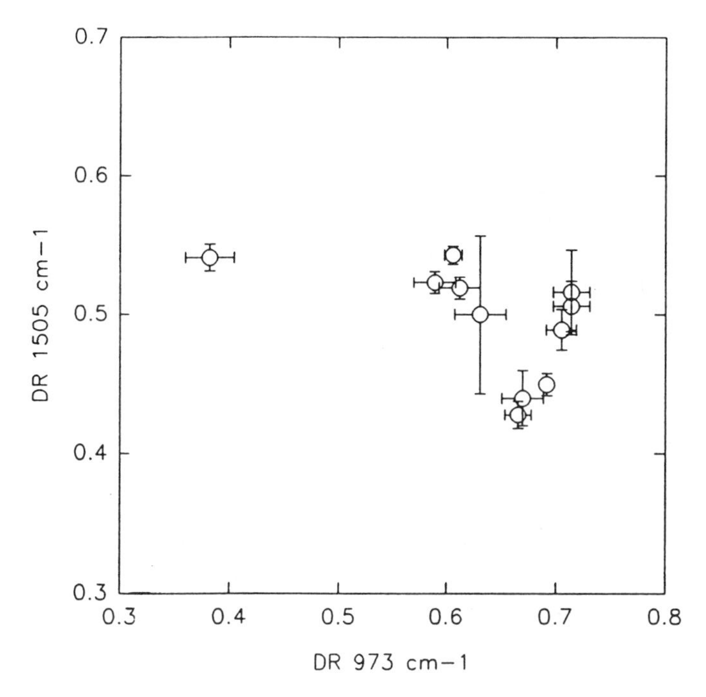

Figure 21. Two-dimensional plot of DR for the 1505- and 973-cm^{-1} bands of 11 types of PET fiber.

expansion of the gas. This expansion is detected as a signal by a sensitive microphone or piezoelectric detector. Spectral range is limited only by the transparency of the sample chamber window. With suitable windows, a detector can operate from the ultraviolet to the far IR. The advantages of Fourier transform (FT)-IR PAS for fiber analysis are as follows: (1) no flattening of the fibers is necessary to reduce optical density; (2) PAS spectra are well suited for distinguishing between similar fibers such as nylons (i.e., bands that are weak in the IR spectrum are more prominent in the PAS spectrum); (3) depth profiling of samples is possible by varying the modulation frequency of the incident IR radiation. Photoacoustic detectors are available as accessories for FT-IR instruments and should see increased use in the forensic analysis of single fibers.

7.3. Microscopic Raman Spectroscopy

Although Raman is not an infrared technique per se, the two techniques are so complementary that they are often considered together. While IR spectra are the result of absorption of radiation, Raman spectra arise from an inelastic scattering effect. If a photon is scattered inelastically by a molecule, it may gain or lose energy, and hence scattering is recorded at higher or lower frequencies. The frequency displacement from the exciting radiation corresponds to a characteristic molecular mode of vibration. The selection rules differ for IR and Raman. Thus bands that appear weak in the IR spectrum will be strong in the Raman spectrum, and vice versa. Both techniques are required to gain a complete picture of the vibrational characteristics of a molecule.

The Raman technique is sensitive enough to analyze single fibers readily with microscopes coupled to spectometers [49]. In addition to the complementary vibrational information obtained, Raman spectroscopy offers several other advantages. First, sample preparation is eliminated because scattering and not transmission is measured. Second, diffraction effects are greatly reduced because the exciting radiation is of a lower wavelength than IR. Third, spatial resolution is enhanced. For example, 10 μm is typically considered to be the diffraction limit for IR microscopy, but 1-μm resolution is easily achieved with a Raman microscope. Finally, narrow-range IR detectors generally have a low-frequency range near 700 cm^{-1}. Raman experiments, on the other hand, are generally performed down to 200 cm^{-1}, and new holographic notch filters are pushing this limit down to 50 cm^{-1}. In this far-IR region one can find information relating primarily to the inorganic components of fibers, such as delusterants. With the rapid growth in FT-Raman spectroscopy, a steady growth in fiber applications is likely.

8. SUMMARY

Infrared analysis should be included in any forensic fiber examination protocol because it is the best general method for identifying the polymer composition of fibers. Infrared microscopy is quick, nondestructive, highly sensitive, and provides a wealth of information from a short piece of single fiber. Every feature of a spectrum contains information concerning the sample. Not every feature, however, is related to the chemical structure of the sample. Certain features, such as interference patterns and stray radiation, may arise from optical phenomena. To minimize the possibility of drawing a false-negtive conclusion, spectra must be acquired and examined carefully.

When making fiber comparisons, the analyst should perform simple, nondestructive tests first. The more obvious dissimilarities will thus be readily evident. Microscopy, visible spectrometry, and IR spectrometry are used to examine the morphology, color, and polymer composition of the fiber. If these exams do not reveal any obvious differences between the fibers, further tests should be performed. These tests may be destructive if sufficient sample exists. Thermal methods such as melting-point determination can be used to examine the secondary structure of the polymer. The dyes present in the fiber may be extracted and examined by thin-layer chromatography (TLC) or high-performance liquid chromatography (HPLC). A final point of comparison may involve an elemental analysis by x-ray microanalysis or inductively coupled plasma analysis to compare additives that may be present, such as metal-based dyes, delusterants, and flame retardants. Finally, the forensic scientist must undertake the most difficult portion of the fiber examination—they must form an opinion as to their interpretation of the analytical results.

REFERENCES

1. Petraco, N. (1985). The occurrence of trace evidence in one examiner's casework, *J. Forensic Sci.*, *30*: 485.
2. Gaudette, B. D. (1988). The forensic aspects of textile fiber examination, in *Forensic Science Handbook*, Vol. 2 (R. Saferstein, ed.), Prentice-Hall, Englewood Cliffs, N.J., p. 209.
3. Fong, W. (1989). Analytical methods for developing fibers as forensic science proof: a review with comments, *J. Forensic Sci.*, *34*: 295.
4. Grieve, M. C. (1990). Fibres and their examination in forensic science, in *Forensic Science Progress*, Vol. 4, Springer-Verlag, Berlin, p. 41.
5. Robertson, J., ed. (1992). *Forensic Examination of Fibres*, Ellis Horwood, New York.
6. ASTM (1987). Identification of textile materials, D 276-87, *Annual Book of ASTM Standards*, American Society for Testing and Materials, Philadelphia, p. 95.
7. Fox, R. H., and Schuetzman, H. I. (1968). The infrared identification of microscopic samples of man-made fibers, *J. Forensic Sci.*, *13*: 397.
8. Smalldon, K. W. (1973). The identification of acrylic fibers by polymer composition as determined by infrared spectroscopy and physical characteristics, *J. Forensic Sci.*, *18*: 69.
9. Grieve, M. C., and Kearns, J. A. (1976). Preparing samples for the recording of infrared spectra from synthetic fibers, *J. Forensic Sci.*, *21*: 307.
10. Grieve, M. C., and Kotowski, T. M. (1977). The identification of polyester fibers in forensic science, *J. Forensic Sci.*, *22*: 390.
11. Read, L. K., and Kopec, R. J. (1978). Analysis of synthetic fibers by diamond cell and sapphire cell infrared spectrophotometry, *J. Assoc. Off. Anal. Chem.*, *61*: 526.
12. Cook R., and Paterson, M. D. (1978). New techniques for the identification of microscopic samples of textile fibres by infrared spectroscopy, *Forensic Sci. Int.*, *12*: 237.
13. Garger, E. F. (1983). An improved technique for preparing solvent cast films from acrylic fibers for the recording of infrared spectra, *J. Forensic Sci.*, *28*: 632.
14. Hartshorne, A. W., and Laing, D. K. (1984). The identification of polyolefin fibres by infrared spectroscopy and melting point determination, *Forensic Sci. Int.*, *26*: 45.

15. Curry, C. J., Whitehouse, M. J., and Chalmers, J. M. (1985). Ultramicrosampling in infrared spectroscopy using small apertures, *Appl. Spectrosc.*, *39*: 174.
16. Messerschmidt. R. G. (1987). Photometric considerations in the design and use of infrared microscope accessories, in *The Design, Sample Handling, and Applications of Infrared Microscopes*, ASTM STP 949 (P. B. Roush, ed.), American Society for Testing and Materials, Philadelphia, p. 12.
17. Bartick, E. G., and Tungol, M. W. (1993). Infrared microscopy and its forensic applications, in *Forensic Science Handbook*, Vol. 3 (R. Saferstein, ed.), Regents/Prentice Hall, Englewood Cliffs, N.J., p. 196.
18. Messerschmidt, R. G. (1988). Minimizing optical nonlinearities in infrared microspectroscopy, in *Infrared Microspectroscopy: Theory and Applications* (R. G. Messerschmidt and M. A. Harthcock, eds.), Marcel Decker, New York, p. 1.
19. Sommer, A. J., and Katon, J. E. (1991). Diffraction induced stray light in infrared microspectroscopy and its effect on spatial resolution, *Appl. Spectrosc.*, *45*: 1633.
20. Bartick, E. G. (1987). Considerations for fiber sampling with infrared microspectroscopy, in *The Design, Sample Handling, and Applications of Infrared Microscopes*, ASTM STP 949 (P. B. Roush, ed.), American Society for Testing and Materials, Philadelphia, p. 64.
21. Hirschfeld, T. (1985). Lens and wedge absorption cells for FT-IR spectroscopy, *Appl. Spectrosc.*, *39*: 426.
22. Daniels, W. W., and Kitson, R. E. (1958). Infrared spectroscopy of polyethylene terephthalate, *J. Polym. Sci.*, *33*: 161.
23. Haslen, J., Willis, H. A., and Squirrel, D. C. M. (1972). *Identification and Analysis of Plastics*, Wiley, New York, p. 313.
24. Schiering, D. W. (1988). A beam condenser/miniature diamond anvil cell accessory for the infrared microspectrometry of paint chips, *Appl. Spectrosc.*, *42*: 903.
25. Tungol, M. W., Bartick, E. G., and Montaser, A. (1990). The development of a spectral data base for the identification of fibers by infrared microscopy, *Appl. Spectrosc.*, *44*: 543.
26. Lin-Vien, D., Bland, B. J., and Spence, V. J. (1990). An improved method of using the diamond anvil cell for infrared microprobe analysis, *Appl. Spectrosc.*, *44*: 1227.
27. Carroll, G. R., LaLonde, W. C., Gaudette, B. D., Hawley, S. L., and Hubert, R. S. (1988). A computerized database for forensic textile fibres, *Can. Soc. Forensic Sci. J.*, *21*: 1.
28. Tungol, M. W., Bartick, E. G., and Montaser, A. (1991). Spectral data base of fibers by infrared microscopy, *Spectrochim. Acta*, *46B*: 1535E.
29. Grieve, M. C. (1983). The use of melting point and refractive index determination to compare colourless polyester fibres, *Forensic Sci. Int.*, *22*: 31.
30. Grieve, M. C., and Cabiness, L. R. (1985). The recognition and identification of modified acrylic fibers, *Forensic Sci. Int.*, *29*: 129.
31. Grieve, M. C. (1993). Another look at the classification of acrylic fibers, using FTIR microscopy, *Proceedings of the 13th Meeting of the International Association of Forensic Sciences*, Dusseldorf, Germany, in press.
32. Bartick, E. G., Tungol, M. W., Carroll, G. R., Carnahan, E. J., and Sprouse, J. F. (1990). A combined infrared spectroscopic and text data base for forensic fiber identification, presented at the *12th Meeting of the International Association of Forensic Sciences*, Adelaide, Australia, Oct. 24–29, unpublished results.

33. Tungol, M. W., Bartick, E. G., and Montaser, A. (1991). Analysis of single polymer fibers by Fourier transform infrared microscopy: the results of case studies, *J. Forensic Sci.*, *36*: 1027.
34. Robinson, G. (1990). The use of fibre data in the Home Office Forensic Science Service, Paper HF459, *12th Meeting of the International Association of Forensic Sciences*, Aldelaide, Australia, Oct. 24–29, 1990
35. Grieve, M. C., Dunlop, J., and Kotowski, T. M. (1988). Bicomponent acrylic fibers: their characterization in the forensic science laboratory, *J. Forensic Sci. Soc.*, *28*: 25.
36. Pandey, G. C. (1989). Fourier transform infrared microscopy for the determination of the composition of copolymer fibres: acrylic fibres, *Analyst*, *114*: 231.
37. Tungol, M. W., Bartick, E. G., and Montaser, A. (1993). Forensic analysis of acrylic copolymer fibers by infrared microscopy, *Appl. Spectrosc.*, *47*: 1655.
38. Federal Trade Commission Rules and Regulations under the Textile Products Identification Act, Title 15, U.S. Code section 70, *et seq.* 16 CFR 303.7.
39. Chase, D. B. (1987). Infrared microscopy: a single-fiber technique, in *The Design, Sample Handling, and Applications of Infrared Microscopes*, ASTM STP 949 (P. B. Roush, ed.), American Society of Testing and Materials, Philadelphia, p. 4.
40. Chase, B. (1988). Dichroic infrared spectroscopy with a microscope, in *Infrared Microspectroscopy: Theory and Applications* (R. G. Messerschmidt and M. A. Harthcock, eds.), Marcel Dekker, New York, p. 93.
41. Young, P. H. (1988). The characterization of high-performance fibers using infrared microscopy, *Spectroscopy*, *3*(9): 25.
42. Reffner, J. A. (1989). Infrared spectral mapping of polymers by FT-IR microscopy, in *Microbeam Analysis—1089* (P. E. Russell, ed.), San Francisco Press, San Francisco, p. 167.
43. Tungol, M. W., Bartick, E. G., and Montaser, A. (1993). Polarized infrared microscopical analysis of single poly(ethylene terephthalate) fibers, *Procedings of the 13th Meeting of the International Association of Forensic Sciences*, Dusseldorf, Germany, in press.
44. Hasenoehrl, E. J., Perkins, J. H., and Griffiths, P. R. (1992). Rapid functional group characterization of gas chromatography/Fourier transform infrared spectra by a principal components analysis based expert system, *Anal. Chem.*, *64*: 705.
45. Mirabella, F. M., Jr., ed. (1993). *Internal Reflection Spectroscopy: Theory and Applications*, Marcel Dekker, New York.
46. Reffner, J. A., Wihlborg, W. T., and Strand, S. W. (1991). Chemical microscopy of surfaces by grazing angle and internal reflection FTIR microscopy, *Am. Lab.*, *23*(April): 46.
47. Bartick, E. G., Tungol, M. W., and Reffner, J. A. (1994). A new approach to forensic analysis with infrared microscopy: internal reflection spectroscopy, *Anal. Chim. Acta*, *288*: 35.
48. McClelland, J. F., Jones, R. W., Luo, S., and Seaverson, L. M. (1993). A practical guide to FT-IR photoacoustic spectroscopy, in *Practical Sampling Techniques for Infrared Analysis* (P. B. Coleman, ed.), CRC Press, Boca Raton, Fla., p. 107.
49. Messerschmidt, R. G., and Chase, D. B. (1989). FT-Raman microscopy: discussion and preliminary results, *Appl. Spectrosc.*, *43*: 11.

8

Infrared Microspectroscopy in the Analysis of Cultural Artifacts

Michele R. Derrick The Getty Conservation Institute, Marina del Rey, California

1. Introduction

Many artifacts and other evidence of previous cultures are slowly disappearing. The lifetime of the art and artifacts that survive is affected by their material composition; their chemical, physical, and biological environments; and by treatments designed to preserve them (i.e., coatings, cleanings, etc.). The field of art conservation works to mitigate or minimize adverse effects in order to maintain and preserve artifacts as material documents of our historic and artistic heritage. Conservation is a complex area because the diversity of materials considered important ranges from seashells to entire city blocks, from rock art to modern paintings, and from ancient bronzes to plastic sculptures.

Whether the object under consideration is metal, wood, stone, or other material, there are some basic factors incorporated in the goal of preservation. These include examination of the object's composition, condition, and deterioration mechanisms. For these evaluations, conservation scientists conduct both laboratory studies and in situ experiments using several types of chemical and physical measuring instruments.

The infrared (IR) microspectrophotometer is a primary analytical tool for the examination and characterization of artifact materials as well as for the evaluation of materials that can be used for repair or protection of artifacts. The knowledge obtained from the analysis of materials in an object can provide a conservator with necessary background information useful in the design of optimal and safe conservation treatment strategies. Also, analysis of materials and their structure in an object can provide an art historian with supporting information for provenance studies and investigations of an artist's technique.

IR instruments have found a number of applications in art conservation for the analysis of paint layers, coatings, and other materials [1–8]. The first application of IR microspectrometry for the analysis of painting materials was published in 1970 by van't Hul-Ehrnreich using an IR dispersive spectrometer [9]. Now, nearly 20 years later, the availability of Fourier transform infrared (FT-IR) instruments have stimulated the revival of the application of IR microspectrometry in art conservation [10–15].

The goal in infrared microspectroscopy analysis of art objects is in many ways similar to that of other fields. It is important to gain as much information as possible from small samples that are often limited in number. Several specific methods and applications are discussed in this chapter. Appropriate natural product reference collections and spectral libraries for these materials are discussed at the end of the chapter.

2. EXPERIMENTAL PROCEDURES

2.1. Sample Collection

Since cultural objects are irreplaceable and their preservation is the ultimate goal in any conservation examination, samples are removed only when absolutely necessary. Nondestructive techniques (x-ray fluorescence, radiography, infrared and ultraviolet photography, etc.) receive priority and are often used to survey the objects. When the removal of sample(s) is permitted, the value of the object and its state of deterioration often dictates very specific restrictions on the number, size, and location of the sample(s) removed.

For example, a typical sampling protocol for a surface-finish sample from a piece of museum furniture is to remove one or two barely visible samples (<50 μg) from arbitrarily chosen, obscure areas near or under metal mounts, in the rear of the carcass, or on its legs. Such a protocol places the representativeness of a sample and its relationship to the object in question. Additional information should be considered when making inferences as to the object's composition, such as very careful visual observations under normal and ultraviolet (UV) light to characterize the uniformity of the object and the sampling area. Historical records of the object are helpful to ascertain previous analyses, treatments, revarnishing, and retouching. Also, since it is the sample that determines the quality and meaningfulness of results, care is taken in each step of its removal, storage, and analysis to prevent contamination or loss.

The selection of a sample removal technique depends on the questions to be answered and on the analytical methods that will be used. For infrared analysis, the samples are removed with a swab or a scalpel to obtain a coating layer, particles, or multilayer cross section.

2.1.1. Swabs

Solvent-dipped swabs are used to conduct solubility studies and to collect samples when the analysis pertains to a solvent-soluble surface of an object, such as a patina or coating on a bronze sculpture. The sampling area can range from a few millimeters to a centimeter in diameter and depends on the size of the swab and the prominence of the area in question.

Commercially purchased cotton swabs or applicators should not be used for this type of sampling because they typically contain an adhesive, such as polyvinyl acetate, to adhere the cotton onto the stick. Swabs for analysis should be prepared by winding a small piece of cotton fiber, gauze, or clean room wipes (additive-free) onto the end of a wooden stick. The fibers, wipes, and sticks selected can be tested before sampling to ensure that they will not be a source of contamination. A sharp, pointed stick is needed to prepare a tiny swab. The sampling stick and the cotton should only be handled with tweezers, to prevent contamination from skin cells and oils. After preparation, the swabs may be wrapped in aluminum foil for easy storage.

When a sample is ready to be taken, the selected area on the object can be swept lightly with an artist's brush to remove any loose particles or dust. A swab is dipped in an appropriate solvent (water, ethanol, acetone, chloroform, hexane, etc.) and rubbed over the small area on the object. The swab collects any solvent-soluble coating from the area for analysis. Multiple swabs, using different solvents, are usually collected from the sampling area. After collection, each sampling swab is placed in a clean glass vial, the stick is removed or its end is broken off, and the vial is sealed and labeled.

2.1.2. Particles and Cross Sections

A scalpel blade is used to remove barely visible particles and multiple-layer cross-section samples. Microsurgical scalpels have sharp, thin blades that work well. Alternatively, microscalpels may be made by adhering 1 to 2 mm broken from a razor blade onto a sharpened wooden applicator stick [16]. After a thorough examination of the object and an evaluation of its problems and the potential sampling sites, samples are removed with the visual aid of a low-power stereomicroscope. Both visible and ultraviolet light are useful in examination of the object.

Particulate samples are collected when the analysis question pertains to a pigment, stone surface, corrosion product, adhesive, or upper coating layer. They are obtained by gently scraping a small area of the object's surface. The scalpel blade is generally held at a low angle, with the blade being pulled backward or away from the cutting edge. This abrades only the very top surface and minimizes the chance of accidentally damaging (slicing or gouging) the object. The particles will often cling to the scalpel, thus aiding in the transfer of the particles from the object's surface to a glass slide. When the sample is nonhorizontal, it is best to hold, or have someone else hold, the slide directly below the sample area to prevent possible loss of valuable particles. Cotton, latex, or other gloves appropriate to the object can be worn to protect the object, sample, and future samples from contamination.

When an object such as a painting has a compound structure with multiple layers, a cross-section sample is taken that incorporates this stratigraphy from the varnish layer down through the paint layers to the support material. Again, using a stereomicroscope, a scalpel is used to remove the sample, cutting from the top down to the bottom. When possible, a sample is taken from the edge of the painting or near a preexisting crack. The resultant sample is often too small to pick up with tweezers and may not have enough static to cling to the scalpel. In these cases, an artist's brush modified to have only a few bristles is used to pick up the particle and transfer it to a glass depression slide. Occasionally, for a few obstinate samples, a short breath of air on the brush (away from the direction of the sample) will supply enough moisture to the bristles to adhere the particle temporarily for transfer. A small sliver of cured silicone, freshly cut from a flexible mold, is also extremely good to pick up and transfer small particles.

2.2. Sample Documentation and Storage

Documentation of samples is crucial for understanding and interpreting the analysis results and relating the sampled area to the conservation problem. Proper documentation starts with a picture of the object (e.g., photocopy or Polaroid) on which the explicit sampling areas can be marked. Corresponding labels should be placed on sample containers with the date of sampling and the initials of sampler. All pertinent information should be recorded in a sampling notebook. This includes a complete description of the object, the sampling area, and the sample. Also noted is the reason for sampling, along with the potential types of analyses.

Each sample is placed in its own well-labeled container. Containers can introduce their own set of problems. Plastic containers (BEEM capsules, Ziploc bags, etc.) often produce static electricity that hinders the addition and removal of samples. Gelatin capsules may hamper the determination of the presence of proteinaceous materials within the sample. A clean, glass container is best for organic analysis samples.

Glass depression slides are ideal for small samples. Normal microscope slides can be placed on top as a lid with tape hinges on one side and a tape

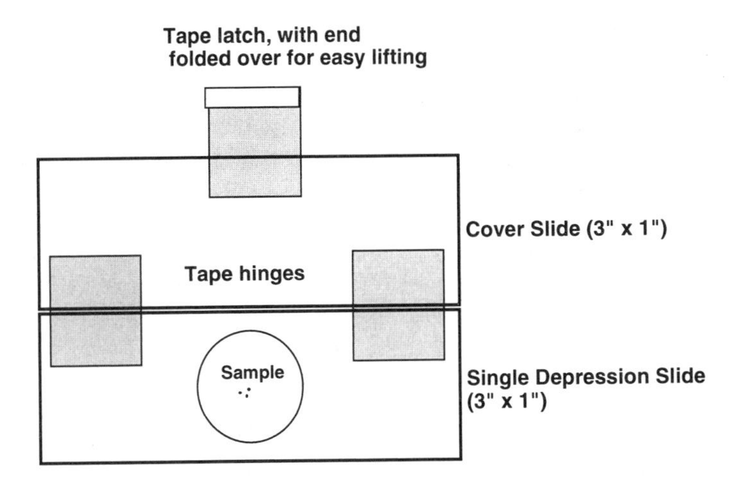

Figure 1. Sample holder using a glass single-depression slide base with microscope slide cover. Tape is placed on the outside of the slides as hinges and a latch. The end of the tape latch is folded over to create a tab for opening.

latch on the other (Figure 1). The cover slide provides suitable flat areas for the required labeling or coding. The primary advantage for using the glass depression slide container is that samples can be examined and photographed with an optical microscope numerous times without opening the container. This minimizes opportunities for contamination or loss. Portions of the sample are readily accessible to tungsten needles, cured silicon rubber slivers, or other probes, for transfer to a salt pellet for infrared analysis. Also, the samples may easily be carried and stored in microscope slide trays.

2.3. Analysis Procedures

2.3.1. Swab Samples

Swab samples contain the evaporation residue of the solvent-soluble portion removed from the sampled area on the object. For analysis, this residue is extracted from the swab using the same type of solvent (water, ethanol, acetone, chloroform, hexane, etc.) with which the sample was originally acquired. Depending on the size of the swab, the reextraction can be done in two ways. The first method is to place the fibers/residue in a microtest tube (5 × 35 mm) and add 1 to 3 drops of solvent to cover the fibers. The test tube can set for about an hour or be briefly agitated in an ultrasonic bath. Using a capillary tube or micropipette, a microdrop (1 to 5 μL) of the extract is then placed on a barium fluoride (BaF_2) pellet for infrared analysis and the solvent is allowed to evaporate. The drop procedure may be repeated to concentrate the amount of sample in the analysis area. Barium fluoride pellets are typically used for this technique because of their stability in most solvents.

A second, more direct procedure is used for tiny swabs or for individual fibers removed from a larger swab. In this case, a fiber or swap tip is placed on the BaF_2 pellet and a small drop (1 to 5 μL) of solvent is placed directly on the fiber(s). As the solvent evaporates, the extracted material will be concentrated in a ring at the outer edge of the solvent drop. For either procedure, a corresponding blank should be prepared to check for contamination from solvents, capillaries, or other contact source.

2.3.2. Particles

Samples of scrapings, powders, and particles are examined under a low-power stereo microscope to characterize the appearance, color, and homogeneity of the sample. If the sample appears homogeneous, one tiny particle is removed and placed on a BaF_2 pellet for analysis. Typically, the particle is flattened by a small stainless steel roller to provide a thin, planar sample on the pellet. A micro diamond cell also may be used for hard samples [17]. If the sample appears inhomogeneous, one particle of each apparent type is removed and analyzed in the manner described above.

Solvent microextractions can be performed on the particles to help define various components. In this procedure, a microdrop (1 to 5 μL) of solvent is placed on the particle while it is on the BaF_2 window (care must be taken that the drop encompasses only the sample of interest). Any soluble portion of the sample will collect at the edge of the solvent ring as it evaporates. Infrared spectra can then be collected from this ring of analyte, after which different solvents may be used on the same sample area to characterize the materials further. This method is especially useful when the spectrum of a coating or consolidant is difficult to identify due to the presence of inorganic components. A solvent will readily separate the organic portion while providing solubility information on the sample.

The selection of solvents for a series of microextractions depends on the components expected to be in the sample. For a sample probably composed of natural products, a typical solvent series would first use a drop of hexane to extract nonpolar components such as waxes. This may be followed by ethyl acetate or chloroform to check for the presence of natural resins, nondrying oils, and some synthetic resins. If multiple components are extracted at this point, ethanol or acetone may be used for further separation. The final solvent is usually a drop of water to check for the presence of carbohydrates or soluble proteins. After the drop of water is placed on the sample, the pellet is positioned under a warm light to heat the water, thereby increasing the solubility of many proteins while also hastening the evaporation of the droplet. After each drop dries, spectra are collected at several positions around the ring. After a series is completed, the insoluble residue is reanalyzed. It is important to analyze a blank for each solvent that is used.

For efficiency, one BaF_2 window can be prepared with multiple samples for infrared transmission analysis. To prevent confusion, detailed drawings of each sample's position and shape should be made before transferring the pellet to the infrared microspectrophotometer. A better system was developed by J. Hill (personal communication, 1993) in which a numbered grid is scratched into the pellet (Figure 2) and a sample is placed within a square. With this method the samples' positions are easily recorded in the notebook, the position numbers are readily discernible under the infrared microspectrophotometer, and chaos does not occur if the pellet is rotated when transferred to the stage.

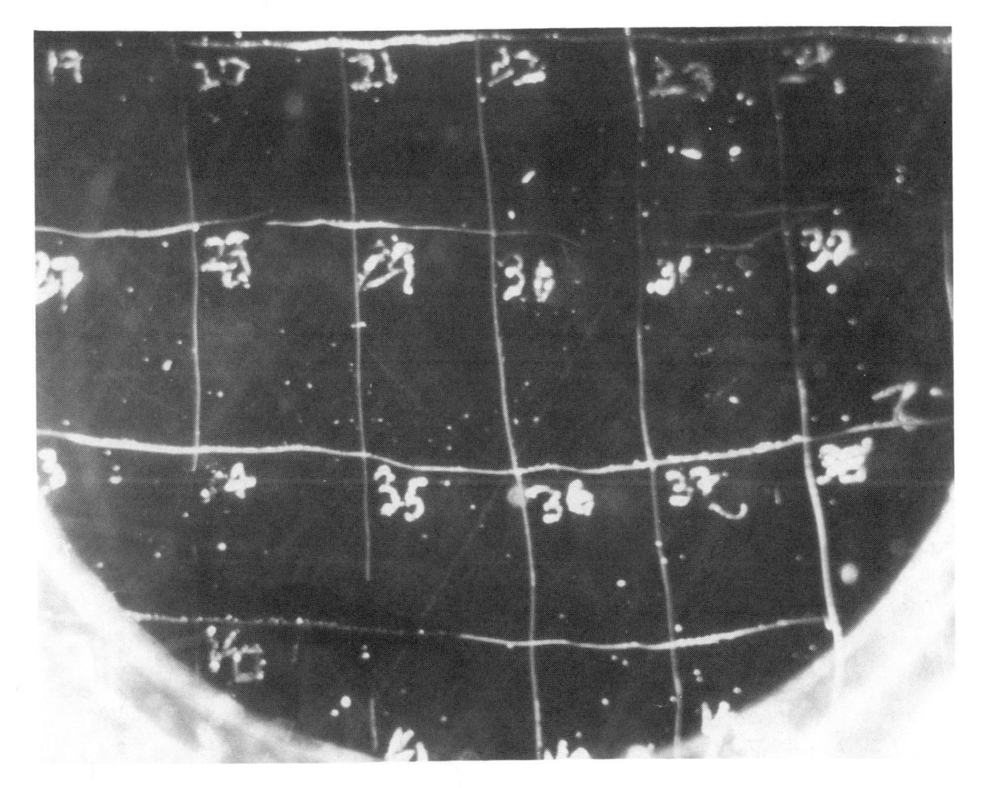

Figure 2. Photomicrograph at 10× magnification of an analysis grid scratched into a NaCl salt pellet. The analysis grid helps in the orientation and placement of multiple samples.

2.3.3. Cross-Section Samples

Before microspectroscopy was commonly used as an infrared analytical technique in conservation science, the two usual sampling techniques for the analysis of paint cross sections consisted of either examining the entire chip or selectively removing visually different sections for analysis. Both methods can misrepresent the composition of the paint. In the first method, when an entire multilayer chip is ground into a homogeneous mixture, usually with KBr salt to make a pellet, any stratigraphic or layering information about the binder is lost. In the second method, trying to sample the separate layers is time consuming and difficult unless the layer is large. Small layers may be missed altogether.

Therefore, to obtain a representative compositional analysis of multilayer media, it is important to look at an intact cross section of the sample. Infrared analysis can be done on cross sections by using reflected or transmitted light. Cross-section samples from paintings contain many materials, such as binders and pigments with various particle sizes. These variations can produce a reflectance spectrum that even after mathematical corrections looks different from the transmission spectrum of the same sample, due to the dissimilar reflectance/absorption characteristics of each material. For the analysis of binders, which generally absorb infrared radiation better than reflect, we chose to concentrate our developmental efforts on transmitted light experiments. To produce good-quality infrared transmission spectra, most paint materials need to be 1 to 10 μm thick. Microtoming is normally used to prepare a thin section of a multiple-layer sample, and an embedding medium is needed for the support of small and fragile paint samples during microtoming.

2.3.4. Embedding Media and Procedures

An ideal embedding medium for paint cross sections obtained from works of art should meet the following requirements:

1. The medium should cure at room temperature. The curing process should not be exothermic, since heat may adversely affect the organic materials or binders in the sample.
2. The mounting medium and its solvent should not react with, soak into, or otherwise interfere with the analysis of the sample.
3. The medium should be clear to ensure proper orientation of the sample in the medium as well as correct positioning of the mold in the microtome jaws.
4. The resulting embedment and sample must be easy to microtome.
5. The medium should not shrink upon curing because this will apply stress to the sample and may cause problems during microtoming.

Several types of embedding media were tested in this study and the best are listed in Table 1 (for more information, Ref. 18). These include acrylics, polyesters, epoxies, and wax. Polyester was determined to be the choice for embedding most paint cross sections from works of art. Some other materials tested, such as silicones, inorganic salts, and malleable metals, did not do well in the microtoming studies.

Paraffin, commonly employed in biological sectioning, was not selected due to its translucent to opaque quality upon hardening. It did slice nicely, although it was somewhat soft on warm days. The only interference noted was with the detection of waxes on the outer edges of samples.

While some epoxies were yellowish in color, they were all transparent upon curing. The epoxies were eliminated as a choice because their hardness was much greater than that of the samples. When microtoming the sample, this differential hardness can cause stress on the sample that makes it "pop out" of the thin section.

Some acrylic resins were found to produce heat as they cured, while others required anaerobic conditions and UV light to cure. All contained toxic chemicals and the acrylic resins were found to readily infiltrate porous samples. In addition, many acrylic polymers shrank significantly upon curing and were generally too hard for samples found in art materials. Samples from car finishes and house paints are typically harder than those found in works of art, and this hardness make the samples more compatible with the epoxy and acrylic media.

Table 1. Types and Brand Names of Polymeric Media Tested for Embedding and Microtoming of Paint Cross Sections

Type	Brands tested	Comments
Paraffin	Paraplast	Opaque, minimal shrinkage, soft, sliced well, elevated temperatures required for preparation
Epoxy	Epon 812 LX-112 Maraglas 655 SPURR	Generally needs elevated temperatures to cure, transparent though sometimes yellow, forms a very hard block that is difficult to slice at >1 micron
Acrylic	Quetol 523M Butylmethylmethacrylate LR White Krazy Glue (cyanoacrylate)	Exothermic cure reactions, transparent, shrinks more than polyesters, cuts well, may infiltrate some samples, some are very toxic
Polyester	Caroplastic Bio-Plastic Castolite	Cures at room temperature, transparent, cuts well, minimal shrinkage, may infiltrate some samples

The polyester resins (see Table 1) appeared to have the most desirable properties and were found to be an ideal embedding media for most samples found in fine art paintings. These polyester–styrene resins were clear, easy to section, cured at room temperature, and did not react with the sample. However, the polyester was found to dissolve wax on furniture finish cross sections [19], some organic dyes on inorganic carriers in modern (post A.D. 1850) pigments (L. Stodulski, personal communication, 1994) as well as fresh natural resin layers [14]. Additionally, it infiltrated porous samples.

Porous, low-binder-content paints, such as found in polychrome sculpture and wall paintings, were found to "wick in" or absorb most embedding media. Since this infiltration of polymer severely inhibits the infrared analysis of the sample, several methods were examined for preventing this occurrence [18]. The most successful method for inhibiting the infiltration was to precoat the sample with a thin layer of acrylic emulsion (Rhoplex AC-33) thickened with fumed silica to form a gel. This gel dries quickly, producing a thin acrylic layer that encapsulates very porous samples, even plaster, thereby preventing polyester infiltration without inhibiting its optimal slicing properties.

To embed a typical sample in polyester media (brands noted in Table 1), 6 drops of catalyst (methyl ethyl ketone ether) are mixed thoroughly with 10 mL of polyester–styrene resin. The resin is initially light blue in color, turns yellow when the catalyst is well mixed, and quickly becomes clear as the reaction proceeds. Excess catalyst will speed up the curing process but will also make the final mount more brittle and difficult to slice. A mold is initially half-filled with the well-mixed embedding medium and cured at room temperature for 3 to 6 h. A representative portion of the sample containing all the layers is transferred to the center of the hardened base layer in the mold with forceps or a probe and positioned in the desired orientation. It is then covered slowly with a small amount of freshly prepared polyester embedding medium and set aside to cure. Although the polyester is cured within 24 h, the best microtoming results are obtained by waiting 36 to 48 h before slicing. The medium continues to cure slowly over time. After 1 month, the microtoming becomes noticeably more difficult and the samples tend to crumble. The bottom half and the top half of the embedment should be prepared within a few days of each other to prevent a hardness differential between halves that would interfere with slicing.

Initially, plastic peel-away molds (2-cm cube) were used for embedding. However, because the presence of excess medium stresses the sample during slicing, most of the excess plastic around the sample needed to be trimmed away to minimize the contact area. To trim, a razor blade or diamond saw is used to remove all but a supporting cone of plastic with at most 1 mm of embedding media surrounding the sample at the cutting surface. Since trimming can be very time consuming, a switch was made to Pelco (Ted Pella, Inc., Tustin, California) silicone rubber molds for embedding (7 × 15 × 3 mm). These molds produce small embedments with trapezoidal tips that do not require much trimming.

2.3.5. Microtoming Procedures

Once the mold is prepared, the sample is ready to be microtomed. Our lab used two rotary microtomes: (a) Reichert-Jung Model 2040 Autocut produces slice thicknesses in the range of 1 to 100 μm, and (b) RMC Model 7000 produces thicknesses in the range of 0.01 to 10 μm. The optimum thickness for infrared transmission analysis is 1 to 10 μm.

Four types of knives (steel, tungsten carbide, glass, and diamond) are readily available for use in any type of rotary-cut microtome. The first three types of knives were evaluated, and both the tungsten and the glass knife routinely produced good, clean slices that were in the range 1 to 10 μm. The diamond knife is also expected to perform well on paint cross sections that contain hard particles, but it is typically used for ultramicrotomy of sections below 1 μm.

The steel blade is an all-purpose, inexpensive microtome blade that can routinely provide slices 15 to 30 μm thick. However, depending on the absorption of the sample, this may be too thick for infrared analysis. The thicker microtome cuts are difficult when the sample contains hard pigments, causing additional stress on the sample, which may result in crumbling. Trimming away excess embedding media, as described above, helps minimize the stresses.

The tungsten carbide blade is the same size as the steel knife and fits in the same holder. Although four times as expensive as the steel blade, it produces high-quality slices a few microns (1 to 10 μm) thick. Samples that were difficult to slice using a steel blade, such as brittle samples with large, hard particles, were sliced easily with the tungsten carbide blades. Both steel and tungsten carbide blades must be sharpened routinely to obtain optimum performance.

Glass knives are made by scoring and breaking a 1-in. square of glass to give two triangular pieces, each with a sharp edge. The cost of a glass knife is small and the knives are disposable. In fact, the glass knives should be used within a few days of generation; otherwise, the sharp edge is lost due to the fluid properties of glass. The one disadvantage of the glass knives is that the apparatus needed to make the knives costs three times the price of a tungsten carbide blade. Still, of the three types of blades, the glass knife was the easiest to learn to use. It cuts the best on samples containing finely ground pigments and routinely produces good, clean slices between 1 and 10 μm thick. The glass knife works best when used in the range 1 to 5 μm, where it creates transparent, cellophane-like thin sections. The thin sections are taken directly from the microtome, placed on a BaF_2 window, and transferred to the sample stage of the infrared microspectrophotometer for analysis. If the analysis is not done immediately, it is best to place a second BaF_2 window on top of the section to keep it flat. This second window can be removed when the sample is ready for analysis, and the thin section will usually remain flat.

2.3.6. Infrared Analysis

A Spectra-Tech IRμS organic microprobe was used for each analytical application presented in this chapter, except for the Roentgen desk furniture finish, which was analyzed using a Spectra-Tech IR-Plan II research infrared microscope positioned in the sample beam of a Perkin-Elmer Model 1760 FT-IR spectrometer. Both microspectrophotometer systems are equipped with a narrowband, cryogenically cooled mercury cadmium telluride (MCT) detector. Except where noted, spectra are the sum of 200 scans collected from 4000 to 800 cm^{-1} at a resolution of 4 cm^{-1}. The IRμS is continually purged with dry, CO_2-free air.

In both microscopes, the analysis area is selected using a dual set of adjustable knife-edge apertures to isolate matching rectangular windows at remote image focal planes above and below the sample. The IRμS has a computer-controlled motorized X-Y stage that may be programmed for automated collection of a set of spectra for mapping studies. The X-Y stage uses two stepping motors with a minimum step size of 1 μm. The IRμS also has three-dimensional graphics software to allow for imaging the absorption intensity of selected bands obtained from linear or rectangular spectral maps of samples.

3. RESULTS AND DISCUSSION

3.1. Degradation Studies

In art conservation, understanding degradation processes, causes, and rates aids in the development of methods and environments that will lengthen the lifetime of an object. This is important because the original materials in an artifact are part of its documentation and need to be retained, not replaced, in any treatment procedure.

Infrared analysis is used to examine the condition of components currently on objects as well as to evaluate changes that occur on test materials after they are subjected to artificial aging procedures.

3.1.1. Dead Sea Scrolls

In 1988, the Getty Conservation Institute began an evaluation of the optimum storage and display conditions for the Dead Sea Scrolls. After determination that controlling the environment is critical to the stabilization of severely degraded parchment samples [20], several methods were used to evaluate the state of degradation of scroll fragments. These included x-ray diffraction, liquid chromatography, and infrared spectroscopy [21].

The collective group of documents known as the Dead Sea Scrolls were written on parchment, papyrus, and copper scrolls. The parchment scrolls were prepared from the skin of sheep and goats about 2000 years ago [22]. Ancient preparation techniques for parchment involved abrasively dehairing the skin and/or using a vegetable matter and enzymatic dehairing bath. After removal from the baths, the skins were stretched tightly on a frame and the water was removed from the skins by scraping with a half-moon knife. When dry, the skins were smoothed by abrasion, usually with pumice stones. Additionally, there is some evidence that the parchment surface was treated with small amounts of vegetable tannage or cedar oil [23].

Nine Dead Sea Scroll parchment samples were analyzed by infrared spectroscopy and microspectroscopy for evaluation of their state of degradation [19]. The analyzed scroll samples consisted of some fragments removed from intact scrolls along with fragments that could not be assigned to any specific scroll. Three types of modern parchment were analyzed as references. The samples were nonhomogeneous and varied significantly from point to point. Areas on the pieces ranged in color from a very dark brown to a light beige.

Small pieces (2×1 mm) from each fragment were removed, embedded in a polyester–styrene medium, and then microtomed to obtain 10-μm thin sections. The thin sections were placed on a BaF_2 window for analysis by infrared transmitted light. Spectra were collected using an analysis window, typically 20×100 μm, isolated on the sample by apertures in the microscope located above and below the sample compartment. The motorized mapping stage was then moved such that the aperture window stepped across the width of the sample in 15-μm increments. This is referred to as a linear spectral map of the sample, and an example is shown in Figure 3 for the cross section from the Cave IV 9A3 scroll. Photomicrographs were then taken of the thin section in normal and polarized light for visual comparison of the spectra analysis (shown in Figure 4).

Collagen, a protein, is the primary component in parchment and can be converted irreversibly to gelatin. The tropocollagen molecule is held in a triple-helical structure via hydrogen bonds that upon the addition of water and/or heat can become structurally disorganized or gelatinized [24]. Additionally, the protein polypeptide chain may degrade via hydrolysis or oxidation.

Collagen denaturation to gelatin has been characterized using infrared spectrophotometry by Brodsky-Doyle et al. [25], Susi et al. [26], and Warren et al. [27], who have shown that the most noticeable change in an infrared spectrum during denaturation is an amide II band position shift from 1550 to 1530 cm^{-1} when the collagen structure is converted to the disordered form found in gelatin.

For the Dead Sea Scroll samples, the shift in amide II band position, in

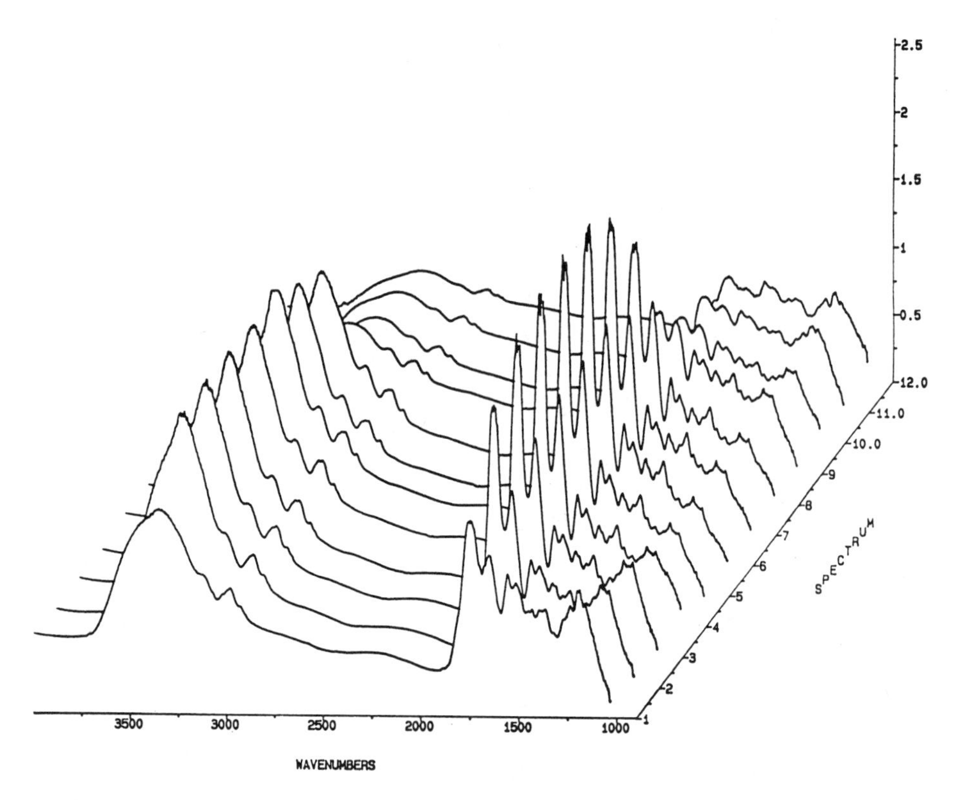

Figure 3. Linear map of 10 infrared spectra collected on a thin section of a sample from the Dead Sea Scrolls. The spectra were collected with transmitted radiation and are the sum of 200 scans at a resolution of 4 cm^{-1}. The aperture size was approximately 20 × 100 μm and the step size as 15 μm.

addition to the change in relative intensities of both amide I and II bands, shows that the denaturation and hydrolysis of the collagen in the parchment is localized in the exterior portions of the samples. The deterioration depth ranged from approximately 20 to 50 μm, depending on the condition of the sample. In general, the skin side of the parchment (smooth upper surface) exhibited degradation to a greater depth than the flesh side (reverse side).

Attenuated total reflectance (ATR) infrared analysis using a 5 × 50 mm KRS-5 crystal was also used to characterize the surface degradation on nonembedded scroll fragments for comparison to modern reference parchment samples. Figure 5 shows a bar chart comparison of the relative differences in position of the amide I and II bands in the spectra. All of the scroll samples exhibit greater degrees of denaturation than the modern parchment reference samples.

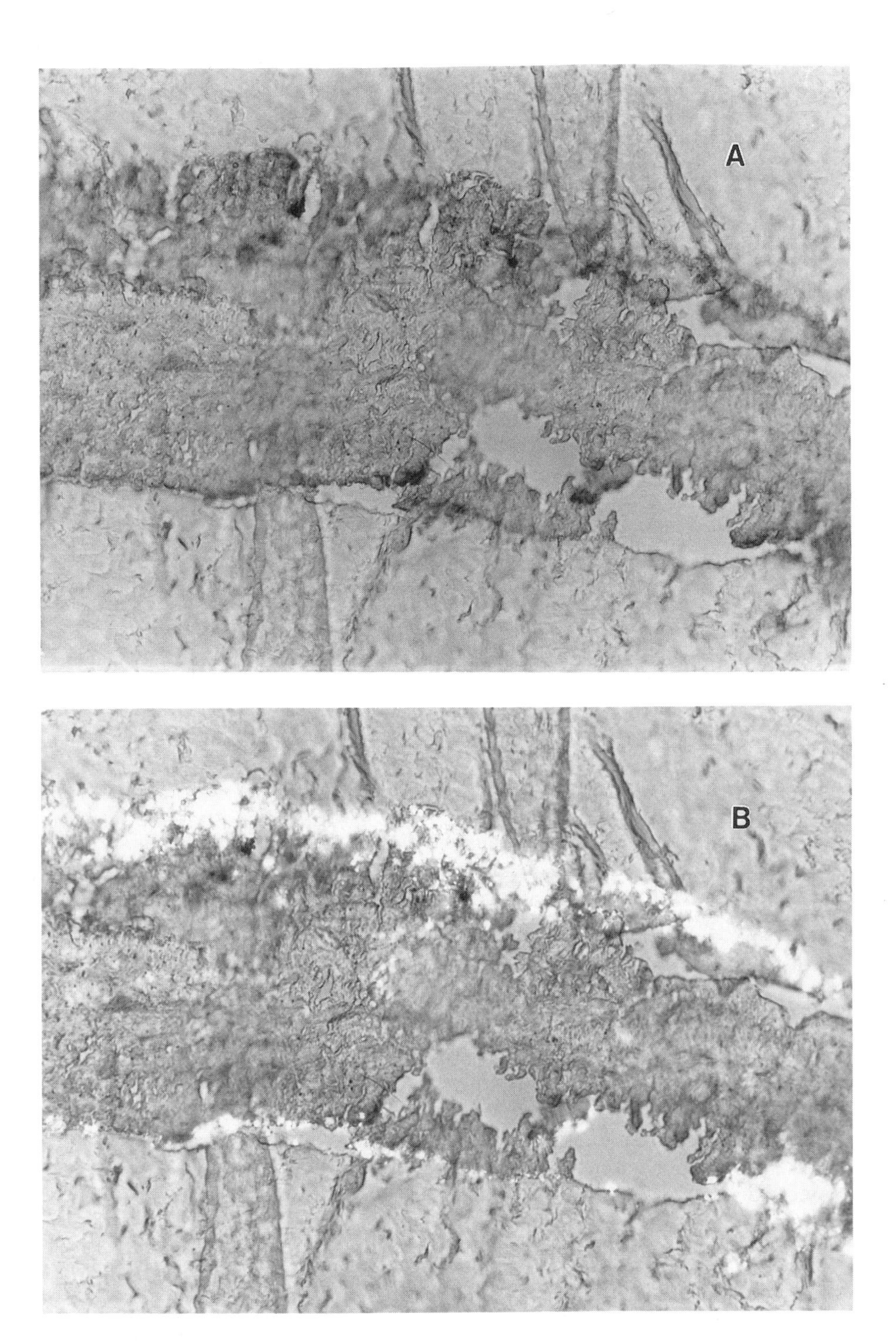

Figure 4. Photomicrographs of the thin section of Dead Sea Scroll analyzed in Figure 3, shown in normal (A) and polarized (B) light at 250×.

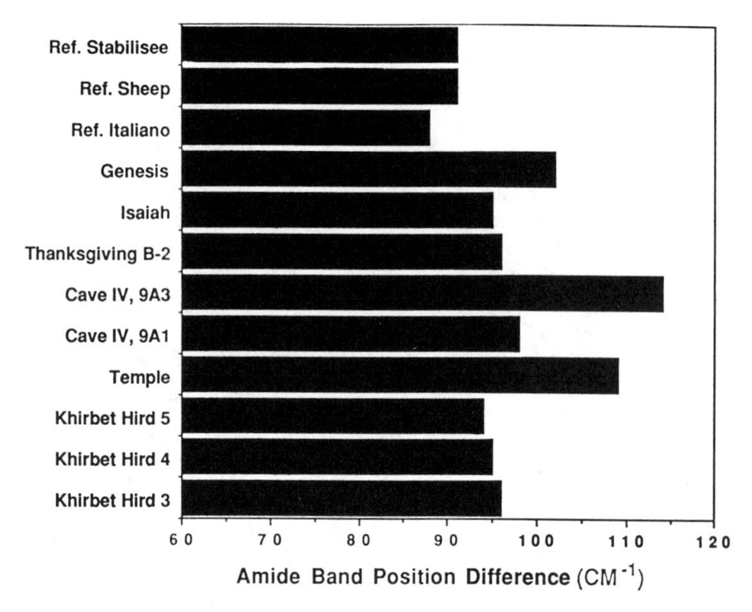

Figure 5. Bar chart to illustrate the relative differences in amide I and amide II band positions obtained from ATR spectra of nine Dead Sea Scroll samples and three modern parchment samples. The greater differences indicate more denaturation of the collagen.

Photomicrographs presented in Figure 4 in normal and cross-polarized light show that the inorganic compounds in the sample exist only near the edges and that the inorganic material appears to be embedded in the parchment rather than just as a surface encrustation. Infrared absorption bands corresponding to non-proteinaceous materials or additives were found in the spectra of the surface areas of all the scroll samples. None were detected in the interior of the samples. Three types of compounds—carbonates, silicates (pumice, talc, or dirt) and alum (aluminum ammonium sulfate)—were found as mixtures or individual components in the samples. Alum, which is used in tawing parchment, was found in two samples. Silicates, possibly from pumice used as an abrasive, were found in the exterior surfaces of all the scroll pieces. Carbonates, possibly used to fill pores or as an abrasive, were present in large amounts in one sample and in very small or nondetectable amounts in the remaining scroll samples. There could be other nonproteinaceous compounds, such as organic tanning agents, present in the samples in amounts too small to detect by infrared spectroscopy.

Microanalysis showed that the samples were nonhomogeneous and that the spectral results varied from area to area within a sample. Thus, while the results themselves were reproducible (as determined by replicate analysis) and are characteristic of the area analyzed, they may not be representative of the entire scroll from which the sample was taken. However, since several samples were analyzed covering a wide range of types, the total set of the infrared results on the pieces are exemplary of many areas of the scrolls. In addition, since some areas of the scrolls were shown to be severely degraded, the recommended storage conditions should be based on the worst (i.e., the most gelatinized) scroll.

3.1.2. Cellulose Nitrate Sculptures

Cellulose esters, the modern plastics of the early twentieth century, furnished a lightweight material that could be transformed into a myriad of shapes and colors. Because of this versatility, the plastic found many uses in everyday products (buttons, billiard balls, and cosmetic cases), in new products (movie film), and in works of art (sculptures [27]). However, it was eventually discovered that plastic products based on cellulose nitrate were inherently unstable. This polymer initially undergoes a very slow, spontaneous decomposition, at normal room conditions, that can progress to faster, autocatalytic degradation in the presence of high humidities, high temperature, or ultraviolet radiation (for a detailed explanation of this chemistry, see Ref. 28).

The Museum of Modern Art (MoMA) in New York City has three early polymer sculptures created by Naum Gabo and Antoine Pevsner in the 1920s that exemplify the constructivist technique. These sculptures were shown by infrared spectroscopy to be composed of cellulose nitrate. Examination showed that they exhibit varying degrees of deterioration, ranging from very good to poor in overall condition. The types of deterioration include crazing, cracking, and discoloration of the cellulose nitrate and corrosion of some metal components that are in contact with the plastic. When two of the pieces began exhibiting drops of clear-to-light brown liquid on the surface, MoMA and the GCI began an examination of these cellulose nitrate sculptures, to obtain a better understanding of the relation between various deterioration mechanisms and the material composition [29].

Samples were obtained from the cellulose nitrate sculptures at MoMA and from other collections. When possible, multiple samples were taken from several areas of each sculpture. The samples were acquired as broken fragments or removed as scrapings, cross sections, and exudates. They were analyzed by IR microspectroscopy at GCI and by scanning electron microscopy with electron dispersive spectroscopy (SEM-EDS), x-ray diffraction (XRD), and optical microscopy at MoMA to identify additives in the polymer as well as to identify the principal component of the plastic. The application presented below will be limited to infrared microanalysis.

Infrared analysis of the liquid exudate, or sweat, showed that it contained an inorganic nitrate. Crystals formed when the liquid was placed in a sealed container and liquified again upon exposure to air. This corresponds to reference samples of nitrate salts (e.g., calcium and zinc) which are hygroscopic and become hydrated to form a liquid at room conditions. In cellulose nitrate, the nitrate salts can be formed by the reaction of nitric acid (a degradation product) with fillers, stabilizers, or colorants in the piece. Zinc oxide was commonly used as a filler and opacifier. It also acts as a stabilizer by rapidly consuming nitric acid to form zinc nitrate.

The samples primarily contained camphor as a plasticizing agent in addition to an unidentified oil. The presence of camphor was confirmed by isolating it from the sample. This was accomplished by placing a tiny sample in the center of a $^1/_4$ in.-diameter glass tube, 4 in. in length. The area of the sample was gently heated with a flame until droplet condensation was noted on the cooler portion of the tube. The tube was then quickly broken open and a droplet was transferred to a BaF_2 pellet for analysis by infrared microspectroscopy. It was imperative to work fast, since the camphor volatilized quickly.

Figure 6 shows infrared transmittance spectra of three samples of cellulose nitrate: one in good condition, one in moderate condition, and one in poor condition, in addition to the spectra of the liquid nitrate salt and camphor. Noticeable in the spectra is the increase in the intensity of the band at 1340 cm^{-1} in the more degraded samples, due to the presence of nitrate salts in the sample. There is also a corresponding decrease in the intensity of the carbonyl band at 1735 cm^{-1} that is due to decreasing amounts of camphor.

The rate of deterioration of cellulose nitrate is dependent on the availability of oxygen and water and exposure to UV radiation [27]. This makes the surface of the polymer the most vulnerable area. To obtain a depth profile of the degradation, a cross-section sample obtained from broken fragments of the sculpture was analyzed. The sample was embedded in a polyester–styrene resin and then polished to a flat, shiny surface. Infrared spectra were collected in reflectance mode. Figure 7 shows three infrared absorbance spectra in the region 1900 to 1500 cm^{-1}; the first was collected at the surface of the cellulose nitrate sample, and the second and third were collected 1 and 2 mm, respectively, deeper into the sample. All three spectra were normalized to the height of the band at 1650 cm^{-1}, which corresponds to the nitrate ester groups in the cellulose nitrate. The spectra show that the surface of the sculpture contains cellulose nitrate, with a diminished amount of camphor (based on carbonyl at 1740 cm^{-1}), compared to the composition at a depth of 2 mm into the sample.

The diminished amounts of camphor in the deteriorated areas of the plastic can be due to sublimation. Since the deteriorated surfaces often have cracks, more area is exposed, which may allow more camphor to sublime. Conversely, the loss of camphor may have a degenerative effect on the plastic. Camphor was the best plasticizer available because it stabilized the cellulose nitrate as well as making it less brittle (for more information on camphor, see Ref. 30).

Infrared microspectroscopy, along with other analysis methods, was used to better understand the various types and degrees of degradation exhibited by each

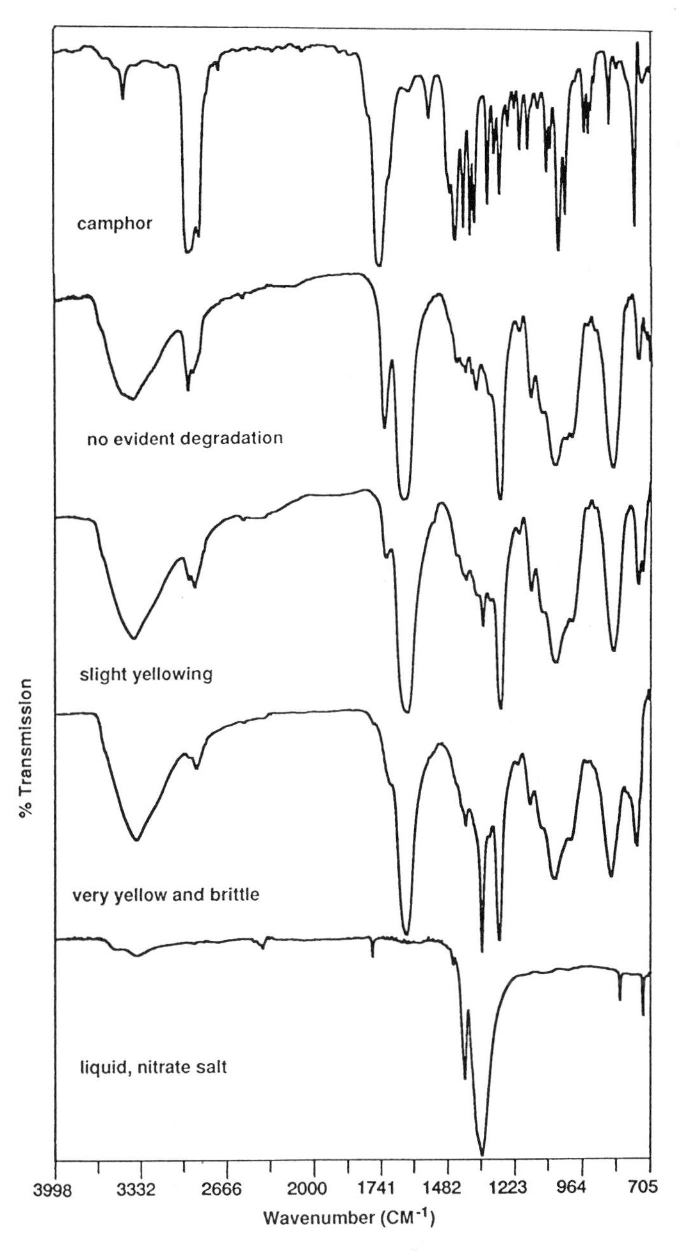

Figure 6. Infrared transmittance spectra for samples obtained from the *Torso* sculpture. Three cellulose nitrate samples: one in visibly good condition, one in moderate condition, and one in poor condition, along with the spectra of the liquid nitrate salt and camphor are shown. All spectra are the sum of 200 scans collected at 4 cm^{-1} resolution.

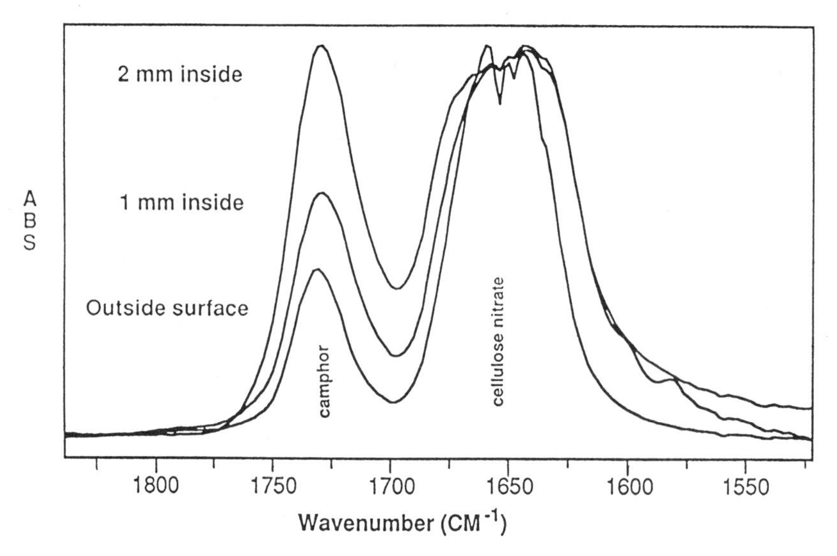

Figure 7. Infrared absorbance spectra for the region 1900 to 1500 cm^{-1} obtained from a cross-section cellulose nitrate sample, at the edge, in 1 and in 2 mm from the surface. All spectra are normalized to the height of the nitrate ester band at 1650 cm^{-1} and are the sum of 200 scans collected at 4 cm^{-1} resolution.

sculpture. At present there are no established treatments for reversing some of the degradation processes that these objects have undergone. Since the deterioration of the polymer is caused by heat, light, acid impurities, and high humidities, conservation treatment of the objects can minimize further degradation by direct control of the environment.

3.2. Identification and Characterization of Materials

The characterization of materials on works of art is one of the most important functions for infrared microspectroscopy. It is often used as the first method in a series of analysis techniques in order to classify the major components in a sample. Then, when needed, more discriminatory analysis techniques, such as gas chromatography–mass spectrometry (GC-MS) or high-performance liquid chromatography (HPLC), are used.

3.2.1. Creosote Lac Resin

Ethnographic data show that native Americans in the southwestern United States used a variety of natural materials as adhesives and coatings on their pots, baskets, and other vessels. These include pitch or sap from several pines, junipers, and brittlebush; animal fats and glues made from horns, skin, or bone from deer or mountain sheep; and resin from creosotebush lac scale insect [31]. Another source states that "holes or cracks in pottery were repaired with creosotebush lac or pitch" [32].

Found on the branches of the creosote bush, this insect resin is sometimes incorrectly labeled as creosote gum. In actuality, the resin is a secretion produced from the female creosote lac scale insect. Lac scale insects comprise a small family of species that exude a resinous substance known as lac. Shellac is the most commonly commercially known product, but exudations from these insects are also used to produce medicines and dyes.

The creosote bush grows in the deserts of California, Arizona, New Mexico, Baja California, and northern Mexico. The resin is found on the outside of branches on infected plants and can easily be removed by twisting it off the branch. The resin is hard and brittle but is thermoplastic and becomes workable when heated and thus must have been applied hot. It hardens on cooling to form a strong bond. Records show that it was used in the past to adhere stone arrowheads in their shafts and to mend broken bowls [33].

In the eastern Mojave desert, several previously undiscovered native American artifacts were recovered by cultural resource specialists of the Bureau of Land Management (BLM). On one ceramic pot, a reddish-brown material was found on the exterior surfaces near a small crack. The repair, on the otherwise intact pot, must have been done when the pot was in use. Thus it was important to identify the adhesive to better understand the culture, technology, materials usage, and perhaps even trade patterns of the native inhabitants in the area.

Infrared spectra were collected on a small sample of the adhesive from the object. Figure 8 shows the spectrum obtain from the sample as well as a reference spectrum for creosote lac. To check the solubility of the sample and to separate any impurities, a small drop of ethyl acetate was placed on the sample, then an additional spectrum was obtained from the soluble portion of the resin that collected in a ring around the sample. A parallel reference spectrum for the ethyl acetate soluble portion of creosote lac is also shown. The solubility test and infrared spectra show that the sample corresponds well with that of creosote lac.

The extraction step used for confirmation in this sample is often necessary for the identificaton of aged, natural products. These materials may contain dirt and insect and animal residues, and some of the material may be slightly chemically altered over time (e.g., oxidized). Separation of the soluble fractions aids in the interpretation of a mixture of materials as well as providing solubility information.

3.2.2. Ultramarine Pigments

Ultramarine blues are composed of complex sodium aluminum sulfosilicates and are one of the oldest blue pigments [34]. Natural ultramarine is an expensive

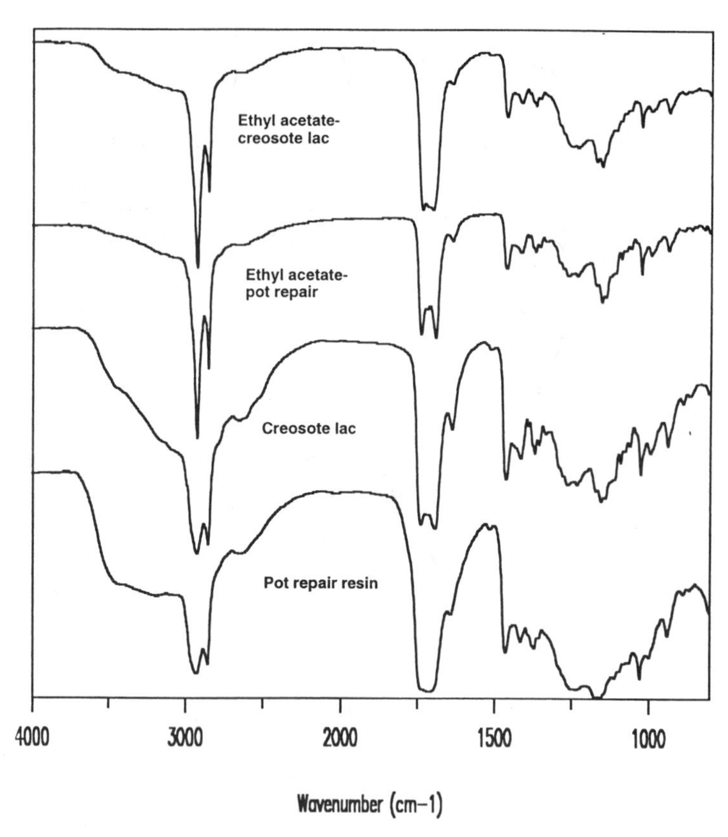

Figure 8. Infrared transmittance spectrum for bulk adhesive sample obtained from a native American ceramic pot along with a reference spectrum of creosote lac. Spectra of the ethyl acetate soluble portion of each sample are also shown. The spectra are the sum of 200 scans collected at a resolution of 4 cm^{-1}.

blue pigment produced from the semiprecious stone lapis lazuli. The colorant can be separated from extraneous minerals in the stone using a time-consuming process; one method was described by Cennino Cennini [39] in the fifteenth century. In this process, the powdered lapis was mixed with a wax–oil–resin mixture and kneaded in a weak lye solution. The dough retained the extraneous particles and the blue particles settled out. In the 1820s, a commercial process for synthetic ultramarine was developed that produced a very pure, deeply colored, fine-particle material. Optical microscopy is used to distinguish between the natural and synthetic ultramarine particles based on the size and shape of the particles. Other analysis methods [x-ray fluorescence (XRF), scanning electron microscopy with electron-dispersive spectroscopy (SEM-EDS), x-ray dif-

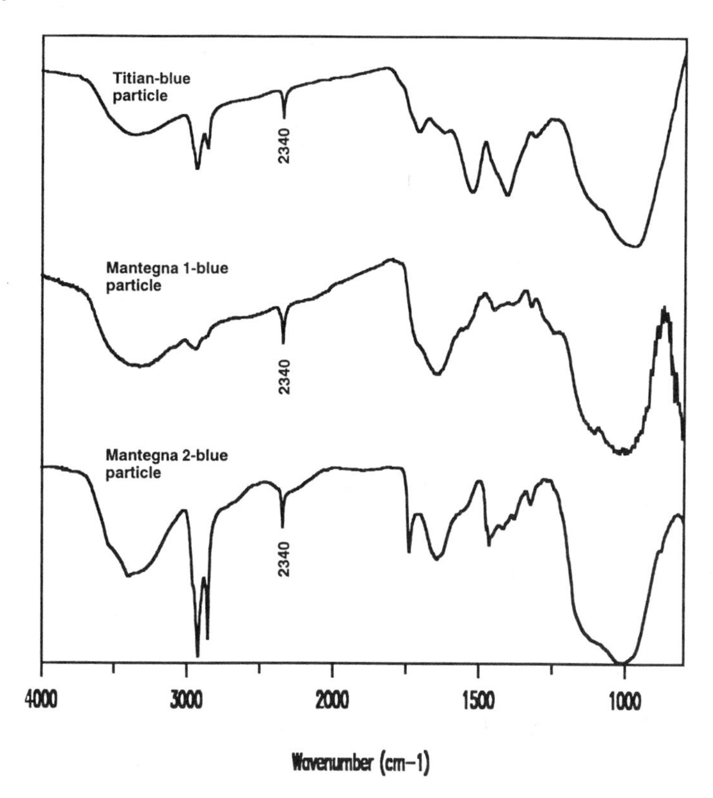

Figure 9. Infrared transmittance spectra for blue particles of natural ultramarine obtained from three fifteenth- and sixteenth-century Italian paintings. Each spectrum exhibits an absorption band at 2340 cm^{-1}. Spectra are the sum of 200 scans collected at 4 cm^{-1} resolution.

fraction (XRD)] commonly used for the identification of pigments cannot readily differentiate between natural and synthetic ultramarine.

The infrared microanalysis of ultramarine particles in this section was done by placing the particles on a BaF_2 window and analyzing with transmitted radiation. The instrument and the sample were continually purged with dry, CO_2-free air. This is important because the absorption band of interest occurs in the same region as the carbon dioxide doublet at 2340 cm^{-1}.

The spectrum obtained from the infrared microanalysis of a natural ultramarine blue pigment sampled from a fifteenth-century Italian painting contained an absorption band at 2340 cm^{-1}. Further examination of other blue particles showed that all the blue particles analyzed in that sample and in a paint sample from a second painting by the same artist contained this absorption band (Figure 9). Comparison with the infrared spectra for pigments containing cyano stretches

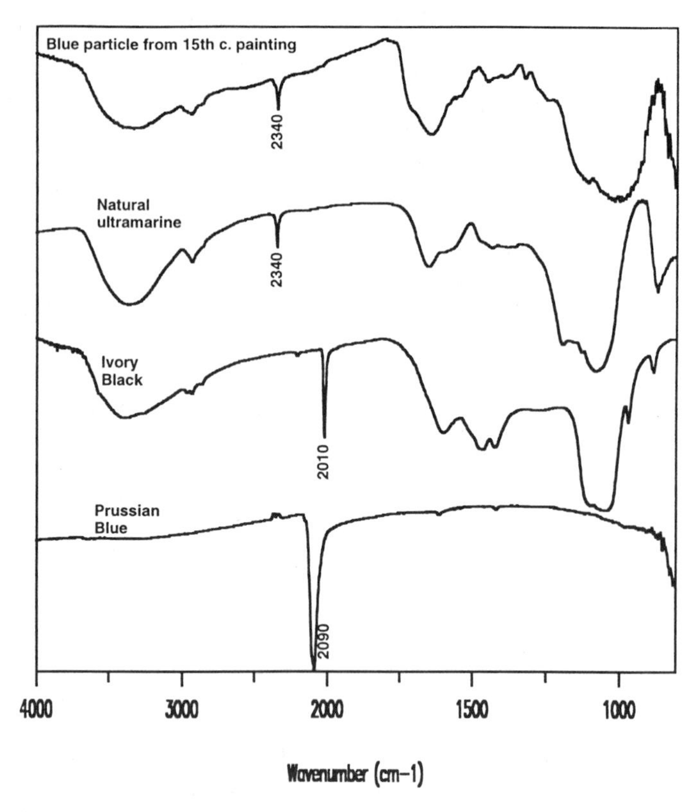

Figure 10. Comparison of infrared transmittance spectrum obtained from a blue particle on a fifteenth-century Italian painting with reference spectra for natural ultramarine, Prussian blue (iron ferrocyanide), and ivory black. Spectra are the sum of 200 scans collected at 4 cm^{-1} resolution.

(Prussian blue and bone black) showed that the small absorption band in the natural ultramarine at 2340 cm^{-1} was distinctly different from the cyano stretch, which occurs near 2100 cm^{-1} (Figure 10). In an analysis of samples from a fourteenth-century Italian manuscript, Orna et al. published an ultramarine spectrum showing this band [36]. Subsequent analyses has identified this absorption band in samples from four other Italian paintings from the fifteenth and sixteenth centuries.

Examination of over 30 samples of ultramarine showed an intriguing trend. The absorption band at 2340 cm^{-1} occurred only in lapis lazuli and natural ultramarine obtained from the Badakshan mines in Afghanistan. The band did not occur in any synthetic ultramarine samples or in samples of lapis lazuli and

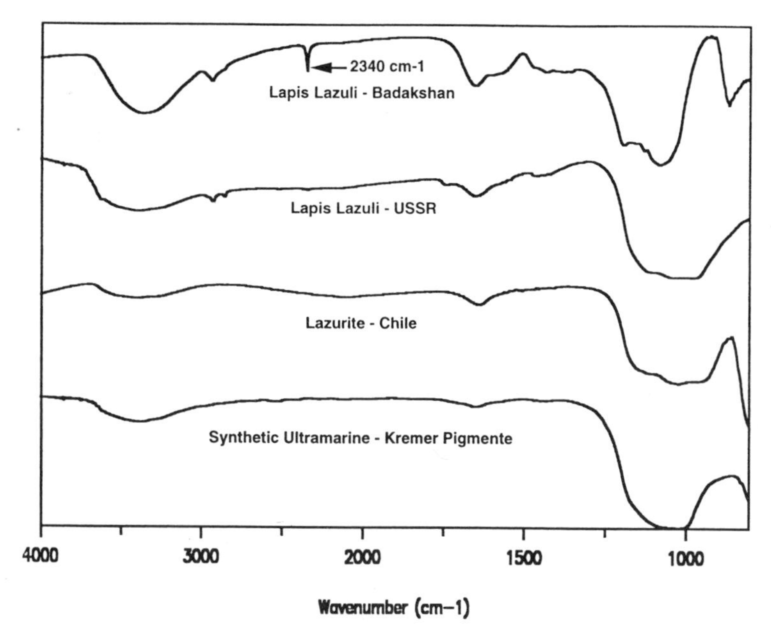

Figure 11. Comparison of infrared transmittance spectra for lapiz lazuli samples obtained from Badakshan, Afghanistan, USSR, and Chile with a spectrum for synthetic ultramarine blue. Only samples from the mines of Badakshan exhibit the absorption band at 2340 cm^{-1}. Spectra are the sum of 200 scans collected at 4 cm^{-1} resolution.

lazurite obtained from known sources in Siberia (former USSR) or in Chile (Figure 11). Thus it appears that the presence of this particular absorption band in an ultramarine sample shows that it is a natural product whose source may be the lapis lazuli mines in Afghanistan.

The Badakshan mines are the most famous source for lapis lazuli and have been worked for over 6000 years [37]. The blue color in the lapis is caused by two minerals, lazurite and hauyne. Geologically, the lapis from the Afghanistan mines contains predominantly hauyne [38]. Hauyne contains a higher concentration of calcium and sulfur than other types of sodalite minerals such as lazurite. It is probably the sulfur, S_6^+, that produces the unique infrared absorption band at 2340 cm^{-1}. While the literature states that the largest European source of lapis was from Afghanistan, there are a few small mines of hauynite found in Italy [39], but samples from the Italian mines have not yet been analyzed.

3.2.3. Finish on Eighteenth-Century German Desk

Identification of resins used for furniture finishes is important for art historical analysis of an artifact as well as for an initial material survey in restoration or conservation treatments. The materials and techniques for furniture surface treatment have, for the most part, been developed empirically throughout the ages. Collections of old recipes provide some insight into the chemistry of varnishes used for furniture surface treatment [40]. Even though the natural resins used for surface finishes (shellac, copal, pine resins, etc.) may be difficult to analyze due to their complex composition, natural variability, and susceptibility to oxidation, infrared microspectroscopy has proved to be a good method for the characterization of many furniture finishes [7].

An unusual mahogany roll-top desk with hidden drawers and elaborate mechanical devices is in the Decorative Arts collection of the J. Paul Getty Museum (Aq# 72.DA.47). The desk, although not stamped with a maker's name, is attributed to David Roentgen, circa 1785. It has a highly polished finish and beautiful filigree gilt-bronze mounts. Although the finish is in good condition,

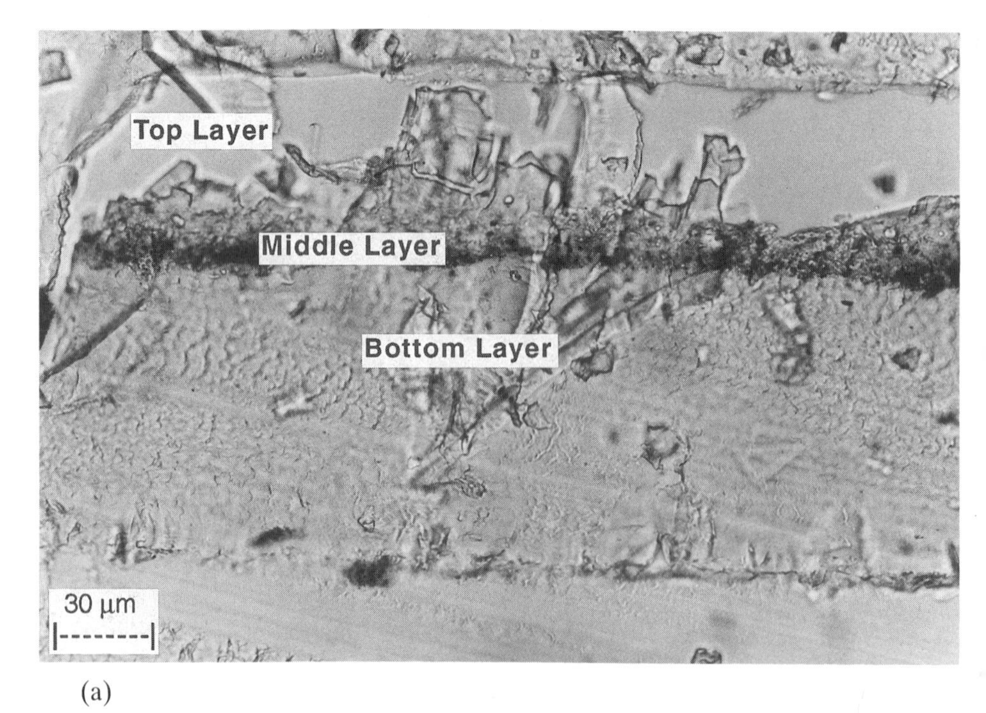

(a)

Figure 12. (a) Thin-section sample from Roentgen desk embedded in polyester resin; (b) infrared spectra (2000 to 700 cm^{-1}) for top layer (cellulose nitrate–shellac), middle layer (shellac–calcium carbonate), bottom layer (shellac), and subtraction spectrum of the middle bottom (calcium carbonate). Spectra are the sum of 120 scans obtained at 4 cm^{-1} resolution.

there was a question about whether it had been refinished, and if so, whether any former finish remained below the new layers.

Previous infrared analysis of the top surface of the German desk had shown that the sample contained shellac, cellulose nitrate, and wax. However, since this sample had been removed by scraping a small amount of finish from the surface of the piece, it was not possible to tell whether these components had been applied as layers or as a single mixture. Thus a new sample was collected as a cross section from the top of the shelf near a small crack.

The cross section was mounted in polyester–styrene embedding media, then microtomed with a glass knife to produce a thin section. A photomicrograph of this cross section after microtoming is shown in Figure 12a. Visually, the thin section contains a lower clear yellow area with a darkened opaque layer, covered with a clear layer. The top clear layer was very brittle, and portions from this layer were lost during the microtoming step. Infrared analysis (Figure 12b) showed that the lower yellow layer was shellac. Further examination of the

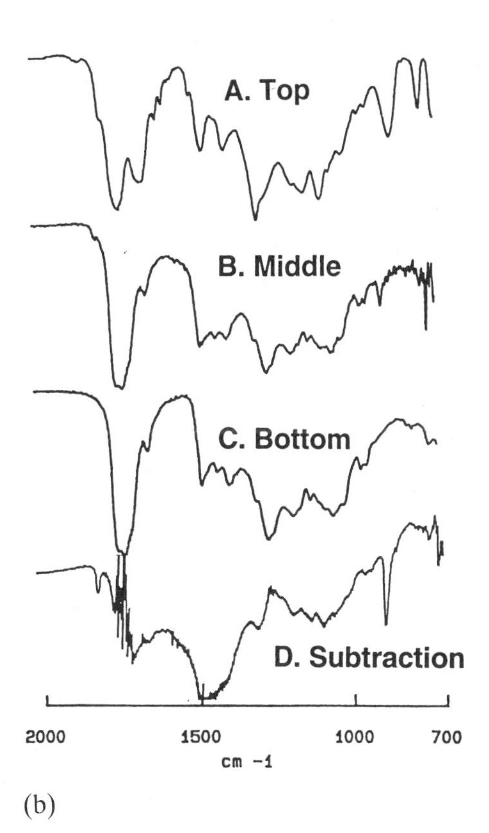

(b)

spectrum obtained from the middle, opaque layer indicated that it contained a few additional bands not present in the spectrum from the lower, clearer region. Subtraction of the spectrum of the lower, clear region from the spectrum of the middle, opaque layer resulted in a spectrum that corresponds to calcium carbonate. It is possible that the chalk was used in the final polishing stages of the shellac finish and becamse embedded in the shellac. Analysis of the top clear layer showed that it consisted of a mixture of shellac and cellulose nitrate. This was a common commercial waterproof varnish sold in the 1930s and 1940s.

Thus, for this sample, infrared microanalysis of the thin cross section provided important information for the conservator. It showed that the desk has been revarnished at least once with a cellulose nitrate–shellac varnish added on top of a thick shellac layer and that the two layers were separated by a layer containing calcium carbonate.

3.2.4. Paint Cross Sections

Andrea Mantegna was an Italian artist in the late fifteenth and early sixteenth centuries. Many of his paintings have the very characteristic matte appearance of glue (distemper) paint on a fine-quality linen support and were never meant to be varnished [41]. The distemper technique, called tuchlein, was popular in the Netherlands and Germany but had not been associated with any other Italian artists.

In the advent of the Mantegna exhibit in London and New York, Andrea Rothe, head painting conservator at the J. Paul Getty Museum (JPGM), began a detailed investigation of techniques in Mantegna's paintings and collected over 30 samples from 10 paintings in museums in Europe and the United States for binding media analysis. Various methods—FT-IR microscopy, SEM-EDS, GC-MS, and HPLC—were used to characterize the components in these samples. The primary reason for analyses was to determine the presence of glue in the media as opposed to an egg medium typical for wood-panel paintings in that time period.

One exemplary cross section from this set of samples came from Mantegna's "Adoration of the Magi" (JPGM, Aq.#85.PA.417). Infrared analysis consistently showed protein as the major component in all the paint layers. SEM-EDS analysis of the samples found that the phosphorus levels were below the detection limit in each layer. GC-MS analysis of the sample was negative for the presence of measurable amounts of cholesterol, but did demonstrate that wax was present in the sample. The absence of phosphorus and cholesterol in a proteinaceous medium is an indication that egg was not used as a binder. HPLC analysis of the amino acids in the protein showed that it contained significant amounts of hydroxyproline, thus confirming that animal glue is present in the sample. Thus all the techniques working together for the analysis of this set of samples were able to substantiate that Mantegna produced several paintings using distemper as a medium.

Computer-controlled X-Y mapping of a sample can be done to characterize the components in a sample and detect components or layers that may not be visually apparent in an optical microscope [42]. This was done on the sample mentioned above. A portion of the sample, a four-layer cross section, was embedded and polished, then placed in the infrared spectrophotometer and an analysis grid of 10 × 15 points was selected. An array of spectra were collected by reflection of the infrared radiation off the surface of the sample at each grid point. The effective resolution of the components in the sample was determined by the size of the analysis aperture and the density of the grid. The size of the aperture for this analysis was 20 × 40 μm. The selection of size was a trade-off between resolution and energy throughput. The step size was approximately 20 μm. The overlap of the windows in the X direction provided an effective increase in the resolution of the components in that direction. Each spectrum was the sum of 50 scans and took approximately 1 min for collection and processing.

From the array of spectra, contour maps were produced that provide information on the concentration and location of compounds in the sample. This was done by selecting a wavelength of interest, such as a hydrocarbon band, and plotting its intensity versus its position in the grid where it was collected. This produced an area map of the intensities of that specific band plotted as a contour map where lines connect the areas of similar value. In these black-and-white plots, variations in line thickness have been used to represent the changes in band intensity. The thickest lines correspond to the areas of strongest band intensity (i.e., highest concentration of the material). The intensities are relative to each other and the background intensity may not be zero due to other absorptions in the region. In this particular sample, because previous extensive analyses has been done to determine its components, the selected infrared absorption band and corresponding functional group can be related specifically to components in the sample. On other samples it would be precarious to identify a material based on only one infrared absorption band.

Figure 13 shows a photomicrograph of the Mantegna sample with four infrared reflectance maps. In the plot for the carbonyl band at 1730 cm^{-1}, the highest intensities shown are due to the polyester embedding media surrounding the sample, and this provides a general indication of the area analyzed. Also, the absence of a carbonyl band in the region of the sample corresponds to the previous analyses that determined that the binder did not contain egg. The plot of the band at 1416 cm^{-1} maps the intensities attributed to carbonates. This indicates that the entire sample contains carbonates with the exception of the surrounding media and a central point in the second layer. The intensity of the band at 1092 cm^{-1} may correspond to sulfates or silicates. The highest concen-

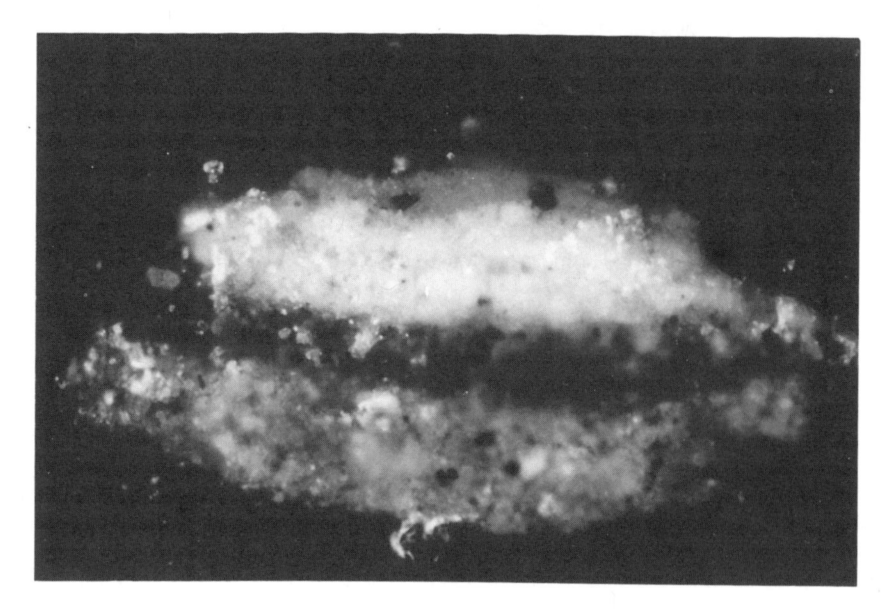

Figure 13. Photomicrograph and infrared reflectance contour maps for painting cross section from "Adoration of the Magi" by Mantegna. Absorption bands mapped are 1730 cm^{-1}, 1416 cm^{-1}, 1092 cm^{-1}, and 2919 cm^{-1}. The thickest lines correspond to the highest-intensity absorptions. Spectra are the sum of 50 scans collected at 4 cm^{-1} resolution.

tration of this material(s) is in the third layer of the sample. The plot of the intensity of the hydrocarbon band at 2919 cm^{-1} indicates the highest concentrations, which occur in the ground (bottom) layer and correspond to wax. Wax, which was found previously in the sample by GC-MS bulk analysis, can now be designated according to its location in the sample. It may be due to the original mixture of components in the ground layer or to a later painting relining procedure that used wax to adhere to a new support to the back of the painting. However, since the infrared map shows a fairly even distribution throughout the lower layer, this tends to give evidence for the wax as an original component in the ground layer.

The method of infrared reflectance mapping shows potential for the determination of materials and their locations within a cross-section sample. At this point, the method has two major limitations. The first limitation is the use of specular reflectance as an analysis method that places a minimum size of approximately 20 × 20 μm on the analysis window due to energy restrictions. Specular reflection can also result in band distortions and shifts for which it is

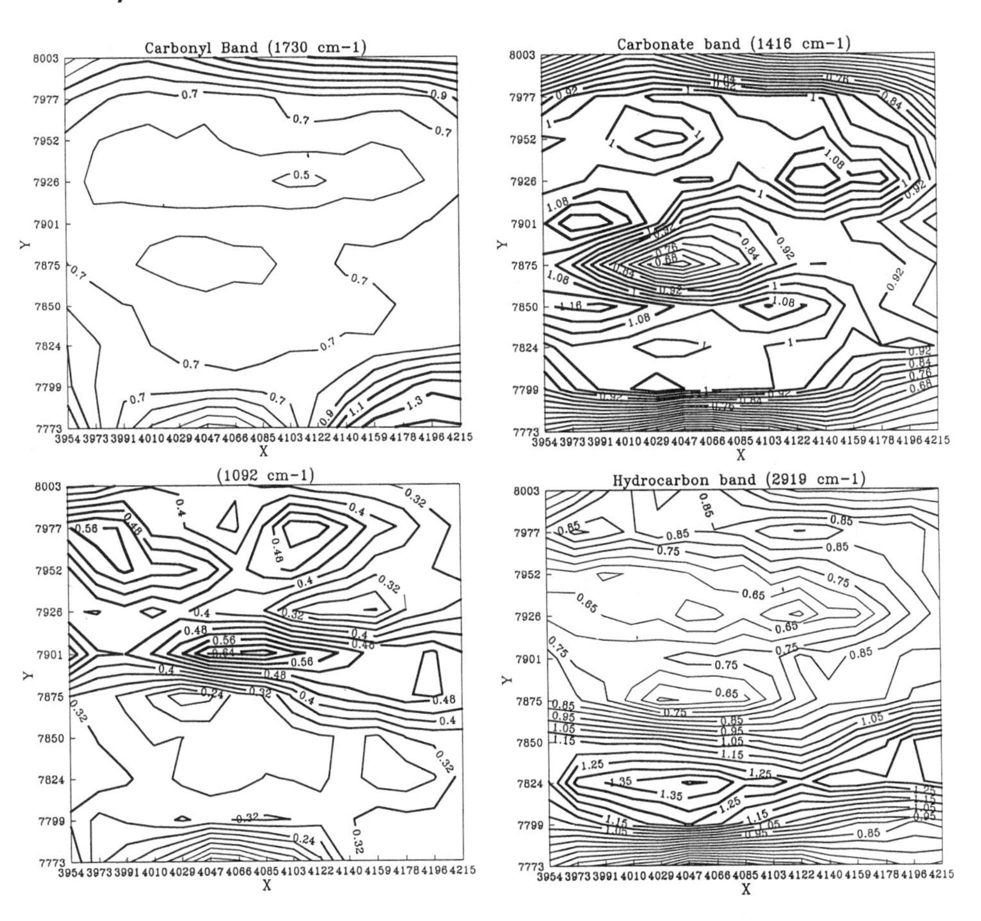

difficult to compensate. The second limitation is that a material cannot be reliably identified based on one infrared absorption band. Thus additional analyses are required to supplement the map and provide interpretation. Future computer programs should allow a map to be created based on the selection of multiple absorption bands that can help identify specific compounds.

3.3. Libraries and Reference Collections

Analysis of natural binding media materials in paints can be difficult because they are often composed of complex products ranging from proteins to oils to sugars. The composition of each of these materials may vary due to geographical location and growth environment. Some artists used more than one material, either as mixtures or as layers (animal glue ground, oil paint layer, watercolor glaze, and natural resin varnish). This diversity, combined with possible additions or changes due to restorations, conservation, and aging, make the analysis of paint media very challenging.

The accurate analysis of these materials depends on the availability of standard materials, test mixtures, and paint facsimiles for comparison. Thus GCI established a Binding Media Collection [43] and through the help of several conservators, scientists, and friends has collected over 1200 reference materials for binding media. This includes eggs from more than 15 sites around the world. Natural resin samples have been obtained from commercial and botanical sources, with materials dated from 1834 to the present. Modern art materials are also part of the collection, so that future scientists will have access to materials used in our generation. Separate collections containing over 2000 materials are also being made of conservation materials and dyes/pigments.

Because many of the materials used in art and ethnographic objects are unusual natural products or unique synthetic products, it is important to characterize the materials in the collection. Infrared spectroscopy is well suited for this purpose. Thus, using this collection, an *Artist and Artisan Materials Infrared Spectral Library* [44] was created. As part of a cooperative effort, 18 conservation and university labs have provided spectra of natural products and pigments, to bring the total number of spectra to over 800.

Figure 14 shows an example page of a wax spectrum from the library, with its corresponding information on sample preparation and analysis method. Use of the spectral library aids in the identification of many materials found in paint samples.

4. CONCLUSIONS

Infrared microspectroscopy is used to characterize the major components found in many types of works of art. Its versatility lends itself well to analysis of materials such as paints, coatings, adhesives, stone, corrosion products, textiles, and photographic materials. For some samples it can serve as a survey method to classify materials prior to other types of analysis, while for other projects, it can be used to quantify changes that occur during deterioration or aging processes.

Infrared mapping microspectroscopy is a valuable tool for the characterization of paint media in multilayered cross sections. It provides complementary information to elemental maps obtained by SEM-EDS analysis and may be done on the same cross section. This allows routine examination of a cross section for organic as well as inorganic components.

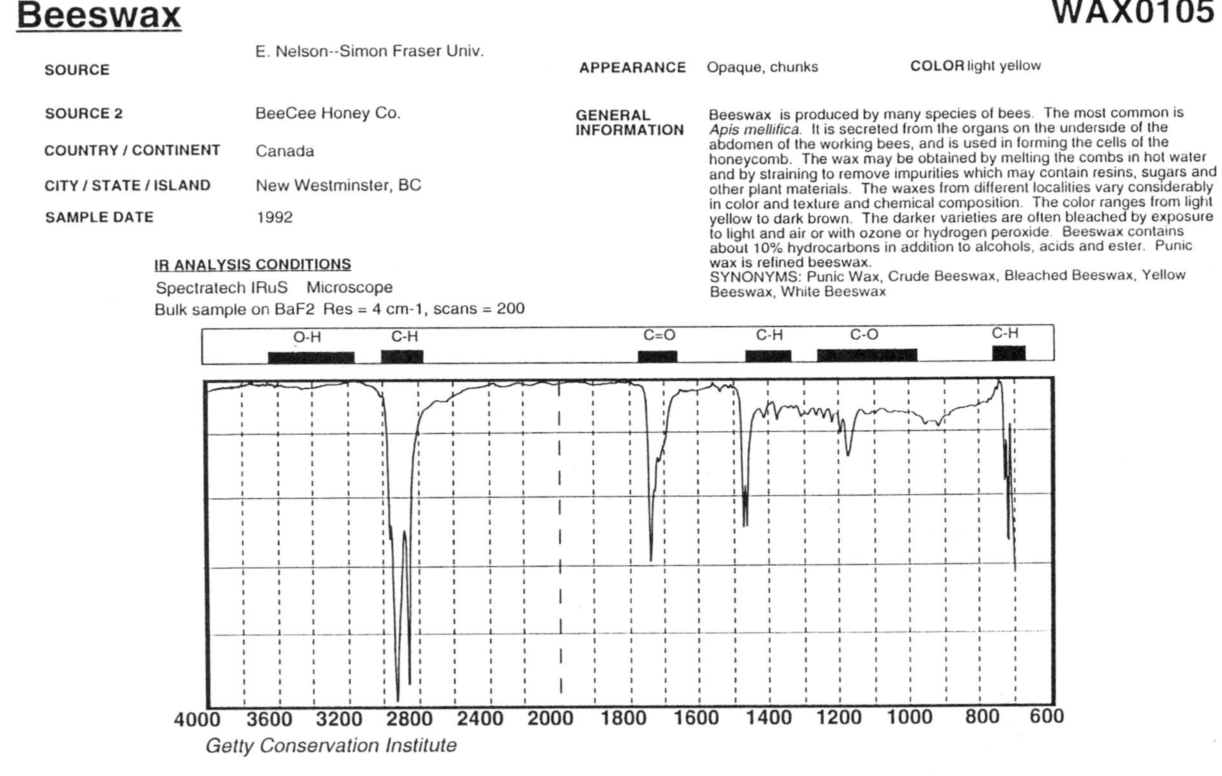

Beeswax **WAX0105**

SOURCE	E. Nelson--Simon Fraser Univ.	APPEARANCE	Opaque, chunks
		COLOR	light yellow
SOURCE 2	BeeCee Honey Co.		
COUNTRY / CONTINENT	Canada		
CITY / STATE / ISLAND	New Westminster, BC		
SAMPLE DATE	1992		

GENERAL INFORMATION

Beeswax is produced by many species of bees. The most common is *Apis mellifica*. It is secreted from the organs on the underside of the abdomen of the working bees, and is used in forming the cells of the honeycomb. The wax may be obtained by melting the combs in hot water and by straining to remove impurities which may contain resins, sugars and other plant materials. The waxes from different localities vary considerably in color and texture and chemical composition. The color ranges from light yellow to dark brown. The darker varieties are often bleached by exposure to light and air or with ozone or hydrogen peroxide. Beeswax contains about 10% hydrocarbons in addition to alcohols, acids and ester. Punic wax is refined beeswax.
SYNONYMS: Punic Wax, Crude Beeswax, Bleached Beeswax, Yellow Beeswax, White Beeswax

IR ANALYSIS CONDITIONS
Spectratech IRuS Microscope
Bulk sample on BaF2 Res = 4 cm-1, scans = 200

Beeswax **Wax**

Figure 14. Example page from *Artist and Artisan Materials Infrared Spectral Library*.

ACKNOWLEDGMENTS

The author is very grateful for the samples received from objects in the collections of the J. Paul Getty Museum, the Museum of Modern Art in New York, the Israel Museum, and the U.S. Bureau of Land Management. She also appreciates the reference samples of lapis lazuli provided by Arie Wallert and V. Golikov and of creosote lac provided by Claire Dean and Tom Holcomb. Much thanks and appreciation go to the following people for providing many helpful comments on this manuscript: Dusan Stulik, Jo Hill, Bill Ginell, Charles Selwitz, Eugenia Ordonez, Jack Katon, Nicholas Stanley Price, Brian Considine, and Andrea Rothe. Many of the procedures for embedding and microtoming were developed in conjunction with Jim Landry at Loyola Marymount University and his students Tanya Kieslich, Margaret Bolton, and Steve Bouffard.

REFERENCES

1. Feller, R. J. (1954). Dammar and Mastic I. R. analysis, *Science, 120*: 1069.
2. Johnson, M., and Packard, E. (1971). Methods used for the identification of binding media in Italian paintings of the fifteenth and sixteenth centuries, *Studies in Conservation, 16*: 145.
3. Low, M. J. D., and Baer, N. S. (1977). Application of infrared Fourier transform spectroscopy to problems in conservation, *Studies in Conservation, 22*: 116.
4. Newman, R. (1980). Some applications of infrared spectroscopy in the examination of painting materials, *Journal of the American Institute for Conservation, 19*: 42.
5. Shearer, J. C., Peters, D. C., Hoepfner, G., and Newton, T. (1983). FTIR in the service of art conservation, *Analytical Chemistry, 55*(8): 874A.
6. Masschelein-Kleiner, L. (1986). Analysis of paint media, varnishes and adhesives, Chapter X in *Scientific Examination of Easel Paintings*, ed. R. Van Schoute and H. Verougstractc-Marco, PACT No. 13, Strasbourg, pp. 185–207.
7. Derrick, M. R. (1989). Fourier transform infrared spectral analysis of natural resins used in furniture finishes, *Journal of the American Institute for Conservation, 28*: 43.
8. Meilunas, R. J., Bentsen, J. G., and Steinberg, A. (1990). Analysis of aged paint binders by FTIR spectroscopy. *Studies in Conservation, 35*: 33.
9. Van't Hul-Ehrnreich, E. H. (1970). Infrared microspectroscopy for the analysis of old painting materials, *Studies in Conservation, 15*: 175.
10. Baker, M., von Endt, D., Hopwood, W., and Erhardt, D. (1988). FTIR microspectrometry: a powerful conservation analysis tool, *AIC Preprints*, 16th annual meeting, New Orleans, La., pp. 1–13.
11. Tsang, J., and Cunningham, R. (1991). Some improvements in the study of cross-sections, *Journal of the American Institute for Conservation, 30*(2): 163.
12. Jakes, K. A., Katon, J. E., and Martoglio, P. A. (1990). Identification of dyes and

characterization of fibers by infrared and visible microspectroscopy: application to Paracas textiles, *Archaeometry '90*, pp. 305–315.
13. Martoglio, P. A., Bouffard, S. P., Sommer, A. J., and Katon, J. E. (1990). Unlocking the secrets of the past, *Analytical Chemistry*, *62*(21): 1123A–1128A.
14. Derrick, M. R., Stulik, D. C., Landry, J. M., and Bouffard, S. P. (1992). Furniture finish layer identification by infrared linear mapping microspectroscopy, *Journal of the American Institute for Conservation*, *31*(2): 225.
15. Derrick, M. R., Landry, J. M., and Stulik, D. C. (1991). *Scientific Examination of Works of Art: Infrared Microspectroscopy*, Getty Conservation Institute, Marina del Rey, Calif.
16. Hill, J. (1989). Micro-scalpel, *Western Association for Art Conservators Newsletter*, *11*(1): 8.
17. Laver, M. E., and Williams, R. S. (1978). The use of a diamond cell microsampling device for infrared spectrophotometric analyses of art and archaeological materials, *Journal for the International Institute for Conservation–Canadian Group*, *3*(2): 34.
18. Derrick, M. R., Souza, L., Florsheim, H., Kieslich, T., and Stulik, D. (1994). Polyester embedding media for paint cross-sections: problems and solutions, *Journal of the American Institute for Conservation*, accepted.
19. Godla, J. (1990). The use of wax finishes on pre-industrial American furniture, M.A. thesis, Antioch University, West Townsend, Mass.
20. Hansen, E. (1992). The effects of relative humidity on some physical properties of modern vellum: implications for the optimum relative humidity for the display and storage of parchment, *Journal of the American Institute for Conservation*, *31*(3): 325.
21. Derrick, M. (1991). Evaluation of the state of degradation of Dead Sea Scroll samples using FT-IR spectroscopy, *Book and Paper Annual*, AIC, pp. 49–65.
22. Ryder, M. L. (1958). *Nature*, *182*: 781.
23. Poole, J. B., and Reed, R. (1962). The preparation of leather and parchment by the Dead Sea Scrolls community, *Technology and Culture*, Winter, p. 1.
24. Weiner, S., Kusanovich, Z., Gil-Av, E., and Traub, W. (1980). Dead Sea Scrolls parchments; unfolding of the collagen molecules and racemization of aspartic acid, *Nature*, *28*: 820.
25. Brodsky-Doyle, B., Bendit, E. G., and Blout, E. R. (1975). Infrafed spectroscopy of collagen and collagen-like peptides, *Biopolymers*, *14*: 937.
26. Susi, H., Ard, J. S., and Carroll, R. J. (1975). Hydration and denaturation of collagen as observed by infrared spectroscopy, *Journal of the American Leather Chemists Association*, *66*(11): 508.
27. Warren, J. R., Smith, W. E., Tillman, W. J. (1969). Internal reflectance spectroscopy and the determination of the degree of denaturation of insoluble collagen, *Journal of the American Leather Chemists Association*, *64*: 4.
28. Selwitz, C. M. (1988). *Cellulose Nitrate in Conservation*, J. Paul Getty Trust, Marina del Rey, Calif.
29. Derrick, M., Stulik, D., and Ordonez, E. (1993). Deterioration of cellulose nitrate sculptures made by Gabo and Pevsner, *Saving the 20th Century: Conservation of Modern Materials*, Symposium '91, Ottawa, Ontario, Canada, pp. 169–182.

30. Reilly, J. A. (1991). Celluloid objects: their chemistry and preservation, *Journal of the American Institute for Conservation, 30*(2): 145.
31. Sutton, M. Q. (1990). Notes on creosote lac scale insect resin as a mastic and sealant in the Southwestern Great Basin. *Journal of California and Great Basin Anthropolgy*, 262–268.
32. Felger, R. S., and Moser, M. B. (1985). *People of the Desert and Sea: Ethnobotany of the Seri Indians*, The University of Arizona Press, Tucson, Ariz., p. 141.
33. Coville, F. V. (1892). The Panamint Indians of California, *American Anthropologist, 5*: 351–361.
34. Moser, F. H. (1973). Ultramarine pigments, in *Pigment Handbook*, Vol. 1, ed. T. C. Patton, Wiley, New York, p. 409.
35. Cennini, C. (1960). *Il Libro dell'Arte*, as translated by D. V. Thompson, Jr. (ed.) in *The Craftsman's Handbook*, Dover Publications, New York.
36. Orna, M. V., Lang, P. L., Katon, J. E., Matthews, T. F., and Nelson, R. S. (1989). Applications of infrared microspectroscopy to art historical questions about medieval manuscripts, *Archaeological Chemistry*, p. 265.
37. Webster, R. (1983). *Gems: Their Sources, Descriptions and Identification*, 4th ed., Butterworth, London.
38. Banerjee, A., and Hager, T. (1992). On some crystals of ''lapis lazuli,'' *Zeitschrift fuer Naturforschung, 47*: 1094.
39. Taylor, D. (1967). The sodalite group of minerals, *Contributions to Mineralogy and Petrology, 16*: 172.
40. Brachert, T. (1978–79). Historic transparent varnishes and furniture polishes I–V, trans. R. Mussey, *Maltechnik-Restauro* (1978) 1: 56, 2: 120, 3: 185, 4: 263; (1979) 2: 131.
41. Rothe, A. (1992). Mantegna's paintings in distemper, in *Andrea Mantegna*, ed. J. Martineau, Royal Academy of Arts, London, Metropolitan Museum of Art, New York.
42. Katon, J. E., Sommer, A. J., and Lang, P. L. (1989). Infrared microspectroscopy. *Applied Spectroscopy Reviews, 25*(3): 173.
43. Stulik, D. C., Derrick, M. R., Druzik, C. M., and Landry, J. M. (1991). Standard materials for the analysis of binding media and the GCI binding media library, *Paintings Group Annual*, AIC meeting, Albuquerque, N. Mex.
44. Derrick, M., and Gergen, M. (1993). *Artist and Artisan Materials Infrared Spectral Library*, Getty Conservation Institute, Marina del Rey, Calif.

9

Pharmaceutical Applications of Infrared Microspectroscopy

D. Scott Aldrich and Mark A. Smith The Upjohn Company, Kalamazoo, Michigan

1. INTRODUCTION

1.1. Background Regarding the Pharmaceutical Industry

Pharmaceutical products consist of a great range of meticulously produced formulations intended for human and veterinary use. From consumer-oriented, over-the-counter products to life-supportive precription items, the modern pharmaceutical product is the result of many years of development and careful testing to ensure safety and efficacy.

The pharmaceutical product ranges from simple sterilized water plus preservative (i.e., Sterile Water for Injection) to complex formulations designed to modulate dosage delivery and maintain the active ingredient potency. For all, the final product forms are scrupulously investigated in terms of efficacy of active ingredient, safety, sterility (or low bio-burden), homogeneity, and stability; consider that release criteria apply at the time of release through the shelf life of the product, a 1- to 5-year period. Buffering agents are utilized to maintain formulation pH and ensure quick solubility, surfactants aid mixing and solubility, preservatives guarantee the biocidal capacity, and packaging is designed to perform multiple tasks such as aiding product use in emergency rooms, pharmacy preparation, label comprehension, and offer the highest degree of product protection. Our attention to quality detail extends beyond our own facility to the vendors that supply us.

Customer expectations of our products are high, and for good reasons. It is presumed that all products work effectively, for many are utilized at critical junctures in the struggle against serious illness. Only those persons not responding to the therapeutic agent should be dissatisfied with the item. Utilization of the products in the hospital and emergency rooms requires the investigation of admixtures with other commercial diluents as well. In recent years, tampering scares have forged a new concern among packaging engineers to incorporate protective devices on the final package and often serial devices on the carton, box, and container. Also common are ''flags'' that allow the customer to judge package consistency and thus product integrity; missing tabs and incomplete or torn seals are typical indicators of compromised integrity.

We employ complex chemical syntheses to produce the active ingredient(s) and assemble the final product in pristine manufacturing facilities, generally as a mixture of excipients, active ingredient, and vehicle, contained in an application-specific package. A variety of scientific disciplines are employed to research, test, fabricate, and maintain this production. Analytical chemists, chemical engineers, process engineers, statisticians, physical chemists, microbiologists, virologists, and synthetic organic chemists are only some of the scientists working cohesively to bring the product candidate to market and assure its quality. The application of analytical microscopy is an essential part of this process, primarily as a problem-solving tool in developing new and maintaining current product lines. Infrared microspectroscopy has become one of the more powerful tools in this analytical microscopy arsenal. In this chapter we review the integration of our microscopic techniques and concentrate on the applications of the infrared microspectrometer.

1.2. Application of Microscopic Methods to Characterization Efforts

Our discipline can be described as materials science, and when applied to pharmaceuticals, provides an evaluation of a specimen in regard to its physical and chemical characteristics. Molecular and atomic characterization has become an essential aspect of the description. The character or ''nature'' of the specimen is very useful in terms of its habit, association, crystalline form, hydration state, and cleanliness. The application of the science provides a process control evaluation tool, as single component or integrated process. It aids consistent production of the active ingredient through final product, alleviating excessive loss through noncompliance or improper form. Similar rationale and methods are applied in problem-solving efforts as well. Our first-pass evaluation concentrates on a full description of the sample ''nature'' in terms of, in sequence, the physical, chemical, and atomic properties.

1.3. Application of Microscopic Methods to Problem-Solving Efforts

The simplest application of the method is description of a new drug form, which comes to us in a homogeneous state. Description of bulk properties in terms of color, powder texture, and particle size range is followed by more quantitative, crystallographic determinations via light microscopy and ultimately, infrared microspectroscopic analysis. A good example of the value of the microscope-based infrared instrument is evaluation of form differences within a sample, which is especially valuable with limited sample quantities. Stepwise evaluation is also applied to process problems, which may occur at any of the various stages of product assembly. Machine wear and disrepair, uncontrolled reactions, contamination, and supplier quality problems are a few examples that have prompted problem-solving efforts. In extreme problem cases, "firefighting" intensity is required to alleviate acute problems, and this is where the stepwise approach is especially valuable.

Problems are evaluated in a progression of descriptive methods, facilitated by light microscopy; moreover, a progression from low to high magnification (human vision through low- to high-power optics) is applied to each aspect of the description. The categorization of the material or the offending isolate may only need to be broad, such as an "animal–vegetable–mineral–crystalline" context. The subject may be isolated or free particles in a medium or contained deep in solid media. A summary of typical particle categories one encounters in the pharmaceutical industry is provided in Appendix A.

Samples we evaluate may consist of whole bulk drug or excipient powders, membrane filter particulate isolates, isolated foreign material, or anomalous material present on or within another material. The quality of the isolation procedure is often monitored with a negative control sample; test samples are evaluated for the presence of materials that are present above negative control levels. The isolated material is qualitatively evaluated through a serial scheme, appropriately deemed a "forensic" light-microscopic scheme, since we approach each incident carefully, striving not to disturb or lose trace or associated evidence that may indicate the ultimate identity and source. The focus of this book is infrared microspectroscopy; however, it is our belief that a comprehensive review of this technique as it applies to the pharmaceutical industry must also include a review of some of the fundamentals and applications of optical microscopic techniques. Although we can apply any of our analytical tools individually, progression through a forensic scheme maximizes the data obtained from a minimal amount of sample in the most efficient manner. The progression of the forensic scheme is typically (1) requester interview, (2) site inspection, (3) visual examination of the sample, (4) low-power stereomicroscopic examination and isolation of the sample, (5) compound light-microscopic examination, (6) thermal analyses, and (7) selected instrumental techniques, such as infrared spectroscopy, scanning electron microscopy, chromatography, mass spectrometry, nuclear magnetic resonance, and so on.

Isolation of the analyte from a matrix is often required. By following the forensic scheme, a method for the isolation often becomes evident. Depending on the sample, isolation steps may be performed at the visual or low-power examination steps. A description of each step in the forensic scheme is detailed in the following sections, with typical isolation techniques and then description of the infrared microspectroscopic evaluation process and examples last.

2. FORENSIC SCHEME

2.1. Requester Interview

Essential to all investigations is a comprehensive interview with the sample submitter. Conditions of formation, isolation, discovery, and so on, may hold clues to the identity of the specimen. Additionally, the submitter may have a ''hunch'' regarding the specimen, often one they have dismissed or ignored, which may be helpful in subsequent work. Coaxing a comprehensive description of the incident or scenario from the requester is an essential contribution to our success.

2.2. Site Inspection

No particulate investigation is complete without a walk-through level of understanding regarding the origin of the specimen. Although not always possible (or practical), this facet of the investigation may reveal to the analyst's eyes a situation, condition, environment, or other factor that has been taken for granted or otherwise ignored by the sample submitter. In this manner the submitter and analyst become partners in the foreign material/defect investigation. Consultant labs and off-site investigators are at a distinct disadvantage in this process, thus must pursue much of this insight from the interview of the submitter, often directing subsequent activities for the submitter to provide supporting information.

2.3. Visual Inspection

With a basic understanding of the origin of the sample, the investigation should proceed with examination of the material through low- to high-magnification light-microscopic methods. The first step, however, is visual examination. Although a visual examination may sound trivial, spending a few minutes observing and documenting the condition or nature of a specimen is an important first step at understanding the context of the material.

Some of the cases involve particle defects noted in the finished products during human or machine inspection; sterile injectable (parenteral) products are

100% inspected for defects by human auditors or human-sensitivity-based machine systems. Defects observed during these inspections and in stability trials of developing product candidates are evaluated for identity toward finding their source(s). In the lab we may simulate the inspection conditions that revealed the defects and thus begin our material evaluation from that point, in order to pinpoint the appropriate analyte. Observing the problem or defect in the same manner as the manufacturing or research method may be important, since other methods can often overlook the true defect. Defects noted during the inspection are isolated for further work, using selected techniques from methods described later. Although nonmagnified human visual limit of detection is near 60 μm [1], grading the sizes of specimens may be useful. A useful qualitative description of particulate matter size via visual detection follows:

Tiny	<60 μm
Small	60 to 100 μm
Medium	>100 to 300 μm
Large	>300 μm

2.4. Low-Power Examination

The progression of analysis should continue with stereomicroscopic examination of the specimen to further categorize and document the nature of the unknown. Each physically different (not necessarily size-different) particle may represent a different category. Categorization is initiated by discrimination of the physical features into general classes. Some useful distinctions are homogeneity, association, transparency, and shape (habit). Color is a subjective determination, which may be diagnostic. The reflective state or luster should be noted, with descending reflectivity, from the brightest (adamantine) to vitreous (glassy), then to greasy, pearly, silky, waxy, and finally, earthy.

The stereomicroscope is also used to aid in numerous other tasks in the lab, including particle transfer and preparation for higher-magnification microscopic (both light and electron) and infrared microspectroscopic preparations. Particle manipulations are dicussed in more detail in Section 3. Many analyses can be performed while viewing at low power (≤ 25×), including solubility determinations on single particles, some microchemical tests, and a simple test for magnetic attraction.

2.5. Polarized-Light-Microscopic Examination

The next step in the forensic scheme is a higher-power examination of the material by polarized-light microscopy (PLM). Detailed review of light microscopy is found in *The Particle Atlas* [2], as the chemical microscopy discipline. The application of these techniques follows sample preparation as simple as immersion in a mounting medium of selected refractive index (or multiple preps in various refractive index media). This may precede or follow isolation efforts, depending on the context or nature of the sample. Often, the unknown can be identified by polarized-light microscopy in seconds by a seasoned microscopist; commonly encountered particles such as cotton fibers, wood, paper fibers, metallic bits, graphite, carbon, paint, rubber, starch grains, and diatomaceous earth are examples of instant recognition. These particle types are referred to as *extrinsic* particles, or those common as environmental contaminants outside the product composition and are described in detail in *The Particle Atlas* [2–4]. Other particle types may easily be recognized by the pharmaceutical scientist as *intrinsic* to the formulation and package, such as rubber closure fragments and fillers, glass fragments and related materials, silicones and other common lubricants, polymers, excipients and their salts, and company-specific active ingredients, their analogous compounds, salts, and degradation products. Some common extrinsic and intrinsic particulates are discussed in Appendix A.

For samples not readily identifiable by PLM, the objective of the examination is to categorize the material further in regard to the following properties: homogeneity (one component or several), color, habit (shape), opacity, reflectivity, relative hardness, size, aggregation, description of surface and interior, description of crystallinity (birefringent or isotropic), refractive index (indices), and solubility. Literature references should be consulted for more information on the determination of crystallographic properties such as birefringence and refractive indices [5–7]; excellent reference tables of crystallographic properties are available [2–4, 8–11].

A system of categorizing by habit employed by our lab relies on an evaluation of the subject particulate matter, beginning with an assignment of relative axial dimensions. Consider the following assignments:

Description	Relative axes (x:y:z)
Equant	1:1:1
Tablet	10:4 to 8:10
Plate	10:2 to 4:10
Flake	10:<1:10
Columnar	$x \approx y$ (2 to 10):z >10
Rod	$x \approx y$ (1 to 2): z > 10
Blade or lath	10:4 to 8:>> 10
Ribbon	2 to 7:<1:>10
Needle or acicular (straight)	$x \approx y$ (<1):>10
Fiber or fibrous (wavy, curled)	$x \approx y$ (<1):>10

Examples of equant particulates are cubes and spheres. Other definitions will depend on personal preference or interpretation. For instance, flat elongated plates may be called blades, laths, or ribbons as a continuum rather than as discrete shapes. Of course, these relationships are useful in our laboratory and may not be universally accepted; however, we find their use extremely helpful in consistent shape description. Figure 1 illustrates the various habits defined above. Additional terms, such as *rounded*, *angular*, and *irregular*, aid habit categorization and the description of surface features. Terms such as *irregular*, *cracked*, *cratered*, *pitted*, *porous*, *smooth*, *rough*, *patterned*, *decrepitated*, or *reticulated* allow more differentiation of particle–particle variation. Careful attention to the individual crystal–particle–semisolid character at this stage of the evaluation may explain or support conclusions drawn at the infrared or other molecular description step (i.e., evidence of phase change or pseudomorphism in otherwise normal steroid drug crystals may be confirmed by the infrared spectral features). Consult Appendix B for definitions of these and other terms used in this chapter.

The opacity and reflectivity of a material can be of value in the choice of further analytical approaches. In an immersion preparation at high magnifications, such as 400× and greater, truly opaque particles that are reflective by top light are probably metallic or graphitic particles. It is essential to evaluate the material in this manner to distinguish truly opaque particles from very dark particles, since a common occurrence in bulk powder "speck" problems is the presence of particles that appear black or gray during visual and stereomicroscopic examinations. Rather than a homogeneous coloration, PLM exam at 400× and greater magnification often reveals the particles to be very small (<5 μm) opaque, reflective particles in a colorless matrix rather than a homogeneous opaque species. These incidents are often the result of metal wear or lubricant contact with the bulk powder; ideally, subsequent microchemical or scanning electron microscopy (SEM) energy-dispersive x-ray spectroscopy (EDX) analyses may be employed to identify the cause. Opaque particles that are not reflective may be carbon black, and a flame test (Section 2.6) may help differentiate them from metals. Contamination with oil or grease which may occur along with the inorganic or metallic particles will cause an otherwise white particle (due to light scatter) to appear gray, due to partial transmission of the illumination. The infrared microspectroscopical methods are quite useful in the elucidation of these matrix contaminants and is dicussed in Section 4.

A measure of solids hardness is appropriate, yet may be difficult with singular grains of material. Mineralogists utilize the Mohs scale of hardness [12], which is a measure of relative hardness scaled numerically 1 to 10. For instance, a common material such as quartz sand has Mohs hardness 7. Therefore, if the unknown material cannot be crushed between glass surfaces, the unknown Mohs > 7. Useful terms for softer solids are, in order: gel-like, sticky, soft (deforms without return to original shape), elastic (deforms and returns to original shape),

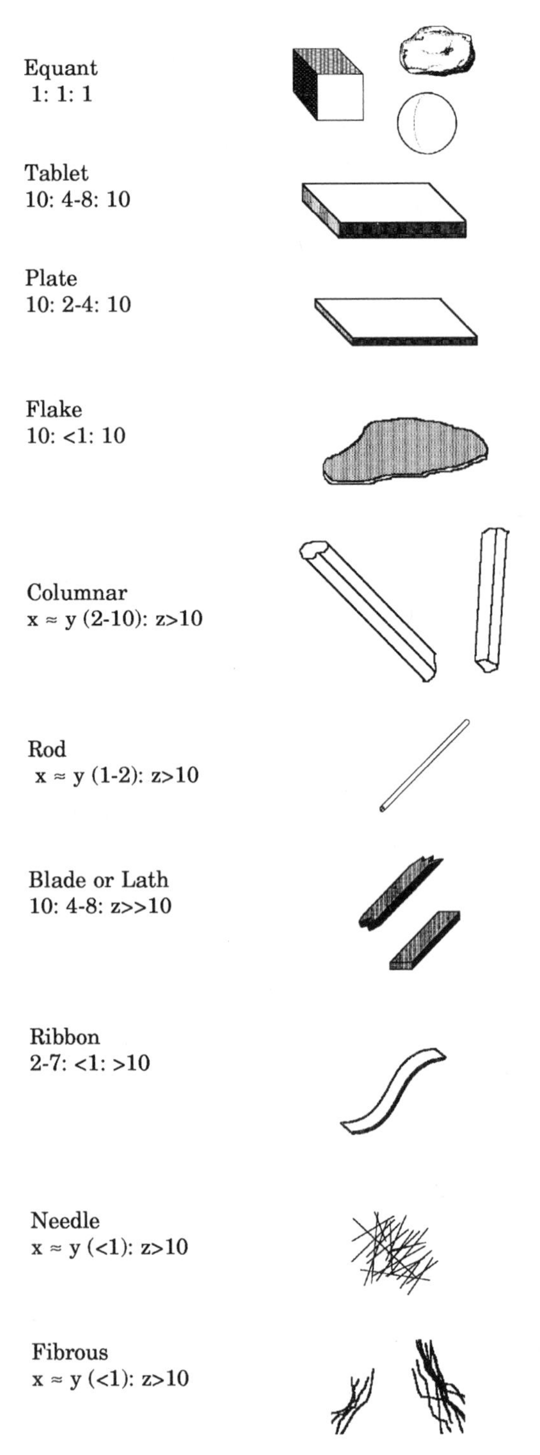

Figure 1. General morphological types: axial comparison.

and pliable (deformable, but firm). Elasticity is a fundamental property of rubber elastomers.

Specimen grain size may be measured in micrometers using a calibrated ocular reticle in the optical microscope. Size distribution, mean size, median size, and other quantitative measures require statistically relevant sampling and validated procedures and are beyond the scope of this document. We have found particulate matter descriptions to be useful; we include colloidal (0.05 to 0.5 μm), fine (to 10 μm), powder (to 100 μm), granular (to 1000 μm), chunk (millimeter sizes), and massive (several millimeter and greater).

2.6. Analysis by Heating

The goal of heating the sample, as with all of the previous steps in this forensic scheme, is to help understand the nature of the sample; questions such as: Is it heterogeneous? Is it organic/inorganic? Is it solvated/hydrated? Does it melt and at what temperature? are common. Thermal behavior may be observed in two general ways: observation of the particle visually in an open-air heating device (flame, on a quartz slide, or platinum foil) or microscopical observation of the sample while using a rheostat-controlled hot-slide device [13]. Commercial devices (Kofler stage, Kofler hot bar, and microfurnaces such as Mettler and Linkam) allow microscopic observation in a controlled-heat microfurnace and are generally termed hot-stage microscopy (HSM). Several physical changes may occur upon heating the sample. The cursory, open-air heating test is a good first-pass technique; possible observations and the substance categories that correspond to them follow:

Observation	Substance
Nothing	Inorganic
Decrepitation	Inorganic
Melting	Inorganic and organic
Decomposition	Organic
Flaming	Organic
Bubbling	Organic
Soot production	Organic
Charring	Organic
Smoking	Organic
Glowing	Elemental carbon
Residual oxide	Metal particles

Observation	Substance
Sublimation	Organic
Residual ash	Inorganic filler, organic salt

If the material is unaffected by flame, it is undoubtedly inorganic and best analyzed by selected elemental testing. Qualitative microchemical tests may be applied to selected samples to indicate elements [14–19], functional groups [15,17,19], or complex molecules [3,16,17,19–22]. Elemental compositional analyses may also be obtained in the solid state by energy-dispersive or wavelength-dispersive x-ray detection systems integrated with electron microscopes [23,24]. Organic or partially organic substances can next be evaluated by infrared (IR) analysis; however, if thermal or PLM analysis indicated a heterogeneous nature, physical or chemical methods for isolating the various components will be necessary (see Section 4.4).

Controlled thermal stages (microfurnaces) provide a more quantitative measure of thermal behavior and will reveal these effects on very small particles (down to 5 μm), given proper optics. Certain literature references [25–28] will provide the pharmaceutical investigator with essential information in thermal behavior deterministic tables and excellent background for the concepts of melts, eutectic forms, polymorph transition, dehydration, desolvation, phase transition, glass transition, sublimation, decomposition, and determination of mixtures. We have applied these thermal techniques to specimens before and after infrared microspectroscopical analyses, representing extension of the microscopists' skills. Compilation and maintenance of an internal library of thermal behavior is an invaluable aid for particle identification, especially in complex pharmaceutical systems with buffers, admixtures, abundant polymorphism, and proprietary chemicals.

2.7. More Complex Analyses

These first six steps in the forensic method should provide a clear course for analysis if the material remains unidentified at this point. IR methods are the next step in the progression if the sample has been determined to be organic. Specific infrared methods are discussed in Section 4. Inorganics can be identified by microchemical testing along with refractive index determinations; x-ray powder diffraction and SEM-EDX/WDX methods with the appropriate standards may also be employed diagnostically. A host of other techniques can be applied to trace substance identification as long as the sample context is well established. A few of these include mass spectrometry, various chromatographic methods, UV-visible spectrophotometry, and nuclear magnetic resonance (NMR) spectroscopy. Comprehensive testing schemes for important pharmaceutical excipients may be found in several texts [21,29–32].

3. ISOLATION TECHNIQUES

Common pharmaceutical dosage forms are sterile solutions (aqueous or oil vehicles), sterile suspensions, sterile powders (freeze-dried drug forms for reconstitution), coated and uncoated tablets, capsules (both hard-filled and soft elastic), aqueous fluids, emulsions, and bulk powders. Isolation techniques commonly used for the analysis of trace components in these dosage forms include direct physical removal of the contaminant with a probe, separation of the specimen by filtration, capillary withdrawal of the specimen from a liquid, centrifugation, zone drying, and extraction. Strategies for utilizing these techniques with various dosage forms are discussed in the following sections.

3.1. Dry Products: Tablets, Capsules, Bulk Powder

3.1.1. Direct Physical Removal

Using a tool designed to separate and remove the solid, the specimen(s) are selected from the matrix visually or aided by the stereomicroscope. First, a description of the type of tools, then a description of the methods.

Tools

MicroTool

Commercially available implements such as the MicroTool have well-balanced metal handles and interchangeable heads, which span a range from fine points to scalpels and flat plates. These implements not only facilitate particle isolation, but certain types aid the infrared preparation of cutting, sectioning, flattening, rolling, and so on.

Tungsten Probe

Developed and promoted by Anna Teetsov at the McCrone Research Institute [33], this is still many microscopists' favorite implement for manual methods. Tungsten wire with a diameter of ≈0.5 mm (available through McCrone Associates, Westmont, Illinois) is chemically etched with molten $NaNO_2$ to a selected tip diameter, generally in the range 1 to 200 μm. The tip diameter dictates the isolate size; particles similar in size to the tip diameter will adhere and can be transferred. Generally, the microscopist keeps a ready supply of three to four tip dimensions on hand and does not let on where they are kept!

Cat's Whisker

Similar in use to the tungsten probe, but using a coarse natural fiber. Particulate analysts at NIST [34] utilize coarse animal fibers for sample preparation for electron microscopy. A cat whisker, thick human eyelash, or similar stiff bristle acts as a reliable probe, having a stiff shaft and a fine natural tip (one may cut the hair within the main shaft for a stiffer probe with a broad tip). Any coarse animal hairs are satisfactory and lend themselves to personalization, which we microscopists favor.

Methods

The following methods for incident and transmitted light observation with a stereomicroscope have proven reliable. Hold the probe in your dominant hand, steady that hand with the other hand or against the base of the stereomicroscope (isometric pressure actually steadies your hand), and at an oblique angle, sweep the selected particle off the substrate (Figure 2). Direct pressure onto the particle may shoot the particle off-stage, frustrating even the calmest microscopist. (Ever play Tiddley-Winks?)

Static effects often complicate fine-particle manipulation efforts. When experienced, one must assure that the transfer probe is suitably sharp, that plastic containers are removed from the isolation stage, and that the room atmosphere is of reasonable relative humidity (above 30%). If still a problem, wet the probe with water or oil, which removes the static and provides some holding power. One must consider the wetting agents' effect on the subsequent microscopy

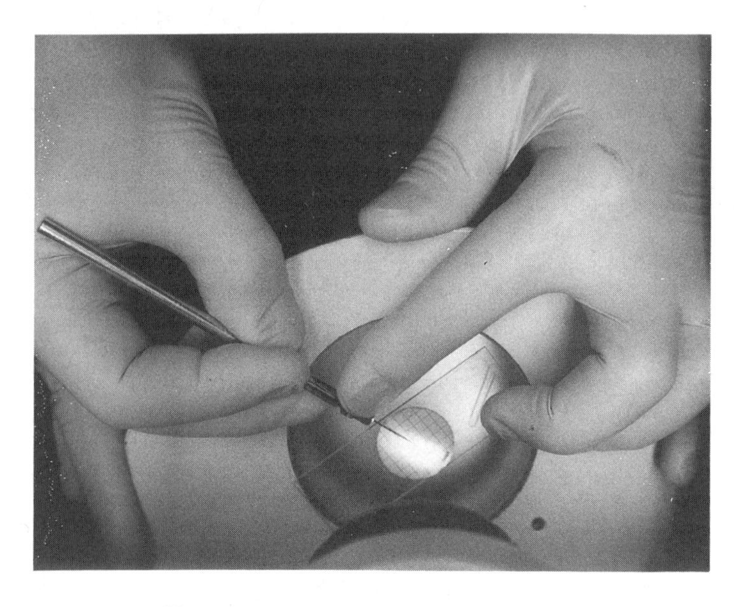

Figure 2. Solid probe isolation technique.

sample or infrared spectrum and your ability to remove it prior to the analyses or interpret the resultant spectrum.

"Goop" the end of the probe in an adhesive that will hold the articles in place. Cargille MeltMount or Aroclor thermoplastic media or adhesive from cellophane tape will work; however, this technique will require its later removal or accommodation in the resultant infrared spectrum. Actually, the adhesive from 3M Magic tape (3M, Minneapolis, Minnesota) is easily manipulated away on the infrared substrate, and as we discuss later, the adhesive matrix can be masked out of the spectrum.

Flatten the particle under a polished tool and transfer to the waiting IR substrate. The best spectra are acquired from thinned specimens, anyway. Flattening offers the "opportunity" to utilize a vexing problem to your advantage. When flattening, one often encounters transfer of much (sometimes all) of the analyte to the flattening device. Choose to use a clean coverslip (for Moh < 7) for flattening; when enough material remains for infrared analysis, use the residue on the coverslip for thermal/PLM analysis.

Place the particles within a "bubble" of clean water and lift the particle out of it—mind that surface tension! This technique has the advantage of rinsing and separating the particles in one operation. Placing a heterogeneous isolate into clean water often provides the analyst with an observation of component separation, which if destructive, still allows the analyst a means to evalaute the nature of the complex assocation.

3.2. Sterile Solutions, Aqueous or Oil Vehicles, Sterile Powders

3.2.1. Filtration

A variety of commercial filters are available in a wide range of porosities and membrane composition. Depth, screen, and track-etched filters may be employed selectively to suit the type of isolation you require; hydrophobic and hydrophilic varieties are even available. Depth membranes are tortuous-path "sponge" organic and metallic media, which provide separation of solids at a nominal porosity rating and are commonly used for solvent cleanup and microbiological removal. The particulate isolate may not be present at the filter surface, however, so our use of this membrane type is often limited to particle counting experiments and is the basis for evolving particulate matter counting methods [35]. Screen filters are often designed as prefiltration devices, capturing foreign particulates within their depth and miscible substances via adsorption, thus do not apply for analytical particulate isolation work unless used as part of a manufacturing simulation. Track-etched filters (Nuclepore) are thin sheets of polymer (polyester and polycarbonate are common) which have regular, consistently sized holes etched into the film. These types provide a relatively hard surface from which manual isolation is simple.

Using a selected membrane porosity to filter the container volume or pooled volumes of multiple containers yields the entire foreign solids population for solids greater in size than the membrane porosity. Careful selection of the membrane type and rinse solvent aids subsequent analyses. This is useful for a "clean" separation of all foreign matter and delivers a fully rinsed isolate (your choice of solvent). However, the following precautions must be considered: insufficiently cleaned glassware may contribute to the filter isolate, and examination of the filter may not directly indicate which particulate species led to the defect observation. Often, the seasoned microscopist will identify far more than the offending species, and even worse, the inexperienced microscopist will identify artifacts of the isolation procedure. The first of these concerns is handled by ensuring (through stereomicroscopic examination) that a blank analysis (negative control) is clean prior to the sample isolation, using the intended sample glassware and same filter type. The second concern requires prudence on the part of the analyst; when examining the filters stereomicroscopically, the use of oblique lighting is recommended; otherwise, soft, semisolid particles that become very thin on the membrane surface during filtration can be overlooked. A membrane isolate that demonstrates the value of varying the illumination is shown in Figures 3 and 4. Incident illumination (acute angle) in Figure 3 reveals only a few large (cellulosic) particles, which might lead the analyst to conclude

Figure 3. Photomicrograph of membrane isolates from a defective product, acute incident lighting (41×; incident light, 40°).

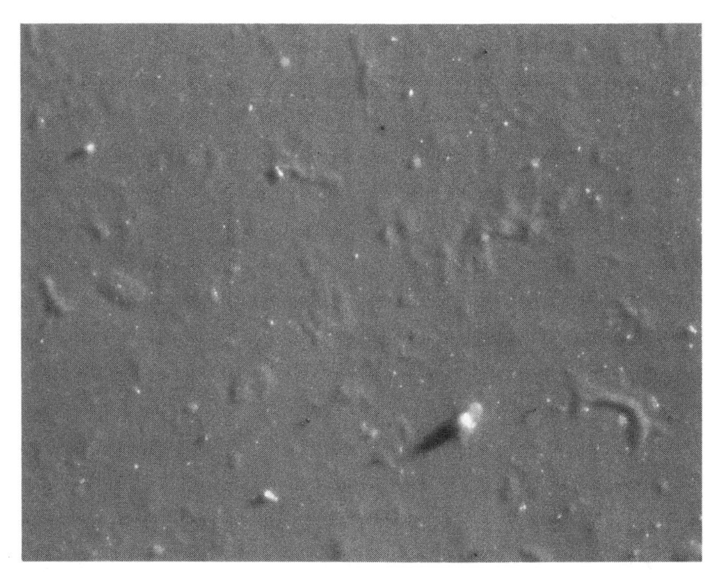

Figure 4. Photomicrograph of membrane isolates from a defective product, oblique incident lighting (41×; incident light, 5°).

that this species was the primary visible defect. Incident (oblique) illumination in Figure 4 would reveal that the cellulose particles are much less prevalent than another particle type, semisolid in consistency and conformed to the filter contours. Quite often a seemingly large particle in the product fluid becomes a thin sheet upon filtration and drying, enhancing the possiblity of overlooking its presence. The capillary withdrawal method of particle isolation (Section 3.2.2.) can be used to ensure that the isolated particles correspond to the visually observed defect.

Once particles have been isolated on a filter, they should be removed manually using the same direct particle isolation techniques described in Section 3.1, then analyzed according to the forensic scheme (light microscopy, thermal analysis, then IR when warranted). Static problems are often encountered with track-etched membranes. Particle removal can more easily be accomplished by laying the membrane on a glass slide and wetting the edge with a drop or two of water or appropriate solvent. Direct analysis of the isolates on the membrane is difficult via reflectance infrared microspectroscopy; transfer of particulates to a salt plate for transmission analysis is more reliable.

Depth-style metallic membranes such as the Selas silver membrane may be appropriate when isolates require subsequent removal via solvent extraction. This type of substrate resists solvent (not exotic pH) differences and is amenable to reflectance IR analyses.

3.2.2. Capillary Withdrawal

A direct sighting and removal of the visual defect provides a distinct isolate for evaluation and eliminates the ambiguity that may exist with filter isolates. Using the inspection method and equipment utilized for terminal inspection, the submitted product containers are examined for defects by viewing/rotating the container to suspend the offending matter. Once the particle is "sighted", it is withdrawn from the opened container using a capillary pipette (Drummond Wiretrol in volumes from 3 to 100 μL are available through VWR Scientific). The foreign material is withdrawn with a minimal amount of the product fluid. Using a stereomicroscope at low power, usually 2× to 20× magnification, the retained isolate is expelled onto a waiting substrate while direct visual contact is maintained. Minimizing the volume of product fluid through choice of capillary size (normally, 3- to 10-μL capillaries are sufficient), and partial expelling of the product fluid will make it easier to eliminate the product solution in the later preparation steps for infrared analyses. The waiting substrate may be a microscope slide, for later examination by polarized light microscopy and thermal analysis. Alternatively, selection of a vehicle-insoluble disk, such as ZnSe, CaF_2, BaF_2, KRS-5, AgCl, AgBr, produces an infrared-ready sample. We prefer zinc selenide plates because they offer the best combination of transmission range, physical stability, and resistance to organic solvents and dilute acids and bases. Zinc selenide plates can be used for direct polarized-light evaluation followed by infrared analysis. Once the particle is on the plate in the liquid droplet, it can be examined and characterized (and sometimes identified) by light microscopy using long-working-distance objectives or conventional objectives if coverslipped; this method obviously is inferior to light-microscopic examinations in a mounting medium with coverslip, but it allows both a light-microscopic and infrared spectroscopic examination of a single particle. Physical properties such as crystallinity (via birefringence), morphology, size, opacity, and homogeneity can be evaluated in this manner, but the refractive index cannot. This direct inspection-microscopy method has been utilized with great success in hundreds of product problem-solving efforts. Techniques for eliminating the product solution are discussed in Section 4, along with examples of small particle isolates.

3.2.3. Centrifugation

Using the entire container or placing an aliquot of the product in a clean centrifugation vessel will provide a separation of solids and liquid phases if needed. The technique is generally useful for larger amounts of materials or large pools of near-micrometer-sized solids. The concentrated material can then be evaluated according to the forensic scheme.

3.2.4. Zone Drying

A curious category for a pharmaceutical product; however, some "defect" incidents may be due to immiscible oils and particulate aggregates, often consisting of multiple components. This mode of separation may be the only successful direct method of analysis. Solutions that appear to be hazy may contain immiscible oils; this can usually be confirmed by the observation of the suspended, tiny droplets with Brownian motion by direct light-microscopic examination (at 400× magnification) of the hazy solution. The method for zone drying follows: The sample is placed on a waiting substrate (again, direct placement onto an IR plate negates the need for a secondary transfer of the separated material) and then is slowly dried while viewing under the stereomicroscope. The drying process may be aided by introducing a nitrogen stream or even directed breath onto the plate. Separation and, usually, coalescence of immiscible and relatively insoluble phases will occur. One such separation is shown in the photomicrograph in Figure 5 and resulting IR spectrum of the immiscible phase (silicone oil) in Figure 6. The silicone oil had been observed upon careful inspection of the vial as small immiscible fluid droplets, which to the unaided vision appeared as particles.

3.2.5. Extractions

Conventional extraction techniques can also be utilized in cases where an immiscible liquid is present, as determined by light-microscopic examination. We often utilize extraction techniques as orthogonal confirmation of the microspectroscopic evaluation, and also in evaluation of product components when searching for the source of a trace contaminant that has been identified in a particle investigation.

3.3. Sterile Suspensions

Any of the methods above may be tried; however, little success is realized trying to filter or centrifuge a product matrix which normally contains crystalline drug solids unless a solvent can be found that dissolves the drug component but not the offending material (a risky procedure if the unknown has not been characterized). Capillary isolation may be useful for obvious dark or clear anomalous solids noted within the vial. Defects are most often visually perceived as "black specks," yet dark, colored, and even large transparent foreign materials or drug crystals will appear as dark foreign matter within a white, opaque granular drug suspension.

3.3.1. Capillary Isolation

Proceed as for aqueous solutions/oils, with optical notation of the selected solid and subsequent withdrawal from the drug solids via stereomicroscopic manipulation.

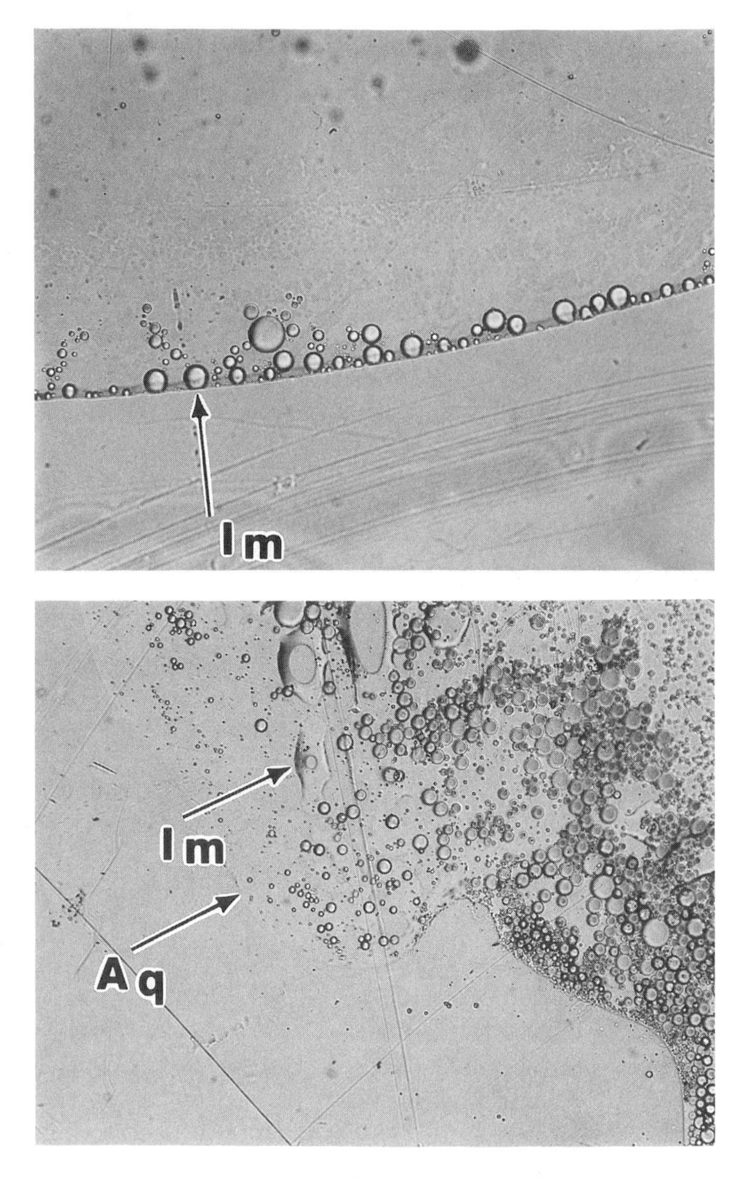

Figure 5. Photomicrograph of a zone-drying preparation. Im, immiscible phase; Aq, aqueous matrix.

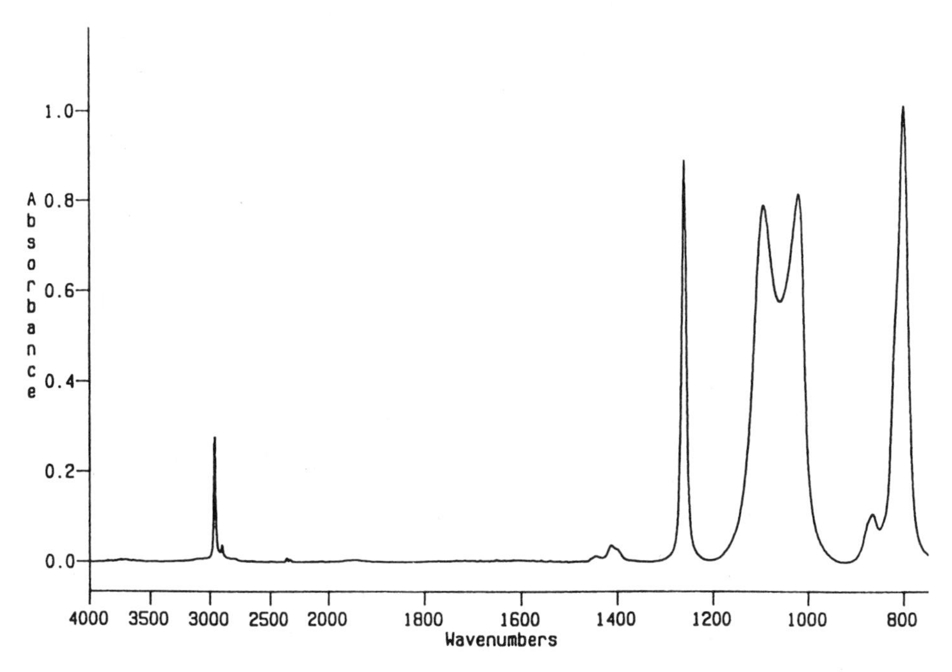

Figure 6. Infrared spectrum of silicone oil, isolated from an aqueous solution via zone drying.

3.3.2. Extraction

Extraction is a minor technique, best employed for coloration incidents, and accomplished using aqueous (acidic/basic) extractions of the separated portion.

3.3.3. Visualization by "Panning"

Visualization of the defect is aided by the surrounding opaque drug suspension. Often, the material is difficult to locate once the suspension has been agitated, even on the trip to the laboratory. We have found reasonable success using clean disposable plastic petri dishes, commercially prepared for microbiological cultures (such as the Falcon 4005 petri dish). The sample is dispersed and poured into a waiting dish in a volume that nearly covers the plastic bottom; a deep fill will not suffice. The pan is then gently rolled about the central axis on a slight angle, 5 to 10°, which "pans" the suspension into a monolayer of solids and usually reveals the foreign material. The proper background must be chosen for the type of foreign material sought, and two or more differing backgrounds, including transmitted light, may need to be employed. Conducted at low power on a stereomicroscope stage, the isolated particles can then be picked out with tungsten probe, fiber probe, MicroTool, fine forceps, or withdrawn with a capillary tube. The capillary is often most helpful for isolating the particle within a small amount of the suspension vehicle, for later separation on microscope slide or infrared disk.

4. INFRARED TECHNIQUES

To use the power of IR microspectroscopy to fullest advantage, the user must become accomplished in both microscopy and spectroscopy techniques and know when to use each. Analysts acquiring an infrared microspectrometer will usually be experienced in only one of these disciplines; you must strive to gain competence in the complementary technique to take full advantage of the capabilities of the instrument. Texts detailing infrared spectral features and interpretation are a lab necessity; since so many excellent reference books are available, one should choose those that best fit his or her own applications and interpretive philosophy. Of course, investing in commercially available spectral libraries is recommended; a few that are particularly helpful for pharmaceutical investigations are Aldrich, Sadtler (standard, monomers/polymers, inorganics, pharmaceuticals), Hummel polymers, and compilations by forensic labs (Georgia State Crime Lab, Canadian forensic). It is important to realize that a close match to a library reference may be a classification rather than an identification; it is also important to confirm possible identifications by a second, orthogonal technique, such as melting range or refractive index data. Final confirmation may be provided by other instrumental techniques, such as chromatography, mass spectrometry, or NMR spectroscopy, where possible.

We cannot stress strongly enough our conviction that all samples should be examined by a progression from simple to complex, as described in Section 2. To reiterate, this starts with a visual exam, proceeds to low-power stereomicroscopy, then compound light-microscopic evaluations. Thermal analyses are usually valuable in early stages of particle identification as well. The objectives of these exams are (1) gaining an understanding of the sample [i.e., what is to be analyzed (not always as easy as it sounds)], (2) determining whether the unknown material is a single component or a mixture, (3) determining whether the sample is organic or inorganic (using refractive index and thermal data) to allow choice of technique (SEM-EDX or microchemical for inorganics, IR for organics), and (4) identification by recognition and/or optical properties. Often, common environmental particles can best be identified by PLM; too often we have seen published spectra of cotton fibers that could be identified more efficiently and reliably by light microscopy. If you are not a microscopist, you are well advised to obtain a copy of *The Particle Atlas* and become familiar with it. Certainly, DeLuca et al. [23], Borchert et al. [24], and Barber [36] provide comprehensive overviews of pharmaceutical particle characterization methods and process control concepts, should any of the preceding discussion prove incomplete.

4.1. Simple Analysis of a Dry Solid

Occasionally, initial light-microscopic and thermal analyses will reveal that your unknown is a single organic entity. In this case IR spectroscopy is the method of choice for the first pass at identifying it. Of course, identification through melting point and refractive index can sometimes be accomplished, but this approach can be time consuming, may require a comprehensive database, or may be limited to well-established compounds. If your unknown is a polymorph or hydrate of a published compound, this may confound identification by physical properties. Again, in the case of a single entity, choice of IR sampling techniques may be related to the amount of material available for characterization. If tens of milligrams of sample are available, the best approach is conventional macroscopic preparation, with the first preparation choice being a KBr pellet. KBr preps allow a full spectral range (down to 400 cm^{-1}), maximize signal-to-noise ratio, and better match the preparation techniques used in most commercial databases. However, even in cases where sample amount seems plentiful, it might be desirable to scale down the IR technique to conserve sample in case additional spectroscopic analyses become necessary. Scaling down to IR microspectroscopy is sometimes selected for efficiency reasons also, since a neat (usually as-is or post-water-rinse, flattened) sample prep is quicker than making a KBr dispersion; signal-to-noise ratio trade-offs can be overcome by longer signal averaging if this is an issue. This is fine, but you should be aware that the IR microspectrometer can sometimes affect the spectral outcome (we discuss dehydration later), so identification of your unknown via library searching (of KBr prepared references) may be thwarted even if the database contains the unknown compound; this problem can offset any gain in sample preparation efficiency. One way to avoid library mismatch is to create your own library of references collected by your specific technique.

More often in trace analysis, more specifically single-particle analysis, there are usually only microgram or smaller quantities available for all analyses. The choices for IR analysis are limited to micro-KBr pellets or IR microspectroscopy. We start with a simple, routine case of analyzing a dry, solid particle with the IR microspectrometer; micro-KBr pellets are discussed later. Assuming that your sample has been physically isolated by manual manipulation or filtration, and that you have determined through light microscopy and thermal analysis that the material is a single organic species, your goal is to present the unknown to the IR so that a spectrum of good quality can be collected to allow you to identify or categorize the material. A transmission spectrum should be attempted first. Particle manipulation skills are essential for successful IR microspectroscopy; use of the techniques described in Section 3.1 is appropriate. Normally, transferral of the particle to an IR-transparent substrate is performed with a fine tungsten probe while viewing at stereomicroscope magnification. Some particles may be difficult to remove from the probe; if you know something about the solubility of the unknown, you can choose a solvent in which it will not dissolve and place a 1 μL droplet of this solvent on the substrate. Of course, if you select water as the solvent, make sure that you use a water-insoluble substrate. The particle should fall off the probe when the tip is placed into the droplet. Unless the particle has a distinctive physical appearance, make sure that you keep the particle in view while the solvent evaporates to assure that you really are analyzing the particle of interest. Your early attempts will center on particles hundreds of micrometers in size. As you hone your skills, subvisible (<60 μm) particles will be no chore.

Once you have placed the particle on the plate, it is a good idea to scribe a mark on the plate to help you locate the sample in the IR microscope and provide assurance that you are analyzing the particle of interest. If you place the particle near the outer edges of the plate, you can scribe a line back to the edge, then line up the line in a north/south/east/west configuration in the plate holder. Finding the particle visually while the preparation is on the stage of the IR microspectrometer is expedited because you can easily see light scattered by the line when it intersects the visible light spot (this is especially pronounced with illumination from above the sample). Although scribes on the surface limit the life of your salt plates, you can squeeze many samples onto a plate; furthermore, sodium chloride windows are cheap compared to your analytical time and have a sufficient transmission range for narrowband MCT detectors.

Most particles will produce better spectra if they are flattened and thinned to less than 10 μm prior to analysis. A variety of techniques have been described to achieve flattening; most particles can be thinned adequately by rolling over them with a probe, either on the salt plate or before transfer. The key to flattening techniques is avoidance of artifacts or secondary contamination while achieving good infrared transmission. As mentioned earlier, pressing the specimen under a glass coverslip also works well. Filled polymers or elastomeric particles may require diamond anvils or other techniques. In any case, inspection of the ratioed spectrum will reveal if any of the stronger bands are unduly broadened. This is an indication of saturation of the features due to excessive sample thickness; the relative intensities of the spectral features will be affected and identification through spectral searching hindered. An example of this problem is evident by comparing the two spectra of ibuprofen in Figure 7. Note the broadened bands and distortion of relative intensities evident in the spectrum of thick (about 10 μm) crystalline drug. Flattening the specimen (to about 2 μm) yields a dramatic

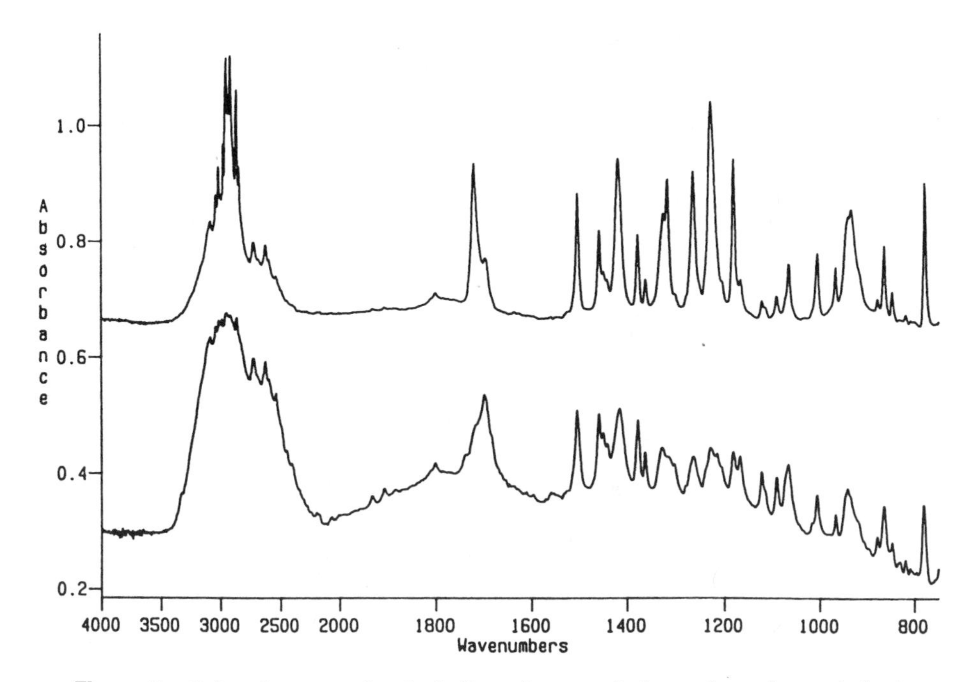

Figure 7. Infrared spectra of a single ibuprofen crystal: (bottom) not flattened; (top) flattened.

improvement, demonstrating the value of thinning the sample or selecting a thinner region (perhaps along an edge) of the sample.

The remote aperture selection (i.e., interactive masking) should be configured so that they generally correspond to the boundaries of the particle or encompass only a portion of the particle. Nothing is gained by setting the mask boundaries larger than the particle. Diffraction effects can be minimized by collecting a background spectrum at the same mask setting as the sample, usually immediately before or following collection of the sample spectrum. This has the added benefit of minimizing atmospheric spectral features and contribution from the substrate disks.

4.2. Micro-KBr (Pinhole) Preparation

We have successfully applied the use of a 500-μm (or smaller) aperture in a stainless steel disk, encapsulating the analyte in a small amount of KBr using a technique reported in the literature [37,38] and refined in our company by staff scientists Chao, Fox, and Duholke. The preparation is by hand pressure, which using the small pin compression anvils (Figure 8) generates enough pressure to

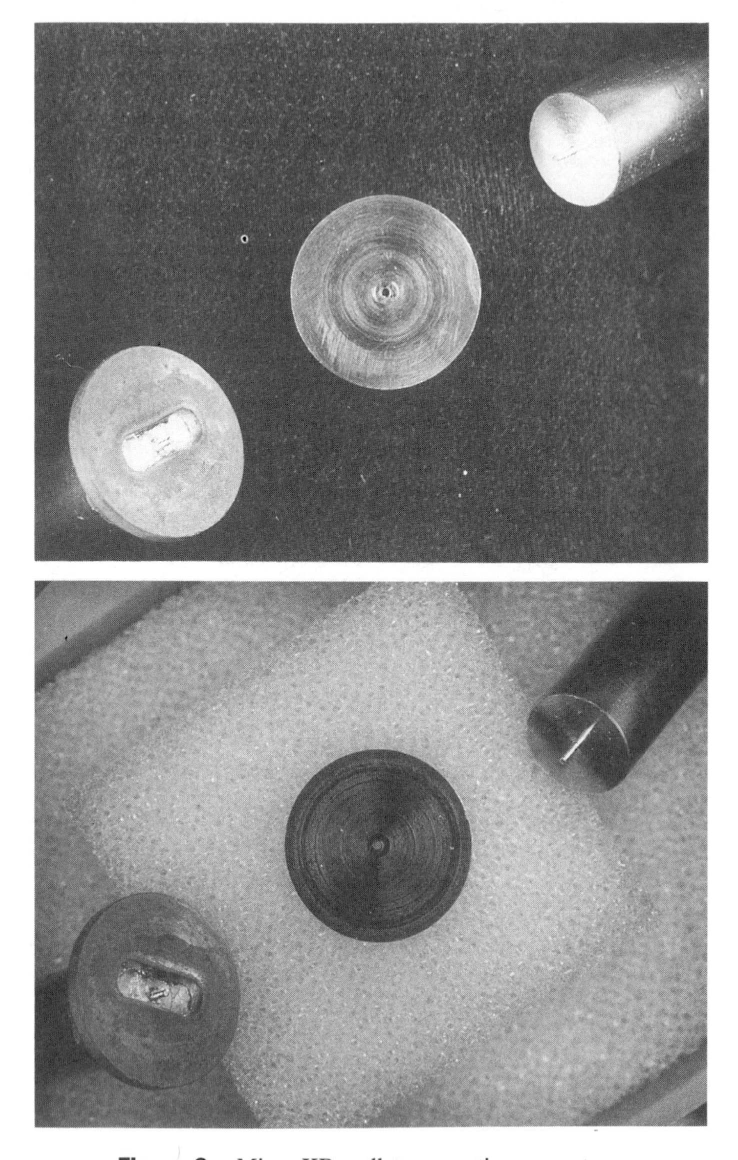

Figure 8. Micro-KBr pellet preparation apparatus.

clear the KBr and fill the voids around the particle. Utilization of the stereomicroscope is essential for this application, since placement of the specimen near the center of the aperture is important, and assuring a clean steel disk prior to encapsulation avoids misdiagnoses. The technique works well for single particles generally greater than 100 μm, or for smaller particles if several can be incorporated into the pellet. Analysis can be performed with an IR microspectrometer or with a standard beam condenser in a conventional spectrometer. This technique may be useful in situations where microspectrometer-induced dehydration of a material is suspected (Section 4.7).

4.3. Analysis of Particles in a Solution

A typical analysis for the pharmaceutical scientist is the identification of particulates in parenteral solutions. An obvious strategy to isolate the particulate defects is filtration followed by transfer of the particles to a salt plate with a probe; however, for a single particle near the visible threshold, more confidence in particle integrity can be achieved by a more direct capillary isolation scheme on a zinc selenide plate, as described in Section 3.2. After light-microscopic examination, the solution must be eliminated; this is accomplished while viewing with a stereomicroscope, and the method will depend on the physical nature of the particle. If the particle consistency will allow, it can be teased out of the droplet with a probe. It is then a simple matter to wick away the excess solution. However, particles will often exist as semisolids in the solution and may break apart when probed (of course, this is frustrating, but may also be diagnostic). The excess solution must be slowly wicked away from around these particle types. This requires some practice, because if the solution is wicked away too rapidly, the particle will be lost. Wicking can be achieved with a wedge of absorbent material such as Whatman filter paper or cellulose ester membrane, and the rate of wicking can be controlled somewhat by the shape and placement of the wedge. Substituting a less-absorbent material (such as Dacron cloth) can also slow down the wicking rate. In either case, the wedge is likely to shed particles, so direct visual contact with the sample is essential.

Occasionally, some particle types will disintegrate during the wicking process. Another practical approach is to expel these types of particles onto a depth membrane of small porosity, such as a 0.8-μm Millipore AABP mixed cellulose esters membrane, while maintaining visual contact with a stereomicroscope. The solution will quickly wick away (be absorbed), leaving particulate residues that can be washed and sampled with a sharpened probe.

After isolation, the particle should be washed (usually with water) to remove residual amounts of the product solution. Our favored washing technique involves flooding the particle with a few microliters of solvent, then wicking away the wash. This is repeated several times. Teetsov has described a similar method

[33] for probe-controlled rinsing, which works as well. The IR spectrum of the particle can be collected as described previously, with flattening if needed. Although this may sound tedious, an experienced analyst can isolate, wash, and analyze several particles an hour. Figure 9 shows a typical small particle isolate, placed onto a ZnSe substrate from a capillary isolation operation, photographed within the product fluid. The particle has two species, a twisted fiber with an adherent amorphous matrix. The fiber was immediately identified by PLM techniques (in situ) as cotton; further characterization by infrared is not necessary. Infrared analysis of the amorphous matrix was conducted after first separating the product fluid (by wicking) from the particle and washing the particle with water (other solvents may be selected). The infrared spectrum of the dried preparation is shown in Figure 10 (identified as polystyrene).

A common item encountered in evaluating particulates in pharmaceutical solutions is methyl silicone oils. This is an inert lubricant that is used extensively in rubber closure processing; invariably, if the product package includes a siliconized rubber closure, a microgram quantity of the silicone will find its way into the product solution. Although this is not considered an inherent particulate defect, the silicone oil may deposit onto other hydrophobic particles present in the solution, hence will show up in the infrared spectrum of the particle. This can be misleading when interpreting the spectrum, since silicone oil features

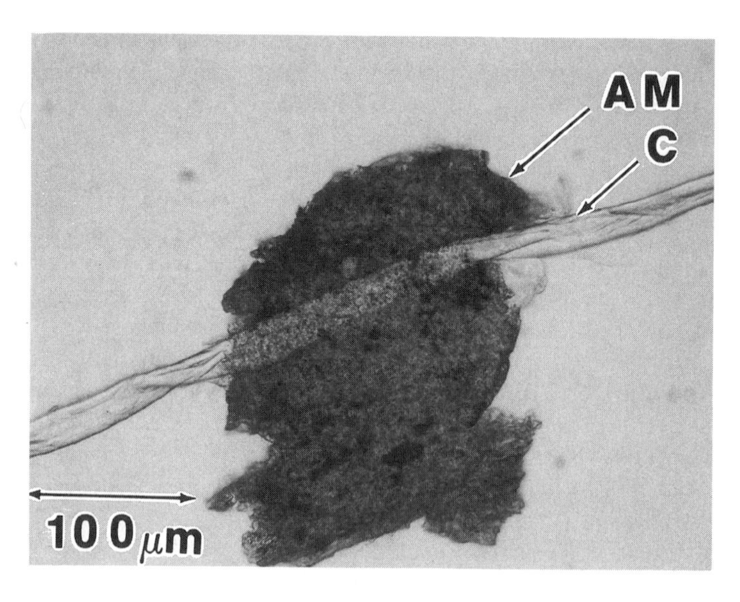

Figure 9. Photomicrograph of a particulate isolate on a ZnSe substrate from a capillary isolate. AM, amorphous matrix; C, cotton.

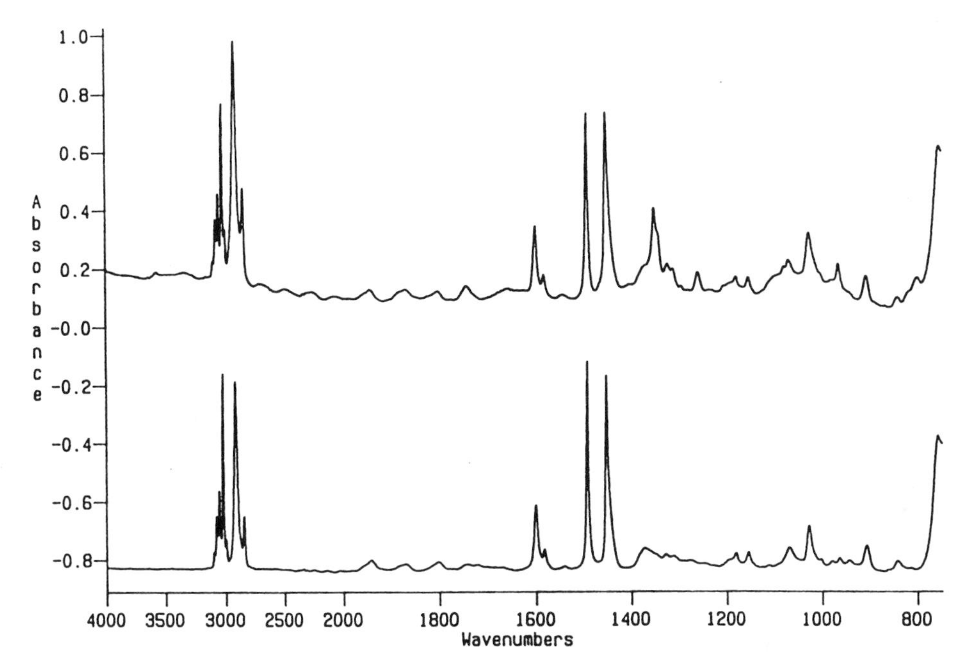

Figure 10. Infrared spectra of the isolate from Figure 9: (bottom) polystyrene reference spectrum; (top) amorphous material on particle isolated via capillary.

tend to absorb strongly. Figure 11 (bottom) shows the spectrum of a particle that was isolated directly from a solution with a Wiretrol capillary. The strongest features, evident at 2963, 1261, 1095, 1020, and 800 cm^{-1} are consistent with a methyl silicone and would appear to constitute the major species present in the particle. However, the particle was washed several times by flooding it with isooctane while viewing with a stereomicroscope. The volume of the particle was not noticeably affected. The spectrum collected from the particle after isooctane washing (top spectrum in Figure 11) shows that the silicone oil has been reduced, and a polyamide material, which is the substance of the particle, is now more apparent. One can also manipulate the extraction solvent on the infrared sample substrate via wire, probe, glass rod, or capillary to a drying "pool" concentrate for later analysis (see Section 4.5 for a discussion of microextraction).

4.4. Mixtures

4.4.1. Optical Separation

If light-microscopic and/or thermal analysis has shown the unknown to be a combination of materials, and particle manipulations have been unable to sep-

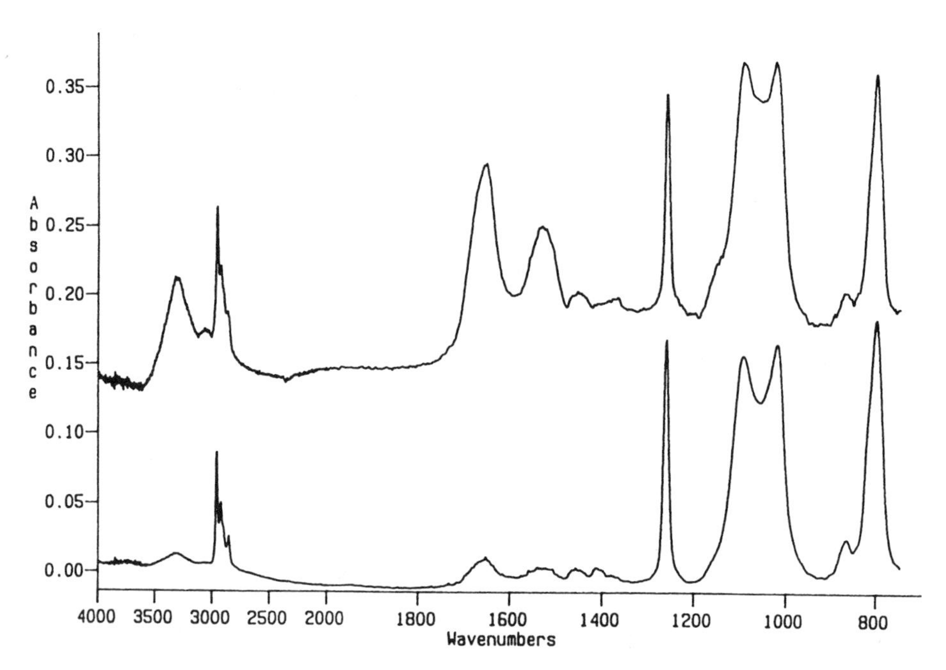

Figure 11. Infrared spectral comparison of a polyamide particle coated with silicone oil before and after isooctane washing: (bottom) particle, water washed; (top) same particle, isooctane washed.

arate the components due to their size or nature, there are some techniques that can help to obtain a "pure" spectrum of each component. The easiest method is to "optically separate" the components on the salt plate with the remote aperture system. This is effective if the particles have morphological or crystallographic differences that can be observed in the IR microscope (remember the value of microscopists' skills). We favor application of PLM at every opportunity; thus polarized-light capabilities on the viewing optics of the infrared instrument are helpful, allowing an evaluation of the crystallographic properties of the infrared analyte. While many instruments are factory equipped with polarizing capability, older or less expensive systems may lack this feature. These systems can often be modified by insertion of inexpensive polarized films (available from Edmund Scientific) in the viewing optics (not in the IR path!) above and below the sample stage. One film should be removable and easily rotatable to allow for viewing with uncrossed, partially crossed, and fully crossed polars.

4.4.2. Spectral Subtraction

When two components cannot be separated physically or optically, a spectrum can often be obtained by spectral subtraction techniques. Figure 12 shows a photomicrograph of remnants that remained after a dissolution test of a sustained-release tablet formulation; light microscopy revealed that birefringent lath-shaped crystals were entrained within an amorphous, sticky matrix of a modified cellulose. Because of the sticky nature of the cellulose matrix, the crystals could not be physically removed without a cellulosic coating. Collection of a spectrum of the crystals involved several steps; first, the sampling area was confined to a single crystal with the remote aperture, and a single-beam spectrum collected. Without changing the remote aperture, a spectrum of the matrix adjacent to the crystal was collected, followed by a background spectrum of the plate. The crystal and matrix spectra were ratioed against the background, and subsequent subtraction of these two gave a "pure" spectrum of the crystal, shown comparatively in Figure 13.

4.4.3. Serial Solvent Washing

Often, typical intrinsic particulate samples are heterogeneous, amorphous masses that cannot be discriminated into components by the spectrometer's microscope

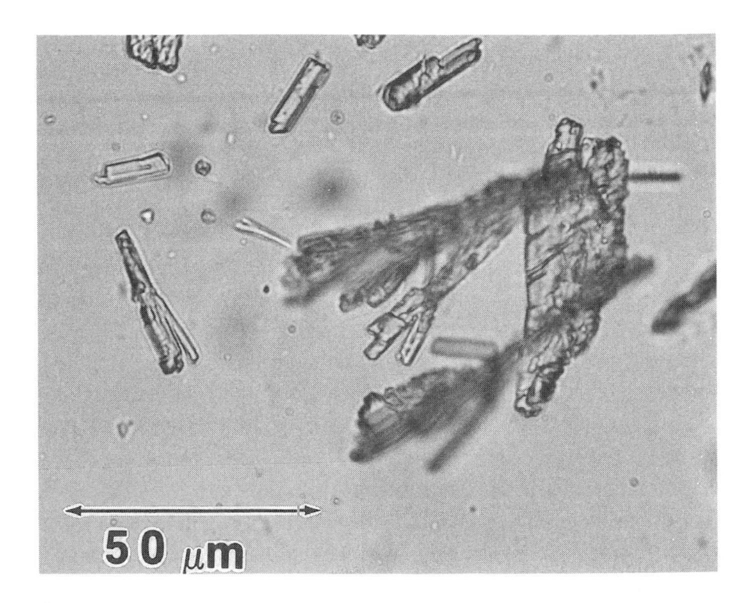

Figure 12. Photomicrograph of crystalline material in tablet remnants.

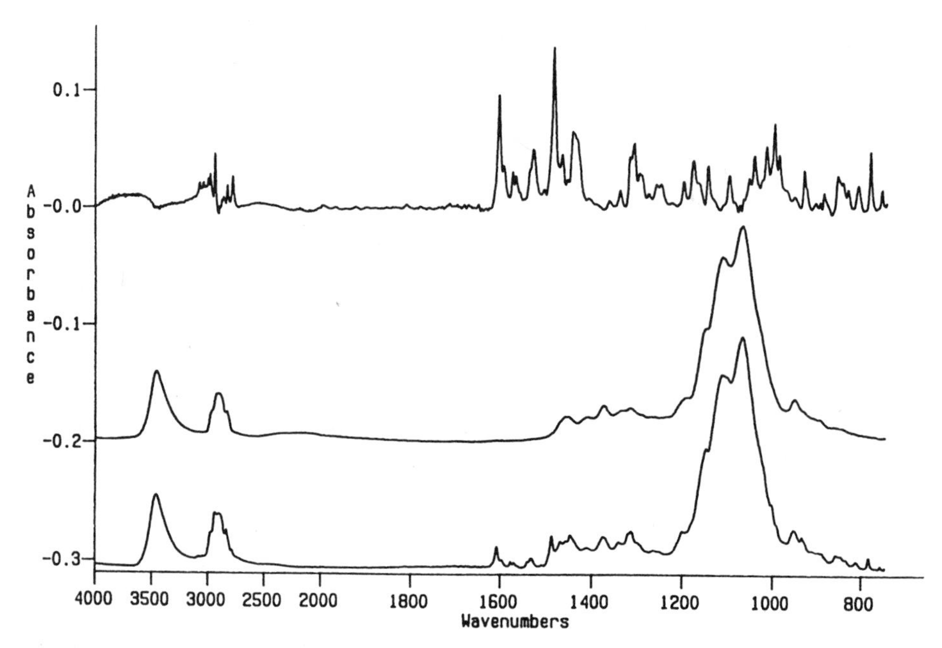

Figure 13. Infrared spectral comparison of the Figure 12 tablet remnant matrix, crystalline material + matrix and the resultant subtraction: (bottom) lath-shaped crystal in cellulosic matrix; (middle) cellulosic matrix adjacent to the lath; (top) difference spectrum (spectrum of the lath).

accessory. Often, but not always, you will observe the heterogeneity in light-microscopic or thermal analyses prior to infrared analysis. There is no trick to obtaining a spectrum of this type of sample, but interpretation of the spectrum can be difficult. If you encounter a spectrum that has an abundance of features which have complicated preliminary spectral interpretation and/or spectral searching is not providing any reasonable matches, you may be dealing with a multicomponent sample. A spectrum of a second preparation may provide a clue if the relative intensities of the features differ from the first preparation, assuming that you have properly flattened both samples. Since we would not base an investigation on the results of one particle isolate, this comparison is part of the characterization process. If there is enough difference between the preps, spectral subtraction may reveal one of the components. More often, however, separation of the components by serial solvent washing (i.e., a sequential solvent washing on one isolate) can be effective. Spectra of components that are washed away can be obtained indirectly by spectral subtraction routines, as well.

The key to obtaining useful data from a serial solvent washing is to perform the analyses on a single particle. Baseline correction routines should not be performed until spectra of the individual components are achieved. After collecting a spectrum of the initial particle mass, flood the particle on the plate with a selected solvent while viewing the preparation with a stereomicroscope. A good choice for start is a nonpolar solvent such as hexane or isooctane. If the particle appears to change its appearance, rerun the spectrum. (A quick survey of 10 to 20 scans may be enough to determine if a component has been removed. If so, run the full number of scans.) If some of the spectral features have disappeared (or diminished), subtracting this spectrum from the first will give you a spectrum of the component that was washed away. If there are residual features of this component in the spectrum of the washed particle, these can be removed by further washing or subtracting away the difference spectrum. If the solvent did not change the spectrum, proceed to another solvent of slightly higher polarity and repeat the process. By gradually increasing the polarity of solvent washes, mixtures of even four components have been resolved. Table 1 shows the progression of solvents we typically use for serial solvent washings; in our application, most mixture analyses involve particles insoluble in aqueous solutions, so washing with water occurs first. The extraction scheme can be extended to acids and bases if necessary.

Table 1. Typical Solvent Progression for Serial Solvent Washings

Solvent	Polarity index
Isooctane	−0.4
(or hexane)	0.0
Toluene	2.4
Diethyl ether	2.8
Methylene chloride	3.1
Tetrahydrofuran	4.0
Ethyl acetate	4.4
Methanol	5.1
Acetone	5.1
Acetonitrile	5.8
Dimethylformamide	6.4
Dimethyl sulfoxide	7.2
Water	9.0
Dilute acids/bases	

Source: Refs. 42 and 43.

An example of this serial solvent washing process begins with Figure 14, which shows the spectrum of a water-insoluble particle and several spectra that resulted from a serial washing of this particle with increasingly more polar solvents. In Figure 14, the top spectrum is of residue that remained after washing with dilute HCl. Comparing the bottom two spectra, a distinct change occurred after an isooctane washing of the particle. A spectral subtraction of the original spectrum minus the isooctane-washed material revealed the spectral features of the material that was removed by isooctane, as shown in Figure 15. This material was characteristic of a primary amide with unsaturation, very similar to erucamide, a common slip agent and antistatic agent in commercial polymers. Referring back to Figure 14, continued washings with solvents of increasing polarity revealed subtle changes (primarily in the region 1100 to 1000 cm^{-1} occurring gradually from nonpolar isooctane through the more polar dimethylformamide (DMF). A second spectral subtraction (methylene chloride washed minus DMF washed, as shown in Figure 16) reveals spectral features similar to a carboxylated cellulose, representing a second component of the particle. Modified cellulose excipients are becoming used more extensively in fluid, dry, and sterile

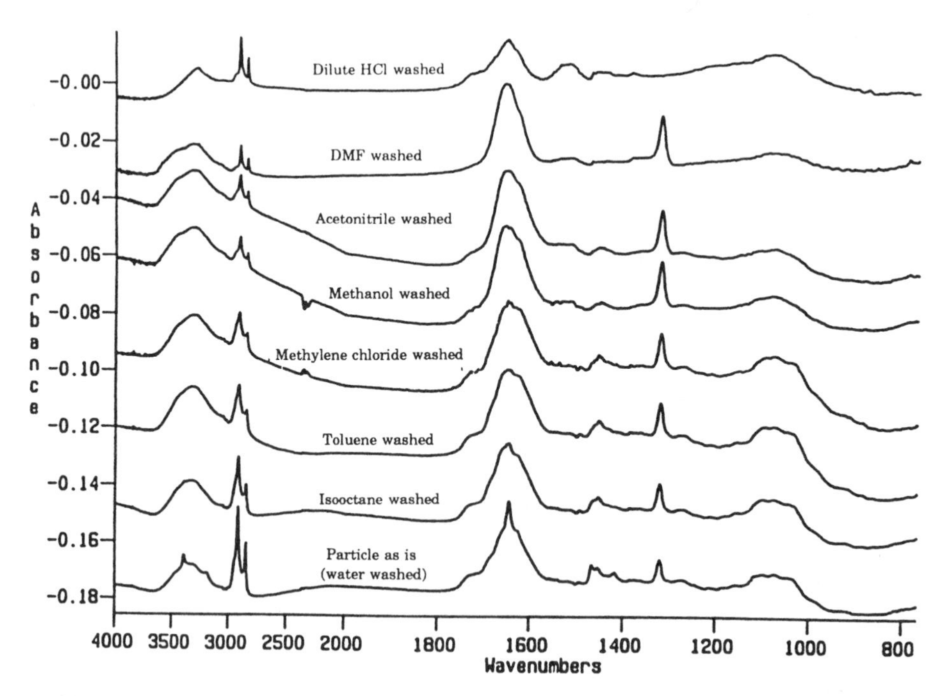

Figure 14. Infrared spectral comparison, particle aggregate (mixture) example: particle as-is through progression of solvents washes of increasing polarity from bottom to top.

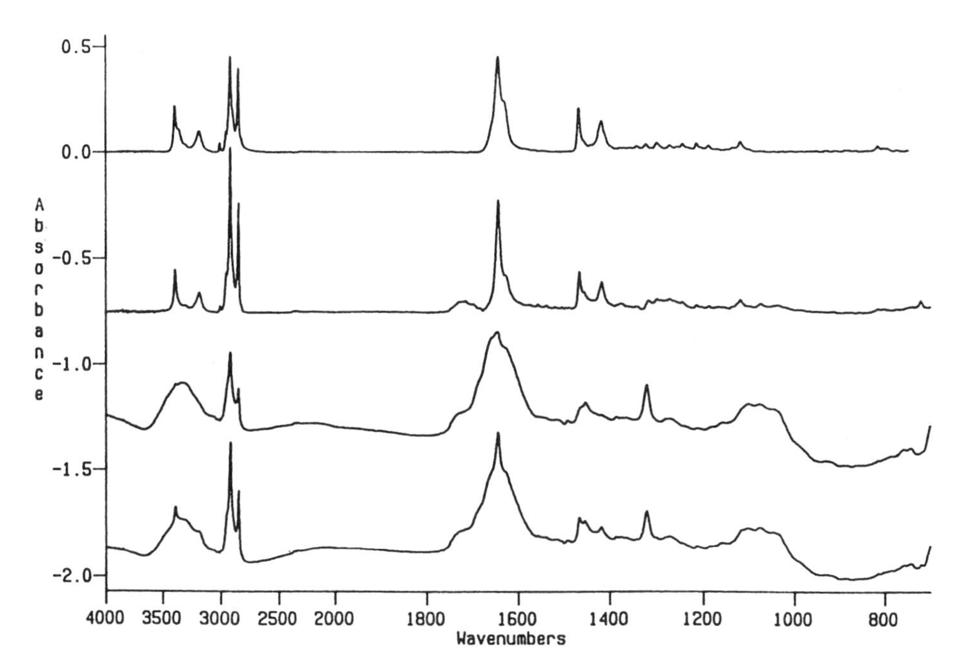

Figure 15. Infrared spectral comparison, Figure 14, residue as-is minus isooctane, parent and difference, plus erucamide reference: (bottom) particle as is; (next to bottom) particle after isooctane wash; (next to top) difference spectrum, as is minus isooctane washed; (top) erucamide reference spectrum.

products. Again referring back to Figure 14, the final wash with dilute HCl resulted in a significant change. Subtraction of the HCl-washed particle from the DMF-washed particle produced a spectrum of the third particle component, which was consistent with an oxalate, as shown in Figure 17. The spectrum of the residue that remained following the acidic wash (Figure 18) was characteristic of a polyamide; this material represents the fourth component that was resolved from the particle mixture by serial washing. Because extinction coefficients are not known, it may be difficult to judge the relative amounts of these components spectroscopically. But if a sufficient amount of sample is available, the proportion of each component can be determined by following a similar solvent washing scheme on a microbalance pan and determining weight loss with each successive washing.

In some instances, the serial solvent washing approach described in this section is not effective if the soluble component is a minor component or is in the presence of a similar compound; in these cases the difference spectrum may be ambiguous. The microextraction approach (see Section 4.5), where the minor

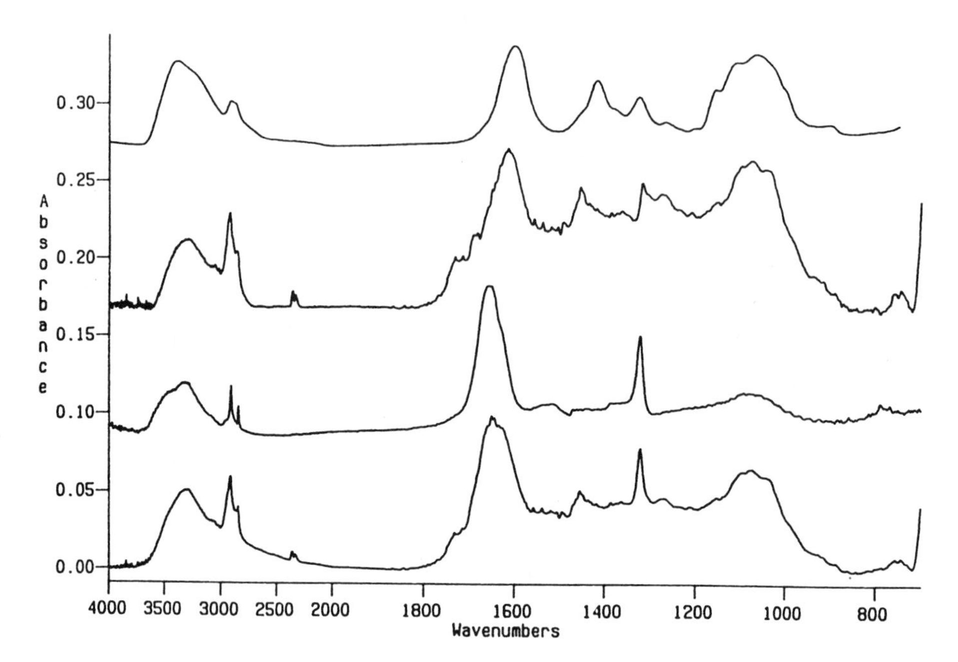

Figure 16. Infrared spectral comparison, Figure 14, residue of methylene chloride–washed minus DMF-washed, parent and difference spectra, plus Na carboxymethylcellulose reference: (bottom) particle following methylene chloride (MeCl) wash; (next to bottom) particle following dimethylformamide (DMF) wash; (next to top) difference spectrum, MeCl washed minus DMF washed; (top) sodium carboxymethylcellulose reference spectrum.

component is analyzed directly following isolation using only a few microliters of solvent, is sometimes more effective.

4.5. Extractions

Extractions are usually justified in situations where a liquid contaminant is present, usually as the result of contact with a grease or oil. This may cause a haze in the case of aqueous solutions or a spot in the case of a tablet formulation. A microextraction can be performed on a salt plate by placing a small (100 to 200 μm) fragment of the tablet spot on the plate and flooding it with 2 or 3 μL of isooctane or hexane. If an oil is present, it will be carried to the edges of the droplet and deposited as it evaporates. Be sure to perform a negative control (solvent on substrate) as well.

On a microscale, as the droplet evaporates, a separation of dissolved components may occur. In one such example, the evaluation of a liquid residue,

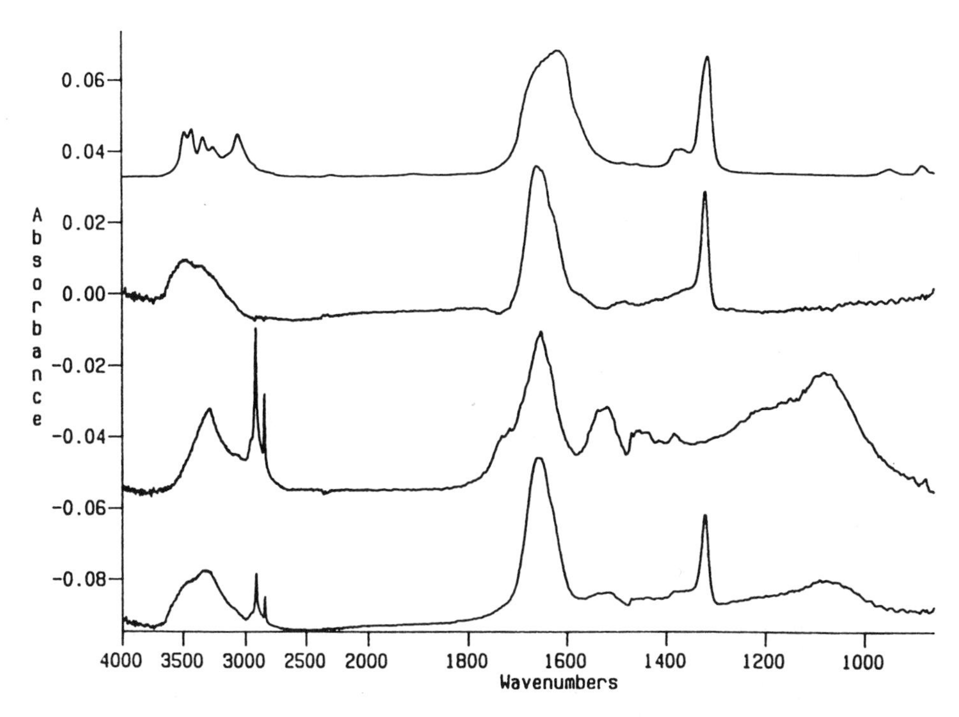

Figure 17. Infrared spectral comparison, Figure 14, residue of DMF-washed minus HCl-washed, parent and difference, plus oxalate reference: (bottom) particle following DMF wash; (next to bottom) particle following dilute HCl wash; (next to top) difference spectrum, DMF washed minus HCl washed; (top) calcium oxalate monohydrate reference spectrum.

rather than discrete "particles," was aided by this technique. Starting with an unknown brown oily liquid (Figure 19, bottom spectrum), a microextraction with isooctane dissolved the liquid, and upon evaporation, both a colorless and a faintly brown-colored residue remained with sufficient spatial separation to allow spectral collection of each. The top two spectra in Figure 19 are the isolated components, which were consistent with a hydrocarbon (top spectrum) and butyl benzoate (middle spectrum).

Microextraction may also be effective as a minimized-solvent volume serial washing technique. Starting with nonpolar solvents and gradually increasing polarity may resolve the components on the salt plate. The spectra in Figure 20 represent the components of a particle that light-microscopic examination had revealed to be a conglomerate of small (<5 μm) birefringent grains and an isotropic, amorphous material. In an attempt to elucidate the mixture by the

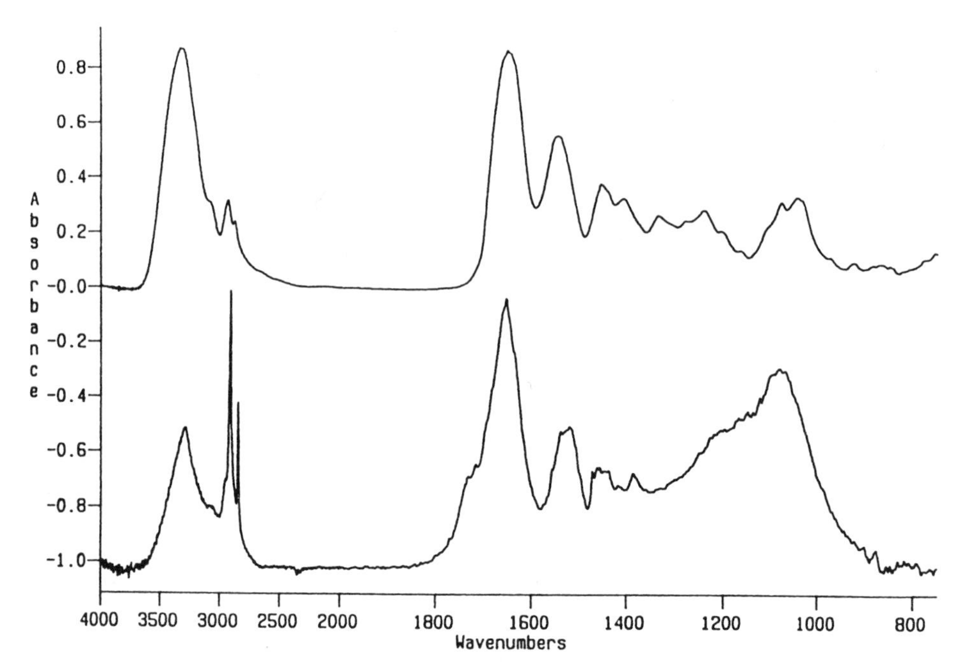

Figure 18. Infrared spectral comparison, Figure 14, residue post HCl-wash plus polyamide reference: (bottom) particle following dilute HCl wash; (top) polyamide material (gelatin) for comparison.

serial solvent washing method, it was found that toluene removed the isotropic material, but the resulting difference spectrum (Figure 20, top) was not clear enough to allow characterization of the isotropic component. A microextraction using only a few microliters of toluene provided a deposit of the isotropic material on the salt plate for direct analysis. Figure 21 shows the spectrum of the toluene isolate (top) from the microextraction and the spectrum of the particle remaining after toluene washing in the serial solvent washing experiment. The spectra are similar to each other and are consistent with amorphous and crystalline forms of the steroidal drug.

Hazy solutions that are the result of an immiscible liquid contamination can be recovered by conventional extraction techniques; a strategy for infrared analysis is to concentrate the organic phase to about 20 to 50 μL, then deposit the extract onto KBr within a diffuse reflectance cup (A Spectra-Tech microdiffuse reflectance cup works well). The remaining solvent is volatilized and the sample analyzed by diffuse reflectance spectroscopy. A negative control is essential since low-level contamination in labware is difficult to avoid. Although this is

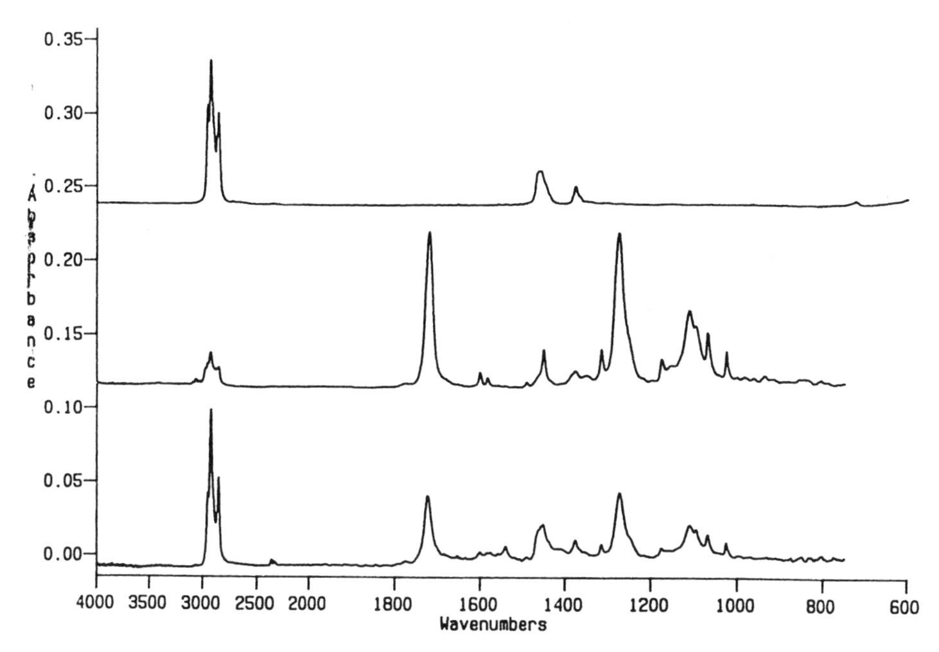

Figure 19. IR spectra of liquid mixture and isolated hydrocarbon and butyl benzoate; (bottom) brown oily liquid mixture; (middle) colorless component isolated by isooctane microextraction (identified as butyl benzoate); (top) faintly brown-colored component isolated by isooctane microextraction (characterized as a hydrocarbon).

not a microtechnique, it has submicrogram sensitivity. Figure 22 shows a spectrum of 100 ng of dioctylphthalate, which was deposited onto KBr in a diffuse reflectance cup from an acetone solution. Consulting more comprehensive discussions of this technique [38,39] is advised for those who find it useful.

Another useful technique for analyzing extracts is the Wick-Stick, sintered KBr wedges available from Spectra-Tech. The technique involves slowly evaporating the solvent, which wicks up the wedge and deposits the analyte at the tip [40]. The tip can then be broken off and analyzed by conventional techniques, or with the Wick-Stick lying on its side on the IR microspectrometer stage, a reflectance spectrum off the tip area may be collected. This is essentially a microdifuse reflectance method. This analysis can be further scaled down by breaking off small (250 to 500 μm) pieces of Wick-Stick, running a single-beam reflectance spectrum of the unused fragment for a background spectrum (on a salt plate), then immersing the fragment in droplets of 1 μL or less of a concentrated (to a few microliters) extract on the salt plate. The solvent will wick quickly into the Wick-Stick fragment and evaporate; the process is repeated until

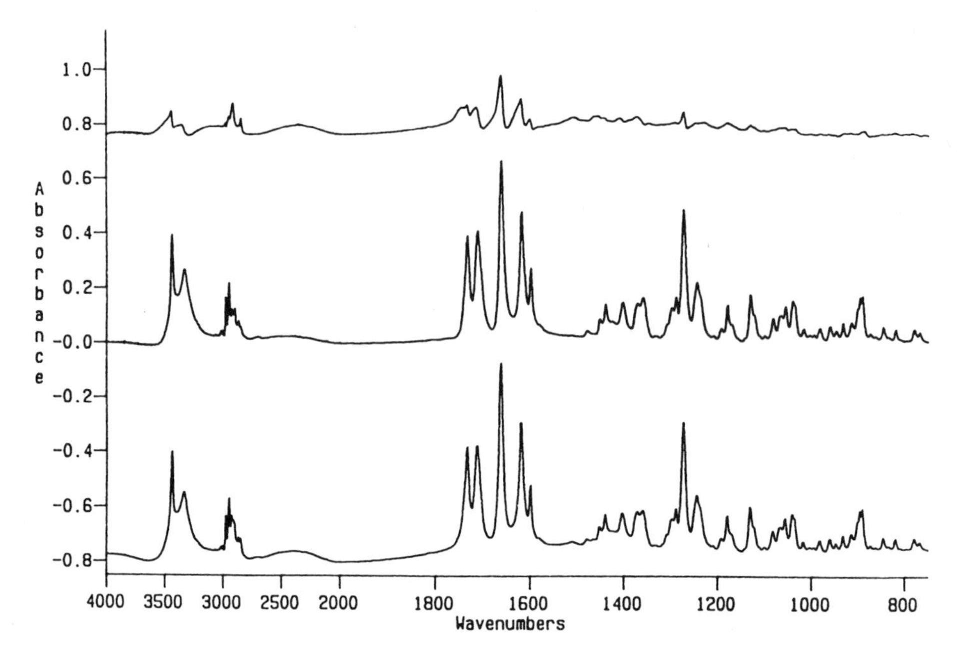

Figure 20. IR spectra of birefringent and isotropic species, solvent washing technique: (bottom) conglomerate particle as is; (middle) same particle after toluene wash; (top) difference spectrum, as is minus toluene-washed particle.

all the extract has been deposited. After the solvent has evaporated, the sample reflectance spectrum is collected and ratioed against the background spectrum. Figure 23 shows the microdiffuse reflectance spectrum of 25 ng of dioctylphthalate that was recovered from a methylene chloride solution by evaporation onto a 500-μm fragment of a Wick-Stick.

4.6. Changing the Form of the Sample (Derivatization, Polymorphs)

Occasionally, particles will require a change in form to permit identification. Examples are carboxylate salts, amine salts, and drug forms that exhibit polymorphism. Although the parent compound may exist in your spectral libraries, identification may be prevented because references are not in the same form as your sample. Carboxylate salts may be recognized by broad bands near 1600 and 1400 cm^{-1} and can often be identified by conversion to the protonated species (this operation can be performed on a ZnSe plate). The bottom spectrum in Figure 24, which was collected on a crystalline aggregate, is typical of car-

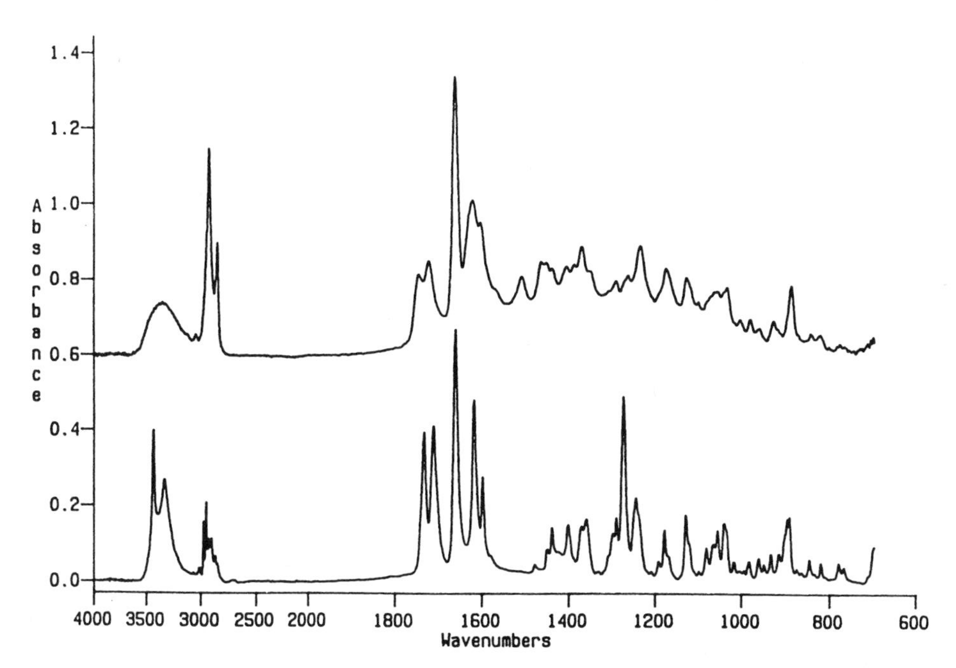

Figure 21. IR spectra of birefringent and isotropic species, microextraction technique: (bottom) birefringent component of conglomerate; (top) isotropic component of conglomerate, recovered from a toluene microextraction.

boxylate salts. On the ZnSe plate, the particle was dissolved in a few microliters of acidified methanol. The IR spectrum of material recovered after evaporation of the acidified methanol was readily identified as adipic acid (a dicarboxylic acid) via library searching (Figure 24, top spectrum), even though the spectral features revealed only partial conversion to the protonated form of the compound. Another example of this methodology is shown in Figure 25; although not as distinctive as the adipate salt in Figure 24 (probably because it contained a variety of cations), the bottom spectrum is a carboxylate salt that was identified as an ion-exchange resin (containing multiple COOH groups) after the particle was dissolved in dilute sodium hydroxide, then precipitated upon acidification with HCl. Similarly, Figure 26 shows before and after spectra of a particle that was converted from the HCl salt form (recognized by the strong absorbance in the 2800 to 2400-cm^{-1} region) to the amine free base with ammonia, which allowed its identification as poly(4-vinylpyridine). Application of conventional wet chemistry techniques on this microscale can be very rewarding.

Polymorphism differences can be more difficult to recognize; however, if your unknown spectrum is similar to but differs somewhat from a reference

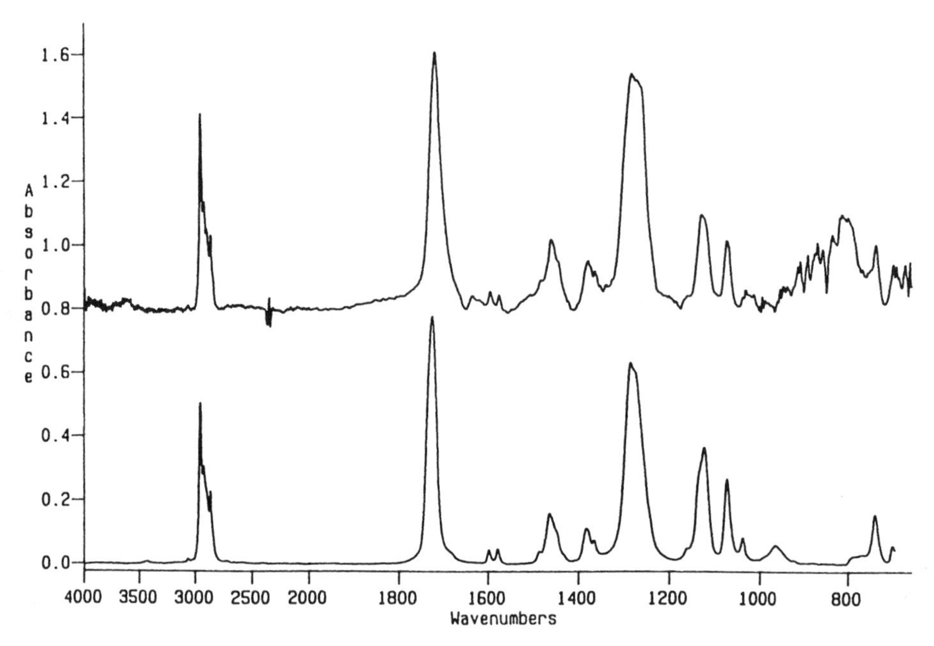

Figure 22. Infrared spectrum: diffuse reflectance of 100 ng of dioctylphthalate. (bottom) dioctylphthalate reference spectrum; (top) diffuse reflectance infrared spectrum of 100 ng of dioctylphthalate collected with a Spectra-Tech Collector, deposited from acetone solution onto KBr in a microcup.

spectrum, it can sometimes be resolved by recrystallizing the unknown from various solvents, or if a sample of the reference is available, recrystallize both reference and unknown from the same solvent. Again, these can be carried out on a microscale on the salt plate. The great value of optical crystallographic determinations and thermal behavior to assess polymorphic states, as discussed in Sections 2.5 and 2.6, respectively, cannot be ignored.

4.7. Dehydration

The focus of the condensers in some IR microspectrometers can provide enough energy to thermally induce dehydration of some compounds; we have observed this occurrence with dozens of compounds. In some cases, the corresponding spectral changes can affect more than just the O—H region. Figure 27 shows spectra of erythromycin dihydrate; the bottom spectrum represents a conventional KBr preparation, while the top spectrum is a microtransmission spectrum, collected on a Digilab UMA-300 IR microspectrometer, which uses a 36× Cas-

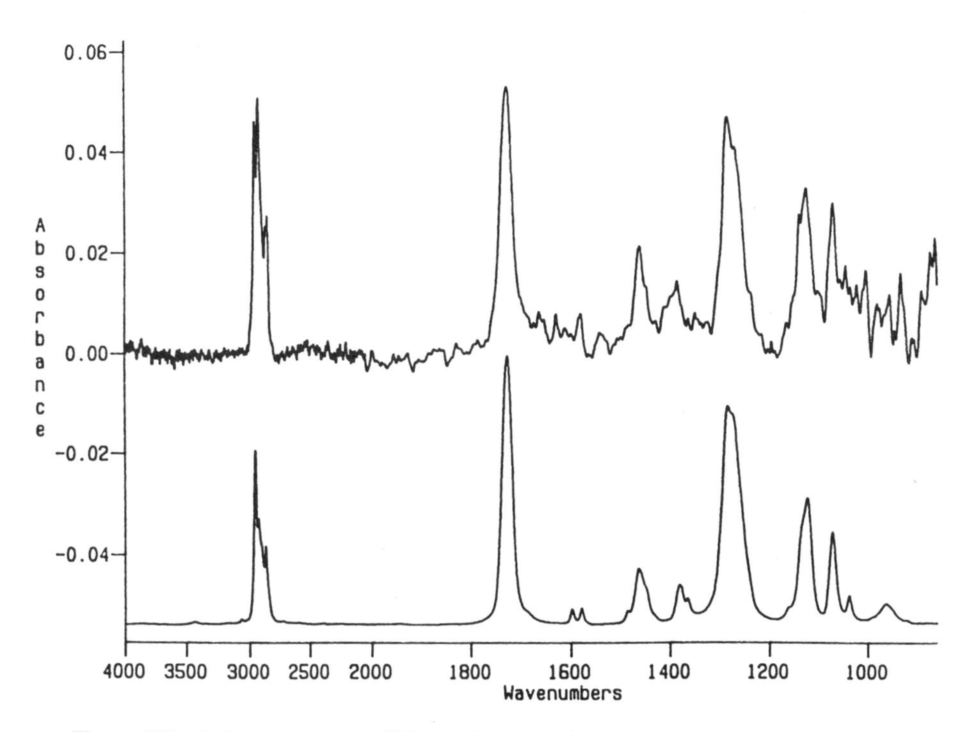

Figure 23. Infrared spectrum: diffuse reflectance of 25 ng of dioctylphthalate: (bottom) dioctylphthalate reference spectrum; (top) microdiffuse reflectance spectrum of 25 ng of dioctylphalate on a 500-μm Wick-Stick fragment, using reflectance mode of an infrared microspectrometer.

segrain objective and a 6× substage beam condenser. The stage is open to the atmosphere, but there is a downward flow of nitrogen over the sample from purge lines on the infrared microspectrometer. The spectrum collected with the IR microspectrometer is consistent with spectra of thermally dehydrated (at 100°C) erythromycin; the spectral differences that occur upon dehydration in Figure 27 span the entire spectral range but are most notable in the carbonyl region, where a lactone band shifts from 1714 to 1732 cm^{-1}, probably due to changes in hydrogen bonding. A ketone carbonyl band at 1704 cm^{-1}, which was present as a shoulder in the hydrated spectrum, becomes more apparent with the shift in the lactone band in the dehydrated spectra.

The propensity for inducing dehydration of a specimen is probably dependent on a number of instrumental parameters, including the intensity of the beam emanating from the interferometer, degree of beam condensation, and stage atmosphere. For erythromycin dihydrate, the spectral differences described pre-

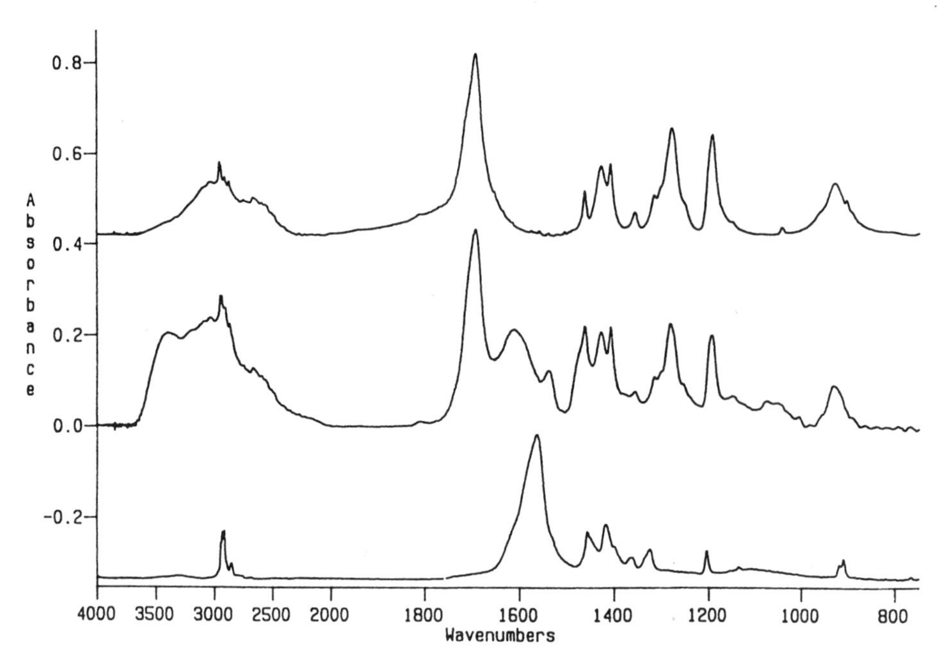

Figure 24. Infrared spectral comparison of sodium adipate particulate matter and the converted free acid form: (bottom) crystalline isolate as is; (middle) isolate after partial conversion to the free acid form; (top) adipic acid reference spectrum.

viously do not occur in micro-IR spectra when the sample stage is not purged (allowing water vapor–sample interaction) but occur rapidly (within a few scans) and reversibly in a dry nitrogen atmosphere [41]; other compounds have been observed to dehydrate gradually over the course of hundreds of scans.

Technique-induced dehydration can impede particle identification attempts during library searches. In the case of erythromycin, the available library references were collected as KBr pellets of the dihydrate form, so a library search of the microtransmission spectrum in Figure 27 would not have resulted in a positive identification. The possibility of dehydration should also be considered in material science characterization of drug forms if hydrated species are a possibility. As mentioned previously, eliminating nitrogen purge to the stage may prevent sample dehydration; limiting the analysis time may also minimize the occurrence. If dehydration is suspected, collect a spectrum of only a few scans, then compare this to a spectrum utilizing the desired number of scans, or collect spectra with and without nitrogen purging; if differences are noted, the specimen may be dehydrating during the analysis. Micro-KBr preparations are an alternative that can circumvent sample dehydration problems.

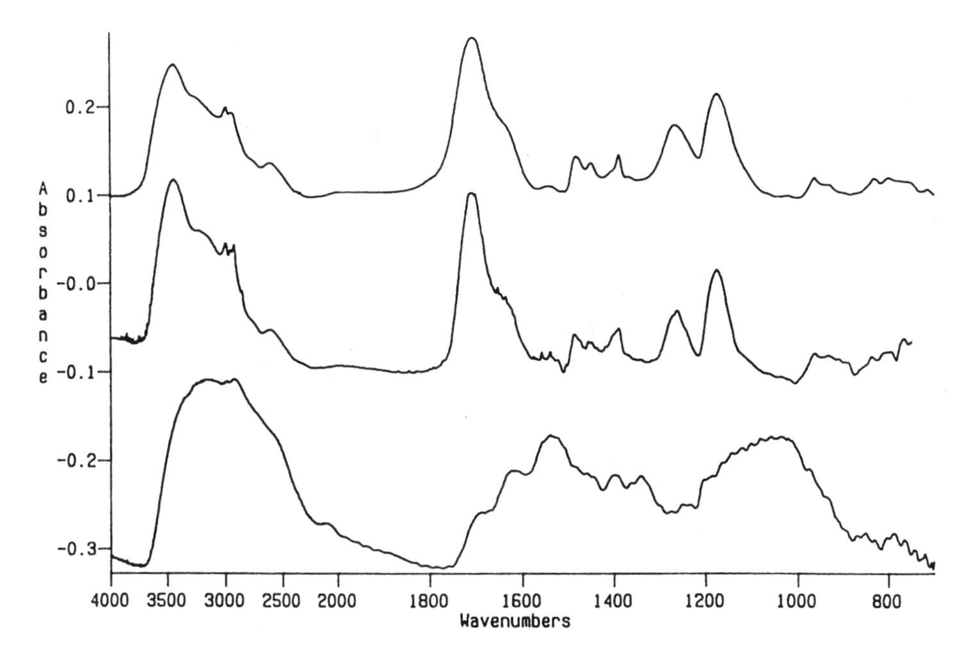

Figure 25. Infrared spectral comparison of carboxylate particulate matter and the converted free acid form: (bottom) isolated particle, as is (carboxylate salt); (middle) particle after conversion to the free acid form; (top) ion-exchange resin reference spectrum.

5. CONCLUSIONS

In our review of infrared microspectroscopic methods for the pharmaceutical scientist, we have described a system of particulate analysis that is an integration of many disciplines and concepts. We have strived to make the point that without building the observational, mechanical, and chemical separation skills of a microscopist, your spectroscopic success will be limited. These skills rely on your studied practice of them. Your application needs and product line will drive the type of talents you develop, although the capabilities of an experienced microscopist will never fail you. The infrared microspectroscopic system is essential for our particle characterization, differentiation of mixtures, confirmation of identity, description of environmental or solvent effects, and demands a high level of infrared interpretive analysis coupled with the microscopical forensic method. As one important twentieth-century microscopist (W. C. McCrone) has taught us, "microscopy is more a way of thinking than a way of looking"; so then is this modern extension of the optical microscope. The skills one brings to infrared microspectroscopy elevates the device to one of the most useful

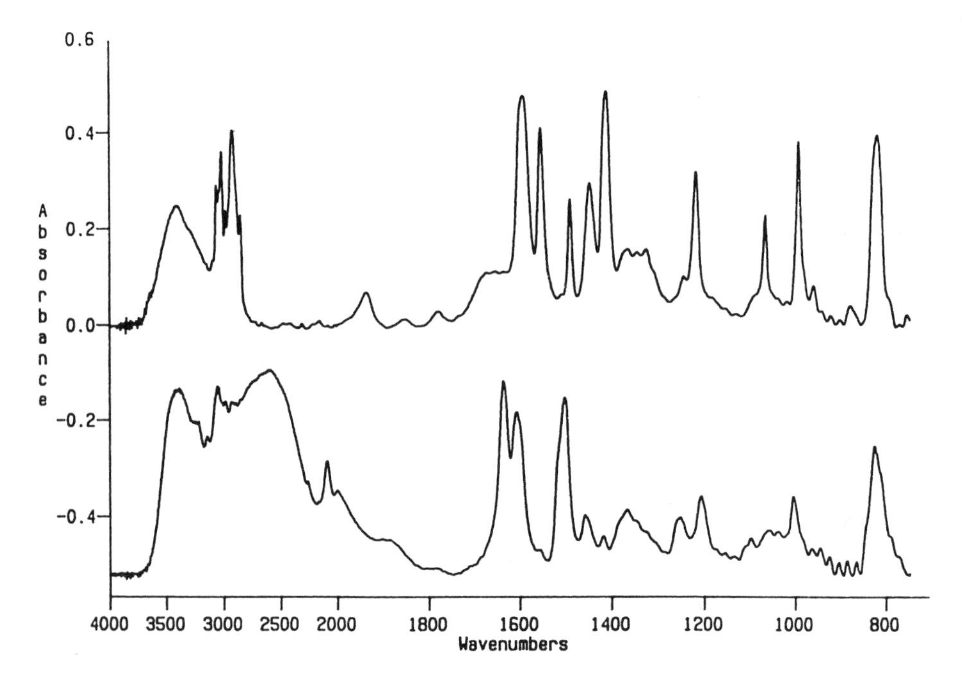

Figure 26. Infrared spectral comparison of the amine salt particulate matter and the converted free base form: (bottom) particle as is; (top) particle following conversion to free base form [identified as poly(4-vinyl)pyridine].

applications for the pharmaceutical investigator. May the examples and methods we have described in this chapter be only a starting point for your evolution as a pharmaceutical microspectroscopist.

APPENDIX A: COMMON PHARMACEUTICAL PARTICLE SOURCES

Environmental Sources

1. *Fibers*: Two classes are seen:
   a. Natural
      (1) Generally as vegetable fibers, which are cellulosic, such as wood, paper, cotton (rayon).
      (2) Hairs, which include animal (horse, camel, cow, hog—originating from carpets, brushes, etc) and human.
   b. Human-made, which occur in a wide range of material types but are primarily personnel- and machine-borne; common are polyamide (ny-

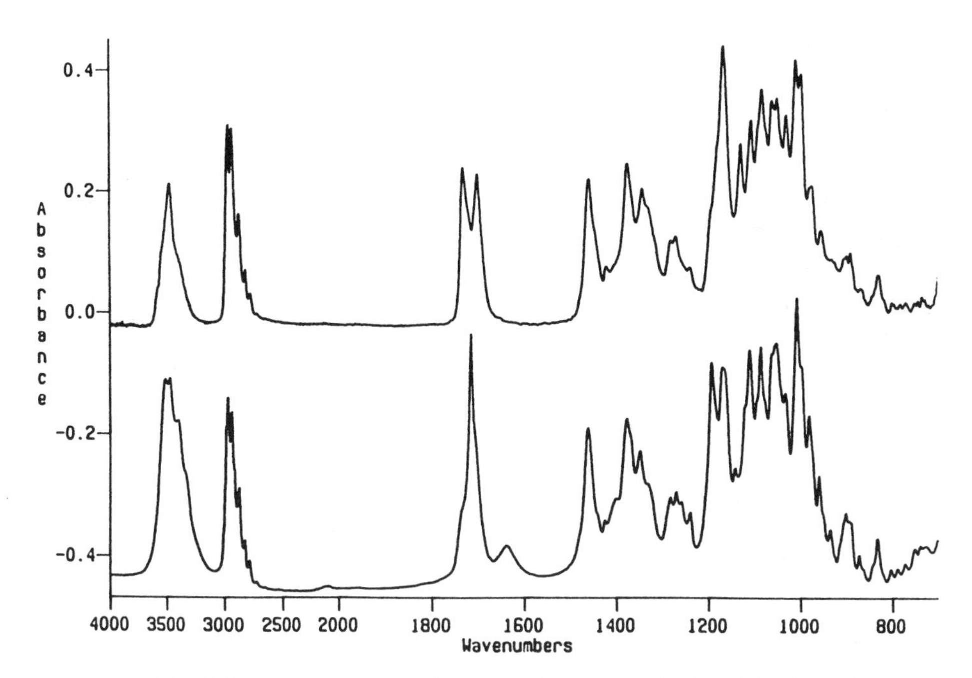

Figure 27. Infrared spectral comparison of erythromycin · $2H_2O$ as-is in KBr with the microtransmittance spectrum: (bottom) KBr pellet, conventional IR; (top) microtransmission spectrum, 10 scans, N_2 purged atmosphere.

lon), aramide (Kevlar), polyester (Dacron), polyethylene (low and high density), polypropylene, polystyrene, and Teflon.

2. *Minerals*: Most are soil-related, but some may be components of equipment or facilities-related; occurring often are quartz, feldspars, clays, mica, hematite, calcite, and gypsum. Often occur with nonspecific greases/oils, which form a "soil" or "dirt." Very site-specific, so know thy geography!

3. *Metals*: Often due to process equipment, but also as a general environmental source. Includes stainless steels 304, 316, 316L, 403, and 303 (Cr/Fe/Ni with minor elemental additives), monel (Ni/Cu), steel (Fe), galvanized steel (Zn on Fe), brass (Cu/Zn), and bronze (Cu/Sn). Solders (Pb/Sn- and Ag-based) are rarely found, yet are distinctive. The metals often occur with or in transition to a corrosion product (see below).

4. *Biologicals*: Primarily airborne matter, such as leaf trichomes (the hairs on the underside of all leaves, specific to the plant), pollen, seeds, vegetative integument, and insect-related particles of exoskeleton (chitin) or wing matrix/scales.

5. *Corrosion products*: These take many forms, such as any transition metal–and product solution–related anion; however, most often occur as Fe_2O_3, hydrated Fe_2O_3, FeO, Fe/Cl species, goethite FeO(OH), and related corrosion products containing Fe-S, plus solid solutions series of other iron and chromium oxides.

6. "*Dark specks*": May be anything relatively dark against a light background (everything mentioned above). Often attributable to charred remnants of the subject matter or graphite (carbon) inclusions in product components. Look for excipients with associated dark regions which provide "macro" color due to dispersed colorants, foreign matter, oils, and coatings.

7. *Paint*: Look for obvious macroopaque, microtranslucent, colored, and laminated flakes, which are somewhat resilient. Often contain TiO_2, $CaCO_3$, and other inorganic or mineral pigments and opacifiers.

Package-Related Sources

1. Cleanliness of the package item is of utmost importance. In the simplest case it is related to environmental debris as particles that fall off the packaging components either through insufficient cleaning or package-quality degradation:

a. All environmental types
b. Rubber particles
c. Glass fragments
d. Inorganic fillers from closures
e. Plastic particles, soft and hard

2. *Glass character*: Atmospheric agents, primarily moisture, act to "weather" glass by extracting alkaline matter and hydrating the extracted layers from replacement of the extracted alkali [44]. This results in physical changes such as cracking and expansion/shrinking. This dynamic process, which starts with container filling, can be described as follows:

a. An excess of sodium oxide (Na_2O) and other alkali oxides leach from the glass, dissolving in the fill.
b. The fill becomes more alkaline.
c. Silica in the glass surface comes under more severe physical stress and chemical exchange.
d. Oxides and silicic acid which have been leached have a progressively retardant effect on the exchange process.

The most common treatment of the glass product to alleviate or minimize this effect is to exchange H^+ for Na^+ in the near-surface, producing Na_2SO_4 deposits on the glass surface following sulfur treatments. These deposits are easily removed with aqueous rinsing. Various treatments such as gaseous SO_2, ammonium sulfate, ammonium chloride, aluminum chloride, and bifluoride treatments of the freshly manufactured glass are effective in preparing the glass for use [45]. Combinations of these treatments have proven to be most effective. The performance of these treatments can be summarized by comparing the Na_2O content at various depths in the inner glass surface.

Other whole-glass defect categories are:

a. Delaminated glass via physical and chemical effects.
b. Glass treatment residues such as Na_2SO_4 from sulfur-treated glass.
c. Excessive or deep pocking of the interior glass, observed during inspection as a haze.
d. Insoluble formation products, primarily inorganic salts, such as $BaSO_4$, which form a "crust" on the interior glass surface.
e. Glass character problems, optical anomalies, such as scratches, nonplanar sidewalls, and glass defects (stones, metal inclusions) may cause "particle" descriptions. These are a minor problem but may not be detected by the pharmaceutical inspection system(s) and thus end up with the customer. All should be removed prior to filling, preferably screened out by the vendor. Shipping the containers should not cause defects through packing material contamination or scratching.

3. *Rubber closure character*: Problems may be caused by filler "debris" as physical loss of material into the product, or as ions and organic compounds leached (extracted) into the product, primarily from colorants, opacifiers, accelerators, and catalysts. Elements included are often Al, Si, S, Ca, K, Mg, Ba, Ti, Fe, Mn, and Zn (older pharmaceutical and device formulations contained Se, Cd, Co, and Cr as well). Leached volatile organics occur from vulcanizing agents with sulfur (thiurams, organodisulfides) and those without sulfur (ZnO, phenolic compounds). Accelerators such as aldehyde–amine products, amines, guanadines, thioureas, thiazoles, thiurams, xanthates, sulfenamides, and dithiocarbamates may also be used. If the item contains porous or cavernous surfaces, it may trap lubricant or other debris, which is later exposed to/released into the product; the best approach is to refrain from or minimize lubrication and to utilize rubber items with low-surface-area or barrier films. Closures may offer a substrate for particle formation; the surface of the rubber provides a tremendous inorganic/organic leachate sink *and* the physical site on which to build particulate matter.

4. *Excipient-related sources*: Excipient particle load does not directly cause particle problems in sterile liquids and lyophilized products, since the parenteral solutions undergo 0.2-μm filtration operations prior to filling. Even terminal sterilization schemes will include filtration for bioburden reduction in these products. Excipient-related particle incidents are insidious, since many involve minor components or impurities which may form particulates during the

shelf life. Lot-to-lot variation of elemental character has the most profound effect, forming drug–salt particulates. In dry products and powder-fill products, where excipients are used directly, particulate matter, texture changes, and color differences may be perceived as "specks" in the final product.

5. *Drug-related sources*: These particulates are present or form from impurities and degradation products and have much to do with the character of the bulk drug lot. The reduction of these particle types must occur at the formulation development stages to alleviate shelf-life losses of product. Visible problems may occur when the species are present in the product solution at ppm and certainly in the low-percent concentrations. Most critical are those that are relatively insoluble in the product solution, given product insert ranges of pH, ion content, temperature, and diluent usage or dilution. Micellar drugs have a critical micelle concentration (CMC), below which the self-solubilizing capabilities of the parenteral are lost and solid forms of impurities and similar, dissolved compounds express (precipitate, nucleate, aggregate, etc.) from the product solution.

APPENDIX B: GLOSSARY OF TERMS

Acicular Needle-shaped.

Acute Steep angle, generally near 45° to the horizontal, used to describe incident illumination.

Adamantine Brilliant reflectance.

Admixture Combination of two or more commercial parenteral products in common container, such as a sterile aqueous antibiotic diluted in a sterile saline vehicle.

Agglomerate Association of homogeneous solids that is tightly compacted, having indistinct margins of the individual solids.

Aggregate Association of homogeneous solids that is loosely compacted, having distinct margins of the individual solids.

Amorphous Shapeless.

Angular Sharp-edged or having roughly polyhedral (many-faces) shape.

Banded Distinct regions of color, opacity, or other feature that form parallel zones.

Birefringence Absolute difference between high and low refractive index: birefringence $= \eta_{high} - \eta_{low}$ (*see* Interference color)

Bumpy Surface feature that consists of regular or irregular spacing of smooth convex forms.

Cemented Smooth or featureless matrix that contains solids.

Chunk Mass of material(s), in millimeter sizes, which may consist of several materials in a variety of forms, yet appears to the unaided vision as a single entity.

Clathrate Crystalline material in which water is physically captured by the crystalline solid and not part of the crystal lattice.

Cleavage Interfacial angles and faces that intersect in a repetitive fashion, due to splitting along crystalline planes.

Cluster Loose association of solids, with little physical integrity.
Coating Layer or cover of material.
Colloidal Dispersion of small solids, 0.05 to 0.5 μm in size, which in a fluid medium do not settle.
Conchoidal Surface fracture lines typical of a glass; generally, fluid and parallel to one another.
Conglomerate Association of heterogeneous solids that is tightly compacted, having indistinct margins of the individual solids.
Congregate Association of heterogeneous solids that is loosely compacted, having distinct margins of the individual solids.
Cracked Surface that looks much like a ''dried riverbed,'' with irregular grooves on the surface.
Cratered Surface feature that consists of regular or irregular spacing of smooth concave forms.
Cryptocrystalline Single form that contains thousands of small crystal centers; usually results from crystallization after solidification (phase transition) such that the exterior form does not change.
Crystalline Symmetrical geometry indicating an atomic or molecular lattice structure.
Decomposition Transition occurring during heating, marked by darkening, which also may include crystallinity loss, glass formation, and outgassing.
Decrepitated Effect of roasting or elevated heating, characterized by increasing opacity and a cracked ''dry riverbed'' appearance.
Dehydration Loss of water, either absorbed or crystalline.
Dendritic Solid state with branched crystalline shape and often a fine structure, much like frost.
Desolvation Loss of solvent of crystallization and/or adsorbed solvent.
Disorder State of crystallinity in which imperfections in the lattice yield a suboptimal crystalline state; evidenced by broadening of refractive indices, imperfect extinction, and x-ray diffraction pattern broadening.
Earthy Nonreflecting, absorbing, dull appearance.
Fibrous Threadlike.
Fine Description of the characteristics of a bulk material with component particles up to 10 μm in size.
Flake/flaky Flat structures, often with x and y axes roughly similar and z axis much smaller.
Foliated Leaflike stacks.
Free Unencumbered, unattached.
Glass/glassy Glasslike, vitreous mass, having bulk, homogeneous character and generally transparent.
Glass transition Metastable solid state in which crystalline materials lose their crystallinity and exist in a disordered array.
Granule/granular Irregularly shaped, equidimensional solid.
Greasy Reflective as if wet, slick-appearing.
Habit Characteristic appearance or shape (*see* Morphology).
Heterogeneous Range of types, dissimilar parts, incongruity.

Homogeneous Uniformity of composition, structure, singularity of type.
Hydrate The state of a substance with molecular water of defined stoichiometry.
Inclusion Any solid, liquid, or gas surrounded by a separate matrix.
Interference color Also polarization colors: formed by the interference of light emergent from crystals that are not in an extinction position, due to the recombination of vector components out of phase from refractive index differences in the crystal and observed when viewed with crossed polarizing filters.
Irregular Lacking symmetry.
Isotropic Substances with a single refractive index and no interference colors in crossed polarized light.
Laminated/lamellar Occurring in distinct layers, stacked plates.
Massive Description of bulk or great size, relative to the comparator.
Melt Change from the solid to the liquid state.
Morphology Structure and form, excluding function (*see* Habit).
Oblique Shallow angle, such as 1 to 10° from the horizontal, used to describe incident illumination.
Occlusion Solid or liquid substances occurring on the exterior of a solid.
Orange peel Description of texture simulating an orange peel; many regular but non-ordered surface points or pits.
Patterned Any flat, concave, or convex surface marking which forms a consistent or periodic figure.
Pearly Smooth, low reflective state which is lustrous or appears to have depth.
Phase transition Temporal state of change, with conversion or change from one form to another.
Pitted Surface covered with many similarly shaped, usually ragged concave surfaces.
Plate/platy Similar to *Flake*, with longer z axis.
Polycrystalline Form that consists of many crystalline entities.
Polymorph Any one of several solid-state forms of a material, being chemically identical, yet filling a different crystalline lattice spacing.
Porous Surfaces that have holes, pores, and voids.
Powder Description of the characteristics of a bulk material with component particles up to 100 μm in size.
Reflectivity Description of the reflective character (e.g., dull, iridescent, shiny, glassy, metallic).
Refractive index Physical property (vector) describing the speed of light through a substance, relative to the speed of light through a vacuum. Expressed as $\eta\alpha_D = 1.567$, where the α refractive index is 1.567, measured with D-line Fraunhofer wavelength (589 nm).
Reticulated Marked with lines resembling a network.
Rough Relative term, dependent on the magnification, indicating an irregular, coarse surface, with a defined, sharp-edged microstructure.
Silky As in pearly, yet more reflectance.
Sintered Formation of a homogeneous, crusty mass.
Smooth Relative term, dependent on the magnification, indicating nonstructured, flat surfaces.

Solvate Crystalline state of a substance of defined stoichiometry with molecular solvent (ethanol, acetone, etc.).
Sphere/spherical Globular, all axes equivalent, least surface area.
Spherulite Association of crystals, in which acicular forms radiate from a common center, often forming spherical masses.
Striated Surface marked with a series of parallel stripes, ridges, grooves, or indentations.
Sublime Transition to the gaseous state directly from the solid state.
Vitreous Glassy, rather shiny appearance; may have positions of intense reflectance from prismatic faces.
Waxy Waxlike, reflective in a soft glow.

REFERENCES

1. Knapp, J. Z. and Kushner, H. K. (1980). Generalized methodology for evaluation of parenteral inspection procedures, *J. Parenteral Drug Assoc., 34*: 14.
2. McCrone, W. C., and Delly, J. G. (1973). *The Particle Atlas*, Vols. I–IV, Ann Arbor Science Publishers, Ann Arbor, MI.
3. McCrone, W. C., Delly, J. G., and Palenik, S. J. (1980). *The Particle Atlas*, Vol. V, Ann Arbor Science Publishers, Ann Arbor, MI.
4. McCrone, W. C., Delly, J. G., and Gavrilovic, J. G. (1980). *The Particle Atlas*, Vol. VI, Ann Arbor Science Publishers, Ann Arbor, MI.
5. Bloss, F. D. (1961). *An Introduction to the Methods of Optical Crystallography*, Holt, Rinehart and Winston, New York.
6. Needham, G. H. (1958). *The Practical use of the Microscope*, Charles C Thomas, Springfield, IL.
7. Chamot E. M., and Mason, C. W. (1958). *Handbook of Chemical Microscopy*, Vol. I, John Wiley & Sons, New York.
8. Winchell, A. N. (1954). *The Optical Properties of Organic Compounds*, 2nd ed., Academic Press, New York.
9. Winchell, A. N., and Winchell, H. (1964). *The Microscopical Characters of Artificial Inorganic Solid Substances: Optical Properties of Artificial Minerals*, Academic Press, New York.
10. Horwitz, W. ed. (1970). *Official Methods of Analysis of the Association of Official Analytical Chemists*, AOAC, Washington, DC.
11. Weast, R. C. ed. (1974). *Handbook of Chemistry and Physics*, CRC Press, Cleveland, OH.
12. Hurlbut, C. S., and Klein, C. (1977). *Manual of Mineralogy (after J. D. Dana)*, John Wiley & Sons, New York.
13. Moran, B. R., and Moran, J. F. (1987). An inexpensive digital temperature monitoring device for the ''poor microscopist's hostage,'' *Microscope 35*: 291.
14. Feigl, F., and Anger, V. (1972). *Spot Tests in Inorganic Analysis*, Elsevier, New York.
15. Feigl, F. and Anger, V. (1966). *Spot Tests in Organic Analysis*, Elsevier, New York.

16. Benedetti-Pichler, A. A. (1964). *Identification of Materials*, Academic Press, New York.
17. Shriner, R. L., Fuson, R. C., Curtin, D. Y., and Morrill, T. C. (1980). *The Systematic Identification of Organic Compounds: A Laboratory Manual*, John Wiley & Sons, New York.
18. Chamot, E. M., and Mason, C. W. (1960). *Handbook of Chemical Microscopy*, Vol. II, John Wiley, & Sons, New York.
19. Schneider, F. L. (1964). *Qualitative Organic Microanalysis*, Academic Press, New York.
20. Stahl, E., ed. (1973). *Drug Analysis by Chromatography and Microscopy: A Practical Supplement to Pharmacopoeias*, Ann Arbor Science Publishers, Ann Arbor, MI.
21. World Health Organization (1986). *Basic Tests for Pharmaceutical Substances*, Macmillan/Spottiswoode, London.
22. Porter, C. C., and Silber, R. H. (1950). A quantitative color reaction for cortisone and related 17,21-dihydroxy-20-ketosteroids, *J. Biol. Chem.*, *185*: 201
23. DeLuca, P. P., Boddapati, S., and Im, S. (1980). Guidelines for the identification of particles in parenterals, *FDA ByLines*, *10*(3), Consecutive No. 94.
24. Borchert, S. J., Abe, A., Aldrich, D. S., Fox. L. E., Freeman, J. E., and White, R. D. (1986). Particulate matter in parenteral products: a review, *J. Pharm. Sci.*, *40* (5): 212.
25. McCrone, W. C. (1957). *Fusion Methods in Chemical Microscopy*, Interscience Publishers, New York.
26. Winchell, A. N. (1954). *The Optical Properties of Organic Compounds*, 2nd ed., Academic Press, New York.
27. Budavari, S., ed. (1989). *The Merck Index: An Encyclopedia of Chemicals, Drugs and Biologicals*, Merck and Co., Rahway, NJ.
28. Kuhnert-Brandstätter, M. (1971). *Thermomicroscopy in the Analysis of Pharmaceuticals*, Pergamon Press, Oxford.
29. Boylan, J. C., Cooper, J., Chowhan, Z. T., Lund, W., Wade, A., Weir, R. F., and Yates, B. J., eds. (1986). *Handbook of Pharmaceutical Excipients*, American Pharmaceutical Association/The Pharmaceutical Society of Great Britain, Washington, DC/London.
30. United States Pharmacopeial Convention, ed. (1989). *The United States Pharmacopeia*. 22nd revision, Mack Printing Company, Easton, PA.
31. The Society of Japanese Pharmacopoeia (1991). *JP XII: The Pharmacopoeia of Japan*, Yakuji Nippo, Ltd., Tokyo.
32. Medicines Commission (1988). *British Pharmacopoeia*, Vol. II, Her Majesty's Stationery Office, London.
33. Teetsov, A. S. (1977). Techniques of small particle manipulation, *Microscope*, *25*: 103.
34. Zeissler, C. J. (1992). Particle preparation for materials analysis (tutorial), *Proceedings of the Electron Microscopy Society of America, Fiftieth Annual Meeting*, Boston, pp. 1788–1789.
35. Health Industry Manufacturers Association Committee (1993). Final report: improved microscopic assay for the enumeration of particulate matter in parenteral solutions, *Pharm. Forum*, *19*(3): pp. 5435.

36. Barber, T. A. (1993). *Pharmaceutical Particulate Matter: Analysis and Control*, InterPharm Press, Buffalo Grove, IL.
37. King, S. T. (1973). Application of infrared Fourier transform spectroscopy to analysis of micro samples, *J. Agric. Food Chem.*, *21*(4): 122.
38. Griffiths, P. R., and deHaseth, J. A. (1986). *Fourier Transform Infrared Spectrometry*, Vol. 83 in Chemical Analysis: A Series of Monographs on Analytical Chemistry and Its Applications P. J. Elving, and J. D. Winefordner, eds.), John Wiley & Sons, New York.
39. Fuller, M. P., and Griffiths, P. R. (1978). Diffuse reflectance measurements by infrared fourier transform spectroscopy, *Anal. Chem.*, *50*: 1906.
40. Garner, H. R., and Packer, H. (1968). New technique for preparation of KBr pellets from microsamples, *Appl. Spectrosc.*, *22*: 526.
41. Smith, M. A., Chao, R. S., and Clark, D. A. (1989). Dehydration of Erythromycin dihydrate: a microscopy–FTIR application, *7th International Conference on Fourier Transform Spectroscopy*, George Mason University, Fairfax, VA, p. P5.55.
42. Higgins, R. S., and Klinger, S. A., (1990). *High Purity Solvent Guide*, Baxter Diagnostics Inc., Burdick and Jackson Division, Muskegon, MI.
43. *Solvent Use Index II*, Waters Associates.
44. Dimbleby, V. (1953). Review article: glass for pharmaceutical purposes, *J. Pharm. Pharmacol.*, *5*: 969.
45. Persson, H. R. (1962). Improvement of the chemical durability of soda-lime-silica glass bottles by treating with various agents, *Glass Tech.*, *3*: 17.

10
Infrared Microspectroscopy in Earth Science

Anne M. Hofmeister Washington University, St. Louis, Missouri

1. INTRODUCTION

Earth science involves the study of our planet as a whole, its evolution and origin. Basic issues concern the physical characteristics of the earth and its constituent materials. Our rocky planet is formed of natural, coherent, aggregate masses of solid material. Each rock contains one or more components called minerals. These chemical compounds vary from the utmost in simplicity and purity (e.g., not-so-rare diamonds consist of a network of tetrahedrally linked C atoms in a cubic array) to solid solutions of incredible complexity in structure and chemistry [e.g., hornblende, a common amphibole (Table 1), consists of paired chains of SiO_4 tetrahedra partially substituted with Al that are cross-linked by various transition metals and other cations in a monoclinic array]. Measurements of the physical properties of minerals, their stability over pressure and temperature ranges, and their structural phase transformations are central to interpreting the formation and history of rocks. Because the accessible surface of our planet is but a thin veneer, the mineral samples we observe do not accurately represent the bulk earth (Table 1). Therefore, indirect measurements (e.g., on meteorites) and comparisons of laboratory data on synthetic high-pressure materials to seismic studies are used to probe the interior of our planet. Furthermore, as the earth's surface has changed dramatically over its 4.5-billion year history due to release of heat through volcanism and convection of its interior, comparisons with the surfaces of other rocky moons and planets are useful in that these bodies provide a series of ''snapshots'' in planetary evolution.

Table 1. Chemistry of the Earth's Mantle and Its Minerals

Bulk mantle and crust composition[a]		Common surficial minerals	Possible interior minerals
SiO_2	48	$(K,Na)AlSi_3O_4$ alkali feldspar	$Mg_{0.9}Fe_{0.1}SiO_3$ perovskite
TiO_2	0.27	$(Na,Ca)(AlSi)_4O_4$ plagioclase	$Mg_{0.9}Fe_{0.1}SiO_3$ majorite
Al_2O_3	5.2	$(Ca,Na,Mg,Fe)(Al,Si)O_3$ pyroxene	$Mg_{0.9}Fe_{0.1}SiO_3$ ilmenite
Cr_2O_3	1.1	$(Ca,Na)_{2-3}(Mg,Fe,Al)_5Si_6(Si,Al)_2O_{22}(OH)_2$ amphibole	$Mg_{1.8}Fe_{0.2}SiO_4$ olivine
MgO	34.3	SiO_2 quartz	$Mg_{1.8}Fe_{0.2}SiO_4$ β- or γ-spinel
FeO	2.1	$KAl_2(AlSi_3O_{10})(OH)_2$ muscovite	$Mg_{0.9}Fe_{0.1}O$ periclase
MnO	0.12	$K(Mg,Fe)_3(AlSi_3O_{10})(OH)_2$ biotite	$CaMgSi_2O_6$ diopside
CaO	3.6	$Al_2Si_2O_5(OH)_4$ kaolin and other clays	$NaAlSi_2O_6$ jadeite
Na_2O	1.5	$(Mg,Fe)_6(Si,Al)_4O_{10}(OH)_8$ chlorite	SiO_2 stishovite
H_2O	0.11	$CaCO_3$ calcite and other carbonates	$Mg_3Al_2Si_4O_{12}$ pyrope

[a]From Ref. 88.

1.1. Problems Amenable to Infrared Spectroscopy

Infrared (IR) spectroscopy is used in earth science for both quantitative and qualitative analysis. Qualitative studies are based on the fact that the vibrational spectrum is a characteristic of the material; hence comparison to a set of standards allows for identification of the phase. Qualitative studies have been applied most frequently by geologists and planetary scientists on a regional scale for reconnaissance missions. Specifically, laboratory data are compared either to reflectance spectra from aircraft to assess economic potential, or to measurements made from spacecraft to model surface compositions of planets and asteroids [1]. Quantitative analysis concerns minerals rather than rocks and now tends toward microsamples due to recent instrumental advances, to the typically small grain sizes of 1 mm or less for chemically homogeneous, untwinned minerals, and to the even smaller (<0.1 mm) crystal size and quantity of synthetic materials formed at high pressures and temperatures inside the earth. Also, the nondestructive nature of reflectance measurements makes this technique amenable to identifying precious gemstones. These types of studies have been made on macroscopic samples.

An ongoing concern in earth science is measurement of the thermodynamic and physical properties of phases that would be stable inside the earth's interior (i.e., those minerals that form at pressures near 10^6 atm and temperatures near 2000°C). Typically, data are generated in the laboratory from various materials that were either synthesized or transformed at high pressures and temperatures. These measurements are then compared to compressibility and sound-speed data extracted from measurements of seismic waves propagating through the earth to ascertain the composition and thermal profile of the earth's interior (e.g., [2]). Spectroscopy is useful in this regard in that entropy, heat capacity [3], and compressibility [4] can be calculated from the measured vibrational frequencies. Moreover, the temperature and pressure dependence of the lattice modes allows for extrapolation of direct measurements at ambient conditions of these and other properties to conditions appropriate to the earth's interior. For this type of study, complete characterization of the sample is necessary, as the accuracy of the calculations depends on the distribution of vibrational modes over frequency. Furthermore, the low-frequency vibrations have a large impact on the calculations, and thus the more difficult to obtain far-IR data are crucial [3].

An entirely different concern is hydrogen. Both OH^- and H_2O are ubiquitous in minerals, either as stoichiometric species or as impurities, so that information on their concentrations is needed for quantitative chemical analysis. Such knowledge is of interest to earth science because incorporation of even small amounts of hydrogen affects rheological properties [5] and the stability field of the phase, and lowers its melting temperature (e.g., [6]). Response to radiation is also strongly influenced by the presence of hydrogen (see the references in

Ref. 7). Conventional methods such as the electron microprobe do not detect this light element. Determinations of very low water concentrations through IR spectroscopy are possible because H vibrates at much higher frequencies than heavy atoms such as Si and the transition metals, which are major constituents of minerals; thus water bands are found in "transparent" regions of most mineral spectra. Furthermore, IR spectroscopy used in conjunction with selection rules is capable of differentiating between hydroxyl, structurally bound water, and fluid inclusions [8]. For the most part, these types of studies have been made on macrosamples, roughly millimeter size [8], but small grain sizes have been investigated [9].

Microscopic fluid inclusions could also be probed by infrared spectroscopy. Minerals such as fluorite (CaF_2) possess hydrocarbons whose speciation can be assessed through comparison of the IR spectrum to standards. These data are of interest because the composition of the fluid inclusions is tied to the aqueous melt from which the minerals formed. So far, these types of studies have been addressed by the Raman technique [10] rather than IR (see Section 5.1). Similarly, glasses are studied as an approximation to the silicate melts, which form a large number of rocks on the earth's surface. Studies of glasses have so far been entirely of macrosamples.

1.2. Background Material and Useful References

Application of IR microscopy to problems in earth science mainly involves minerals. Description of the vibrations of such crystalline structures and the molecular units contained within them is simplified through use of group theory. A variety of chemistry books concern the nomenclature and basic principles (e.g., [11]). One of the most useful reference books is Cotton [12]. Although an occasional mineral is mentioned, applications to earth science are not covered.

Interaction of light with solids is covered by several textbooks on solid-state physics (e.g., [13]; and in detail in Ref. 14). A few aspects are mentioned here. For crystalline solids, the number of modes is determined by the space group. Interaction of IR light with a material occurs only if the dipole moment of the vibrating atoms changes during the cycle. Thus not all vibrations of a crystal or molecule are IR active. In addition, the interaction depends on the orientation of the dipole. For a cubic solid, vibrations perpendicular to the beam are directly affected, whereas vibrations parellel to the beam are indirectly stimulated, leading to slightly different frequencies. The symmetry of the cube gives two transverse optic frequencies (the TO components) and one longitudinal (LO). For less symmetric solids, the splitting is one of each type. Reflectivity measurements contain both TO and LO components (Figure 1). Deciphering their positions requires mathematical analysis of the spectra, discussed in detail in later sections. It is generally thought that absorptivity measurements involve

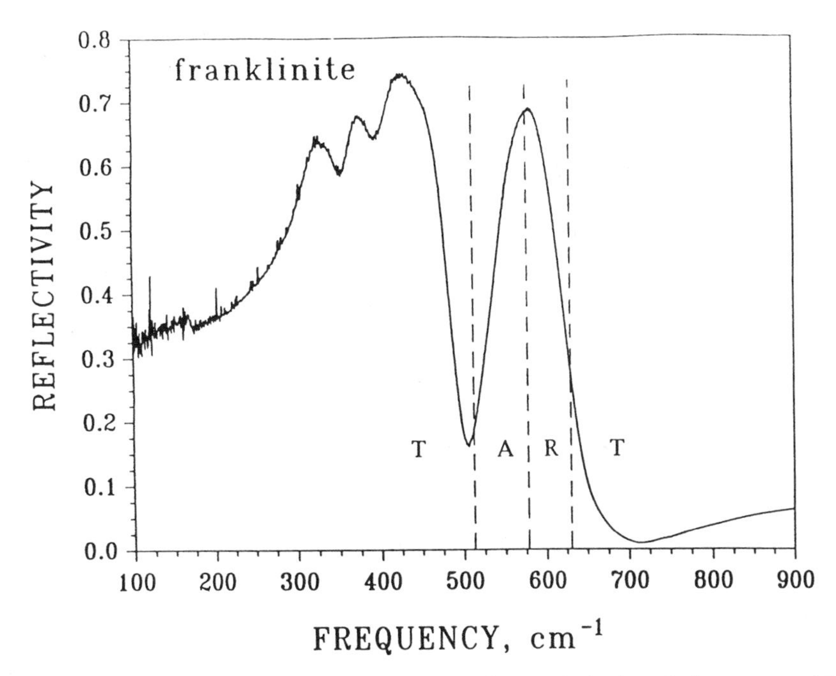

Figure 1. Schematic of a reflectivity spectrum. The sample is the spinel structure with composition $ZnFe_2O_4$. T, transmitting region; A, absorbing region; R, reflecting region.

only the TO component. However, realities such as convergence of light in the spectrometer, wedging of samples, and nonuniform thickness can induce obvious manifestations of the LO component in the spectra [15]. Furthermore, a small amount of the LO component is predicted as shoulders on asymmetric peaks even in the "infinitely thin" samples from the Fresnel equations or from calculations of spectra for ideal harmonic oscilators.

Vibrational spectroscopy of minerals is reviewed in a variety of books and articles. Reference 16 offers a reasonable compromise between an introductory approach and comprehensive coverage and contains an extensive list of references. The book edited by Farmer and Lazarev [17] is excellent and not yet outdated.

Background information on minerals and their properties can be found in numerous texts. Reference 18 is an excellent and inexpensive introduction, whereas Ref. 19 is comprehensive and thorough. In brief, the most common and important rock-forming minerals are the silicates (Table 1), which are classified according to the polymerization of their SiO_4 tetrahedra (e.g., chains, sheets, networks). Additional cations are present in various proportions to balance charge and to link the different polymers. The standard classification scheme is useful for spectroscopists in that the structure is manifest as the number of peaks in the vibrational spectra. Furthermore, the low mass of the Si atom (and of Al, which commonly substitutes for it) and its tight bonding with oxygen places vibrational modes involving Si at high frequency. The much weaker bonding of oxygen with the transition metals and alkali ions makes classification of vibrations into internal (Si—O stretches near 1000 cm^{-1} and bends near 500 cm^{-1}) and external modes (translations of cations from 50 to 400 cm^{-1}) a useful and reasonable first approximation.

2. EXPERIMENTAL PROCEDURES

2.1. Special Problems with Minerals

One intrinsic limitation of IR spectroscopy is that light diffracts when it encounters an aperture with a diameter comparable to the wavelength. The minimum size of samples that can be examined depends on the frequency region of interest. Study of micron-sized fluid inclusions does not present significant problems with diffraction because their IR bands exist at high frequency (near- or perhaps midranges). In contrast, diffraction effects [20] are a severe problem in the acquisition of single-crystal data from the relatively larger high-pressure phases, because data near 100 cm^{-1} is needed to characterize these samples completely. These materials are usually synthesized in microgram quantities and with grain sizes typically 10 to 50 μm; only on rare occasions is a 200-μm sample encountered. Thus data are usually limited to above 200 or 400 cm^{-1}. One compromise is use of thin-film rather than single-crystal techniques, discussed below. The loss is, of course, information on orientational dependence of the IR modes.

A new (albeit expensive) approach to problems with diffraction from small samples is use of synchrotron radiation as the source [21] such that the intense radiation serves to compensate for the loss of light. Currently, the sample sizes of 10 μm measured with this approach in the region above 1000 cm^{-1} and 1.5 μm measured above 4000 cm^{-1} [21] are only slightly smaller than those accessible with conventional instrumentation, as described above.

One unavoidable problem with minerals is that their IR peaks vary considerably in intensity. In general, the lower-frequency bands are much less intense. This necessitates acquisition of multiple spectra from different thickness of samples to establish the positions of all peaks. Sometimes the weaker peaks occur as shoulders on intense bands. Spectral fitting routines are helpful to address this problem. Also, the low symmetry and large unit cells of many minerals result in a large number of bands, 20s to 100s. Accidental degeneracies occur and can only be proven through comparison to spectra from structural analogs or to evolution of spectra with chemical substitutions (solid solutions).

Many important accessory minerals in earth science are iron-bearing oxides or sulfides, which are opaque at visible wavelengths in standard 30-μm thin sections [19]. Depending on the degree of the electronic interactions, these samples can be opaque in the IR also. Typically, Fe–Fe or Fe–Ti charge transfer and semimetals yield reflectivities near 60 to 80%, with indication of only the major vibrational bands; hence a combination of reflection and absorption measurements and investigation of chemical substitutions are needed to locate the IR bands [22]. Because of this intrinsic difficulty, relatively few quantitative studies have been of oxides, sulfides, and many other classes of nonsilica bearing minerals.

Glasses present less of a problem in technique, but the weak, broad nature of their bands and existence of overlapping bands requires use of Fourier deconvolution [23] or peak fitting [24]. The interpretation of the results and the accuracy of the analyses have resulted in controversies.

The prohibitively high cost of gemstones and the uniqueness of some samples requires nondestructive analysis. For the most part, gemstones are sufficiently large that specular reflection from available facets is possible. For zoned samples, microanalysis is needed, and this again requires a compromise between minimization of diffraction effects and collection of a complete data set.

2.2. Instrumentation

Infrared spectra can be measured using dispersive or Fourier transform spectrometers through a variety of technqiues, including (1) by direct transmission through thin (0.1 μm to a few micrometers) layers created by compression, (2) by transmission through pellets of powdered specimen embedded in a dielectric medium (typically, KBr or petroleum jelly), (3) by specular reflectance from polished surfaces, (4) by attenuated total reflectance, and (5) by diffuse reflectance. Spectra obtained by direct transmission and by deconvolution of specular reflectance spectra have proper line shapes. Both of these methods are amenable to microsamples. Spectra of polar materials (virtually all minerals) have distorted line shapes and shifted band positions when measured on powder dispersions (see, e.g., [17]).

The large number of research problems in earth science involving small samples has led to the almost exclusive use of Fourier transform infrared (FTIR) spectrometers. Griffiths and de Haseth [25] offer up-to-date and throughout discussions on the principles and operation of these types of instrument. Hirschfeld [20] provides a useful and concise synopsis of the problems encountered with FTIR spectroscopy.

For small samples, high-sensitivity detectors are essential. Mid-IR quantum detectors suffice as these have a high sensitivity. However, the standard DTGS (deuterated triglycine sulfate) detector for the far-IR range is insufficient for microanalysis. Commercially available liquid-helium-cooled bolometers[1] with Si or Ge elements are necessary. These offer reasonably broad detection ranges, although for the narrow-range varieties, the detectors are more sensitive, so there is a trade-off in capabilities. The main drawback of sensitive detectors is the expense. Principles of operation and sensitivity of the various types of detectors, quantum and thermal, are covered by Griffiths and de Haseth [25].

An additional consideration is use of a purged or evacuated instrument. A vacuum is needed for study of minute water contents and for measurements in the far-IR. Otherwise, the absorption of water on the sample surface may be larger than the intrinsic bands (this is often a problem when KBr pellets are used [8]) or the small amounts of unpurged water vapor contribute a rotational spectrum that is comparable in strength to the weak, low-frequency absorptions of minerals [16].

2.3. Specular Reflectance

The sample surface needs to be polished smooth and flat. Standard techniques for preparation of thin sections suffice, with the final grinding involving submicron-sized alumina or diamond polishing compound [26]. For microsamples, Crystal-Bond[2] is advantageous for adhering the sample to the mount, because this glue dissolves in acetone or relaxes by heating to about 100°C. Only one side should be polished because reflections from the back of a doubly polished microscopic sample can occur if the sample is thin enough [27].

Commercially available IR microscopes allow measurement of reflection spectra at wavelengths as low as the diffraction limit [28,29]. We use the Spectrascope[3] for mineralogical studies because this can be inserted in the evacuated compartment. Interference fringes can obscure the bands at wavelengths slightly above the size of the limiting aperture [30]. Quantitative analysis is still possible if one uses a combination of Kramers–Kronig and classical dispersion [29]. The main disadvantage of IR microscopes is steep convergence of the beam, giving a significant component of nonnormal incidence. Articles in Ref. 31 cover design and use of IR microscopes in detail. For unpolarized measurements this effect is inconsequential. The reflectance data collected using a FTIR microscope are identical to data collected on the same sample with a noncondensing specular reflection device. For polarized measurements, this results in a significant pro-

[1]Infrared Laboratories, Inc., 1808 East 17th Street, Tucson, AZ 85719.
[2]Aremco Products, Inc., P.O. Box 429, Ossining, NY 10562-0429.
[3]Spectra-Tech, Inc., 652 Glenbrook Road, Stamford, CT 06906.

portion of *P* polarization in the beam (Figure 2), and as a result, the symmetries of the crystal are mixed (e.g., [32]). The amount is not terribly large (ca. 5% [28]), but this can bring in spurious bands and can alter the positions of peaks which overlap with modes present in the other symmetries [33].

One approach in obtaining information on polarizations with a microscope is to measure crystals oriented in different ways. For example, comparing reflectance data from a polycrystal to that of a single crystal allows polarizations to be inferred from intensity differences between the two samples [30].

Vibrations from different symmetries can be fully separated by using specular reflectance devices that do not involve condensing mirrors or lenses, but

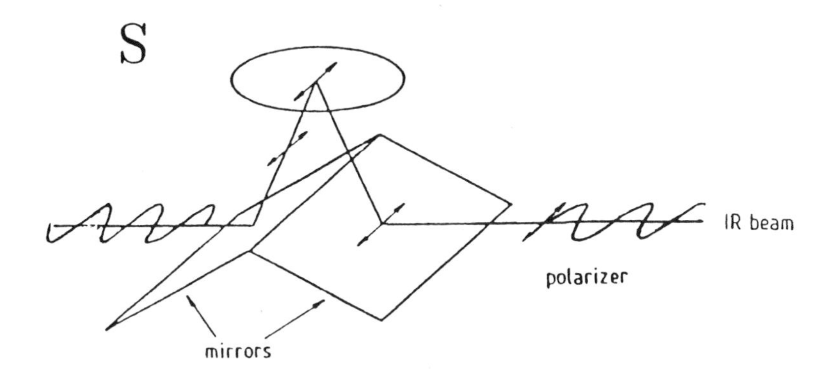

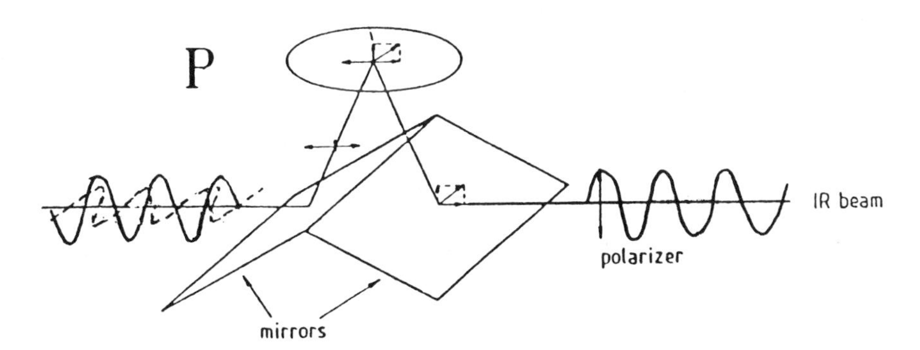

Figure 2. S and P polarizations. The wavy line lies in the propagation direction of the light and shows whether or not the light is polarized. The arrows indicate the polarization of the **E** vector. Polarizations are maintained for the *S* orientation but are mixed in the *P* orientation.

these devices cannot probe as small a sample. Samples as small as 1 mm in diameter can be measured with commercially available equipment.[3] It should be possible to construct a device to make polarized measurements on smaller samples.

2.4. Absorption Spectroscopy of Single Crystals

Absorption spectroscopy has two drawbacks: the difficulty in preparing geologic samples that are "thin enough" and artifacts induced by a mismatch of refractive indices of the sample with the reference. Samples should be doubly polished, but this is impossible to achieve for the smallest samples. The maximum thickness for which all IR modes are "on scale" depends on the material and the wavelength region. For example, examination of mid-IR lattice modes involving Si—O stretching and bending in silicates requires thicknesses on the order of tenths of microns (Section 3.2). In contrast, weaker lattice modes in the far-IR that arise from translations of metal ions are best studied at 1 to 5 μm thickness [34]. Even greated thickneses (10 μm to centimeter size, depending on the concentration) are required to study absorptions from impurities such as OH^- [8]. Thickness as small as 5 μm can be achieved through polishing by hand, and those near 20 μm are amenable to use of an automated polishing apparatus,[4] but the minute dimensions needed to examine the intense fundamental lattice modes require ion thinning [27].

Commercially available machines[5] for preparation of sections for the transmitting electron microscopes serve well to prepare sections for IR spectroscopy. The literature descriptions of the procedures [35] apply directly with only minor modification. The ion guns should be aimed to cover as wide an area as possible instead of focusing on one spot. This technique produces an unavoidably "lumpy" surface (Figure 3) which could produce some artifacts in relative intensity of the peaks. We also tried dimpling[6] the sample but found that the thinned area was too small for thicknesses needed for mid-IR measurements and the sample too difficult to detach for spectroscopic measurements [27]. This approach would work for thicker samples (e.g., for the study of fluid inclusions or of far-IR modes) and offers the advantage of lower cost.

Quantitative analysis requires exact knowledge of thickness. Thicker samples, on the order of 10 μm or larger, can be determined directly in a standard petrographic microscope or a high-power binocular microscope. Thinner sam-

---

[4]Buehler, 41 Waukegan Road, Lake Bluff, IL 60044.

[5]Model 600 Dual Ion Mill, for example: Gatan, Inc., 6678 Owens Drive, Pleasanton, CA 94566.

[6]Model 656/3 Dimple Grinder: Gatan, Inc., 6678 Owens Drive, Pleasanton, CA 94566.

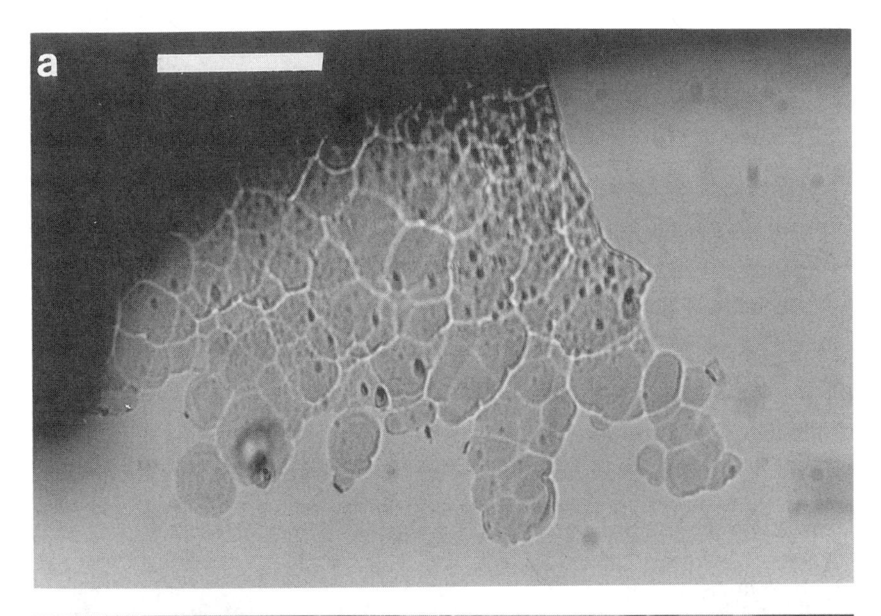

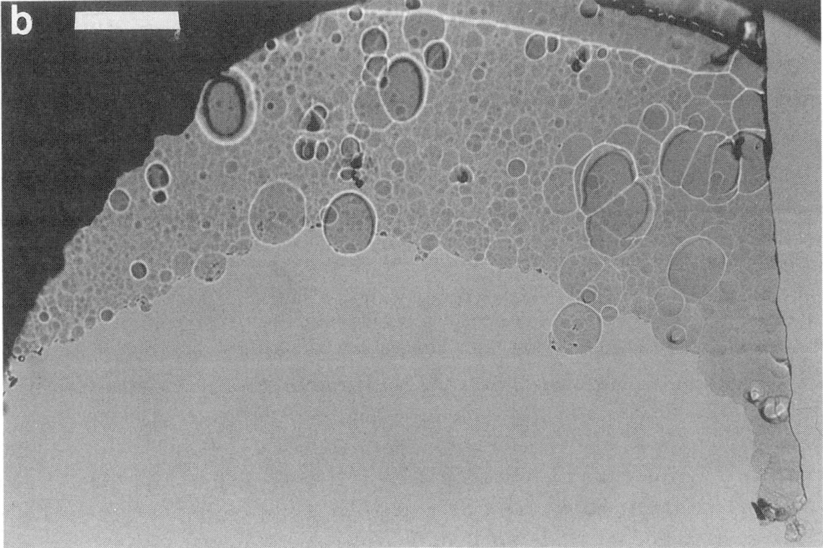

Figure 3. Photomicrograph of ion-thinned single crystals of garnets. (a) Natural grossular ($Ca_3Al_2Si_3O_{12}$) from Asbestos, Quebec. The field of view is about 0.5 mm. (b) Natural grossular with Fe^{3+} substituting for 6 mol % of the Al, and Fe^{2+} substituting for 4 mol % of the Ca. The field of view is about 0.8 mm. Scale bars are 0.1 mm.

ples require use of a scanning electron microscope (SEM).[7] Thickness (Figure 4) can be determined through comparison to a wire grid of known spacing. We used a grid with 2160 lines/mm. Alternatively, one would need to measure accurately the tilt of the sample in the SEM and the distance from the sample to the objective to measure the thickness precisely. Interference fringes produced with a laser may be a feasible method of measurement, but this has not been attempted. Use a profilometer is probably not feasible, due to the fragile nature of submicron crystals.

An indirect determination of thickness is possible if reflectivity data are available on a similar sample. A Kramers–Kronig analysis of the reflectance (Section 3.1) gives the absorptivity (in mm^{-1}). Ratioing the calculated absorptivity to the measured absorbance of various peaks (Figure 5) gives the thickness of the thin film. However, since the films are not perfectly uniform, intensities for various peaks can yield differences in estimated thicknesses (see Section 3.2).

An IR microscope is essential for collecting absorption spectra from single crystals because of their small size. The lesser of the two intrinsic problems involves mixing of polarizations. This mixing is induced by the converging beam, in that the curvature of the condensing mirrors changes the polarization of the light (as in Figure 2) during the reflections. If the polarizer is positioned such that polarization is in the S orientation (Figure 2) for most of the mirrors, mixing is minimized (to about 5% for the Spectrascope). Intensities should not be affected (according to classical dispersion analysis; see Section 3.1) for typical angles of convergence less than about 25°, and because the amounts of LO components in absorption spectra are small.

The major problem with use of microscopes for minerals is mismatch between the index of refraction of the crystal and the index of refraction adopted in designing the microscope (G. R. Rossman, personal communication, 1993). Newer designs of microscope deal with this effect but cannot remove the problem completely because the index of refraction is not a constant but depends on the strength of the absorbance (see, e.g., [14] and Section 3). In turn, this aspect changes with wavelength region and the nature of the material being studied. This problem is magnified for earth science applications because silicate minerals have high indices of refraction, and as a result, the peak positions differ from the true measurements (see Section 3.2). One possibility is to use a high-index-of-refraction substrate for both the reference spectrum and as a support for the sample. Wafers of type II diamond would serve this purpose, and scrap pieces are very inexpensive.[8]

---

[7]The JEOL Model 840, for example.

[8]For example, 5-mm pieces are about $10 each from Dubbledee Harris Diamond Corp., 100 Steirli Court, Mount Arlington, NJ 07856.

Figure 4. SEM images. (a) Portion of the same single crystal of Asbestos, Quebec, grossular that was shown in Figure 2a. This was taken at 6.0 kV with a magnification of 4000. The scale bar is 1 μm. (b) Portion of the same single crystal of Eden Mills grossular that was shown in Figure 2b. This was taken at 4.4 kV with a magnification of 4000 times. The scale bar is 1 μm. (c) Piece of a thin film formed in a diamond anvil cell from synthetic grossular. This was taken at 3.0 kV with a magnification of 4000 times. The scale bar is 1 μm.

2.5. Thin-Film Technique

Diamond anvil cells (DACs) come in a variety of types and designs. All DACs consist of two brilliant cut stones with the large facets (tables) pointing away from each other and smaller facets (culets) in contact (Figure 6). For IR studies, type II diamonds are usable for all wavelength ranges. The nitrogen impurities in type I diamonds absorb light near 1000 cm^{-1} so that these cannot be used for mid-IR measurements (e.g., [36]). Other high-strength, ''transparent'' materials, such as cubic zirconia and sapphire, can be used for various wavelength ranges [37,38]. The cell itself offers a means of alignment of the stones and as a lever to create pressure. Generation of pressures of 10^6 atm are possible with small facets (0.5 mm) and a moderate applied only load of about 1000 kg. Various designs and their uses are covered in the literature [39–42]; some of

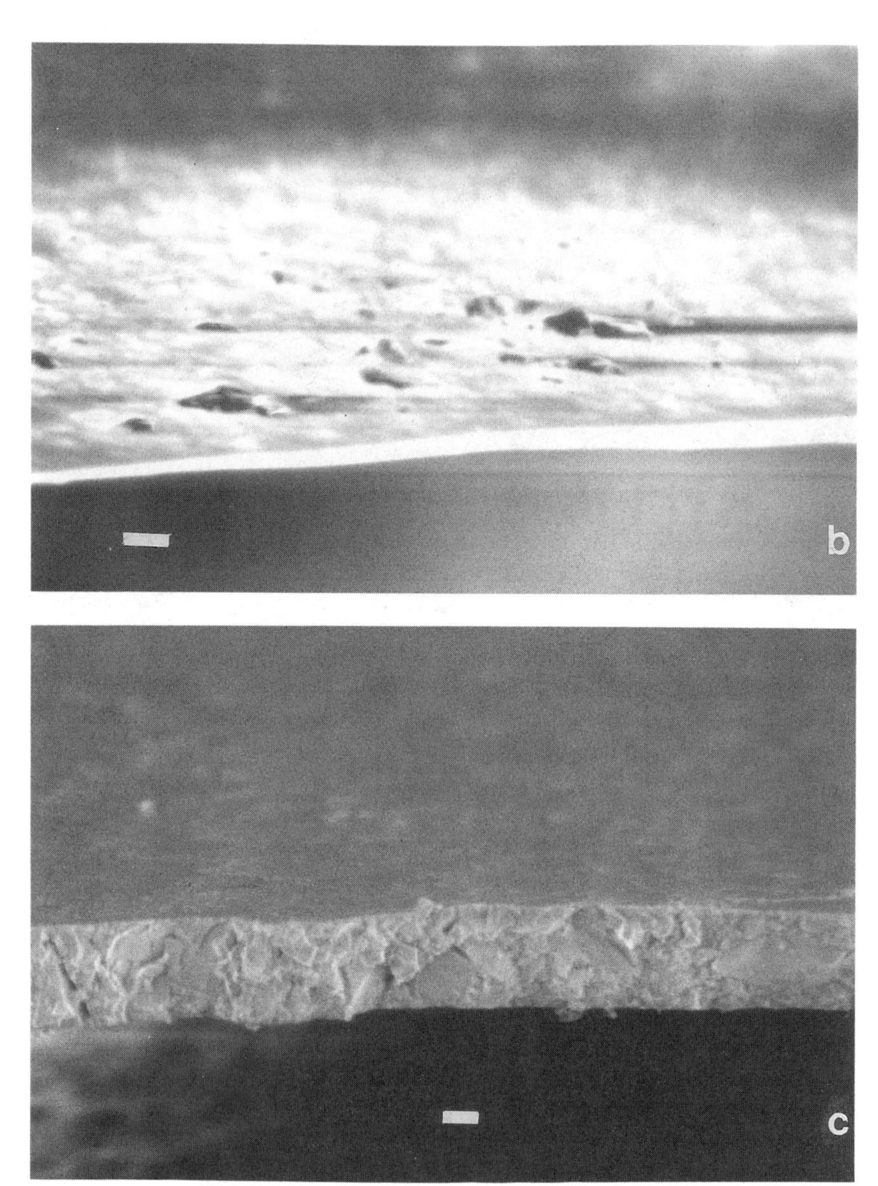

Figure 4 Continued

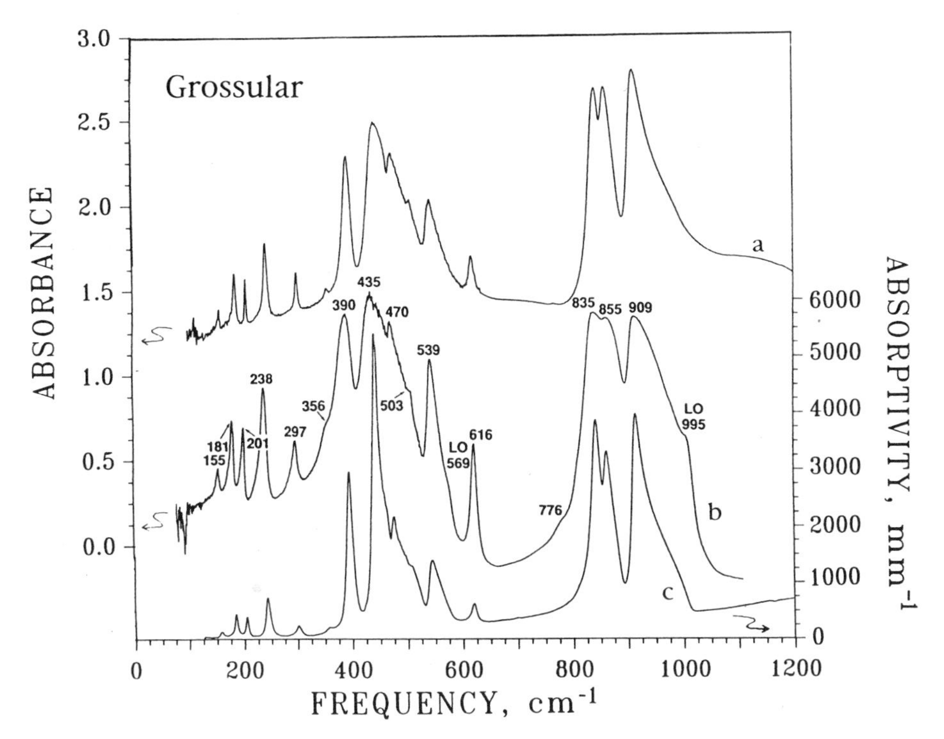

Figure 5. Absorbance spectra of grossular garnets. Spectra are offset for clarity. From top to bottom: (a) thin-film absorbance of synthetic grossular (provided by P. Richet, IGP, Paris IV); (b) single-crystal absorbance from the natural grossular from Asbestos, Quebec, shown in Figures 2a and 3a; (c) calculated absorptivity from reflectivity data on the same natural grossular. The scale to the right side pertains to spectra (a) and (b), whereas the scale to the left pertains to spectra (c).

these cells are commercially available.[9] Construction details are available for the Merrill–Bassett design [40]. For the purpose of this article, the DAC serves as a fancy sample holder, and pressures are applied to produce a thin film.

Thin films can be prepared from single crystals, polycrystals, or finely ground powders. The sample is picked up with a needle through surface tension or with forceps and placed on one of the diamonds (usually, on the piston). A binocular microscope with a long working distance (ca. 20 mm) is essential in

[9]For example, High Pressure Diamond Optics., Inc., 7400 North Oracle Road, Suite 372, Tucson, AZ 85704.

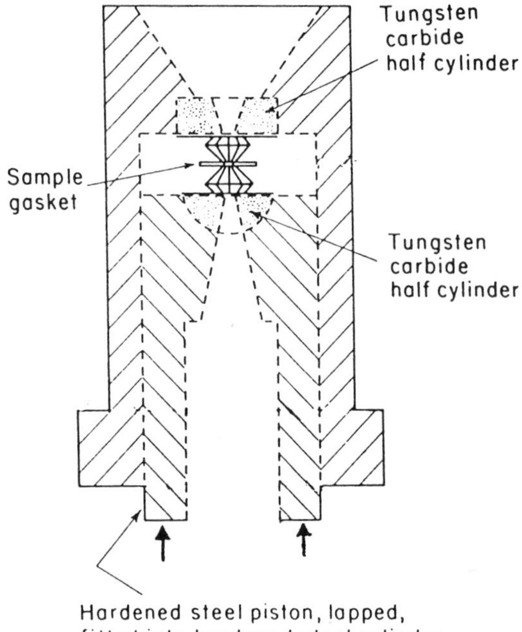

Figure 6. Schematic of a diamond anvil cell.

this process.[10] The ungasketed DAC is then assembled and pressure is generated by squeezing the piston and cylinder together using one's hands instead of the lever arm. The roughly 5 GPa attained in this manner is more than sufficient to crush any crystals larger than a few microns. The production of the film is monitored under the microscope and with experience, roughly correct thicknesses can be attained. Generally, a rough spectrum is acquired to ascertain whether the peaks are on scale and sample is removed or added until the desired absorptivities are produced.

The films obtained in this manner are not perfectly uniform (Figure 4c). Incompressible materials such as silicates tend to form films that are thicker in the middle of the diamond, or at the sides, if the diamonds are not properly aligned. Light leakage around the edges of the film or from cracks, if the diamond anvils are not completely covered, can also affect the quality of the spectra.

---

[10]For example, the SZ series of Olympus Optical Corporation, 4 Nevada Drive, Lake Success, NY 11042-1179.

Thicknesses of thin films are measured with most of the same approaches as taken with single crystals. Because the films are opaque due to scattering when the pressure is released (Figure 4c), thicknesses cannot be measured easily using interference phenomena. Nor is profilometry possible, due to the fragile nature of the films.

The DAC can be placed directly in the beam for mid- or near-IR experiments, although the low throughput of the diamond anvil cell, due its small aperture (ca. 0.6-mm-diameter diamonds) and small limiting angles (ca. 7°), makes use of an all-reflecting beam condenser advantageous. These are available commercially for some designs of DAC (such as that of Spectra-Tech). The smaller Merrill–Bassett design can also be viewed through an FTIR microscope. Lenses are possible for measurements at high frequencies, but the hard materials suitable for a wide range of far IR are lacking. For instance, Si is opaque in the visible, making alignment difficult.

2.6. Powder Dispersions

Powder dispersion, a popular and long-standing technique, involves grinding a small amount of sample and mixing this with a diluent [43]. For the mid-IR, typically 0.7 mg of a mineral is mixed with about 200 mg of KBr or CsI and then pressed into a pellet. For the far IR, sample is dispersed in petroleum jelly on a polyethylene card with roughly the sample proportions as above. For microsamples, the amounts of sample and medium can be reduced. The limiting factor for small samples is loss of material during the grinding. Typically, this method is used for samples that are found in ample amounts but with grain sizes too small for single-crystal work.

3. QUANTITATIVE ANALYSIS

3.1. Kramers–Kronig and Classical Dispersion Analyses of Reflectance Data

The measured reflectivity $R(\upsilon)$ is analyzed by casting the equations in terms of complex numbers [44] to give the magnitude

$$r(\upsilon) = [R(\upsilon)]^{1/2} \tag{1}$$

and the phase angle

$$\Theta(\upsilon_i) = \frac{2\upsilon_i}{\pi} \int_0^\infty \frac{\ln r(\upsilon) - \ln r(\upsilon_i)}{\upsilon_i^2 - \upsilon^2} d\upsilon = \frac{\upsilon_i}{\pi} \int_0^\infty \frac{\ln R(\upsilon) - \ln R(\upsilon_i)}{\upsilon_i^2 - \upsilon^2} \tag{2}$$

Application of Maxwell's equations and the boundary conditions (light traversing a vacuum impinging on a crystal) leads to a relationship with the real and imaginary parts of the complex index of refraction:

$$n(\upsilon) = \frac{1 - r^2(\upsilon)}{1 + r^2(\upsilon) - 2r(\upsilon)\cos(\Theta(\upsilon))} \tag{3}$$

and

$$k(\upsilon) = \frac{2r(\upsilon)\sin(\Theta(\upsilon))}{1 + r^2(\upsilon) - 2r(\upsilon)\cos(\Theta(\upsilon))} \tag{4}$$

Conversely,

$$R = \frac{(n - 1)^2 + k^2}{(n + 1)^2 + k^2} \tag{5}$$

The complex dielectric functions are then

$$\epsilon(\upsilon) = \epsilon_1(\upsilon) + i\epsilon_2(\upsilon) \tag{6}$$

where

$$\epsilon_1(\upsilon) = n^2 + k^2 \tag{7}$$

and

$$\epsilon_2(\upsilon) = 2nk \tag{8}$$

Absorptivity α is obtained from [14]

$$\alpha(\upsilon) = \frac{2\pi\upsilon\epsilon_2(\upsilon)}{n(\upsilon)} \tag{9}$$

The positions of the transverse optic modes υ_{TO} are defined by the maxima in the dielectric function ϵ_2. The positions of the longitudinal optic modes υ_{LO} are taken from the minima of imaginary part of $1/\epsilon$.

Evaluating the integral of equation (2) requires data over all frequencies from zero to infinity, an impossible experimental task. Therefore, it is necessary to split the integral of (2) into three parts: The middle phase angle is evaluated between the lower cutoff frequency υ_{low} and the upper limit of measurements υ_{up} by applying the trapezoidal rule. For frequencies above υ_{up} to infinity, we use Wooten's [14] approximation, and for frequencies from 0 to υ_{low}, reflectivity is assumed to be constant in computing the phase angle.

Classical dispersion analysis (e.g., [45,46]) involves construction of a synthetic spectrum from a set of damped harmonic oscillators that are described by TO peak positions υ_i, full widths at half height γ_i, and oscillator strengths f_i.

The method approximates vibrations as Lorentzian modes:

$$\epsilon_1(\upsilon) = n^2 - k^2 = \epsilon_\infty + \sum_i f_i \upsilon^2 \frac{\upsilon_i^2 - \upsilon^2}{(\upsilon_i^2 - \upsilon^2)^2 + \gamma_i^2 \upsilon^2} \tag{10}$$

$$\epsilon_2(\upsilon) = 2nk = \sum_i \frac{f_i \upsilon_i^2 \gamma_i \upsilon}{(\upsilon_i^2 - \upsilon_i)^2 + \gamma_i^2 \upsilon^2} \tag{11}$$

The dielectric constant ϵ_∞ is approximated from the indices of refraction. Fits to ϵ_1 and ϵ_2 are made by varying the oscillator strengths taken from the Kramers–Kronig analysis. Reflectivity is then calculated from inverting equations (7) and (8) and inserting this into (5). Absorptivity is calculated similarly from (9).

Nonnormal incidence requires that the Fresnel equations [equations (3) to (8)] be modified to analyze reflectance data and that the classical dispersion analysis be recast slightly. Details are given in Refs. 32 and 47–49. The small departure from normal incidence encountered in a microscope for unpolarized measurements does not perturb the results significantly, and thus the equations above are relevant.

3.2. Comparison of Thin Film with Single-Crystal Absorption and Reflection Techniques

In this section we compare absorbance spectra from thin films of synthetic garnets [grossular = $Ca_3Al_2Si_3O_{12}$ (Figure 7), pyrope = $Mg_3Al_2Si_3O_{12}$ (Figure 8)] to single-crystal measurements of identical or similar samples. Differences are due primarily to variations in thickness and how absorbance peak positions are perturbed by use of a microscope.

Spectra should be taken from as thin a film as possible. Thick films (Figure 7) have rounded peak shapes, positions shifted to slightly higher wavenumbers, and incorrect relative intensities among the peaks. The slight shifts shown in Figure 7 are due at least partially to difficulties in pinpointing the maxima on a rounded peak. The rounded shapes are due to (1) light leakage through cracks or around the edge of the film, (2) nonuniform thickness, and (3) an increase in the proportion of LO to TO components with increasing thickness. It is the latter that moves the peak position to higher frequencies with increasing thickness. Once the film is "reasonably thin," the spectra change only in absorptivity. From Figure 7, peak positions remain constant over a wide range of thickness. Thus accurate peak positions are obtained more easily than accurate peak shapes. The most intense peaks in the spectrum are the most affected because obtaining these on scale requires the thinnest samples. Relative intensities are strongly affected by the film thickness. Weaker modes, such as the peak near 620 cm^{-1} in Figure 7, are enhanced relative to the intense modes as thickness increases. Overtones can be strongly enhanced (e.g., the peaks near 717 and 1100 cm^{-1} in Figure 7).

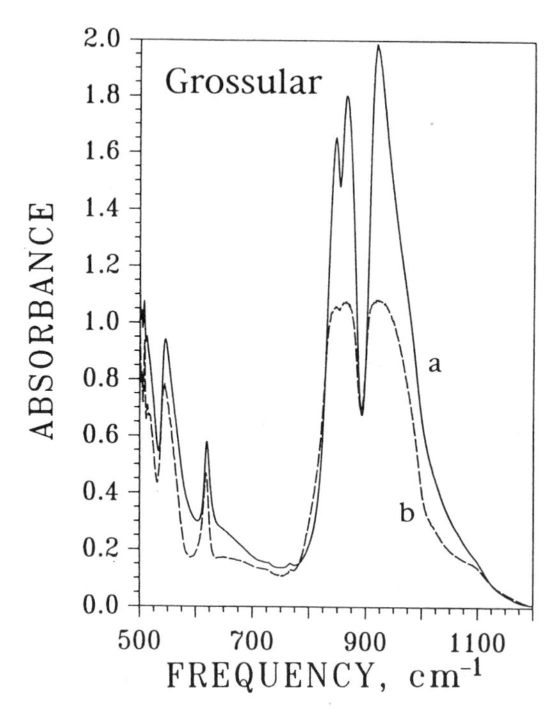

Figure 7. Absorption spectra of thin films of synthetic grossular: (a) spectra from the entire thin film formed in a DAC; (b) spectra acquired in the microscope from the thickest portion of the thin film of part (a) supported on a low index of refraction medium (KBr). The thin film of (b) is shown in Figure 4c. The synthetic sample was provided by T. Haselton (U.S. Geological Survey, Reston, VA).

The number, shape, and intensity of peaks from thin-film spectra of grossular closely compare closely with the spectra obtained from both single-crystal absorption measurements and absorptivity calculated [via equation (9)] from single-crystal reflectance (Figure 4). The greatest differences occur among the most intense peaks near 450 to 500 cm^{-1}. The intense modes near 430 to 500 cm^{-1} are on scale for the thin-film spectra but not for the thicker single-crystal data because light leakage from cracks and edges can be avoided with the microscope and because a uniform thickness can be selected by masking the image in the microscope. Thicker films (not shown) have an additional shoulder in the high-intensity region near 435 cm^{-1}, due to the increased proportion of an LO mode.

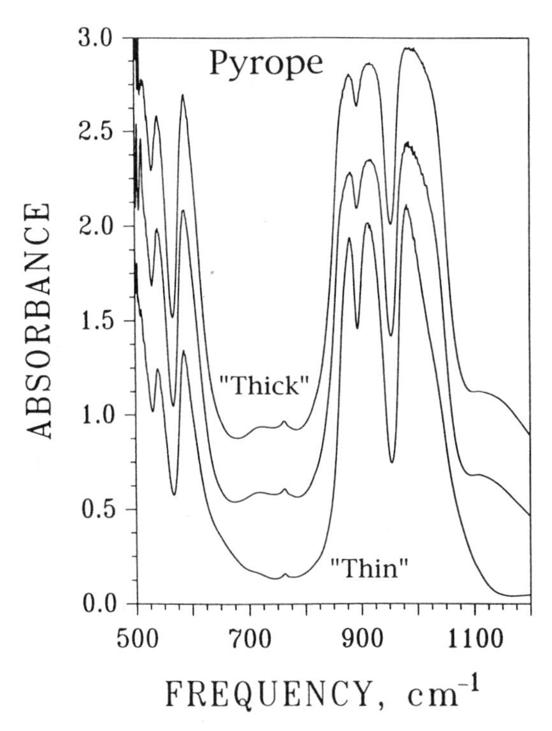

Figure 8. Pyrope ($Mg_3Al_2Si_3O_{12}$) spectra from progressive thinner samples (thickness was not measured). The "sharpness" of the peaks increases and their frequencies gradually shift to lower wavenumber with thinning. This synthetic sample was provided by T. Haselton (U.S. Geological Survey, Reston, VA).

Peak positions from thin-film spectra (Table 2) are slightly higher than those obtained from the calculated absorptivity spectrum, with thin-film frequencies differing by -0.4 to $+8.4$ cm^{-1} from calculations. The deviations roughly correlate with peak intensity, showing that the differences depend on thickness and are created by the departure from the ideally infinitely thin sample. The average deviation is $+1.17$ cm^{-1}, which is close to the accuracy of determining frequencies for the broader peaks of ± 1 cm^{-1} (Table 2).

Frequencies obtained from single-crystal absorbance spectra lie below those calculated from reflectance measurements. Deviations range from $+0.3$ to -7.5 cm^{-1}. The average deviation is -2.78 cm^{-1}, with the greatest differences associated with the intensely absorbing bands. Because the calculations represent

Table 2. Infrared Absorption Frequencies of Grossular Garnets

Mode and assignment[a]		Synthetic thin film[b]	Natural reflectivity[c]	Natural single crystal[d]
υ3[e]	(1B)	912.9	911.7	910
υ3[e]	(3C)	860.4	859.6	859.9
υ3[e]	(4D)	842.4	840.8	833.3
υ4	(7E)	619.5	619.3	617.3
υ4	(9F)	543.4	542.8	540.6
υ4	(10G)	508.3	508.7	—
T(Al)[e]	(19H)	475.4	437.1	—
υ2[e]	(13I)	445.5	437.1	430
T(Al)	(20J)	420.5	—	—
$R(SiO_4)$	(15K)	394.6	391.8	387.5
$R(SiO_4)$	(18L)	355.3	355.5	353.2
$T(SiO_4)$	(27M)	301.3	300.9	298.1
T(Al)	(21N)	242.8	242.6	240.6
T(M)	(22O)	242.8	242.6	240.6
T(M)	(23P)	205.2	205.4	203.8
T(M)	(24Q)	185.5	185.3	183
$T(SiO_4)$ + T(M)	(29R)	156.6	156.3	154.9
Overtones		767		689
LO modes		976 ± 1		995
		455		572
Film thickness[f] (μm)		0.213 ± 0.040	n/a	1.38 ± 0.81[g]
Film thickness[h] (μm)		0.176 ± 0.033	n/a	1.60 ± 0.94[g]

[a]All frequencies are in wavenumbers (cm^{-1}). Nomenclature is from Refs. 89 and 90; vibrational assignments are from Ref. 91.
[b]Thin-film measurements of synthetic grossular provided by Pascal Richett (IGP, Paris).
[c]Absorptivity calculated from single-crystal reflectance of natural $Gr_{99}An_1$ from Asbestos, Quebec [27].
[d]Values from single-crystal absorbance of the same natural $Gr_{99}An_1$ [27].
[e]Intense modes.
[f]Mid-IR, covering modes B to F.
[g]From Figure 4a, thickness is about 0.1 μm.
[h]Far-IR, covering modes G to Q.

the ideal sample, the incorrectly low frequencies from single-crystal spectra are attributed to problems with microscope optics. This inference was confirmed by comparing thin-film spectra in the DAC to that obtained from the same film in the microscope (Figure 7). The shift of the thicker film in the wrong (negative) direction shows that optics are the problem.

The correlation of the deviations in Figures 5 and 8 and Table 2 with peak height indicates that the variations in index of refraction are the key issue in that the weaker peaks have lower indices of refraction. A possible solution to this problem is changing the path length by using a diamond substrate as the reference spectrum, as well as to support the sample. This approach is suggested by reflectivity measurements being unaffected by use of a microscope. Unfortunately, we could not make a direct comparison of absorption from a single crystal with and without a diamond substrate because the samples were destroyed in obtaining the SEM images in Figure 3.

Given the time and effort required for sample preparation of single crystals for absorption spectroscopy, the difficulties intrinsic in obtaining polarized data with a microscope, and the ease with which one can obtain quantitative measures of vibrational spectra from thin films, at this time we recommend use of a diamond anvil cell to obtain absorption spectra from microminerals.

3.3. Difficulties in Extracting Quantitative Measurements from Dispersions

The ease and convenience of the powder technique are offset by inaccuracies in peak positions, incorrect relative intensities, and absorption of water onto the substrate [16,50]. The latter problem can easily be avoided by comparison to reference spectra, but the others are problematic, as follows.

Surface modes cannot be avoided because of scattering from the particulates [43], and for highly polarizable materials such as silicates, the surface modes can be stronger than the bulk lattice modes [17], leading to a bogus spectrum. For all materials, scattering induces a substantial amount of the LO component (Figure 9). If the LO-TO splitting is large enough, the LO components are manifest as separate peaks. These have been mistaken for TO modes. For moderate LO-TO splittings, the IR peaks contain partial contributions from both components, resulting in a shift from the true absorption position.

Incorrect relative intensities result because grinding produces a range of sizes, typically 1 to 3 μm. Because the thickness at which light is transmitted depends on the intensity of the modes, and because the strong mid-IR modes require a thickness of 0.1 μm, edges are sampled for the strong peaks and the entire particles are sampled for the weak peaks (Figure 9). Thus weak peaks and overtones have highly exaggerated intensities in powder spectra. Owing to

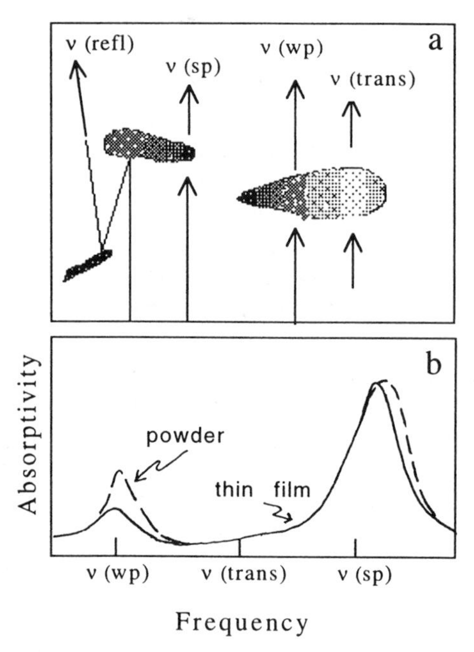

Figure 9. Interaction of light with crystals dispersed in a medium: (a) physical phenomena; (b) schematic of differences between the thin-film and powder techniques. υ(refl) indicates a component induced by scattering which shifts peak positions to higher frequency. υ(wp) and υ(trans) indicate that thicker portions of crystals are sampled for weak peaks and transparent regions of the spectra. υ(sp) indicates that the strongest peaks are produced by absorptions only from the edges of the crystallites. Thicker sections are opaque.

this effect, overtones have sometimes been assigned incorrectly as fundamentals, as discussed in Ref. 27. Hence powder spectra are not reliable for quantitative analysis, and care should be taken with qualitative analysis.

4. APPLICATIONS TO MINERALS

In this section we summarize examples of various types of problems in earth science that have been addressed through IR microspectroscopy. The list is not all-inclusive but should be illustrative.

4.1. Structural Studies

4.1.1. Tridymite

The SiO_2 polymorph tridymite commonly occurs in silicic volcanic rocks, yet its structure at low tempertures remains debated [51]. Spectroscopic as well as crystallographic determinations have been complicated by the existence of multiple low-temperature polymorphs, the likelihood of stacking disorders [52], and the ubiquitous presence of twinning on various scales [53]. The structure of high-temperture (β-)tridymite is well characterized (e.g., [54]), consisting of layers of interlocking undistorted six-membered rings of tetrahedra that alternately point up and down with a mirror plane between the layers. This highly symmetry structure ($D_{3h}^2 = P\bar{6}c2$) is expected to possess seven IR modes. β-Tridymite forms between 1470 and 870°C at ambient pressure, and exists metastably to about 163°C, where a displacive transformation occurs to the α′ (or M) form, which appears to be orthorhombic, $D_2^5 = C222_1$ [55]. Symmetry analysis gives 25 IR bands for this polymorph [51]. At 117°C, a second displacive transformation to α (or S)-tridymite occurs (e.g., [52]). Crystallographic refinements yield a monoclinic structure, $C_s^2 = Cc$ [56], which would have 213 IR modes. This transition can be driven by grinding [57] as well as by heating.

As twinning and some types of stacking disorder have little or no effect on vibrational spectra, IR data were collected to determine independently the space groups for the various displacive transformations. Infrared absorption spectra at room temperature from thin films of natural α-tridymite in the diamond anvil cell show roughly 16 IR bands [57], far too few for the monoclinic or even the orthorhombic symmetries. A concern is whether grinding during sample preparation or the pressure applied to make the thin film transformed the sample to one of the higher-symmetry polymorphs. However, the number of bands observed is double that expected for the β-polymorph, and single-crystal reflectance spectra at 1 atm (Figure 10) showed a similarly small number of bands. From these data, from heating experiments performed on larger amounts of sample which show a transition at 220°C to a structure with eight IR bands [58], and from symmetry analysis of potential pathways of displacive transformation [51], we propose that a hexagonal phase of tridymite (space group $D_6^6 = P6_322$ with eight IR bands) is stable from about 450 to 220°C, and that another phase exists from about 220 to 100°C (space group $D_3 = R32$ with 19 IR bands). A lower-symmetry space group could occur below 100°C, but the IR data cannot address this problem at present. Assignment of different space groups through spectroscopy and crystallography can be attributed to differences in scale, in that macroscopic characterization may be affected by ubiquitous twinning, strain, and stacking disorder.

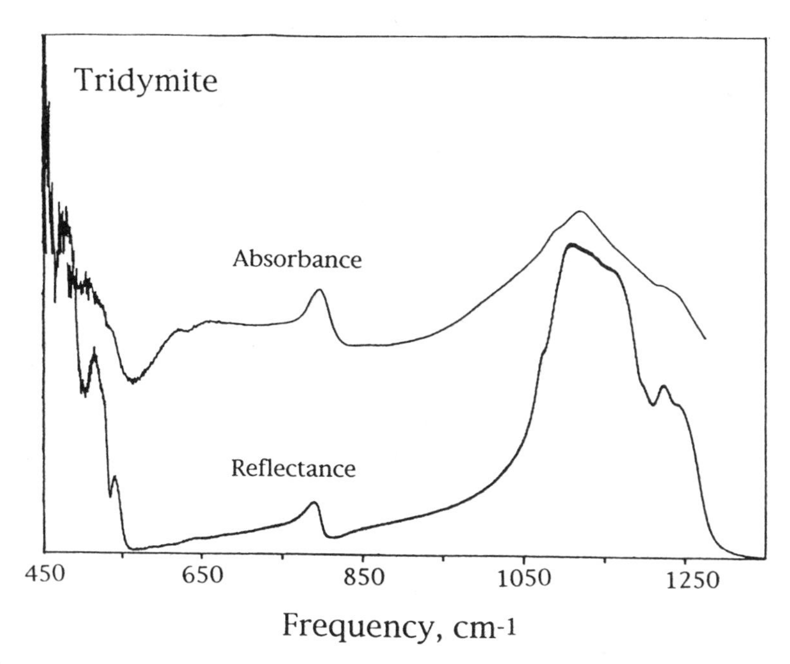

Figure 10. Mid-IR spectra of natural tridymite from Topaz Mountains, Utah. The reflectance spectra was collected from a gemmy, polished section of the crystal. The absorbance spectra were collected from a thin film ground to submicron size and compressed. All features occur in both spectra.

4.1.2. Optically Anomalous Garnets

The cubic symmetry of garnets implies optical isotropy, yet grossular-andradites ($Ca_3Al_2Si_3O_{12}$-$Ca_3Fe_2Si_3O_{12}$) from skarns commonly exhibit weak birefringence. Such anomalous optical properties could either originate in strain or by reduction of symmetry through partial ordering of Al and Fe^{3+} in the octahedral sites (or of Ca and Fe^{2+} in the dodecahedral sites), as indicated by crystallographic refinements [59–62]. Analysis of infrared spectra should distinguish between these two hypotheses because the suggested low symmetries of $I\bar{1}$ or *Fddd* require a large number of IR vibrations (129 or 98), compared to 17 IR peaks observed for cubic garnets.

Samples nearly identical to those crystallographically refined were acquired. Single-crystal absorption and reflection measurements (polarized and unpolarized) obtained from optically oriented sections of four birefringent garnets showed no extra bands and an intensity pattern characteristic of isotropic cubic garnets [27]. Data for one sample are shown in Figure 5. The contrasting results from spectroscopy and crystallography indicate that strain causes anomalous optical anisotropy in garnet (as occurs in diamond [63]) in that x-ray diffraction requires juxtaposition of about 1000 unit cells, whereas lattice vibrations can be produced by isolated molecules.

4.2. Lattice Modes

4.2.1. Samples Synthesized at High Pressure and Temperature

Characterization of the thermodynamic properties of minerals at pressure requires intimate knowledge of their vibrational properties. Stishovite, the high-pressure polymorph of SiO_2, has been problematic in that previous infrared spectra are inconsistent with expectations for the ideal rutile structure (e.g., [3]). Raman spectra [64] show that synthetic stishovite possesses the rutile structure, although the origin of the extra bands in the natural sample was not clear.

Unpolarized IR reflectance spectra were obtained from a single crystal of synthetic stishovite [65] which had dimensions of about $10 \times 40 \times 100$ μm. Absorption spectra were collected from polycrystalline natural samples that were extracted from Coconino sandstone using concentrated HF and HCl [66] as well as from powdered synthetic samples by compressing them into a submicron thin film with a diamond anvil cell.

The IR data for synthetic stishovite [29] corroborate the x-ray diffraction [65] and Raman [64] results, which show that stishovite crystallizes in the ideal rutile structure. As predicted by symmetry analysis, three strong E_u peaks were observed at 1020 to 820, 700 to 580, and 565 to 470 cm^{-1}, and one weak A_{2u} mode was found at 950 to 675 cm^{-1}.

Natural stishovite from Meteor Crater, Arizona, had fundamentals similar to those of synthetic samples and an additional strong feature near 740 cm^{-1}. Excess bands in the IR and the Raman spectra are best explained as $(SiF_6)^{2-}$ contamination that is charge compensated by alkali ions. This proposal is supported by the following observations:

1. The intense extra bands seen in natural samples occur at the same positions as octahedral Si—F vibrations observed in other compounds (at 741 and 483 cm^{-1} in IR spectra and at 663, 477, and 408 cm^{-1} in Raman specta [67]).
2. The presence of multiple, weak far-IR bands in natural stishovite are consistent with translations of alkali ions.
3. Treatment with strong acids could yield some surface exchange of O^{2-} with F^- and could easily extract alkalis from feldspar in the Coconino sandstone, for later deposition in defects on the surface of stishovite crystals.

4. Surface modes in such fine-grained polycrystalline samples are expected to perturb the IR spectrum, resulting in high-intensity bands. The positions of the impurity bands in IR data are not consistent with significant contamination with SiO_2 glass, although glass could be produced through laser heating in the Raman experiments.

IR frequencies for stishovite are considerably higher than those of the other oxides, as expected, because stishovite has the lighest mass and shortest metal–oxygen bond distance of the four compounds. However, frequencies of SiO_2 cannot be predicted from rutile TiO_2 spectrum by accouting for differences in mass and bond length. Neither are rigid ion calculations for stishovite particularly close: model values [68] are low but within 3 to 20% of the measured TO and Raman frequencies.

The LO-TO range (1020 to 820 cm^{-1}) of the highest-energy Si—O octahedral stretch overlaps considerably with that of Si—O tetrahedral stretches which occur from 1250 to 850 cm^{-1}. The placement is considerably higher (Figure 11) than that of other compounds containing octahedrally coordinated silicon (e.g., $MgSiO_3$ with the ilmenite structure has an LO-TO stretching vibration at 940 to 740 cm^{-1} [30], and similarly, perovskite has its Si—O octahedral stretch at 943 to 771 cm^{-1} [69]).

The relationships above imply that frequency alone is insufficient to establish coordination numbers for silicon, and that octahedrally coordinated Si may bond differently in stishovite than in other high-pressure silicates. The difference is due likely to the high degree of edge sharing of Si octahedra that is present in stishovite compared to other high-pressure silicates.

4.2.2. Hydrogen Gas Under Pressure

The behavior of hydrogen at pressure is pertinent to understanding the interiors of the gas giant planets. IR microspectroscopy on H_2 has been performed up to pressures of 216 GPa at temperatures of 77 to 295 K using synchrotron radiation [70,71]. Data were acquired on samples as small as 1.5 μm for the wavelength range of 4000 to 6400 cm^{-1} and revealed a previously undetected phase transformation. An increase in absorption of three orders of magnitude was also observed at the 150-GPa phase transformation. This change is attributed to symmetry breaking, which gives rise to new electronic states as the band gap of the material closes with increasing pressure [71].

4.3. Hydrous Impurities in Synthetic and Natural High-Pressure Samples

Near-IR spectroscopy is useful for ascertaining speciation as well as concentration of hydrogen impurities [8]. The OH^- groups yield one sharp stretching

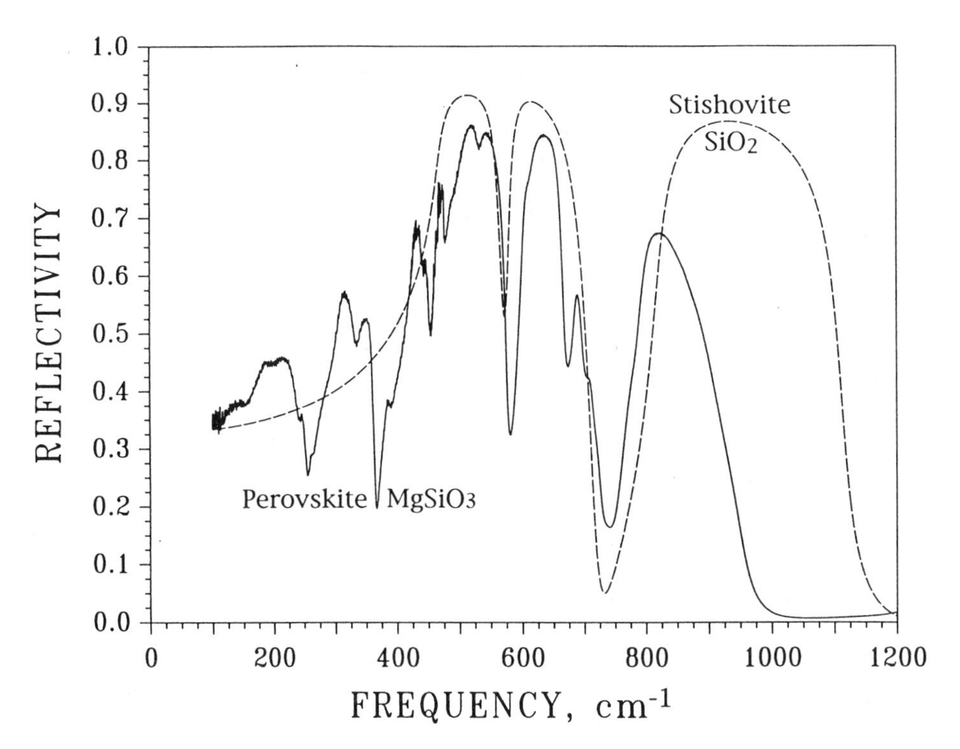

Figure 11. Lattice modes in high-pressure silicates. The dashed line is a calculation of the E_u symmetry based on measurements of a single crystal of synthetic stishovite. The perovksite spectrum was taken from a synthetic polycrystal. Interestingly, the correspondence between the O—Si—O bending frequencies near 400 to 700 cm^{-1} is excellent, and that between the Si—O stretching modes is poor. Bands below 400 cm^{-1} in perovskite are connected with translations of the Mg cation.

band near 3600 cm^{-1}. In contrast, the bent H_2O molecule has two IR stretching modes (symmetric and asymmetric) in the region 3200 to 3600 cm^{-1} and one bending mode near 1600 cm^{-1}. A combination mode is also found near 5200 cm^{-1}. Sharp bands are expected and observed for water oriented within the mineral structure, whereas broad bands are related to liquid water (fluid inclusions consisting of more than a few hundred molecules). Concentrations can be determined from Beer's law,

$$A = \epsilon t C \tag{12}$$

where the extinction coefficient ϵ is a constant determined from standards, A the absorption coefficient, t the thickness, and C the concentration [8].

Evaluation of the solubility of hydrous species at pressures and temperatures found in the earth's interior is needed to constrain mechanisms for water recycling in the mantle as well as to set limits on the water budget of the entire earth. Moreover, important physical and chemical properties of rocks and minerals, such as mechanical strength, phase equilibria, and the chemistry of magma generation, commonly depend on water content. The following four examples indicate that ample storage sites exist for water in phases that are stable under deep mantle conditions, and the final example considers hydrogen present in the shallow mantle.

4.3.1. Wadsleyite

Near-IR spectra were obtained from crystals and polycrystals of β-$(Mg,Fe)_2SiO_4$ using a Spectra-Tech Spectrascope [9]. Regions as small as 10 μm were examined on samples having thicknesses of 5 to 30 μm. Three distinct bands observed at 3329, 3580, and 3615 cm^{-1} are attributed to hydroxyl. The maximum concentrations are 1400 H atoms per 10^6 Si atoms. Comparison with olivine α-$(Mg,Fe)_2SiO_4$ coexisting in the same sample shows that hydrogen is preferentially incorporated at a 40:1 ratio in the β-$(Mg,Fe)_2SiO_4$ structure due to wadsleyite having regular sites for protonation [61,72], whereas the mechanism for olivine is defects or grain boundaries [73].

4.3.2. Stishovite

Near-IR spectra were obtained from synthetic crystals of SiO_2 which varied from 25 to 1000 μm across. A sharp polarized band seen at 3111 cm^{-1} is clearly due to hydroxyl [74]. Its low frequency indicates that strong hydrogen bonding is present [75], as expected for a structure with edge-sharing octahedra. The concentration of H depends on the Al impurity content and reaches a maximum of 550 H per 10^6 Si for samples with 1.5 wt % Al_2O_3 [74].

4.3.3. Perovskite

Absorption spectra were obtained from a 230-μm-thick polycrystalline sample and a 50-μm-thick single crystal of roughly 100 μm dimensions [69]. The polycrystal possessed no near-IR bands and thus anhydrous within the detection limit of <0.001 wt % H_2O. However, the single crystal had a prominent band centered at about 3450 cm^{-1}. Hydroxyl is inferred because the observed frequency is too low and the band is too narrow for speciation as H_2O. A concentration of 0.06 wt % H_2O was estimated from extinction coefficients of water in sanidine [7].

4.3.4. Garnets

Upper mantle minerals such as garnet $(Mg,Ca,Fe)_3Al_2Si_3O_{12}$ often have trace amounts of hydrogen. Of interest is the total amount of water stored in the mantle via such nominally anhydrous phases. Although these crystals are macroscopic (millimeter to centimeter sized), the samples are often heavily included with other minerals and fluids, or are altered along cracks. The pristine, gemmy regions representing the mineral which formed at depth are considerably smaller, requiring micro-IR techniques to assess water contents. Typically, regions of 10 to 50 μm are studied with the aid of an FTIR microscope [76,77]. Grossulars $Ca_3Al_2Si_3O_{12}$ contain 0.0003 to 3.6 wt % water as H_2O, such that for high water contents the hydrogarnet substitution ($4H^+ = Si^{4+}$) serves as the incorporation mechanism, with H probably substituting for the other cations at low concentrations [77]. Pyrope $(Mg_3Al_2Si_3O_{12})$-rich garnets (Figure 12) have lower

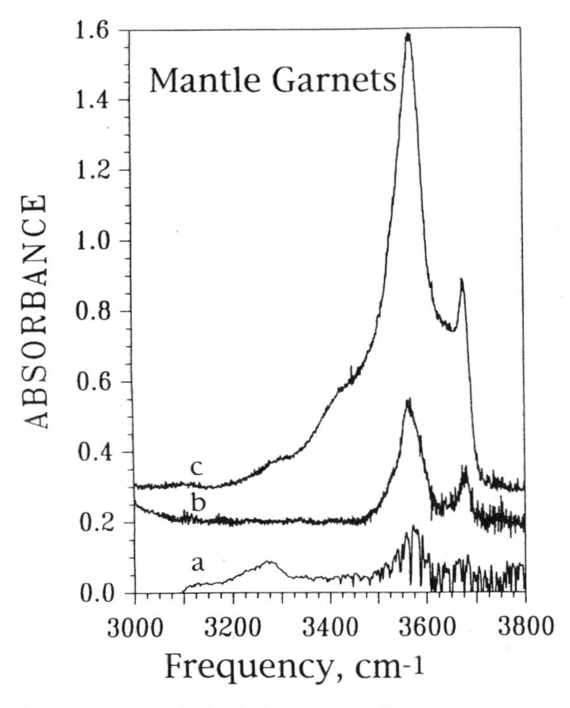

Figure 12. Water in garnets derived from the shallow upper mantle near the Four Corners region. These were collected from Moses Rock diatreme on the Navajo Indian reservation. (a), (b), and (c) are three garnets with slight chemical differences. OH^- is present in all three to varying concentrations (0.01 to 0.24 wt %).

amounts, 0.00016 to 0.03 wt % as H_2O [78,79] incorporated through the hydrogarnet substitution [76]. Even such minor concentrations of hydrogen in garnets and other mantle minerals suggest that an amount of water roughly equal to that in earth's oceans resides hidden in crystals deep within the earth's interior (George Rossman, personal communication, 1994).

5. COMPARISON WITH OTHER TECHNIQUES

In this section, IR microscopy is compared with various techniques that are routinely or widely applied in earth science to measurement of physical properties from microscopically sized samples. For the most part, the techniques listed below provide complementary information. Comparison studies using two or more approaches have prooved most advantageous.

5.1. Micro-Raman Spectroscopy

The Raman technique stimulates vibrations with a laser, generally an argon-ion gas type. The vibrational frequencies equal the difference in position between the Raman peaks and the frequency of the incident laser light. Vibrations are active if the polarizability changes during the atomic excursions. A given vibrational mode can be active in either the Raman or the IR or both or neither. Thus IR and Raman techniques sample similar properties of the solid and are used to address the similar problems.

A Raman spectrum is collected in the visible region. Two approaches are possible for microstudies. Usually, the laser beam is directed through a microscope, but simple lenses have also been used successfully. Because measurements are made in the visible, diffraction is much less of a problem than with micro-IR studies and all vibrational frequencies are affected equally. Samples as small as 1 μm are studied routinely [10] and polarized measurements present few difficulties.

Differences between the IR and Raman methods frequently make one approach preferable for specific projects. Several examples follow:

1. Opaque and deeply colored materials are best studied through IR spectroscopy because these samples absorb the laser light. The consequent increase in temperature shifts the vibrational frequencies, creating problems for quantitative studies. In extreme cases, the sample melts.
2. Glasses can be characterized by either method, because weak peak intensities in the Raman are compensated by narrow peaks and conversely for the IR method.

3. Concentrations of impurities can be determined with the IR technique, but not through Raman spectroscopy, although determinations of speciation can be done with either technique.
4. Raman spectra of samples held at high pressure in a diamond anvil cell [80–82] are more easily obtained and often give better results because the small spot size allows sampling at one pressure rather than across the entire cell, which has a pressure gradient, and because the Raman bands are narrower, and thus pressure broadening has a much smaller effect.

In summary, the ease with which small samples can be studied and the similarities in the problems that can be addressed have made micro-Raman spectroscopy more popular than micro-IR spectroscopy among earth scientists.

5.2. X-Ray Diffraction

The high brilliance of synchrotron sources makes it possible to make crystallographic measurements on extremely small samples, for example, 30-μg quantities of high-pressure phases [83,84]. With conventional sources and commercially available diffractometers, samples 100 μm across are routinely measured. Thus sample sizes are comparible to and even smaller than those studied through micro-IR spectroscopy. However, the type of problems addressable by these two technqiues differ considerably. Only crystallography can give atomic coordinates and bond lengths, whereas information from IR spectroscopy is limited to determination of the space group. However, the IR technique samples the unit cell, whereas crystallography requires 100 to 1000 adjacent unit cells, and hence strain will not affect IR data and can serve as an independent check for crystallographic refinements. Glasses are the one type of material where more structural information can be gleened through spectroscopy than through crystallography; see Ref. 85 for a discussion.

5.3. Electron and Ion Microprobes

Knowledge of the chemical composition of the natural and synthetic species studied in earth science is essential and cannot be obtained accurately with IR spectroscopy. Roughly speaking, variations in 5 wt % are difficult to detect with IR spectroscopy. Chemical analyses of major elements (>0.5 wt %) are routinely performed from spot sizes of 1 to 5 μm using the electron microprobe, which compares x-rays emitted from the sample via electron bombardment to those emitted from a set of standards. A related class of instruments have been developed in the past 20 years whereby secondary ions are produced by ion bombardment. These ion microprobes are used to study isotopic variations and trace element concentrations in samples 20 μm and below [86].

5.4. TEM and SEM

Transmission electron microscopes (TEMs) record scattering of an electron beam from a small (micrometer-sized) and very thin (0.05- to 0.5-μm) samples. Structural information obtained from the different pattern pertains to a much smaller area than conventional x-ray crystallography. Production of images from the diffracted and undiffracted beams reveals twinning, exsolution, and inclusions not observable by any other method. This technique has greatly promoted research on mineral transformations with temperature (e.g., [87]). Semiquantitative chemical analyses can also be obtained.

Scanning electron microscopes (SEMs) measure reflected electron beams from similarly small samples. The information is mainly morphological but allows identification of samples through approaches similar to that of optical microscopy (see below). In addition, semiquantitative chemical analyses are routinely done.

5.5. Optical Methods

Polarizing petrographic microscopes have historically been a major tool for identification of minerals. Numerous instruments are available commercially for study of minerals larger than about 5 μm. Observation of cleavage and interfacial angles on a crystal allows one to infer the symmetry of atomic arrangements at the unit cell scale. The interaction with light provides further clues. Measurement of the index of refraction and its dependence on crystallographic orientation allow identification of the mineral and in many cases give constraints on the chemical composition. The accuracy of the chemistry is comparable to that determined via IR spectroscopy for most minerals. Instruments capable of imaging much smaller samples are available, but applications to study of minerals have not yet been published to my knowledge.

6. FUTURE TRENDS

6.1. Temperature and Pressure Studies

To date, many studies of vibrational spectra have been made of compressed or of heated samples, but collection of IR data during simultaneous application of high temperatures and pressures has not been realized. The motivation for such difficult experiments is obvious: the interior of the earth is hot (ca. 2000°C). Direct measurements would yield information on stability of phases, details on their structures, and thermodynamic data. Laser heating is the most promising technique to raise the temperature of samples compressed in a diamond anvil cell during spectroscopic data acquisition. The temperature could be calculated from the emission of blackbody radiation from the sample with the same instrument that collects the vibrational data.

6.2. Microreflectance of Smaller Samples

Large, gemmy materials are the exception rather than the rule. Thus application of microtechniques is needed to investigate the majority of mineral samples. Current micromeasurements are made of unpolarized samples with FTIR microscopes. Generation of complete data for the majority of geological samples will require modification of the optics to prevent polarization mixing or development of specular reflectance devices for microsamples.

6.3. Need for a Spectral Database

Existing IR spectral databases for minerals are comprised of data collected from powder dispersions. The current data set suffices for identification purposes but not for quantitative or semiquantitative study. The recent growth of the field has produced a large number of reflectance spectra from macro-sized samples and of thin-film spectra from microsamples. These are scattered among geology, chemistry, and materials science journals. Continued expansion of the field is likely to create a need for a compilation of single-crystal IR spectra in the not-too-distant future.

7. CONCLUSIONS

Application of infrared microspectroscopy to problems in earth science is in its infancy. Only a small number of problems that can be addressed by this technique have been investigated.

ACKNOWLEDGMENTS

A critical review was provided by R. E. Criss (Washington University). I thank B. McAloon (University of California–Davis, Geology) for acquiring the spectra in Figures 5, 7, and 8 and the photos in Figure 3; P. Richet (IGP, Paris) and H. Haselton (USGS) for supplying samples of synthetic grossular and pyrope garnets; and Brenda Zimny (University of California–Davis, Engineering Department) for obtaining SEM photos in Figure 4. Karla Campbell (University of California–Davis, Geology) helped greatly with illustrations. Support was provided by the David and Lucile Packard Foundation and by NSF-INT 90-16362.

REFERENCES

1. Abrams, M., and Siegal, B. (1980). Lithologic mapping, in *Remote Sensing in Geology* (B. Siegal and A. Gillespie, eds.), John Wiley & Sons, New York, p. 702.
2. Anderson, D. L. (1989). *Theory of the Earth*, Blackwell Scientific Publications, Boston, p. 366.
3. Kieffer, S. W. (1979). Thermodynamics and lattice vibrations of mineral: 3. Lattice dynamics and an approximation for minerals with application to simple substances and framework silicates, *Rev. Geophys. Space Phys.*, *17*: 35.
4. Hofmeister, A. M. (1991). Calculation of bulk moduli and their pressure derivatives from vibrational frequencies and mode Gruneisen parameters: solids with cubic symmetry or one-nearest-neighbor distance. *J. Geophys. Res.*, *96*: 16.
5. Griggs, D. T., and Blacic, J. D. (1965). Quartz: anomalous weakness of single crystals, *Science*, *147*: 292.
6. Fyfe, W. S. (1964). *Geochemistry of Solids*, McGraw-Hill, New York, p. 199.
7. Hofmeister, A. M., and Rossman, G. R. (1985). The inhibiting role of water in irradiative coloring of smoky feldspar, *Phys. Chem. Minerals*, *12*: 324.
8. Rossman, G. R. (1988). Vibrational spectroscopy of hydrous components, in *Spectroscopic Methods in Mineralogy and Geology* (F. C. Hawthorne, ed.), Mineralogical Society of America, Washington, D.C., p. 193.
9. Young, T. E., Green, H. W., II, Hofmeister, A. M., and Walker, D. (1993). Infrared spectroscopic investigation of hydroxyl in β-$(Mg,Fe)_2SiO_4$ and coexisting olivine: implications for mantle evolution and dynamics, *Phys. Chem. Minerals*, *19*: 409.
10. Pasteris, J. D., Wopenka, B., and Seitz, J. C. (1988). Practical aspects of quantitative laser Raman spectroscopy for the study of fluid inclusions, *Geochim. Cosmochim. Acta*, *52*: 979–988.
11. Wilson, E. B., Jr., Decius, J. C., and Cross, P. C. (1955). *Molecular Vibrations: The Theory of Infrared and Raman Vibrational Spectra*, Dover Publications, New York, p. 388.
12. Cotton, F. A. (1971). *Chemical Applications of Group Theory*, John Wily & Sons, New York, p. 386.
13. Burns, G. (1990). *Solid State Physics*, Academic Press, San Diego, p. 810.
14. Wooten, F. (1972). *Optical Properties of Solids*, Academic Press, New York, p. 260.
15. Berreman, D. W. (1963). Infrared absorption at longitudinal optic frequency in cubic crystal films, *Phys. Rev.*, *130*: 2193.
16. McMillan, P., and Hofmeister, A. M. (1988). Infrared and Raman spectroscopy, in *Spectroscopic Methods in Mineralogy and Geology* (F. C. Hawthorne, ed.), Mineralogical Society of America, Washington, D.C., p. 99.
17. Farmer, V. C., and Lazarev, A. N. (1974). Symmetry and crystal vibrations, in *The Infrared Spectra of Minerals* (V. C. Farmer, ed.), Mineralogical Society, London, p. 51.
18. Mottana, A., Crespi, R., and Liborio, G. (1978). *Simon and Schuster's Guide to Rocks and Minerals*, Simon and Schuster, New York, p. 607.
19. Zoltai, T., and Stout, J. H. (1984). *Mineralogy: Concepts and Principles*, Burgess Publishing Company, Minneapolis, Minn., p. 505.

20. Hirschfeld, T. (1979). Quantitative FTIR: a detailed look at the problems involved, in *Fourier Transform Infrared Spectroscopy: Applications to Chemical Systems* (J. R. Ferraro and L. J. Basile, eds.), Academic Press, New York, Chapter 6.
21. Reffner, J., Carr, G. L., Sutton, S., Hemley, R. J., and Williams, G. P. (1994). Infrared microspectroscopy at the NSLS, *Synchrotron Radiation News*, *7*: 30.
22. Guerrera, A. J., and Hofmeister, A. M. (in preparation). IR reflectance and absorbance spectra of Fe^{3+} bearing spinels, *Phys. Chem. Minerals*,
23. Kauppinen, J. K., Moffatt, D. J., Mantsch, H. H., and Cameron, D. G. (1981). Fourier self-deconvolution: a method for resolving intrinsically overlapped bands, *Applied Spectroscopy*, *35*: 271.
24. Hawthorne, F. C., and Waychunas, G. A. (1988). Spectrum-fitting methods, in *Spectroscopic Methods in Mineralogy and Geology* (F. C. Hawthorne, ed.), Mineralogical Society of America, Washington, D.C., p. 63.
25. Griffiths, P. R., and Haseth, J. A. (1986). *Fourier Transform Infrared Spectrometry*, John Wiley & Sons, New York.
26. Hutchinson, C. S. (1974). *Laboratory Handbook of Petrographic Techniques*, John Wily & Sons, New York, Chapter 1.
27. McAloon, B. P., and Hofmeister, A. M. (1993). Single-crystal absorption and reflection infrared spectroscopy of birefringent grossular-andradite garnets, *Am. Mineralogist*, *78*: 957.
28. Hofmeister, A. M. (1987). Single-crystal absorption and reflection infrared spectroscopy of forsterite and fayalite, *Phys. Chem. Minerals*, *14*: 499.
29. Hofmeister, A. M., Xu, J., and Akimoto, S. (1990). Infrared spectroscopy of synthetic and natural stishovite, *Am. Mineralogist*, *75*: 951.
30. Hofmeister, A. M., and Ito, E. (1992). Thermodynamic properties of $MgSiO_3$ ilmenite from vibrational spectra, *Phys. Chem. Minerals*, *18*: 423.
31. Roush, P. B. (1987). *The Design, Sample Handling, and Applications of Infrared Microscopes*, ASTM, Philadelphia.
32. Jackson, J. D. (1975). *Classical Electrodynamics*, John Wiley & Sons, New York, 848 pp.
33. Reynard, B. (1991). Single-crystal infrared reflectivity of pure Mg_2SiO_4 forsterite and $(Mg_{0.86}Fe_{0.14})_2SiO_4$ olivine, *Phys. Chem. Minerals*, *18*: 19.
34. Hofmeister, A. M., Xu, J., Mao, H.-K., Bell, P. M., and Hoering, T. C. (1989). Thermodynamics of Fe-Mg olivines at mantle pressures: mid- and far-infrared spectroscopy at high pressure, *Am. Mineralogist*, *74*: 281.
35. Tighe, N. J. (1976). Experimental techniques, in *Electron Microscopy in Mineralogy* (H.-R. Wenk, ed.), Springer-Verlag, Berlin, p. 144.
36. Seal, M., and van Enckevort, W. J. P. (1988). Applications of diamond in optics, *Diamond Optics*, *969*: 144.
37. Sherman, W. F., and Wilkinson, G. R. (1980). The sapphire high pressure cell, *Adv. IR Raman Spectrosc.*, *6*: 158.
38. Patterson, D. E., and Margrave, J. L. (1990). Use of a gem-cut cubic zirconia in the diamond anvil cell, *J. Phys. Chem.*, *94*: 1094.
39. Weir, C. E., Lippincott, E. R., Van Valkenburg, A., and Bunting, E. N. (1959). Infrared studies in the 1- to 15-micron region to 30,000 atmospheres, *Journal of Research of the National Bureau of Standards, Section A*, *63*: 55.

40. Hazen, R. M., and Finger, L. W. (1982). *Comparative Crystal Chemistry*, Wiley-Interscience, New York, p. 231.
41. Jayaraman, A. (1983). Diamond anvil cell and high-pressure physical investigations, *Rev. Modern Phys., 55*: p. 65.
42. Jephcoat, A. P., Mao, H.-K., and Bell, P. M. (1987). Operation of the megabar diamond-anvil cell, in *Hydrothermal Experimental Techniques* (G. C. Ulmer and H. L. Barnes, eds.), John Wiley & Sons, New York, p. 469.
43. Horak, M., and Vitek, A. (1978). *Interpretation and Processing of Vibrational Spectra*, John Wiley & Sons, New York, p. 425.
44. Andermann, G., Caron, A., and Dows, D. A. (1965). Kramers–Kronig dispersion analysis of infrared reflectance bands. *J. Opt. Soc. Am., 55*: 1210.
45. Spitzer, W. G., Miller, R. C., Kleinman, D. A., and Howarth, L. E. (1962). Far-infrared dielectric dispersion in $BaTiO_3$, $SrTiO_3$, and TiO_2, *Phys. Rev., 126*: 1710.
46. LeDuc, H. G., Coleman, L. B., and Chandrashekhar, G. V. (1984). Infrared and far-infrared studies of zinc-compressed β-gallates, *Phys. Rev. B, 30*: 7206.
47. Fahrenfort, J. (1961). Attenuated total reflection: a new principal for the production of useful infra-red reflection spectra of organic compounds, *Spectrochimica Acta, 17*: 698.
48. Roessler, D. M. (1965). Kramers–Kronig analysis of non-normal incidence reflecton, *Br. J. Appl. Phys., 16*: 1359.
49. Hofmeister, A. M. (1993). IR reflectance spectra of natural ilmenite: comparison with isostructural compounds and calcultion of thermodynamic properties. *Eur. J. Mineral., 5*: 281.
50. Hofmeister, A. M. (1991). Comment on ''Infrared spectroscopy of the polymorphic series (enstatite, ilmenite, perovskite) of $MgSiO_3$, $MgGeO_3$ and $MnGeO_3$'' by M. Madon and G. D. Prive, *J. Geophys. Res., 96*: 21959.
51. Hofmeister, A. M., Rose, T. P., Hoering, T. C., and Kushiro, I. (1992). Infrared spectroscopy of natural, synthetic, and ^{18}O-substituted and α-tridymite: structural implications, *J. Phys. Chem., 96*: 10213.
52. Sosman, R. B. (1965). *The Phases of Silica*, Rutgers University Press, New Brunswick, N.J., 388 pp.
53. Goldschmidt. V. (1923). *Atlas der Krystallformen, Tafeln*, Vol. 9, Trechmannit-Zoisit, Carl Winters Universitätsbuchhandlung, Heidelberg.
54. Wyckoff, R. W. G. (1965). *Crystal Structures*, John Wily & Sons, New York.
55. Dollase, W. A. (1967). The crystal structure at 230°C of orthorhombic high tridymite from the Steinbach Meteorite, *Acta Crystallogr., 23*: 617.
56. Kato, K., and Nukui, A. (1976). Die Kristallstruktur des monoklinen Tief-Tridymits, *Acta Crystallogr., B32*: 2486.
57. Xiao, Y., Kirkpatrick, R. J., and Kim, Y. J. (1993). Structural phase transitions of tridymite: a ^{29}Si MAS NMR investigation. *Am. Mineralogist, 78*: 241.
58. Work in preparation.
59. Takeuchi, Y., Haga, N., Umizu, S., and Sato, G. (1982). The derivative structure of silicate garnets in grandite, *Z. Kristallogr., 158*: 53.
60. Allen, F. M., and Buseck, P. R. (1988). XRD, FTIR, and TEM studies of optically anisotropic grossular garnets, *Am. Mineralogist, 73*: 568.

61. Kingma, K. J., and Downs, J. W. (1989). Crystal-structure analysis of a birefringent andradite, *Am. Mineralogist, 74*: 1307.
62. Griffen, D. T., Hatch, D. M., Phillips, W. R., and Kulaksiz, S. (1992). Crystal chemistry and symmetry of a birefringent tetragonal pyralspite-grandite garnet, *Am. Mineralogist, 77*: 399.
63. Tolansky, S. (1966). Birefringence of diamond, *Nature, 211*: 158.
64. Hemley, R. J., Mao, H. K., and Chao, E. C. T. (1986). Raman spectrum of natural and synthetic stishovite, *Phys. Chem. Minerals, 13*: 285.
65. Sinclair, W., and Ringwood, A. E. (1978). Single crystal analysis of the structure of stishovite, *Nature, 272*: 714.
66. Fahey, J. J. (1964). Recovery of coesite and stishovite from Coconino sandstone of Meteor Crater, Arizona, *Am. Mineralogist, 49*: 1643.
67. Begun, G. M., and Rutenberg, A. C. (1967). Vibrational frequencies and force constants of some group IVa and group Va hexafluoride ions, *Inorg. Chem., 6*: 2212.
68. Striefler, M. E., and Barsch, G. R. (1976). Elastic and optical properties of stishovite, *J. Geophys. Res., 81*: 2453.
69. Lu, R., and Hofmesiter, A. M. (1994). Thermodynamic properties of ferromagnesium silicate perovskites from vibrational spectroscopy, *J. Geophys. Res., 99*: 11795–11804.
70. Hanfland, M., Hemley, R. J., Mao, H. K., and Williams, G. P. (1992). Synchrotron infrared spectroscopy at megabar pressures: vibrational dynamics of hydrogen to 180 GPa, *Phys. Rev. Lett., 69*: 1129.
71. Hanfland, M., Hemley, R. J., and Mao, H. (1993). Novel infrared vibrom absorption of solid hydrogen at megabar pressures, *Phys. Rev. Lett., 70*: 3760.
72. Smyth, J. R. (1987). β-Mg_2SiO_4: a potential host for water in the mantle? *Am. Mineralogist, 72*: 1051.
73. Miller, G. H., Rossman, G. R., and Harlow, G. E. (1987). The natural occurrence of hydoxide in olivine, *Phys. Chem. Mineral., 14*: 461.
74. Pawley, A. R., McMillan, P. F., and Holloway, J. R. (1993). Hydrogen in stishovite, with implications for mantle water content, *Science, 261*: 1024.
75. Nakamoto, K. (1978). *Infrared and Raman Spectra of Inorganic and Coordination Compounds*, John Wiley & Sons, New York, p. 448.
76. Geiger, C. A., Langer, K., Bell, D. R., Rossman, G. R., and Winkler, B. (1991). The hydroxide component in synthetic pyrope, *Am. Mineralogist, 76*: 49.
77. Rossman, G. R., and Aines, R. D. (1991). The hydrous components in garnets: grossular-hydrogrossular, *Am. Mineralogist, 76*: 1153.
78. Rossman, G. R., Beran, A., and Langer, K. (1989). The hydrous component of pyrope from the Dora Maira Massif, Western Alps, *Eur. J. Mineral., 1*: 151.
79. Aines, R. D., and Rossmann, G. R. (1984). Water in minerals? A peak in the infrared, *J. Geophys., 88*: 4059.
80. Chopelas, A. (1990). Thermal properties of forsterite at mantle pressures derived from vibrational spectroscopy, *Phys. Chem. Minerals, 17*: 249.
81. Jayaraman, A., Sharma, S. K., Wang, Z., Wang, S. Y., and othes (1993). Pressure-induced amorphization of Tb, *J. Phys. Chem. Solids, 54*: 827.
82. Hemley, R. J. (1987). Pressure dependence of Raman spectra of SiO_2 polymorphs: α-quartz, coesite, and stishovite, in *High Pressure Research in Mineral Physics*

(M. H. Manghani and Y. Syono, eds.), Tetra Scientific Publishing Company, Tokyo, p. 347.

83. Finger, L. W. (1989). Synchrontron powder diffraction, in *Modern Powder Diffraction* (D. L. Bish and J. E. Post, eds.), Mineralogical Society of America, Washington, D.C., p. 309.
84. Jackson, W. E., Knittle, E., Brown, G. E., Jr., and Jeanloz, R. (1987). Partitioning of Fe within high-pressure silicate perovskite: evidence for unusual geochemistry in the lower mantle. *Geophys. Res. Lett.*, *14*: 224.
85. Wong, J., and Angell, C. A. (1976). *Glass Structure by Spectroscopy*, Marcel Dekker, New York, p. 864.
86. Zinner, E. (1988). Isotopic measurements with the ion microprobe, in *New Frontiers in Stable Isotopic Research* (W. C. Shanks and R. E. Criss, eds.), U.S. Government Printing Office, Washington, D.C., p. 145.
87. Putnis, A., and McConnell, J. D. C. (1980). *Principles of Mineral Behavior*, Blackwell Scientific Publications, Oxford, p. 257.
88. Ganapathy, R., and Anders, E. (1974). Bulk compositions of the Moon and Earth estimated from meteroites, *Proc. Lunar Sci. Conf.*, *5*: 1181.
89. Tarte, P. (1965). Étude expérimentale et interprétation du spectre infra-rouge des silicates et germanates. Application a des problèmes structuraux relatifs á l'état solide. Memoires de l'Académie Royale de Belgique 35, Prats 4(a) and 4(b), p. 259.
90. Moore, R. K., White, W. B., and Long, T. V. (1971). Vibrational spectra of the common silicates: I. The garnets, *Am. Mineralogist*, *56*: 54.
91. Hofmeister, A. M., and Copelas, A. (1991). Vibrational spectroscopy of end-member silicate garnets, *Phys. Chem. Minerals*, *17*: 503.

11
Unique Preparation Techniques for Nanogram Samples

Anna S. Teetsov McCrone Associates, Inc., Westmont, Illinois

1. INTRODUCTION

By applying methods used by microscopists to the preparation of samples for infrared microspectroscopy, one can analyze particles less than 1 ng in size. For example, a dispersion of particles too small to analyze individually can be transferred to a KBr crystal and assembled to cover a larger area. Excellent spectra can be obtained for an array of 1- to 2-μm particles by careful sample preparation. Techniques for solving similar problems have been carefully documented in this chapter. These techniques require practice and proper tools. These techniques are extremely useful in isolating and identifying contaminants encountered in the pharmaceuticals, electronics, forensics, and manufacturing industries.

It is a useful concept to consider 1 ng as the lower limit and strive for larger specimens whenever possible, always keeping in mind just how many 1-μm particles are needed to make a 1-ng sample. The relative volumes of 1 μL, 1 nL, 1 pL, and 1 fL are shown in Figure 1 along with the relationship between size and weight assuming a density of 1. Table 1 lists the number of spheres of various diameters that are necessary to make a nanogram sample.

To document and prepare samples properly for infrared microspectroscopy, it is essential to have a polarizing microscope and a stereomicroscope equipped with transmitted, oblique, and coaxial illuminations. With the exception of a few filtering steps, all of the techniques described in the following section were performed using a stereomicroscope at 12× to 25×.

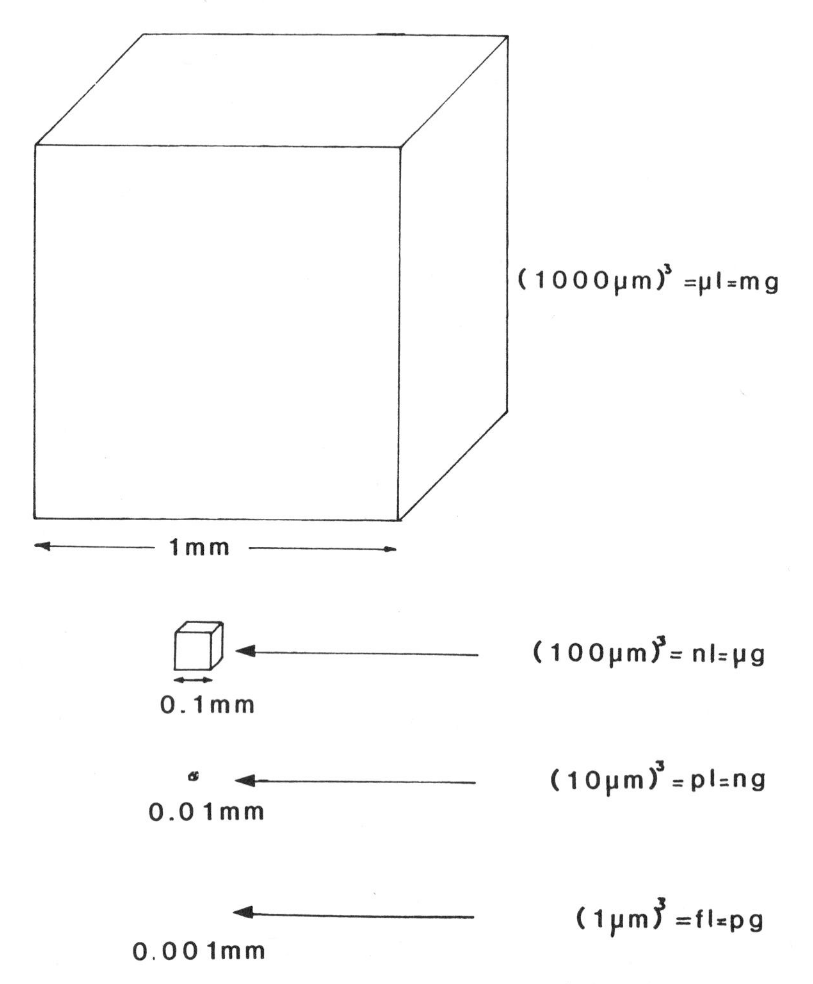

Figure 1. Relative volume of a microliter (μL), nano liter (nL), pico liter (pL), and femtoliter (fL).

2. MICROTOOLS AND SUPPLIES

The effectiveness of a microtechnique depends largely on the quality of the analyst's microtools. Fine tungsten needles, micropipettes, diamond knives, KBr crystals, small salt cubes, polyester filter squares, and various solvents are needed as discussed below.

Table 1. Relationship Between Size, Number, and Volume for Spheres

Spheres		
Diameter (μm)	Number[a]	Volume (pL)
1	~2000	1
2	~250	1
3	~10	1

[a]Number required to equal 1 ng or 1 pL.

2.1. Tungsten Needles

Tungsten needles made of 1-in. lengths of 24-gauge tungsten wire and sharpened with sodium nitrite [1, pp. 226–227] are essential to small sample isolation and preparation. Figure 2 shows how the tungsten wire is heated until the tip is red hot and then passed through molten sodium nitrite, which etches it to a very sharp point. Finished needles are graded into three broad categories—coarse, medium, and fine—as shown in Figure 3. About 30 of them can be stored in a shallow plastic box, as shown in Figure 4. It is useful to have a large supply of the various grades of needles.

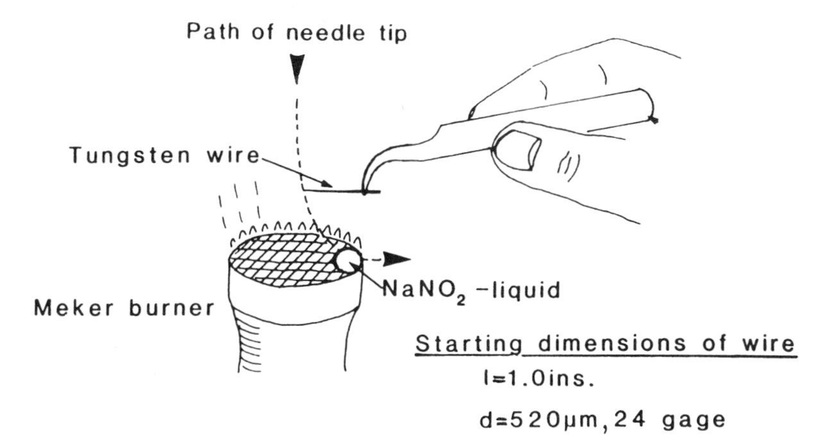

Figure 2. Sharpening the tungsten wire.

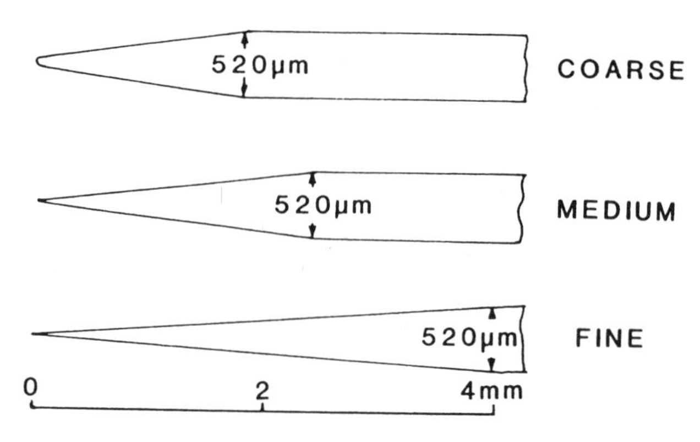

Figure 3. Grading the finished needles.

2.2. Micropipettes

Micropipettes are unique in that they can be filled, discharged, and cleaned only by capillary action. When the tip of a filled pipette is touched to any surface, it will deliver a drop that can be 100 to 1000 μm in diameter. These small drops of solvent are essential in extracting, concentrating, and washing nanogram samples for infrared analysis. The pipettes can be made from polyethylene tubing [1, pp. 228–229] or they can be purchased and modified slightly as shown in Figure 5, in which the tip of an Eppendorf GELoader is bent over a heat source. Micropipettes made in-house are stronger and last longer than purchased pipettes because they are made from a higher-density polyethylene. They are soft, unbreakable, and will not scratch delicate surfaces. A colored rubber band wound

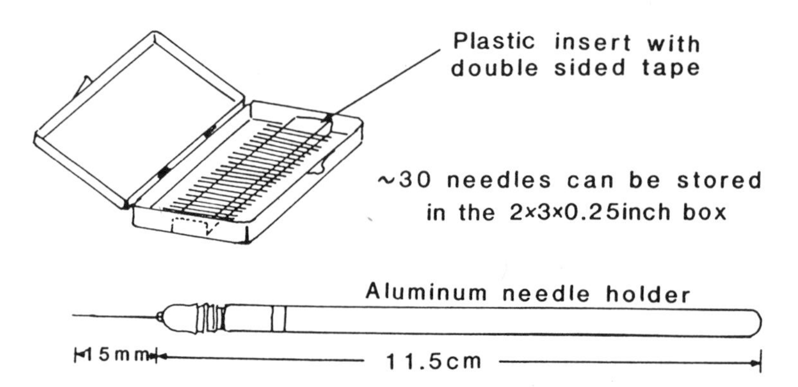

Figure 4. Storage box and needle holder.

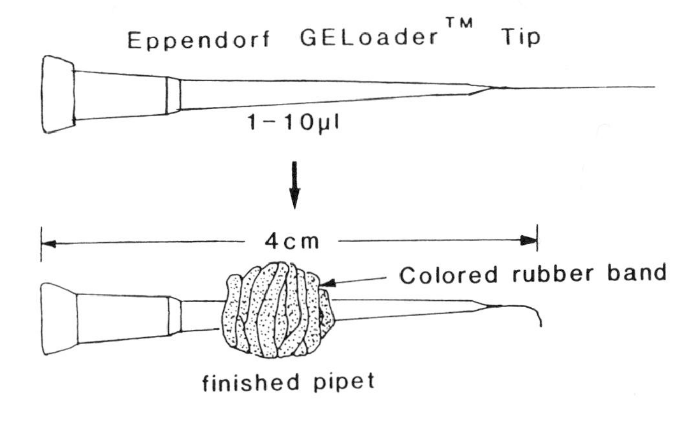

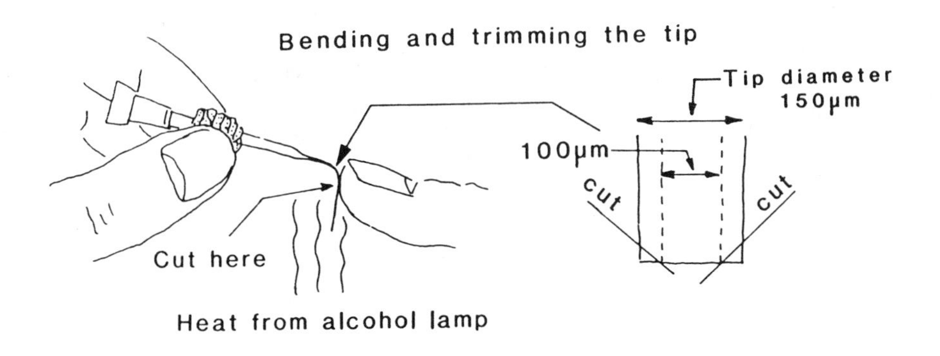

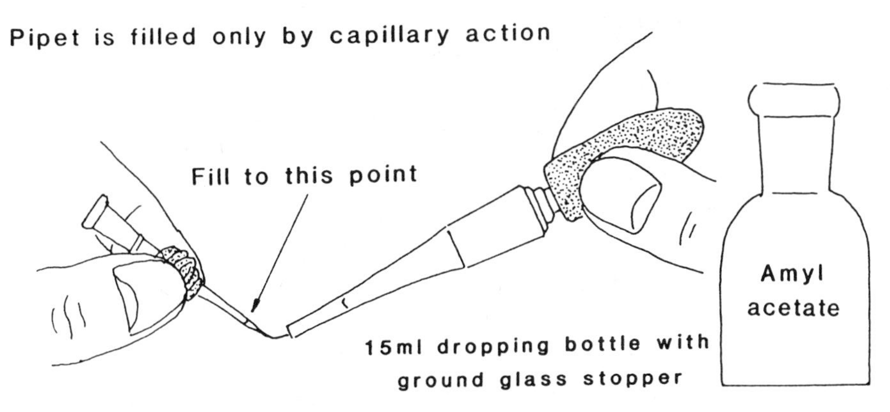

Figure 5. When the tip of the pipette is touched to any surface, it will deposit a drop of solvent, which will spread to 100 to 1000 µm in diameter. The spread of the solvent depends on the surface tension of the solvent and the surface on which it is deposited.

around the center of the micropipette will prevent the tip from touching the bench and facilitate handling. Different-colored bands can be used to indicate different sizes.

2.3. Diamond Log

Small diamond logs can be purchased (Edge Technologies, Indianapolis, Indiana) and attached with epoxy to the top of a split 26-gauge tungsten wire as shown in Figure 6a. The diamond edges are excellent for scribing or cutting, while the two front faces can be used for pressing soft microsamples into a thin film directly on a KBr crystal. Harder samples can be pressed first on a glass slide, then transferred to a KBr crystal as illustrated in Figure 6b.

The diamond log can also be used to remove a thin film in a hard-to-get-at place as shown in Figure 6c. Figure 6d illustrates how the diamond's flat surface can be used to transfer small volumes of a nonvolatile liquid to a KBr crystal for infrared analysis. The diamond log is also useful for scraping thin organic coatings or defects from metal or plastic surfaces. When sufficient material has been scraped onto the log, it is rotated 180°. A KBr crystal is placed under the diamond and the scrapings are pressed onto the crystal to form a thin clear film. These steps are repeated until sufficient material is accumulated for infrared analysis.

2.4. KBr Crystals

In a laboratory where large numbers of microsamples are analyzed, it is economical to prepare your own polished KBr crystals (Figure 7). The crystals are cut, approximately 15 × 5 × 1.5 mm, from a larger crystal. The crystals are first sanded on both sides with 2000-grit lapping paper, then they are picked up with a small piece of double-sided sticky tape, attached to the end of a glass slide, polished for a few seconds on a wet synthetic wipe, and then polished for a few more seconds on a dry paper wipe (Figure 8). The crystal is then checked for flaws under the stereomicroscope using coaxial illumination before polishing the reverse side. Used KBr crystals can be repolished using a paper wipe moistened with amyl acetate and water as shown in Figure 9a.

2.5. KBr Crystal Holders

KBr crystals must be supported and held securely while preparing a sample. A preparation slide or "working slide" can be made from a glass slide with a 2-cm piece of double-sided sticky tape. A piece of polystyrene and a fragment of a membrane filter can be used for cleaning the diamond knife and the tungsten needle. They are arranged on the double-sided tape on the working slide alongside the KBr crystal (Figure 9b). To facilitate easy removal, the crystal is at-

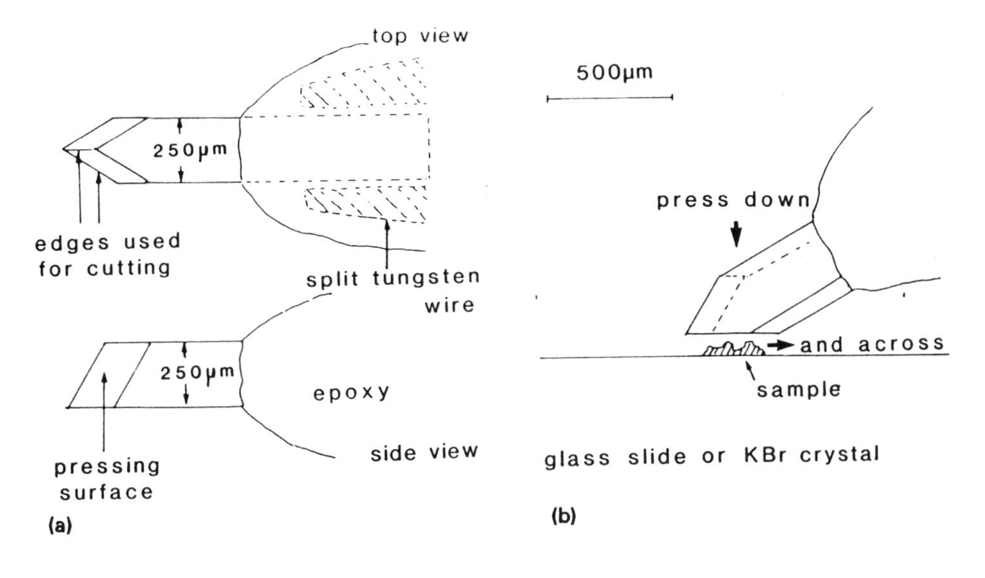

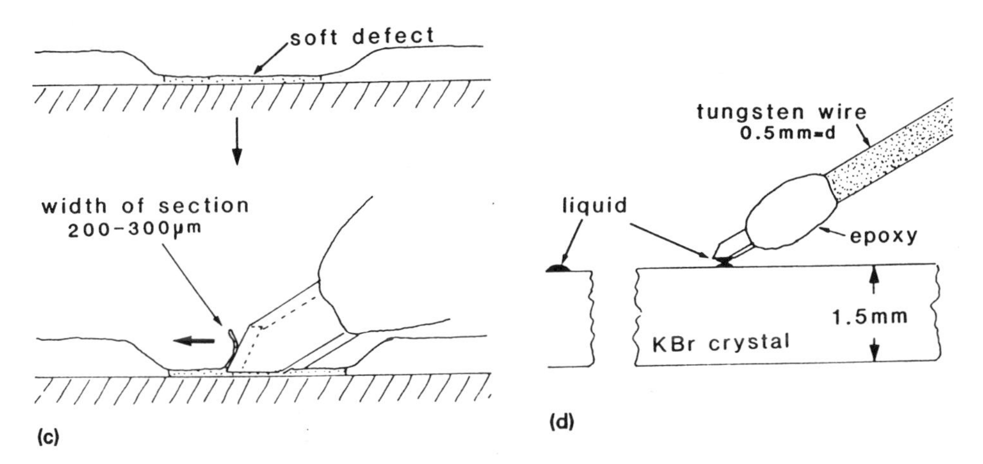

Figure 6. (a) Finished diamond log; (b) pressing hard or soft samples; (c) isolating a thin film from a crater; (d) transferring a small volume of a nonvolatile liquid to a salt plate.

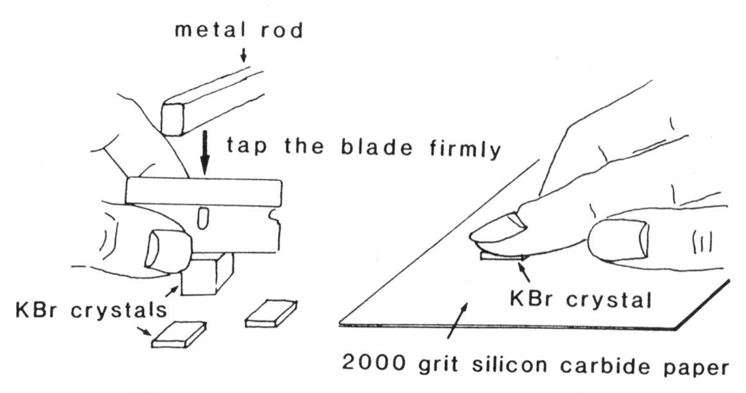

Figure 7. Cutting and polishing KBr crystals.

tached only to the edge of the double-sided tape during sample preparation and is transferred using blunt tweezers to an aluminum or stainless steel holder that fits the mechanical stage of the light microscope and the infrared microspectrometer. As shown in Figure 9b, this aluminum or stainless steel plate has a variable opening slit, running lengthwise along the plate, that can accommodate KBr crystals from 5 to 15 mm wide. Both the working slide and the holder are reusable. It is suggested that working slides be kept in a cleanroom or clean

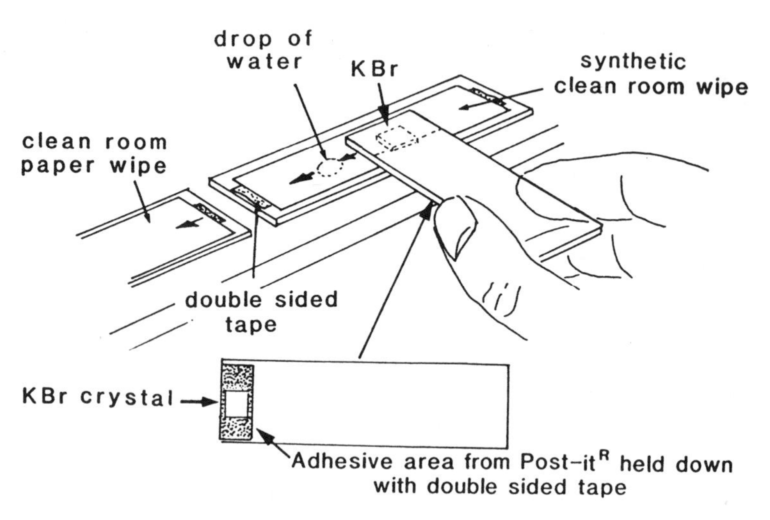

Figure 8. Putting the final polish on a KBr crystal.

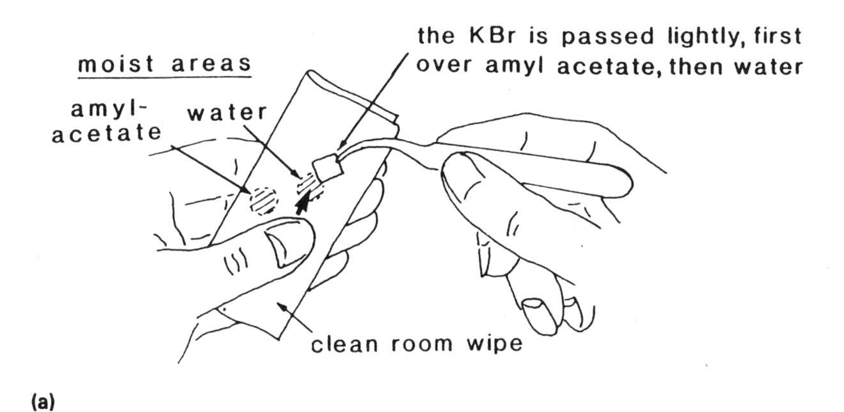

(a)

PREPARATION SLIDE

polystyrene

KBr crystal 5–15mm across

membrane filter

glass slide

double sided tape

CRYSTAL HOLDER

Aluminum 1mm thick

variable width slit

SLIDE STORAGE

(b)

Figure 9. (a) Repolishing crystals of KBr; (b) crystal holders and storage box.

area while holders with prepared samples on the KBr crystal are sent for analysis.

2.6. KBr Microcubes

Microcubes of KBr, approximately 0.1 to 0.4 mm on edge, are used for pressing out elastomeric samples when compression cells or diamond anvils are not available. They are also useful for removing nonvolatile liquids from liquid–solid mixtures. An example would be the separation of the oil from an oily wear product as shown in Figure 10.

These tiny microcubes are made from a large crystal of KBr using a stereomicroscope. The large crystal (<5 mm) is placed inside a shallow plastic box and covered with a piece of thin plastic film such as Saran. The large crystal is cut into 0.1- to 0.4-mm microcubes using a razor blade. The cuts are made directly through the thin plastic film to keep the small cubes from scattering. Unlike larger KBr crystals, these microcubes will be relatively clear and will not require polishing. Some of the cut pieces will not be perfect cubes, which are necessary for pressing out elastomeric samples, but the irregularly shaped ones can be used to remove oil from oily wear particles or to press out small soft specimens.

2.7. Polyester Filter Squares

Small, 0.2- to 0.4-mm polyester filter squares are used mainly to collect very small volumes of a nonvolatile liquid from a surface as illustrated in Figures 14, 30, and 31. A large supply can be made from a single 0.4-μm-pore-size Nuclepore polyester filter. The Nuclepore filter is placed on a plastic surface and cut into small squares with a razor blade under a stereomicroscope. Unlike a fragile membrane filter, the polyester squares do not break when handled with a tungsten needle during sample preparation.

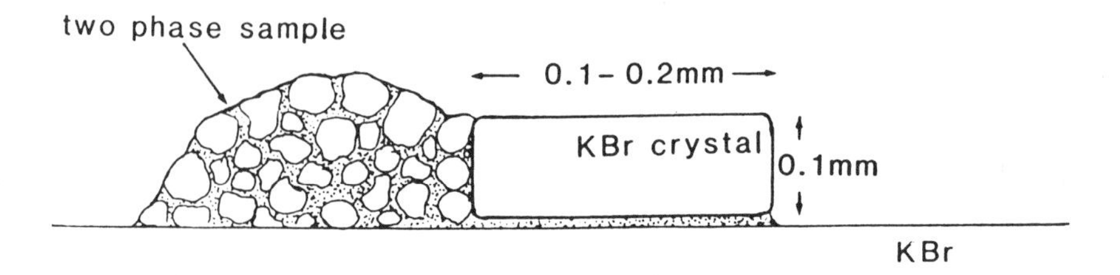

Figure 10. Separating a two-phase sample.

2.8. Solvents

High-purity, low-residue solvents are a requirement for sample preparation. A drop of solvent placed on a glass slide and spread over an area 1 mm in diameter should leave no visible residue upon evaporation when viewed with coaxial illumination. Useful solvents must also have a low evaporation rate to give the analyst sufficient time to perform the technique before the 1-mm drop evaporates. When viewed under the stereomicroscope, a 1-mm drop of solvent that is held between a needle and a glass slide should not evaporate in less than 12 s. Table 2 lists the evaporation rates of some useful solvents.

The two solvents most frequently used in preparing small samples are amyl acetate and water. Amyl acetate has a low evaporation rate and is also a good polar solvent for thin films of flexible collodion used in some sample preparation steps by microscopists. Drops of water are used to separate the thin collodion film from the glass slide. Solvents such as ethanol and 1,2-dichloroethane with fast evaporation rates can be used by more experienced analysts.

Since large volumes of solvents are not required, a useful storage container is a 15-mL ground-glass dropping bottle. To prevent contamination of the solvent by degrading rubber bulbs, the droppers can be fused at the tip. The amount of solvent retained on the fused tip is sufficient to fill a micropipette.

3. BASIC PARTICLE AND SOLVENT MANIPULATION TECHNIQUES

The techniques described in this section require familiarity with small-sample and solvent manipulation using a tungsten needle. If the tungsten needle vibrates more than +10 μm when hand-held, Figure 11 shows the hand and needle position that best helps reduce vibration. The hand and arm (if possible) should rest on the microscope stage, with the hand always relaxed and the needle never firmly gripped. The needle rests on the middle finger, and the thumb simply stops the needle from slipping off. Large movements involve the index and middle fingers, and fine manipulation requires slight pressure from the index finger only. Vibration is also reduced if the needle is kept short (13 to 15 mm) and the fingertips are kept just outside the field of view of the stereomicroscope.

Although particles smaller than 20 μm can be picked directly from the surface of filters or contaminated objects with tungsten needles, larger particles may fall from the needle tip. If the needle tip is made tacky with a soluble gum, larger particles can be transferred to the KBr crystal, where a drop of amyl acetate will dissolve the gum and leave the particle free and clean. For a more detailed coverage of particle-handling techniques, see *The Particle Atlas* [1, pp. 233–244].

Table 2. Evaporation Time for a 1-mm Drop of Solvent Held Beneath a Coarse Tungsten Needle and a Glass Slide at 72°F

Solvent	Evaporation time (s)
Dimethylformamide	65
Amyl acetate	28
Isopropyl alcohol	28
Water	17
Xylene	13
Ethanol	6
1,2-Dichloroethane	5
Acetone	1.5

Many organic solvents have low surface tension and tend to spread out on the surface of a slide and evaporate quickly. Figure 12 illustrates how a solvent deposited with a micropipette and drawn simultaneously beneath a needle can create a longer-lasting and more easily manipulated drop. If the needle is raised more than about $\frac{1}{10}$ mm, the surface tension between the solvent and needle will be broken and the solvent will again spread out.

A solubility check on a nanogram-sized sample is shown in Figure 13. The edge of the drop is moved back and forth over the particles, dissolving those that are soluble and leaving those that are insoluble. Only the edges of the droplet should be used, because the center portion of the drop will dislodge all the particles from the glass and make it difficult to tell if the particles have dissolved or simply been moved out of the field of view.

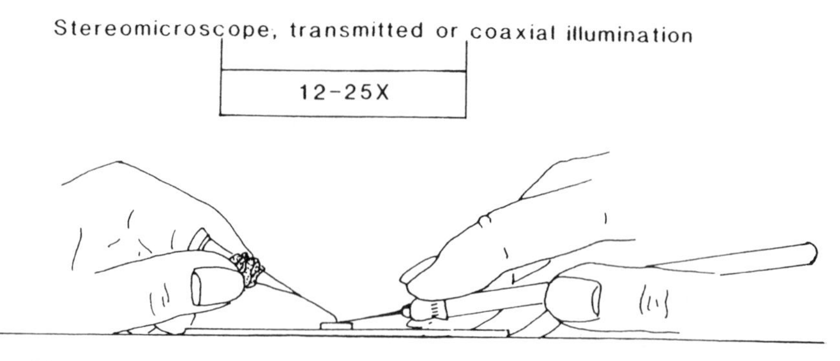

Figure 11. Position of hand needle and pipette for particle and solvent manipulation.

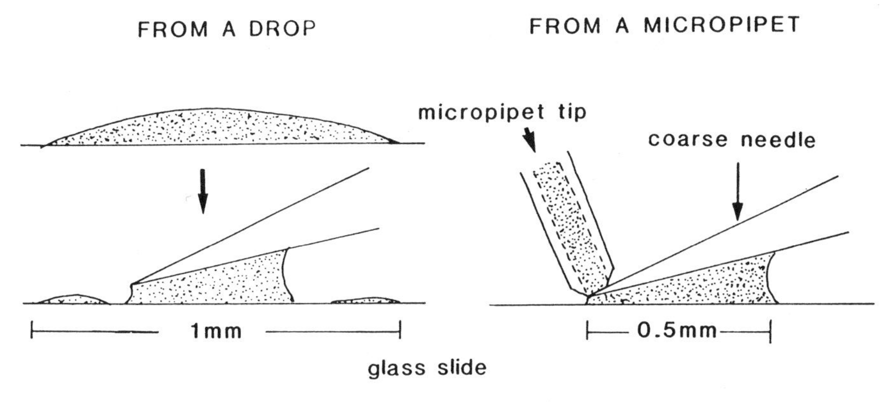

Figure 12. Pulling a solvent beneath a coarse needle.

Occasionally, one encounters surface contamination in which minute droplets of liquid are dispersed over the surface of an object. If these droplets are within the normal size range for reflection spectroscopy and the object has a shiny reflecting surface, reflectance microspectroscopy can be carried out. There are occasions, however, when the droplets are too small in diameter and too thin to be analyzed directly. Figure 14 illustrates how a liquid (e.g., a silicone or a

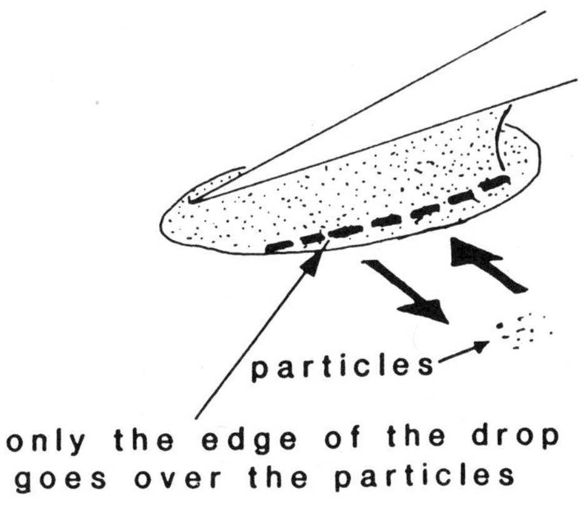

Figure 13. Checking the solubility of nanogram samples.

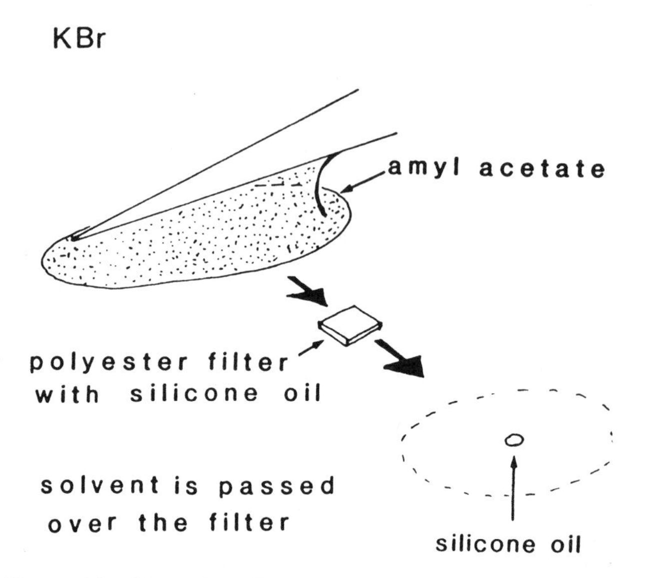

Figure 14. Extracting silicone oil from a filter with amyl acetate.

hydrocarbon oil) can be swabbed up with a 0.2-mm square of polyester membrane and then extracted with a drop of amyl acetate onto a KBr crystal. The amyl acetate droplet is pulled beneath the tungsten needle, then passed quickly over the contaminated filter and onto a clear area of the KBr crystal, where it is held for 5 to 10 s or until it is reduced in size to <100 μm in diameter. The needle is then lifted, leaving a neat drop of oil, as illustrated in Figure 14.

3.1. Pressing Nanogram Samples into Thin Films

Nanogram samples are often too thick or too opaque to be analyzed directly by transmission techniques. Therefore, they must be pressed out to a thickness of 2 to 8 μm, taking care not to reduce the sample size by smearing it or losing it altogether. A number of simple, fast, and reliable procedures are described below.

3.1.1. Soft Samples

Specimens that are soft can be pressed directly onto a KBr crystal with a needle or a microcube until they have the desired thickness. This is illustrated in Figure 15.

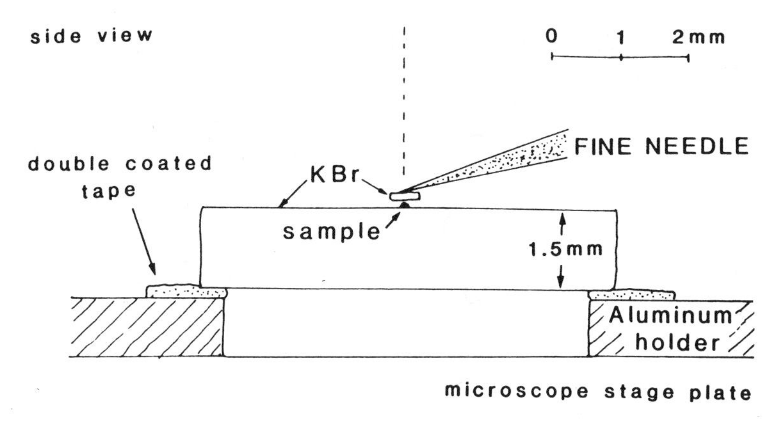

Figure 15. Pressing out a soft sample.

3.1.2. Hard Particles

Because a hard particle cannot be pressed out directly on a KBr crystal, it should be placed inside a marked area on a glass slide and covered with a small (5 × 10 mm) piece of glass slide. Unlike a coverglass, the small piece of glass slide will not easily shatter. Pressure is applied to the small piece of glass with a tungsten carbide scribe while observing how far the particle has spread, as shown in Figure 16. The particle is removed from the glass slide with a tungsten

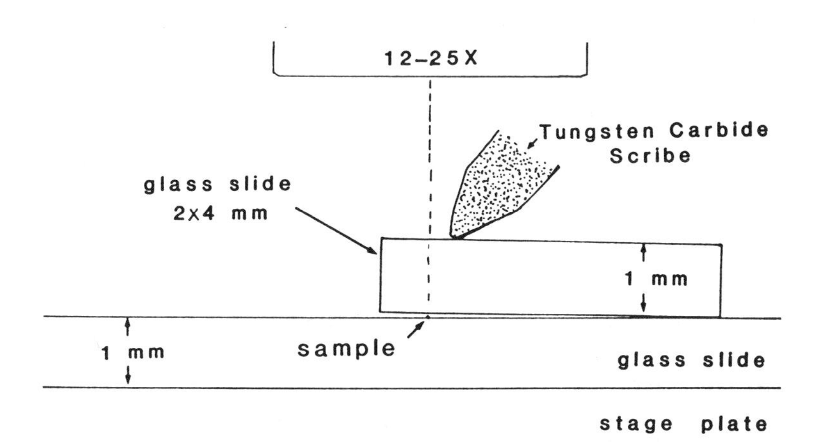

Figure 16. First step in pressing out a hard sample.

needle and placed on a KBr crystal. Extremely hard particles require a diamond micropress; however, what makes a particle difficult to press out is often not its hardness but its size. For example, a 25-μm piece cut from a 200-μm particle is much easier to press out than the original 200-μm particle.

Hard particles can also be pressed with a diamond log as follows. A few small, hard particles are placed on a clean slide and with the flat end of the diamond log are pressed and smeared until they form a thin, translucent film on the glass slide. This film is then scraped off the slide with the tip of a clean razor blade held at a relatively low angle. These scrapings are transferred from the razor blade to a KBr crystal with the aid of a drop of solvent from a micropipette.

Softer particles can be pressed directly onto KBr crystals with a clear hard plastic such as CR-39, which is the trade name of an allyl diglycol carbonate. Most particles will not stick to CR-39, although they may stick to glass or diamonds instead of the KBr crystal. Good control over the movement of the CR-39 during the pressing step is possible because the tungsten carbide scribe works itself slightly into the top surface of the plastic. Thus the particle to be analyzed can be pressed out, and if it is large enough, it can be smeared, producing a thickness gradient as shown in Figure 17.

3.2. Elastomeric Particles

Elastomeric specimens are generally too thick to be analyzed directly and, consequently, must be analyzed in a compressed state. The move convenient way of achieving this is with a compression cell or with a micro-ATR (attenuated

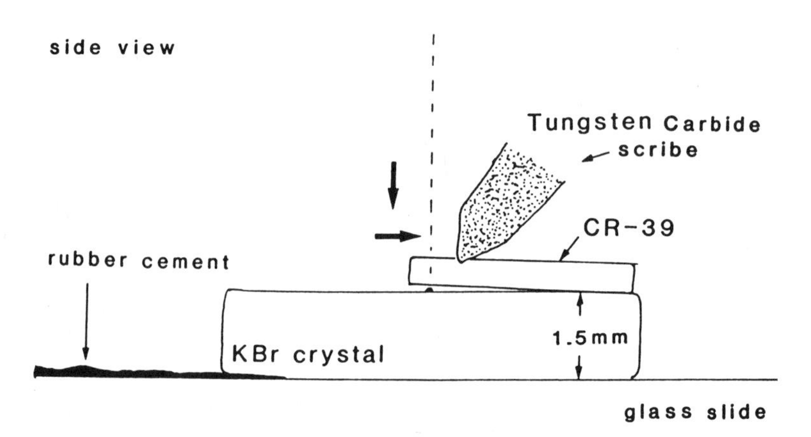

Figure 17. Pressing out medium-hard samples.

total reflectance) objective. In the absence of these devices, we have found a simple technique that works on many samples and requires only a large KBr crystal, a micro-KBr cube and a piece of resin such as CR-39. The technique is illustrated in Figure 18. The microcube and particle are pressed completely into the larger KBr crystal by firmly pushing on the CR-39 with a tungsten carbide scribe, then sliding the CR-39 away from the pressed sample. If done correctly, the elastomeric particle as well as the microcube will be several times thinner than the original. This technique should be used to press out opaque and strongly elastomeric samples <50 μm in size, since these samples give poor results in a compression cell. Larger particles should be cut before they are processed in the same manner. This technique requires patience; however, no solvents are necessary and the particle can be recovered intact by popping off the small KBr cube. Figures 19 and 20 are micrographs of a 20-μm opaque rubber particle before and after it was pressed out with a microcube of KBr.

4. ISOLATING NANOGRAM SAMPLES FROM A LIQUID MATRIX

Frequently, it is necessary to identify haze-producing particles in a small volume (<2 mL) of fluid. These particles may be visible only when a beam of light is passed through the solution. Using filtration, the contaminant is concentrated on a 1- to 4-mm area of a ≤0.4-μm Nucleopore polyester filter using a disposable pipette and no upper funnel as illustrated in Figure 21. The filter is rinsed with particle free water and examined wet with coaxial illumination. If the 1- to 4-mm area is dark, the haze is produced by solid particles (see Figure 22). Interference colors in the central portion of the filter indicate a fine suspension of oil droplets. A residue of solid particles can be removed with a fine tungsten needle. If the residue does not separate from the filter or if it is highly charged, a drop of water is added to the filter. This softens and separates the residue from the filter. Now it can be transferred to a KBr crystal and pressed out with a piece of CR-39 or a diamond knife. Oil residues (see Figure 23) can be extracted from the filter directly onto a KBr crystal as shown in Figure 14.

5. ISOLATING NANOGRAM SAMPLES IN A MALLEABLE SOLID MATRIX

Small contaminants or defects in plastics are common and many approaches can be considered to isolate them depending on their size, hardness, opacity, and the thickness of the matrix. Inclusions in soft films <200 μm thick are not difficult to remove and can be cut out with a sharp razor blade.

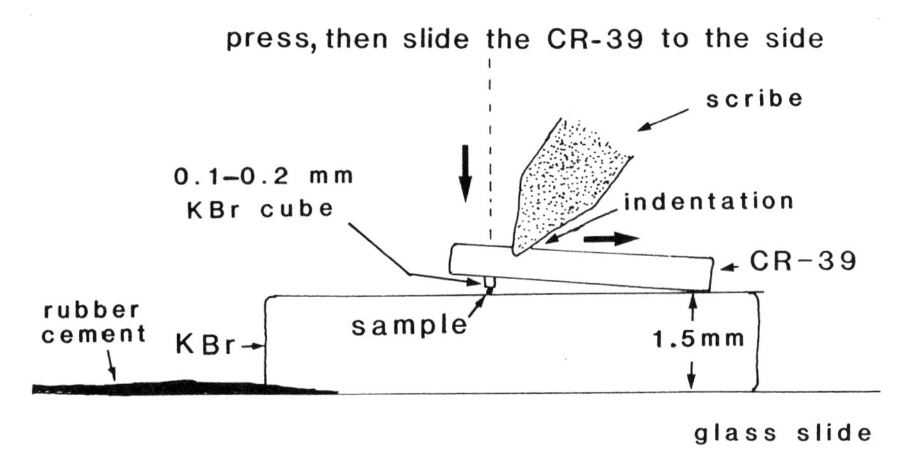

Figure 18. The small cube is pressed into the large KBr crystal.

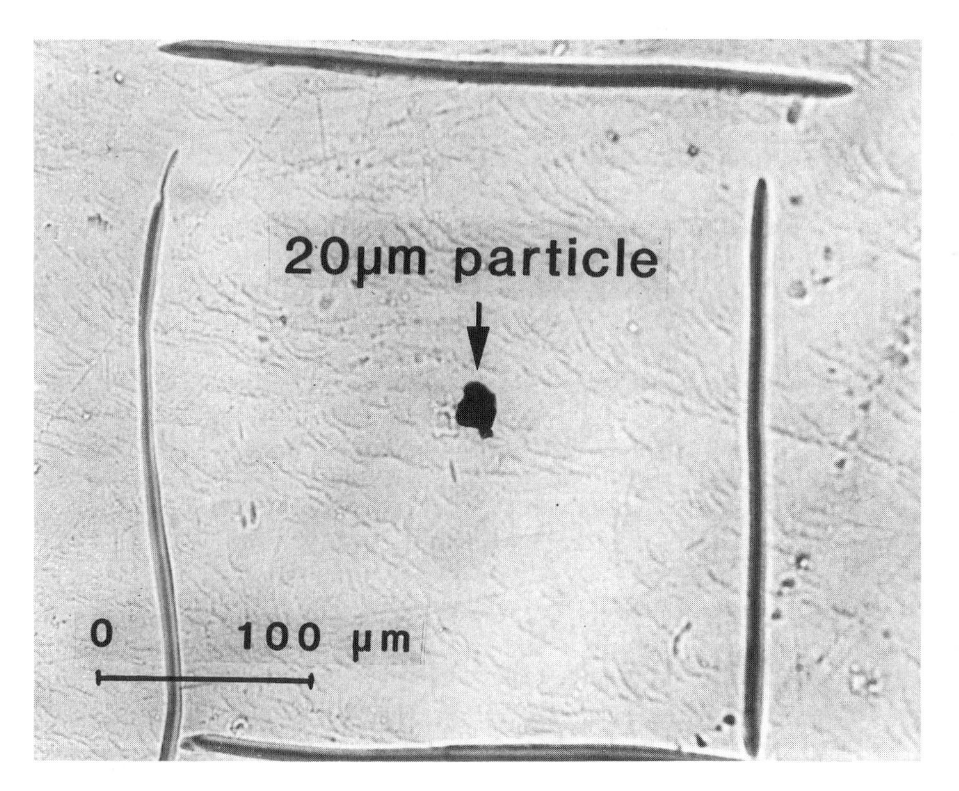

Figure 19. A 20-μm opaque elastomeric particle on a KBr crystal (200×).

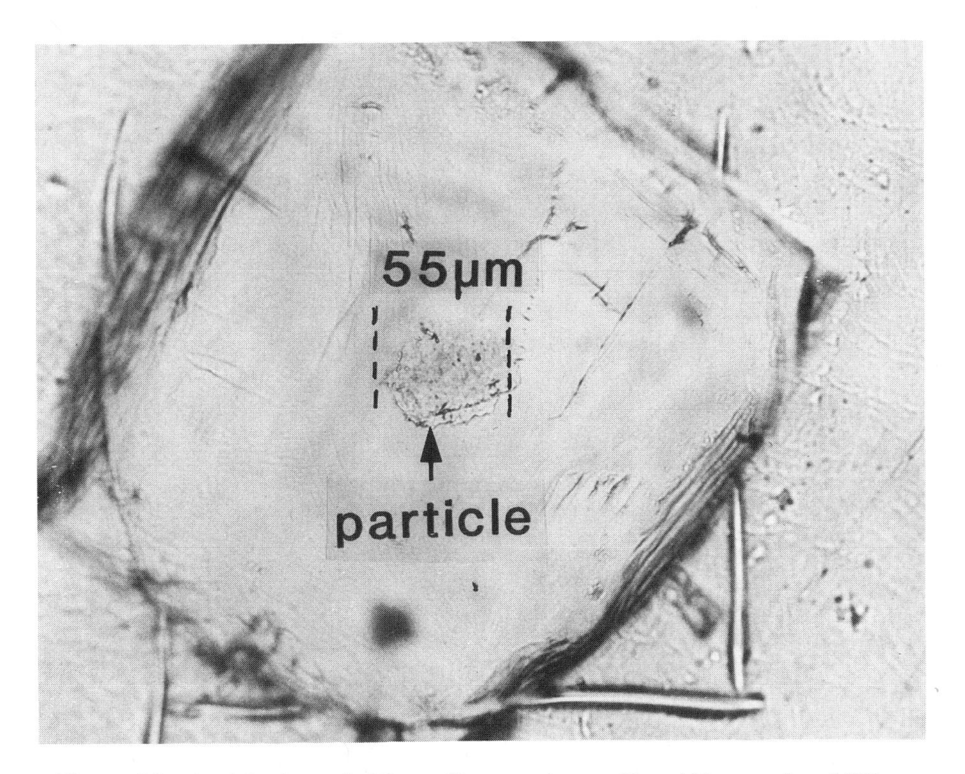

Figure 20. Particle shown in Figure 19, pressed out with a 100-μm cube of KBr (200×).

Inclusions in thick, hard plastics are, however, more difficult to remove. The defect is cut out using a sharp razor blade. The sample is first raised about 1 cm from the surface of the microscope stage, so the razor blade can be held comfortably at about a 20° angle as shown in Figure 24. The matrix surrounding the defect is cut away by making four cuts symmetrically around the defect at a 45° angle away from it, and four more cuts are made at a 45° angle toward the defect as shown in Figure 25. The defect is now isolated in a small volume of the matrix, which is then cut into approximately 10-μm-thick sections as shown in Figure 26. If a solvent for the matrix is known, the entire center area can be isolated and placed in the solvent, while the inclusion is observed continuously with the stereomicroscope. Sections with defects thicker than 10 μm can be pressed between two glass slides before they are analyzed as shown in Figure 27. Other defects that are at or near the surface of the film can be isolated with a diamond log as shown in Figure 6c. The diamond log works well on defects in polyethylene and other soft plastics.

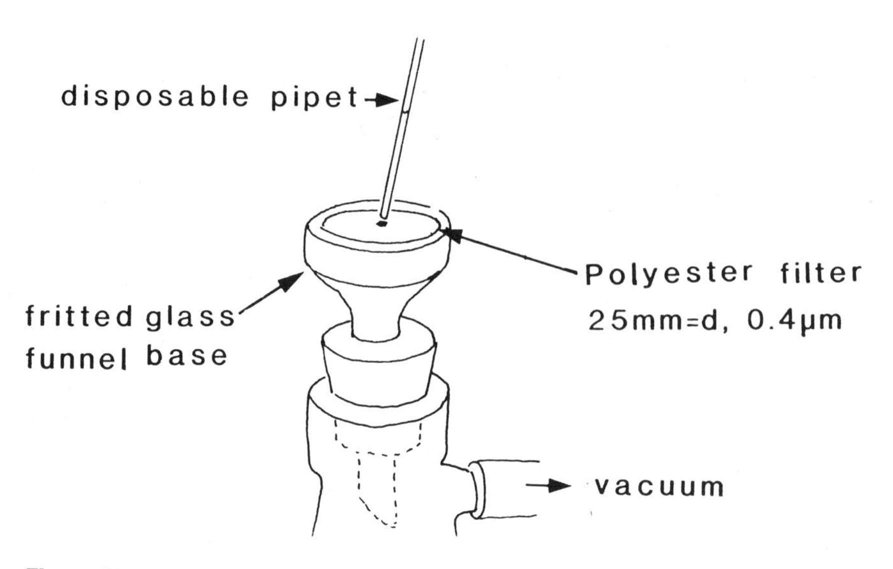

Figure 21. Setup for collecting nanogram quantities of contaminants in a small volume of liquid.

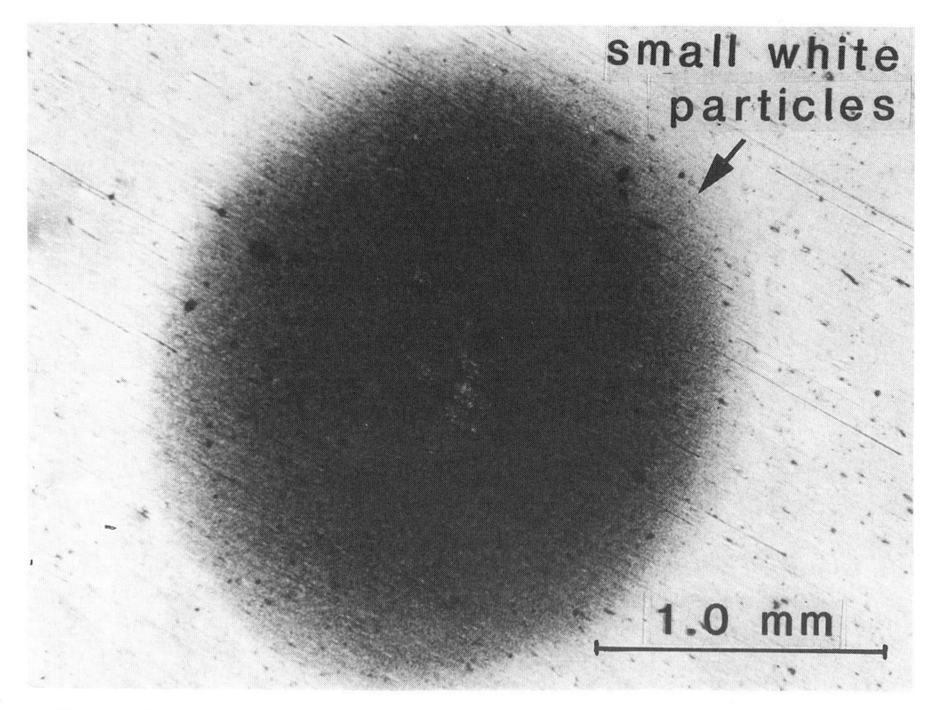

Figure 22. Appearance of filter when the contaminant is a solid (coaxial illumination 30×).

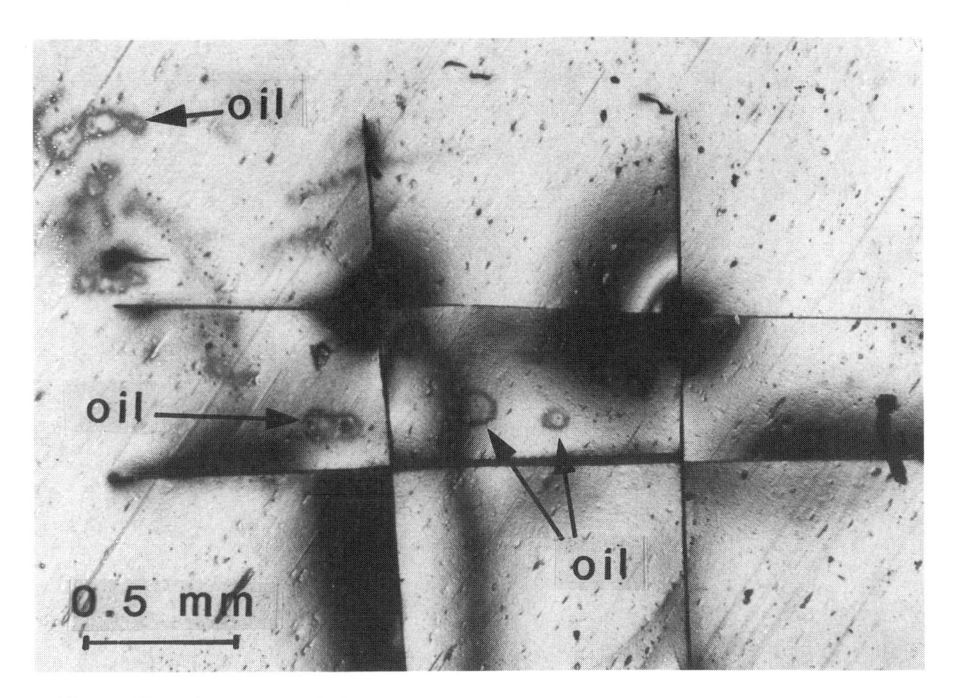

Figure 23. Appearance of filter when the contaminant is a fine suspension of oil droplets. Filter cut for extraction as shown in Figure 14 (coaxial illumination 30×).

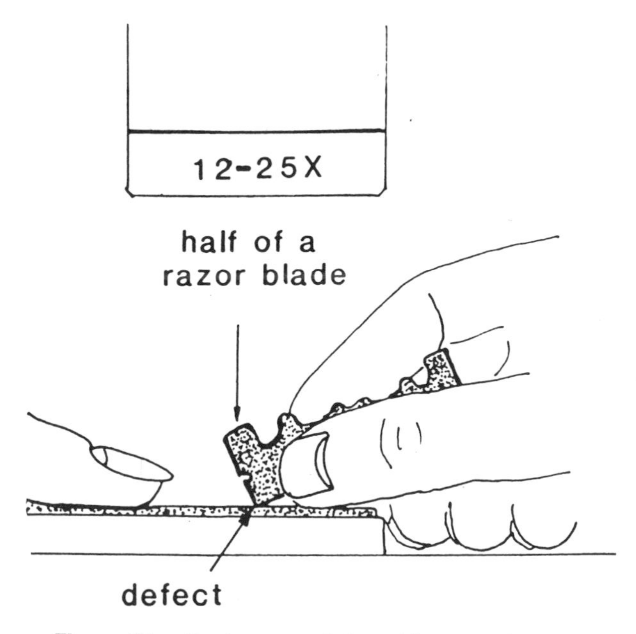

Figure 24. Cutting out a defect with a razor blade.

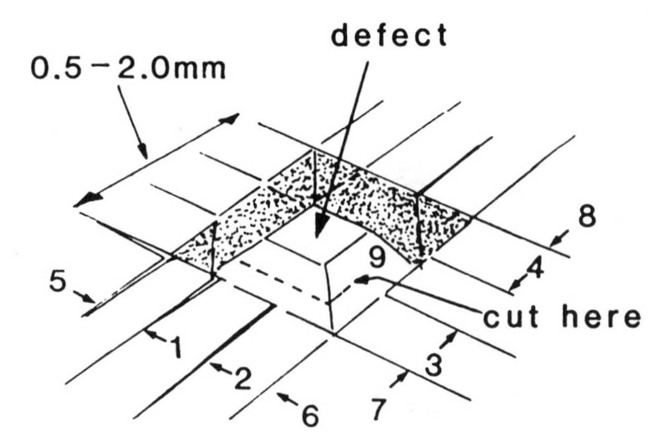

Figure 25. Small isolated volume of matrix with a defect.

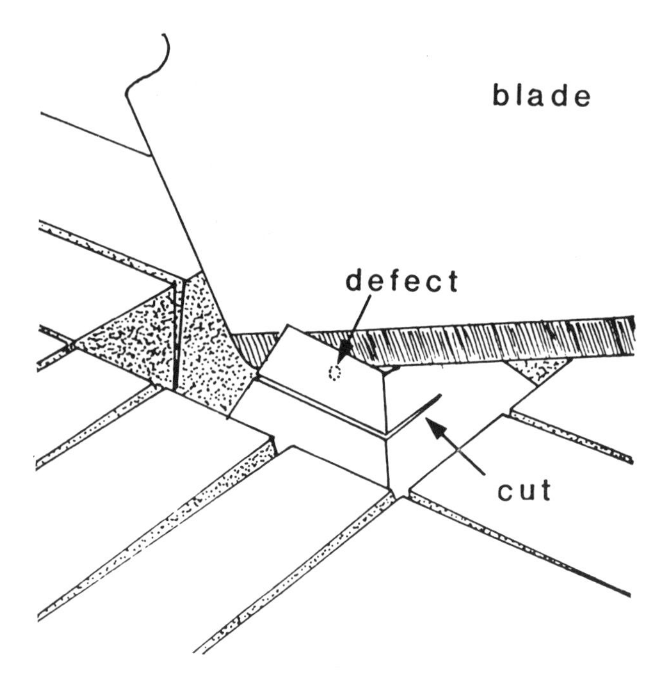

Figure 26. One- to 2-mm-long sections, 10 μm thick, are cut with a fresh blade.

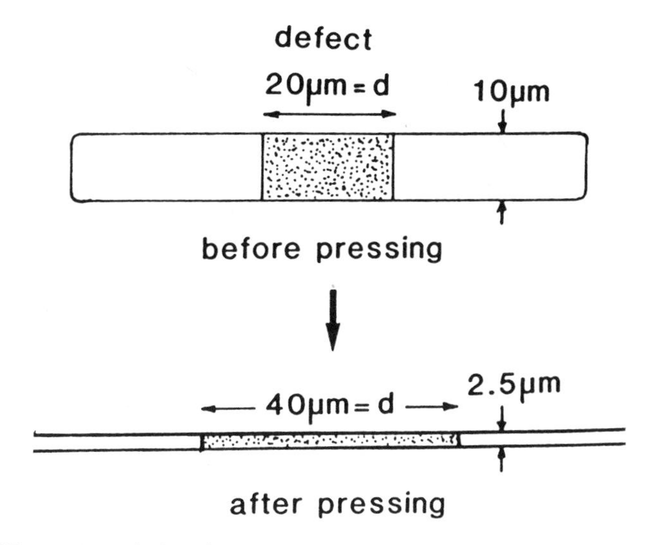

Figure 27. Defects thicker than 6 μm are pressed to about 3 μm between two glass slides as shown in Figure 16.

6. GATHERING MICROMETER PARTICLES FROM A SURFACE

Frequently, a small piece of glass, metal, or plastic will have micrometer-sized particles scattered over the surface as shown in Figure 28. The particles may be too small to scrape off with a needle, but if a suitable solvent is found, they can be concentrated as follows. A solubility check is done on a 1-mm area of the surface as shown in Figure 13. If the particles do not dissolve in amyl acetate, water is tried next, followed by xylene, nonane, ethanol, and 1,2-dichloroethane. Most substances will dissolve in one or more of these solvents. When a suitable solvent has been found, a 1- to 2-mm drop is pulled under a coarse needle and as much of the contaminated surface is covered with the drop as is possible in the 10 to 30 s that it will take for the solvent to evaporate. Frequently, it is easier to keep the needle stationary, 100 to 200 μm above the surface, and move the contaminated object instead, as shown in Figure 29. The needle is removed from the drop a few seconds before the solvent evaporates, leaving a circular area of solid or liquid residue. It may be possible to analyze the residue directly on the substrate, or it can be transferred to a KBr crystal with a needle if it is a solid. A liquid residue can be transferred to a KBr crystal with a diamond log as shown in Figure 6d.

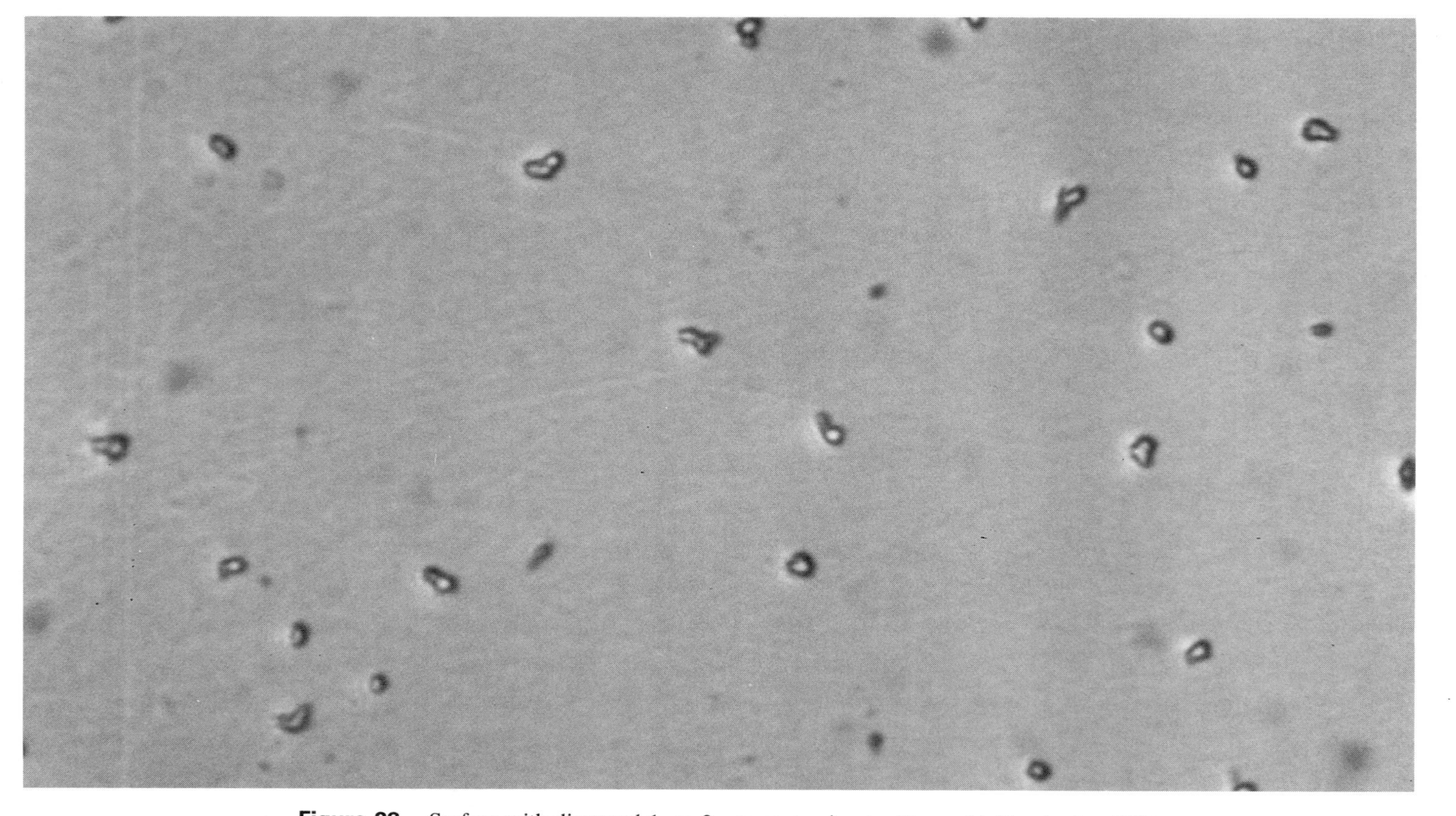

Figure 28. Surface with dispersed 1- to 2-μm contaminants. Nomarski illumination 500×.

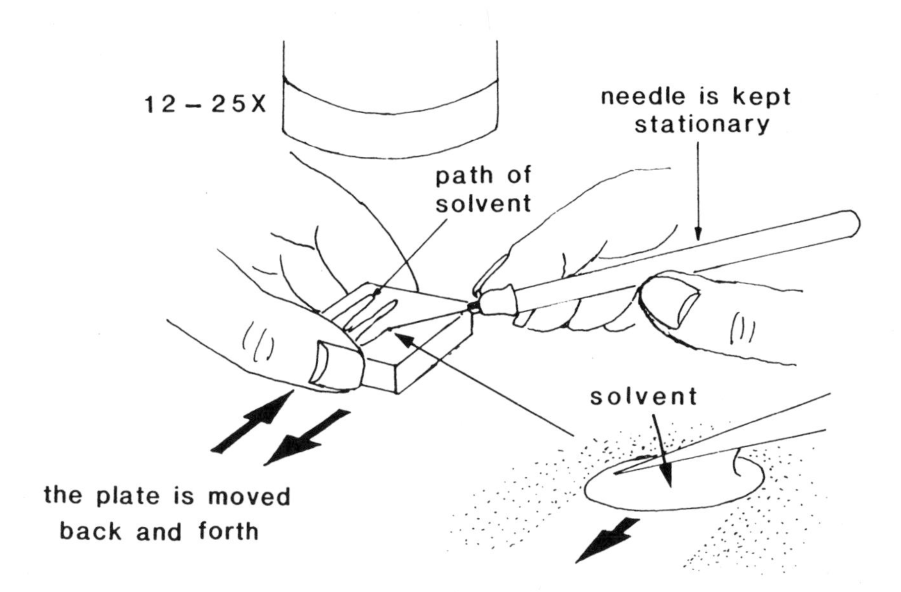

Figure 29. Concentrating the contaminants into a drop of solvent.

7. ISOLATING NANOGRAM LIQUID DROPLETS SCATTERED ON A SURFACE

This technique illustrates how a small number of droplets that are selected over a large or hard-to-get-at surface can be combined into a sufficiently large sample for analysis. The droplets are absorbed by a small square of polyester filter as shown in Figure 30. The filter works well on smooth or rough surfaces because

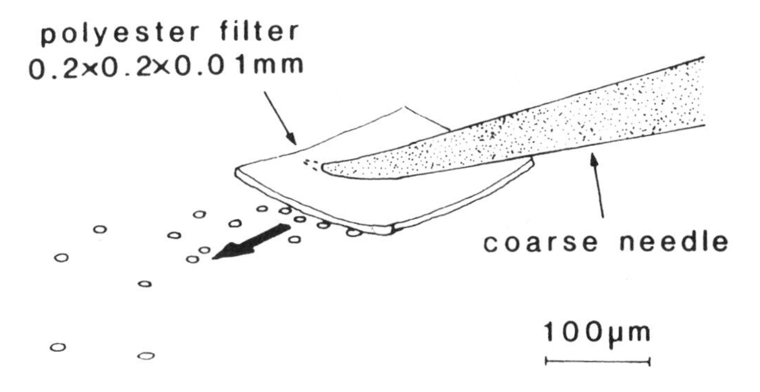

Figure 30. Concentrating the drops of liquid.

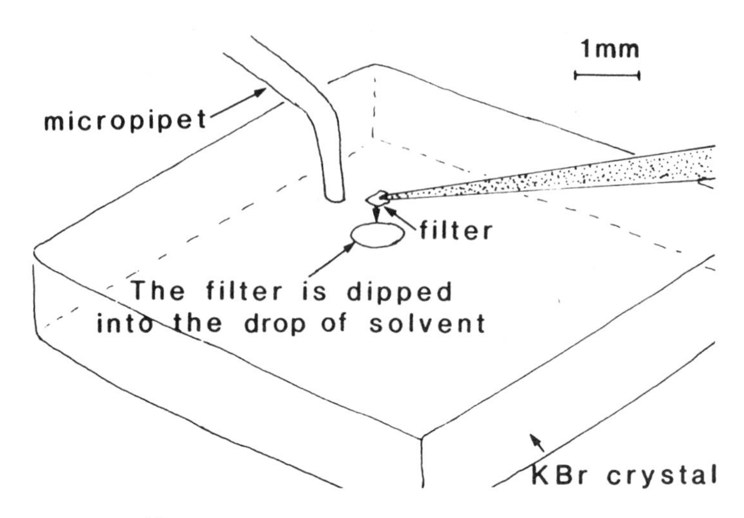

Figure 31. Extracting the liquid from the filter.

it is strong and can be maneuvered into difficult-to-reach places. After the liquid has been absorbed by the filter, it is extracted either by dipping the filter in a 1-mm drop of an appropriate solvent on a KBr crystal as shown in Figure 31 or by passing the solvent over the filter as illustrated in Figure 14. When the solvent evaporates in 10 to 15 s, the unknown liquid will remain as a thin film, or ring of small droplets, or a single large drop as shown in Figure 32. The

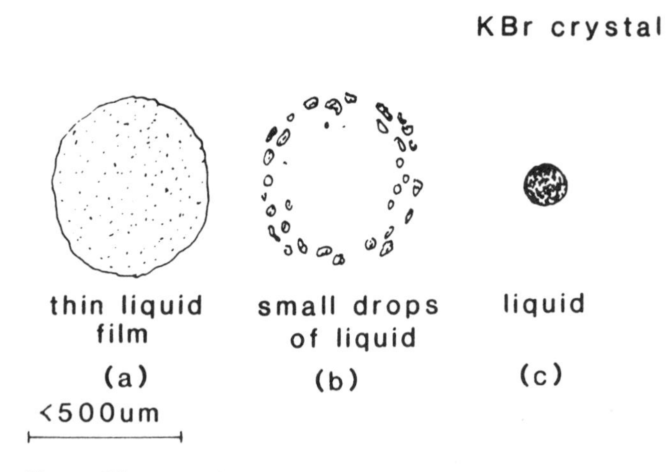

Figure 32. How the extracted liquid can appear on the salt plate.

liquid in Figure 32a and c can be analyzed directly after its location has been marked. The small droplets in Figure 32b can be combined with a suitable solvent. Amyl acetate and nonane are two commonly used solvents for extracting silicone and hydrocarbon oils, respectively. The 1-mm drop used for extraction must be pure. To ensure the purity of a solvent, a drop should be deposited on a KBr crystal prior to extraction. If a residue is observed, the micropipette must be flushed a number of times to remove any contamination. If the problem persists, the solvent must be purified.

8. CONCLUSIONS

In this chapter we have presented some general microtools, techniques, and procedures used in preparing nanogram samples for infrared microspectroscopy. With these tools and procedures, a little ingenuity, and lots of practice, almost every nanogram-sized sample can be located, isolated, and analyzed.

REFERENCE

1. McCrone, W. C., and Delly, J. G. (1973). *The Particle Atlas*, 2nd ed., Vol. 1, Ann Arbor Science Publishers, Ann Arbor, MI.

12

Microsample Preparation Techniques

Howard J. Humecki McCrone Associates, Inc., Westmont, Illinois

1. INTRODUCTION

"What you see is what you get!" This is a response that often comes to mind when we attempt to decipher infrared spectra of complex specimens. One of the advantages of modern instruments is the ability to exclude unwanted portions of our specimen through the use of apertures. Yet the components of many samples are too intimately mixed to permit optical isolation of pure or even enriched phases. The result is that the spectrum contains the total absorption of all the ingredients present in the aperture path. If we are comparing one specimen to another—for example, a reference material from a known source to a specimen found at the scene of a crime—a visual comparison of a spectral pair may be sufficient. A detailed interpretation of the spectra in terms of molecular composition may be less important than showing sameness. But this is not always the case.

When determining the source or sources of a contaminant, it is useful to be able to identify as many components as possible. One substance may be used in several locations in the manufacturing plant, but it may be possible to pinpoint its exact source if it is found in conjunction with some other material. Sometimes separations can be made by physically isolating particles or by solvent extractions using one of the methods described in Chapter 11. In the case of paints or other filled polymers, these techniques are limited to matrices soluble in some solvent. Supporting evidence may be obtained through polarized-light microscopy (PLM), scanning electron microscopy (SEM), x-ray diffraction (XRD), energy-dispersive x-ray spectrometry (EDX), micro-Raman spectroscopy, or microchemical tests.

Pyrolysis, followed by infrared spectroscopy, has been employed to identify polymer matrices of highly filled intractable materials when milligram or larger specimens were available for sacrifice. Although situations exist where destructive tests may not be employed, the sample size has now been reduced to the microgram range, so that even if the samples are relatively small, it is not necessary to sacrifice a major portion. In our laboratories, we occasionally have need to identify the polymer system of a 1-μg sample of a paint or plastic fragment (equivalent to a cube about 100 μm on edge). The spectra prepared from the pyrolyzates of these specimens are clean and free of interferences and baseline irregularities. Contrary to what one might expect, the reproducibility of spectral data is better for these tiny particles than for specimens thousands of times larger.

Controlled degradation can lead to recovery and definitive identification of the hydrolysis products of many polymers. Nylons, for example, cannot be distinguished from one another by their infrared spectra. Raman spectra are unique, but few laboratories have Raman microspectrometers. Hot-stage microscopy is also useful, but skilled microscopists are often not available. Furthermore, the results are not always unambiguous. The hydrolysis products of nylons result in the generation of dicarboxylic or ω-amino acids. These are recoverable and for most nylons, have spectra that are different enough to be useful for identification purposes. Nylon particles smaller than 1 μg can be identified from their hydrolysis products.

Polyesters and polyurethanes can be hydrolyzed in similar quantities, but the identification process understandably becomes more difficult as the polymer system becomes more complex. Acid components are most easily recovered because they usually constitute a greater percentage of the polymer and are stable to hydrolysis conditions. In addition, the ease with which they can be converted from acid to salt form aids in their extraction, precipitation, and subsequent recovery. Polyols have not been recovered and identified successfully and are probably best dealt with by gas chromatographic analysis of their acetates or other derivatives. Hydrolysis of some polyurethanes has been successful at the microgram level, but recovery of their diamines in form pure enough for identification has been unsuccessful thus far. It is unknown at this time if this is because of the degradation of the amines, or because of the low concentrations of the amines, or a combination of both factors. Perhaps HPLC is the solution; non-UV absorbing diamines may be derivitized to UV absorbers or fluorescent substances to improve sensitivity.

The analysis of samples by micropyrolysis or microhydrolysis is simple, requiring only a few steps. It does require some degree of patience and steady hands, and experience with a microscope is a definitive advantage. The major tools required are a stereomicroscope with transmitted and reflected light capabilities and a bench at a height such that the arms and hands can rest comfortably. Do not attempt to use a microspectrometer as a microscope for viewing and preparing samples. None are at a convenient height for this purpose.

In addition to a stereomicroscope, fine tungsten needles, scalpels, a variety of borosilicate capillary tubes, a hobbyist's microtorch, and some fine stainless steel wires, such as needle plungers, are useful. All solvents must be checked to ensure that they leave no residue upon evaporation.

2. MICROPYROLYSIS OF POLYMERS

The identification of polymers by infrared analysis of their pyrolyzates was reported at least 40 years ago [1]. Many polyesters unzip to semivolatile oligomers, having spectra virtually identical to the original polymer; others generate predictable products. Phthalic anhydride crystals are frequently recognized in the condensate of phthalate polyesters, for example. Even those polymers that produce a multitude of degradation products do so in a reproducible manner. Libraries of pyrolyzate spectra are available for identification purposes.

Specimens no larger than 150 μm in two dimensions must be separated from other debris. If a plasticizer is suspected, it must be extracted. Colored samples that are easily located can be sonicated in a suitable solvent for 5 min in a 0.3-mL minivial. If it is necessary to recover the plasticizer, it is best to place the particle in a 0.5 × 30 mm test tube made by quickly sealing one end of an open 0.5-mm borosilicate tube containing solvent with a minitorch. Transferring solvent to a tube having one end closed can be difficult but can be accomplished easily by expelling the solvent from a 0.3-mm-OD capillary into the larger capillary. As low-boiling solvents, such as ether, are lost quickly, higher-boiling-point solvents, such as heptane, dichloroethane, toluene, or amyl acetate, are preferred as the extractant.

After sonication, the end of the sample tube is scratched with a scribe and broken off. The solvent is then slowly expelled onto a small spot on a warmed microscope slide. The extracted plasticizer can be transferred with the tip of a pointed scalpel to a salt window for analysis. The polymer fragment should then be removed from the extraction tube and inserted into a capillary brush for pyrolysis.

Capillary brushes consist of borosilicate glass wool fused to the inside surface of one end of a 0.2-mm-ID borosilicate capillary tube in a manner that allows the flow of the solvent in the capillary to exit from the brush when it is touched to a microscope slide or salt window. When the brush is pulled away from the surface, the flow of liquid stops.

The advantage of the capillary brush over an open capillary tube lies in the ability to control the deposition of fluid and thus avoid spreading the deposit into ever-widening circles. The use of a capillary brush was first described by McCrone and Delly [2] and adapted, we are told, from an idea from Fred Schneider of IBM. The construction of capillary brushes for use in micropyrolysis has been described elsewhere [3] but bears repetition because of its simplicity and usefulness.

A small bundle of borosilicate glass fibers is wet and pulled to a fine point. The pointed tip of fibers is inserted into the end of a 0.2-mm-ID borosilicate glass capillary to a depth of several millimeters. The end of the capillary is then played quickly over the flame of a hobbyist's butane microtorch. The flame must touch the capillary where the ends of the fiberglass touch the tube but not close enough to the protruding fibers to melt them. With some practice, the fibers can be fused to the inside surface of the capillary without too much distortion. The exposed fibers are then snipped with small scissors, at an angle, to complete construction of the microbrush. A microbrush suitable for deposition of liquids is shown in Figure 1. The cleanliness of the capillary is obviously important. I recommend treatment with hot hydrochloric acid followed by a wash with water and sonication to free any particles and broken fibers. The use

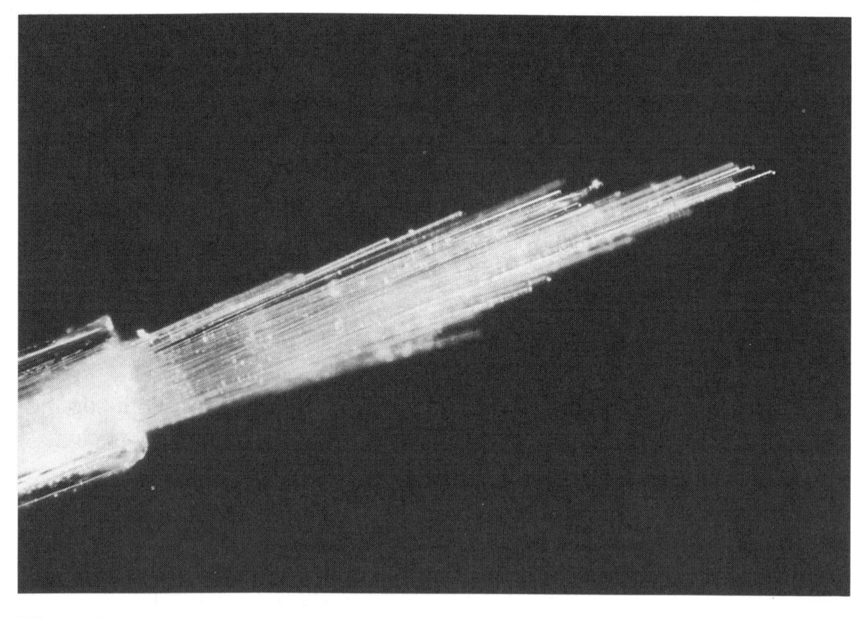

Figure 1. Microbrush prepared by fusing borosilicate glass-wool fibers to the inside surface of 0.2-mm-ID borosilicate glass capillary.

of larger-diameter capillaries is possible, but their value diminishes as the sample size increases. Capillaries of 100 μm ID are not easily constructed and are even more difficult to use. Sample volumes in the range of 0.5 nL to several nanoliters are most conveniently handled with a 0.2-mm-ID capillary.

The particle to be pyrolyzed is inserted into the open end of the capillary opposite the brush. This can best be accomplished by holding the capillary horizontally on a microscope slide and pushing the particle into the open end with a fine wire. The wires found to be most convenient are those supplied as needle plungers for microsyringes. After the particle or particles have been pushed into a depth of about 3 mm, the open end of the tube is quickly fused closed with the microtorch. The sample is pyrolyzed by playing the flame under the sample portion of the capillary. The capillary is then scored with a sharp instrument such as those employed for cutting glass capillaries by gas chromatographers. The glass fragments are wiped from the outside surface, and the tip of the capillary containing the ash is snapped off, holding the end with a forceps.

A drop of solvent is quickly applied to the broken end of the capillary and drawn in to a length of about 5 to 10 mm. Ethylene dichloride, heptane, amyl acetate, and toluene are useful solvents for this purpose. Water-soluble solvents or those that evaporate very quickly are not advisable because they result in pitting of the salt plate or evaporation of the solvent in the glass fiber brush before it reaches the salt plate.

The capillary brush is held vertically until the solvent reaches the brush end. The brush is then touched to the salt window and the liquid is allowed to flow onto the salt plate. If the brush has been constructed properly and if not too much liquid is in the capillary, a small bead of pyrolyzate will accumulate just beyond the tip of the brush. The brush can then be used with or without additional solvent, to push the small bead onto a convenient spot. It has not been found necessary to oven dry or air dry the deposit before analysis.

The following examples were prepared to illustrate the usefulness of the micropyrolysis technique. The instrument employed was a Spectra-Tech IRμs equipped with a 15× objective. A total of 50 interferograms were coadded with the resolution set at 8 cm^{-1}. Baseline corrections were made, but no smoothing or other manipulations were made unless specifically noted. Sample sizes were approximately 1 μg (approximately 1 nL).

Figure 2 shows the spectra of three generally available white flat paints. These are transmission spectra of a sample prepared by pressing, with a hydraulic press, a small flake placed between two steel disks. With the exception of the spectrum having the strong carbonyl band, it would be difficult to speculate as to the nature of the binder although the ''rabbit ears'' appearing at 1602 and 1583 cm^{-1} in the lower spectrum suggest this to be of an alkyd paint. The

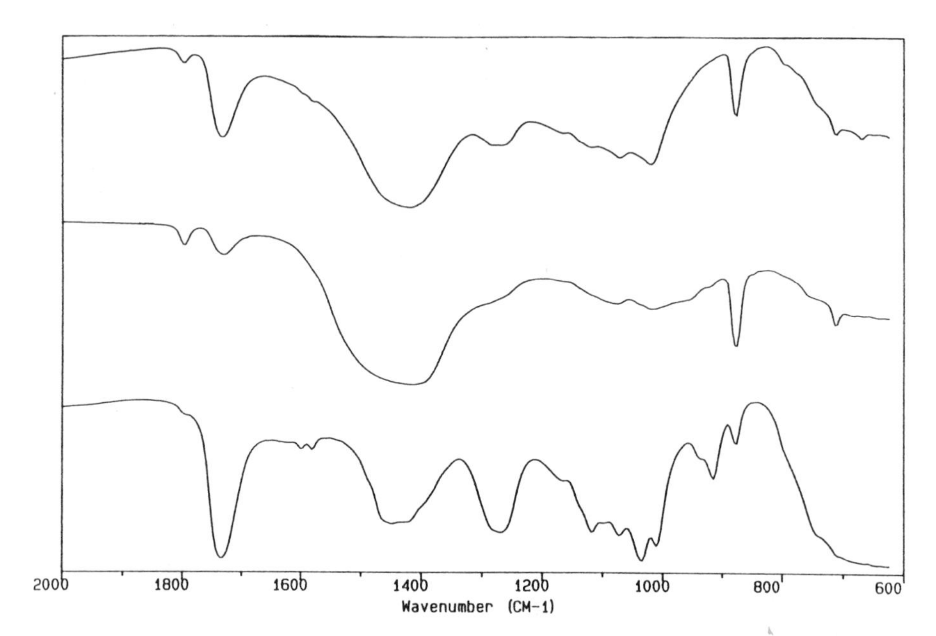

Figure 2. Transmission spectra of pressed flakes of flat, white paint from three different manufacturers.

pyrolyzates of these same paints, shown in Figure 3, clearly indicate that the polymers are similar in all three cases.

Titania and siliceous clays mask the identity of the binder in the spectrum of the white paint chips shown in Figure 4 (top), but the pyrolysis spectrum reveals it to be a latex (bottom). The acrylic base is easily identified in the spectrum of the pyrolyzate of the red artist's paint in Figure 5 (bottom). Similarly, the identity of the drying-oil-modified alkyd paint is more obvious after pyrolysis (Figure 6).

The spectra of pyrolyzates from microgram-sized samples appear to be more reproducible than for larger specimens. It has been suggested that this is because the entire micropyrolyzate is analyzed rather than only a small portion as is sometimes the case for milligram-sized samples, and also because micropyrolysis conditions are more reproducible (L. Nylander, personal communication). Microgram samples pyrolyze virtually instantly, whereas the destructive distillation of large specimens takes place over a longer period of time, during which further degradation may occur under less reproducible conditions. Figure 7 shows three replicate micropyrolyzates of a polyurethane paint. Although there are differences in relative peak intensities, they are readily identifiable as being of the

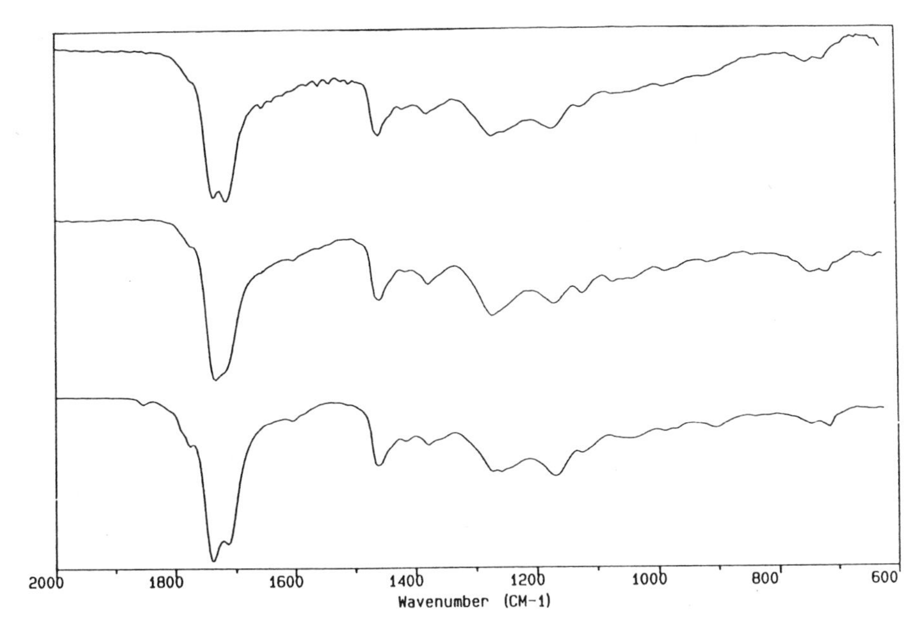

Figure 3. Spectra of the pyrolyzates of the three paints shown in Figure 2.

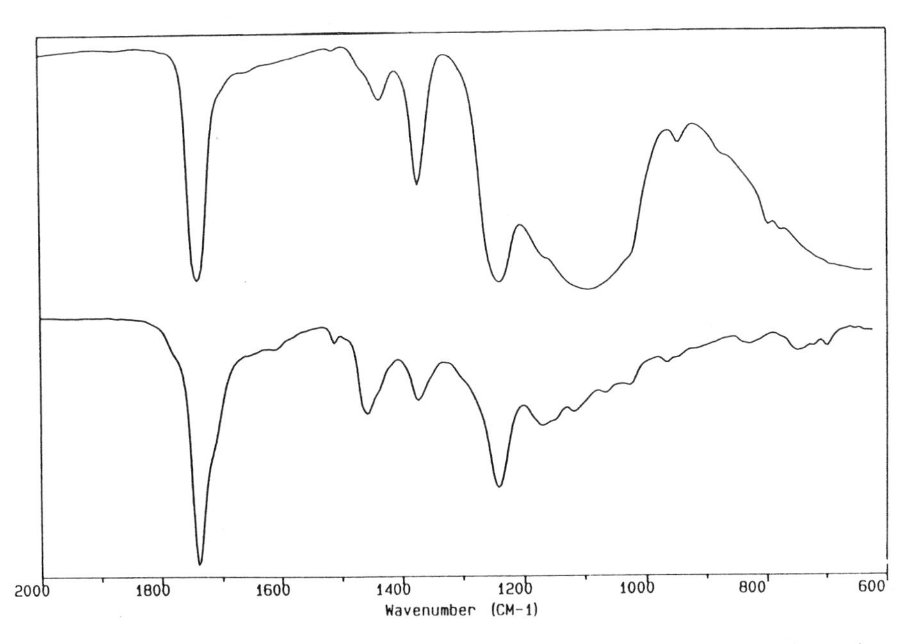

Figure 4. (Top) transmission spectrum of a thin flake of a white, latex paint; (bottom) spectrum of the pyrolyzate of the paint above.

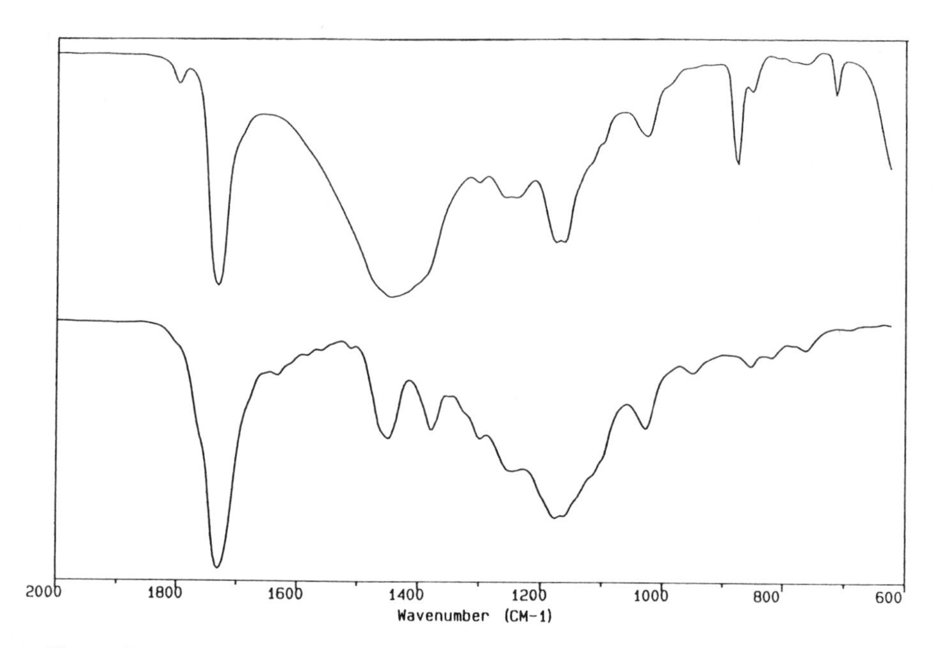

Figure 5. (Top) transmission spectrum of a pressed flake of red, acrylic, artist's paint; (bottom) spectrum of the pyrolyzate of the paint above.

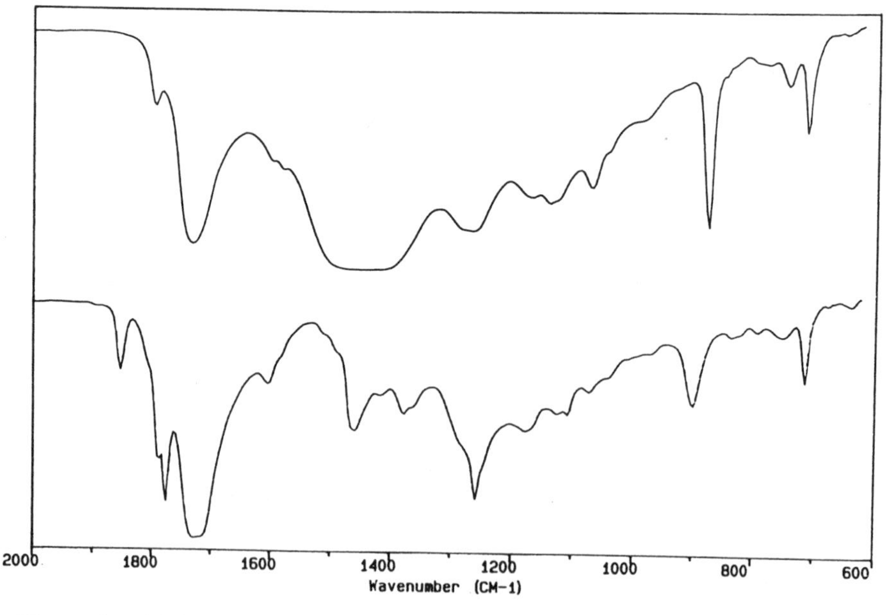

Figure 6. (Top) transmission spectrum of a pressed flake of a black, carbonate-filled paint; (bottom) spectrum of the pyrolyzate of the paint flake above, showing oil-modified alkyd character.

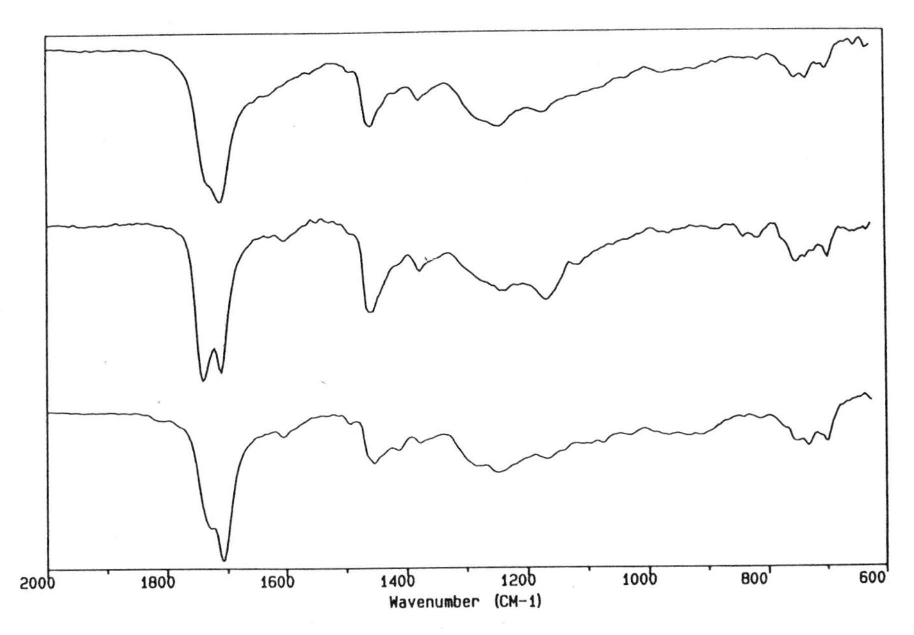

Figure 7. Spectra of pyrolyzates of three submicrogram specimens of the same polyurethane paint.

same polymer type. Based on the measured size and ash content of 30%, it is estimated that the polymer in each case represents about 500 ng.

Other products having high levels of strengtheners and carbon black are gaskets, and automobile and truck tires. A unique problem is the characterization of the polymer in tire-wear particles such as might be found at vehicular accident sites. Figure 8 shows the infrared spectra of a tire-wear particle from the author's car by both transmission and ATR. The transmission spectrum was prepared by flattening the particle, as much as possible, between two salt plates. The attenuated total reflectance (ATR) spectrum was obtained with a Spectra-Tech micro-ATR attachment to the Spectra-Tech IRμs. Considerable information concerning the elastomer is revealed by the transmission spectrum, although the strong absorption peaking at 1115 cm^{-1} completely washes out large areas of the useful range. Figure 9 shows the spectrum of the pyrolyzate of a tire-wear particle from the same tire. The cigar-shaped tire-wear particle was approximately 100 μm in diameter and 400 μm in length. It was cleaned by sonicating in toluene for several minutes. The particle was then removed and allowed to dry before pyrolysis. A spectrum of a 0.5-mg slice from the same rubber tire is also shown in Figure 9.

Figure 8. (Top) ATR spectrum of a tire-wear particle, 1000 interferograms coadded; (bottom) transmission spectrum of the particle above pressed between two KCl disks, 50 interferograms.

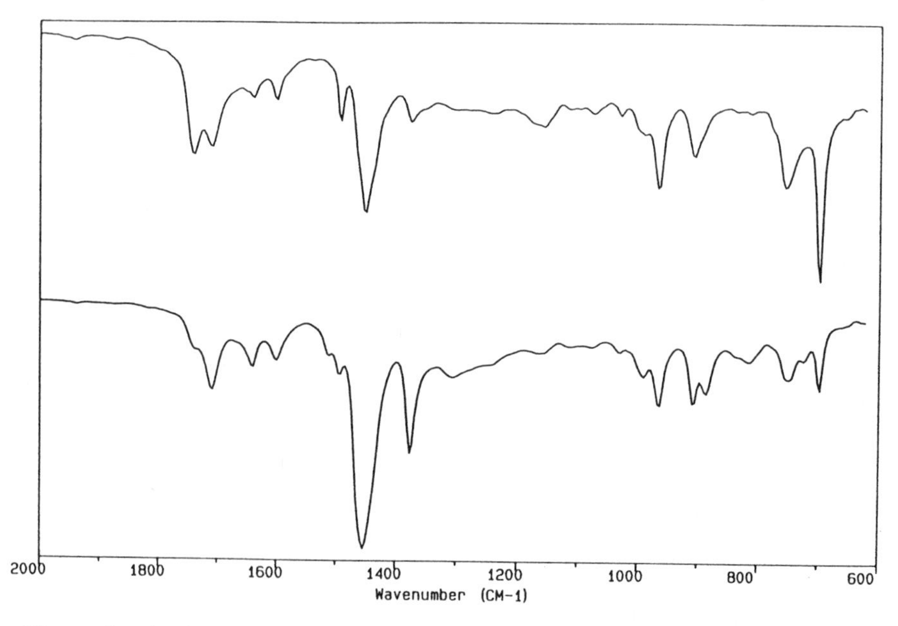

Figure 9. (Top) spectrum of the pyrolyzate of a 100-μm-diameter by 400-μm-long tire-wear particle; (bottom) spectrum of the pyrolyzate of a 0.5-mg specimen. Both specimens are from the same tire as Figure 8.

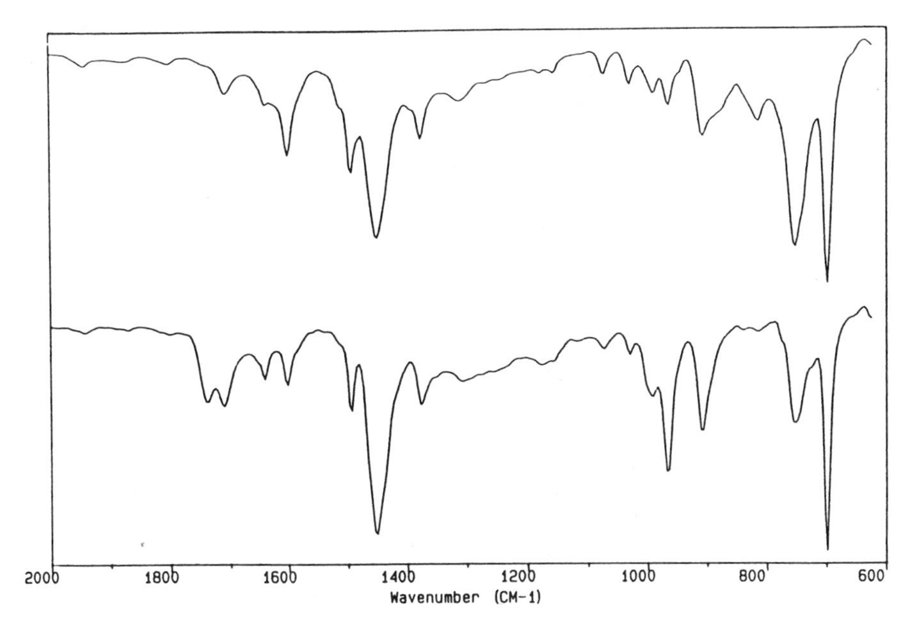

Figure 10. (Top) pyrolyzate of a 5-mg tire specimen; (bottom) pyrolyzate of 150- × 400-μm tire-wear particle. Samples are from a different manufacturer than those shown in Figure 9.

Differences in the relative intensities of some absorption bands appear. Figure 10 shows the spectrum of the pyrolyzate from a 150- × 400-μm tire-wear particle (bottom) from another tire manufacturer compared to the spectrum of a milligram-sized sample pyrolyzate (top). Both specimens were sonicated with toluene to remove possible contaminants. Variations are obvious. In an attempt to shed some light on the cause for these discrepancies, spectra of pyrolyzates of three separate tire-wear particles from this second manufacturer were compared. They are virtually identical not only to each other (Figure 11) but to that of the tire-wear particle shown in Figure 9. This confirms that reproducibility among small-sample pyrolyzate spectra is quite good. One must be cautious when comparing the spectra of pyrolyzates of small samples to those of larger ones if the possibility exists that the sample may have been altered by prepyrolysis treatment.

Micropyrolysis of highly filled and pigmented polymers can reveal considerable information concerning the nature of the polymer. The technique does require skill and practice, and in addition, the sample specimen is destroyed.

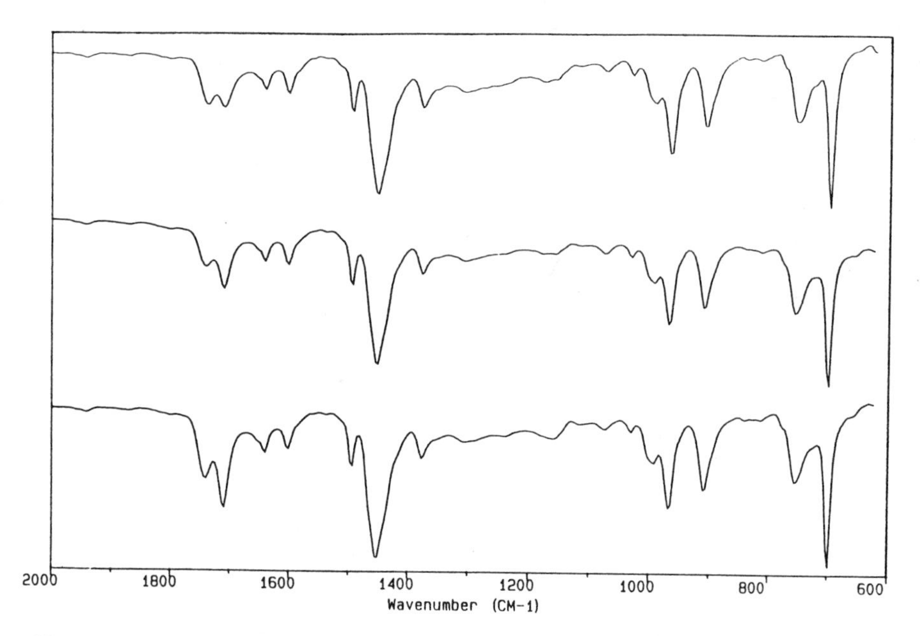

Figure 11. Comparison of the pyrolyzates of three tire-wear particles from the same tire as shown in Figure 10: (top) 400 × 150 μm; (center) 300 × 100 μm; (bottom) 275 × 100 μm.

3. HYDROLYSIS

3.1. Introduction

Polymer technology is becoming more complex and sophisticated. Additives and modifiers are difficult to deal with even under normal analytical circumstances. With very small samples, the identification of even major components and ingredients can be difficult or impossible. A reasonable degree of success has been achieved by subjecting the polymer system to conditions of controlled degradation, separation, and identification of the products. The difficulties increase with decreasing sample size, with the presence of substances similar in chemical composition to those of interest, and to the decreasing proportion of the ingredient in the total sample.

A single procedure cannot be recommended for degrading all types of polymers yielding to hydrolysis. Nylon-type polyamides are easily hydrolyzed with aqueous hydrochloric acid, but epoxides, polyurethanes, and some polyesters are not. Epoxy resins and polyurethanes are resistant to all but the most aggressive media. My advice to those attempting to follow this course of action would be to use volatile reagents wherever possible, to use the minimum amount of solvents and reactants, and to employ as few steps as possible in isolating the specimen. As in the case when analyzing any specimen, one should have as much information about the specimen as possible before starting and be prepared to modify the procedure if the first attempt fails.

3.2. Nylons

The direct identification of small nylon particles can be difficult if one does not have available a Raman microspectrometer or a microscope equipped with a programmable hot stage. Identification of nylons from their infrared spectra is, with few exceptions, impossible. It is fortuitous that the infrared spectra of the acid hydrolysis products are unique and readily distinguishable from one another in all but a few less common nylons. Since nylons are distinguishable from one another on the basis of the diacids rather than their amine components, the process is straightforward. Amines and their hydrochlorides are comparatively weak absorbers and generally do not present serious interference problems.

The sample is prepared by inserting it into the end of a 0.5-mm-ID borosilicate glass capillary approximately 1 in. in length. The tube, held in a horizontal position, is touched to a drop of 50% hydrochloric acid in water. Several millimeters of acid solution are sufficient. Much more than that may result in difficulties recovering some of the acid components which are slightly soluble in aqueous acid. These tend to crystallize nicely from small volumes, making them easier to recover. Rocking the tube will roll the acid back and forth so that it wets the nylon particle. Holding the tube with the fingers, one end can be sealed with a microtorch. Holding the tube with the fingers rather than forceps will prevent overheating the tube and expelling the sample and acid from the opposite end. The opposite end can be sealed by warming it with the torch about 3 mm from the end to dry it and preheating the air in the end, thus preventing it from bursting as the tip is quickly heated and sealed. If the liquid is too close to the end to allow the end to be sealed, it can be moved down by placing the tube in a centrifuge for a few minutes. Hydrolysis of the lower nylons such as nylon 6 and nylon 6,6 can be accomplished by placing the reaction tube in an oven at 110°C overnight. Nylon 6,9 and higher require higher temperatures. Raising the oven temperature to 175°C has been found adequate for all of the nylons through nylon 6,12.

The tube may be centrifuged to move all of the liquid to one end. The opposite end is scored and broken to remove the tip. The capillary is held on a microscope slide and the opposite end is scored and broken off to prevent the fluid from being expelled by any pressure buildup. The contents are expelled onto a clean portion of the slide using a mouthpiece and rubber tube as supplied by some manufacturers of micropipettes. The droplet is allowed to dry. A small quantity of the residue is removed with the tip of a scalpel and pressed onto a salt plate for analysis. The hydrolysis products from nylons comprised of short-chain diacids are relatively soluble, so the hydrolysis will yield a single phase. The higher nylon products may show a highly crystalline phase, and in fact, this may be evident in the reaction tube before the contents are expelled. Since these represent the diacids, it is this portion of the sample that is of most interest.

Some typical spectra are shown in Figures 12 through 17. All of the microsample sizes were estimated to be between $\frac{1}{2}$ and 1 μg. The macrosamples (20 mg) were prepared in the same way and from the same series of reference standards. Figure 12 shows the spectrum of the hydrolyzate of a microgram-sized particle of nylon 6 compared to one of milligram size. The agreement is excellent. Figure 13 is a comparison of the spectra of a microgram and a large sample of nylon 6,6 hydrolyzates. Again, there are no significant differences. Furthermore, comparing the spectra of Figure 12 to those of Figure 13 demonstrates that one can easily distinguish between the hydrolyzates of nylon 6 and 6,6. Comparable spectra of nylon 6,9, shown in Figure 14a and b, appear to show discrepancies. These discrepancies are actually caused by the presence of the amine hexamethylenediamine dihydrochloride shown in Figure 14c. Nylon 6,10 is less readily distinguished from the 6,9, as it differs in only a CH_2

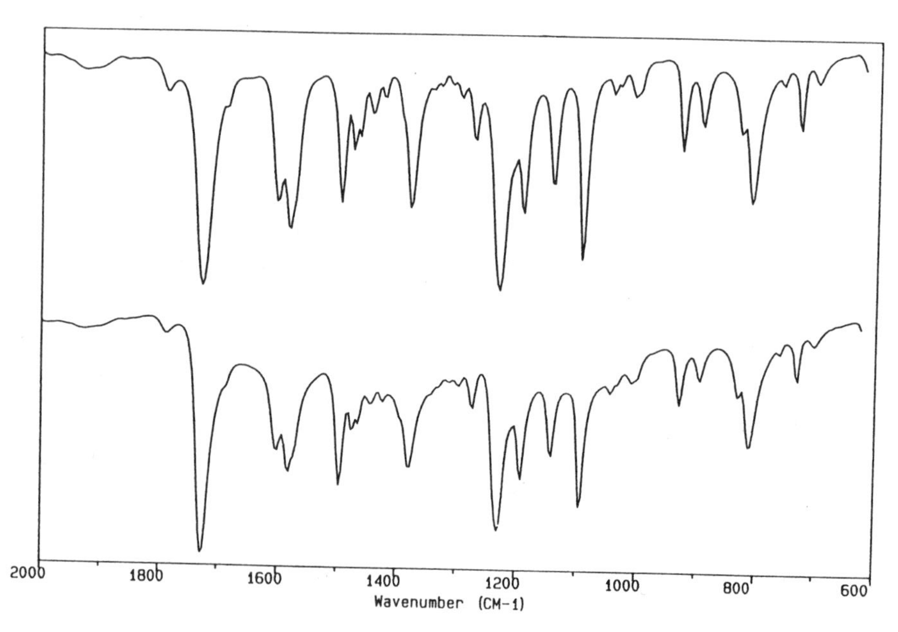

Figure 12. Acid hydrolyzates of nylon 6: (top) macrosample; (bottom) microsample.

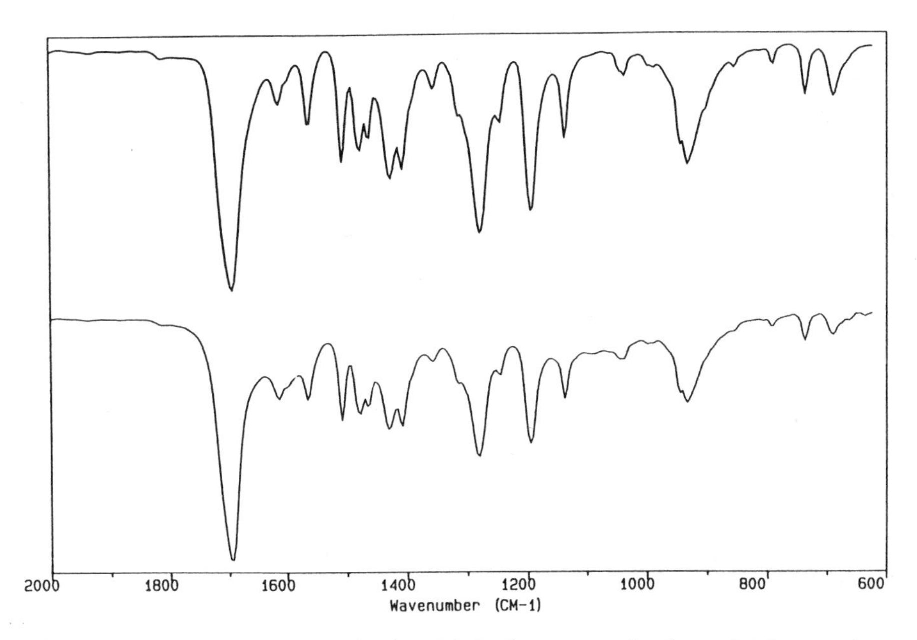

Figure 13. Acid hydrolyzates of nylon 6,6: (top) macrosample; (bottom) microsample.

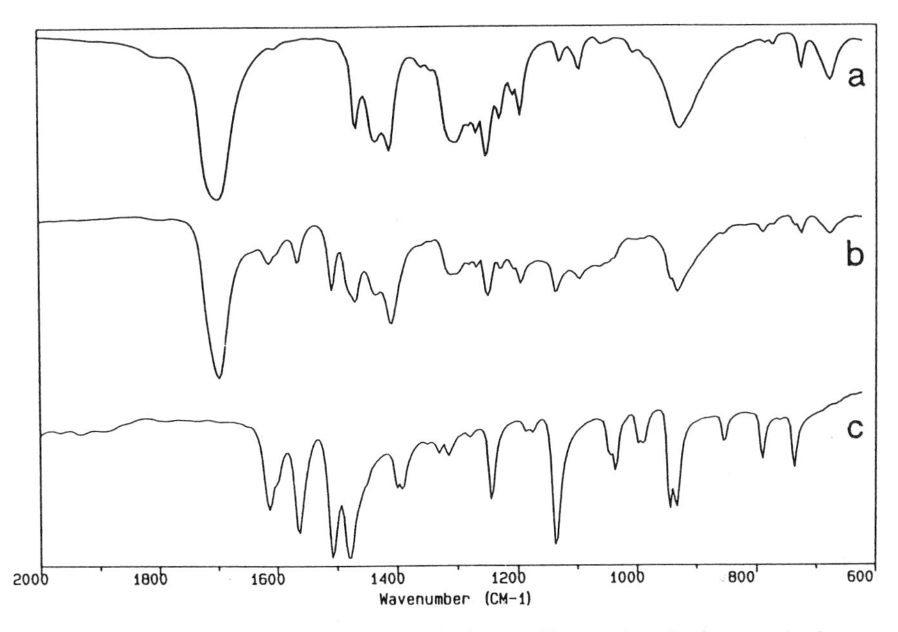

Figure 14. Acid hydrolysis of nylon 6,9: (top) crystalline portion of macrosample; (center) total hydrolyzate of microsample; (bottom) hexamethylenediamine dihydrochloride.

unit in going from azelaic to sebacic acid (Figure 15). Terephthalic acid is readily recognized from the spectrum of the insoluble crystals generated from nylon 6T (Figure 16). Similarly, nylon 11 can be identified from the spectrum of aminoundecanoic acid hydrochloride shown in Figure 17.

In summary, microgram-sized samples of the common polyamides can be identified from their hydrolysis products.

3.3. Polyesters

Polyesters can be hydrolyzed by either acids or bases and their diacids can be isolated and identified. If acids are to be employed, 50% aqueous hydrochloric acid may be used and the sample prepared in a manner identical to that of the polyamides as described in Section 3.2. The temperature of 110°C has been found to be insufficient and 175°C is recommended. The phthalic acids form crystalline solids and can usually be recognized and isolated for anlysis.

In some instances, it may be advisable to carry out an alkaline hydrolysis. Figure 18 shows the spectrum of dipotassium terephthalate isolated upon hydrolysis of polyethylene terephthalate with 2 *N* potassium hydroxide in isopropyl alcohol. This spectrum is that of dipotassium terephthalate.

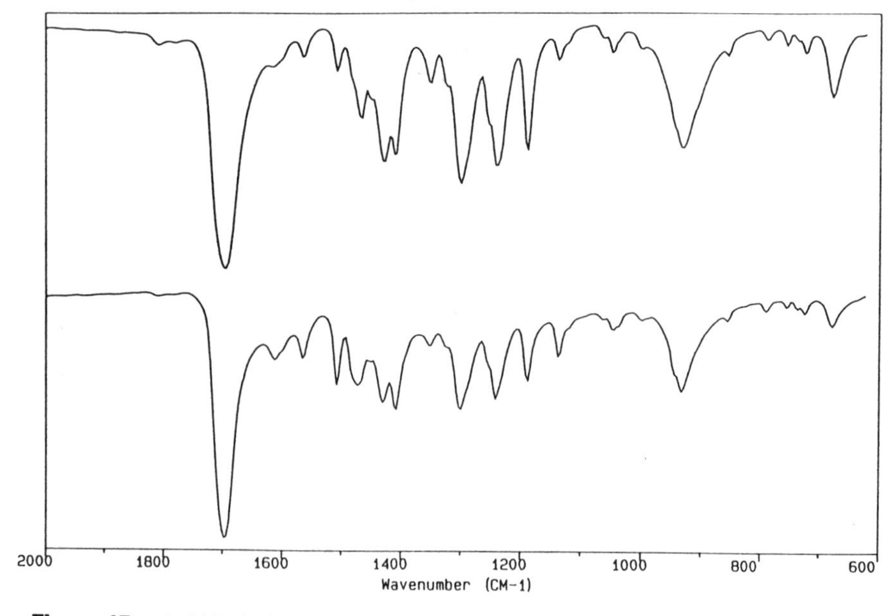

Figure 15. Acid hydrolyzate of nylon 6,10: (top) mostly crystalline portion of macrosample; (bottom) microsample.

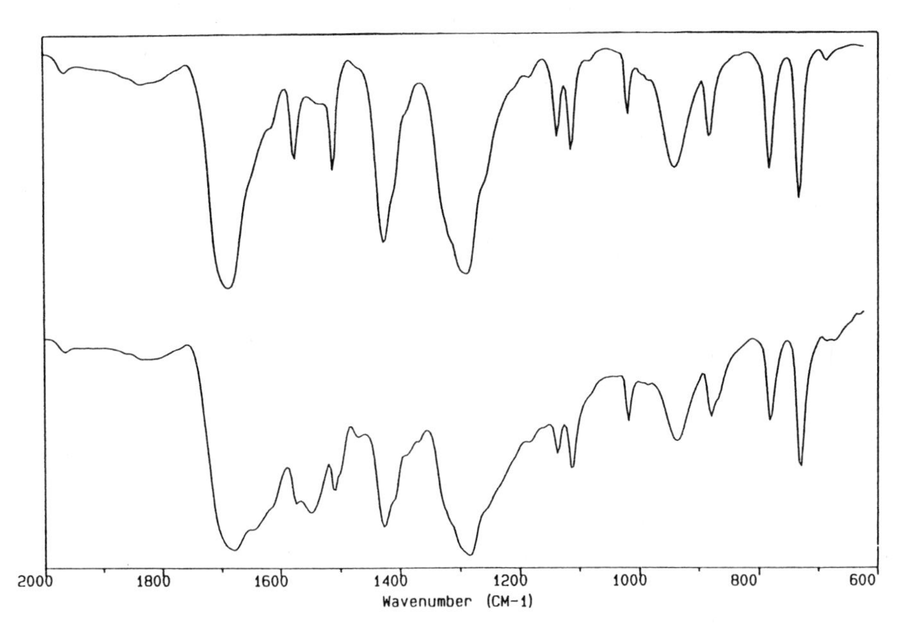

Figure 16. Acid hydrolyzate of nylon 6T: (top) crystals from macrosample; (bottom) crystals from microsample.

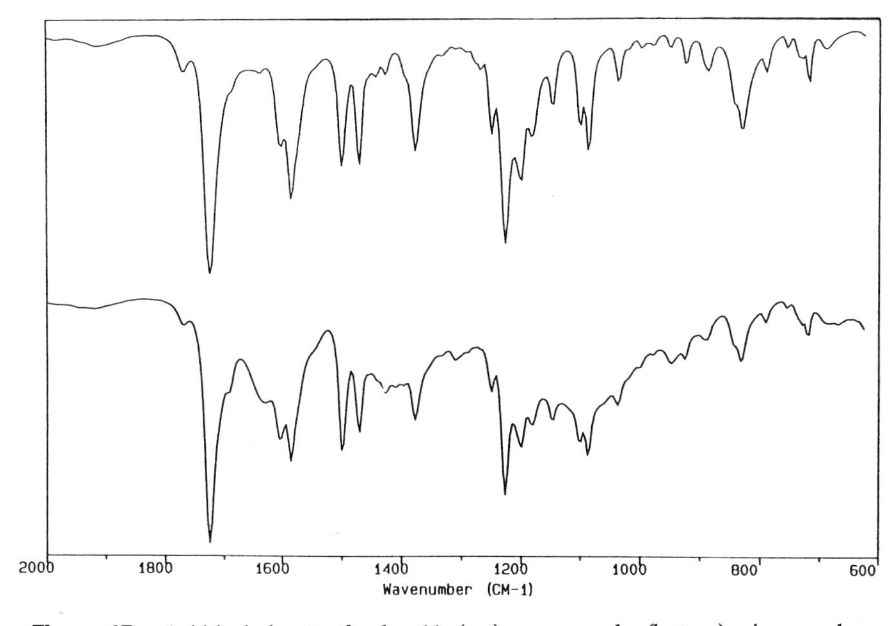

Figure 17. Acid hydrolyzate of nylon 11: (top) macrosample; (bottom) microsample.

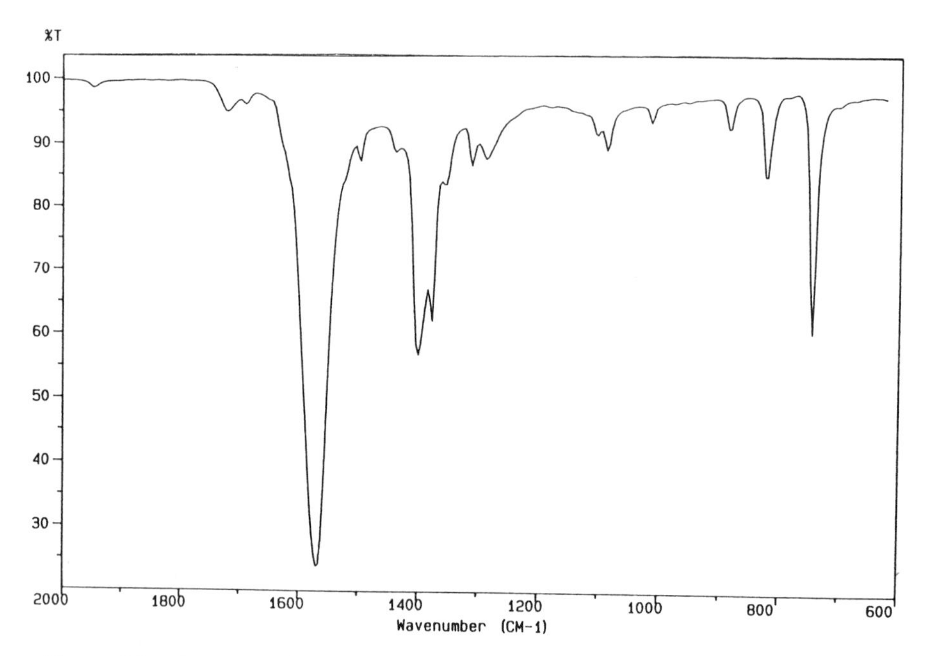

Figure 18. Spectrum of dipotassium terephthalate isolated from micro hydrolyzate of polyethylene terephthalate.

Upon treatment of an alkaline hydrolyzate with a drop of dilute hydrochloric acid, the free acids can be regenerated (Figure 19). Figure 20 shows the spectrum of the phthalic acids recovered from the hydrolyzates of 1-μg quantities of polydiallyl isophthalate, polydiallyl phthalate, and polybutylene terephthalate. The spectra conform closely to those of the free acids with the exception of the phthalate where some interference is evident.

Other polyesters can be hydrolyzed and their acid components identified. This is demonstrated by the recovery of dipotassium fumarate from the alkaline hydrolysis of poly(4,4-dipropoxy-2,2-diphenylpropane fumarate) shown by the spectrum in Figure 21. Had the double bond been lost through cross-linking, this would not be the case. The same would be true of any acid component whose integrity had been lost through an irreversible reaction.

The presence of two or more acids will certainly complicate matters. The identification of these acids will depend not only on the complexity of the spectrum but also on the skill and experience of the analyst. A complete history of the sample and its environment can produce valuable clues. The preparation of derivatives of the acids may permit qualitative and quantitative information by chromatographic techniques. For example, gas chromatography of the methyl

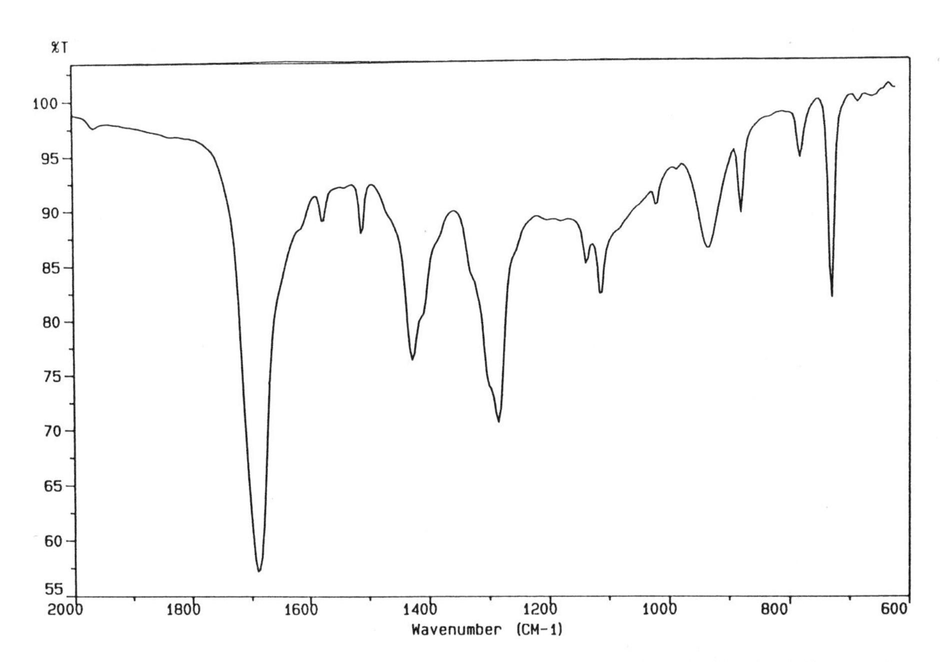

Figure 19. Spectrum of terephthalic acid recovered from the alkaline hydrolyzate of polybutylene terephthalate after treatment with dilute hydrochloric acid.

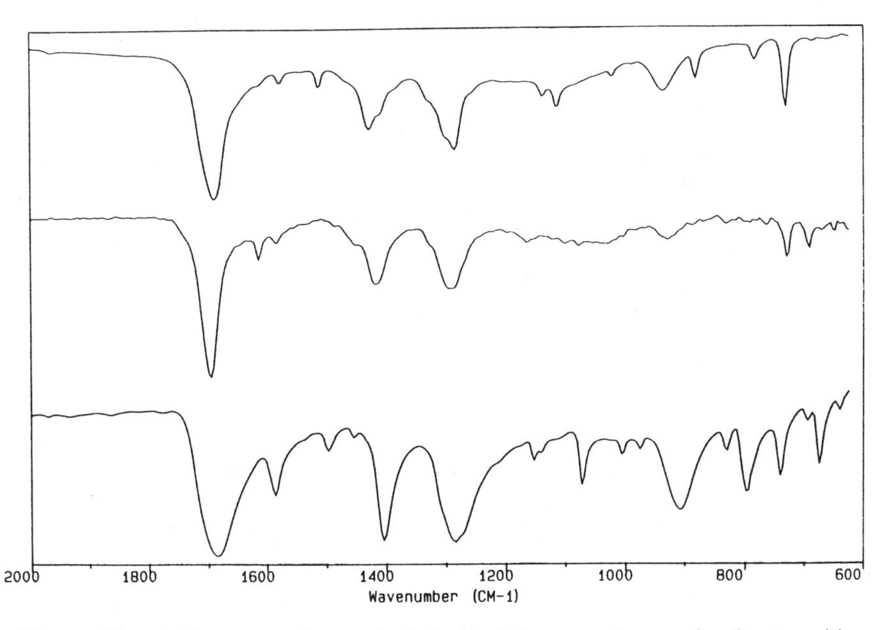

Figure 20. Acids recovered upon hydrolysis of 1-μg specimens of polyester with KOH/isopropanol followed by treatment with dilute hydrochloric acid: (top) terephthalic acid; (center) isophthalic acid; (bottom) *o*-phthalic acid.

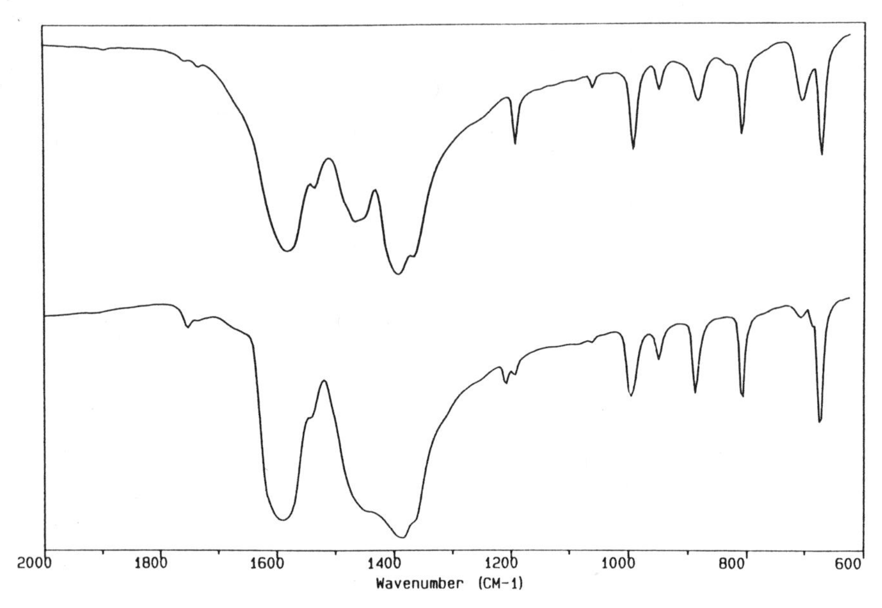

Figure 21. Alkaline hydrolysis of poly(4,4-dipropoxy-2,2-diphenylpropane fumarate): (top) reference dipotassium fumarate; (bottom) dipotassium fumarate recovered from microhydrolysis.

esters or direct analysis of hydrolyzates by HPLC can be used, provided that sufficient sample is available. The gas chromatographic identification of the acetate derivatives of the glycols is another course of action, although this has not been attempted by this author. A combination of techniques may be required to obtain maximum information.

3.4. Polyurethanes

The controlled degradation of microgram quantities of polyurethanes can be accomplished quickly and easily, although the writer has not been very successful in the recovery and identification of these products. Several reasons have been identified. Polyurethanes and epoxides are not as easily hydrolyzed as polyesters. Consequently, it is necessary to resort to more aggressive solvents. The paper by Tindall et al. [4] presents detailed methods for quickly hydrolyzing polyester samples. These authors discuss a variety of hydrolytic solvents and conditions. We have made several attempts to apply these to polyurethanes. The most promising procedure appeared to be 1 M potassium hydroxide in isopropanol for 1 h at 120°C. This had very little effect on a polyurethane of unknown composition. When the conditions were changed to 175°C for 16 h, the polyurethane had degraded completely, but attempts to recover the amine were unsuccessful. A polyethylene glycol was recovered, but this information was of little value because it was evident from a transmission spectrum of the original sample that the polyurethane was a polyethylene glycol derivative. Tindall studied four polar aprotic solvents in some detail. The most promising was a mixture of TMAH (tetramethylammonium hydroxide) in pyridine. This combination was selected because the pyridine could be evaporated, leaving only a residue of the hydrolyzate and the TMAH. This combination was found to be successful in hydrolyzing the few polyurethanes to which it was applied. The pyridine was evaporated and the water-soluble components were extracted. The insoluble free base was treated with dilute hydrochloric acid and washed free of any residue. The amine hydrochloride was analyzed and its infrared spectrum was recorded. The initial sample size in this case was approximately 19 mg. Attempts to reproduce this experiment on a 2-μg sample were unsuccessful. The major problem seemed to be elimination of interfering substances in the TMAH. The TMAH contaminants appeared to be carried through to the final step. Although these residues are insignificant at the milligram range, they are significant and difficult to remove from microgram-sized preparations. For the milligram-size samples, extractions were carried out in a 0.3-cc minivial. It is suspected that the several extractions which are carried out on a microscope slide for the microgram samples are not as effective as those carried out for the milligram samples. Until these difficulties can be overcome, this procedure cannot be recommended for such small samples. However, there is no reason to suspect that these interferences would prevent successful chromatographic analyses of the amines generated from the polyisocyanates.

4. SUMMARY

When direct infrared analyses such as transmission, reflection, and ATR do not provide sufficient information about a specimen, additional sample treatment may be helpful. This could include extraction with a series of solvents to selectively concentrate or isolate species of interest, pyrolysis of polymeric substances to eliminate interferences from pigments and fillers, controlled degradation such as hydrolysis of polyesters and polyamides, and distillation or sublimation to remove semivolatile substances from a nonvolatile matrix. These procedures are practiced routinely to allow identification of milligram-sized and larger specimens, but little has been reported on extending these treatments to microgram-sized specimens. This is largely because, until recently, a prepared specimen of this size would have been useless as a subject for analysis by molecular spectroscopy because instruments were not available to analyze them.

To the microscopist, a 1-μg sample may seem enormous, yet I frequently struggle when faced with a sample mixture of this size—sometimes even with a relatively heterogeneous sample. I have taken some time-tested methods and adapted them to these small specimens. The results have been good for micropyrolysis of polymers and the hydrolysis of polyamides. Success has been achieved for the hydrolysis of a polyesters, but only the acidic components have been recovered and identified by infrared spectroscopy. Very limited success has been achieved in the isolation and infrared identification of the hydrolysis products of polyurethanes.

I do not believe that much more can be gained by applying numerous extractions in an attempt, for example, to isolate amines generated by the hydrolysis of polyurethanes. This is not to say that microhydrolysis is unsuccessfull, only that the steps following the hydrolysis need to be refined.

Amines and polyols can be derivitized for analysis by gas chromatography. With careful sample preparation, sensitivity should be more than adequate. HPLC offers an alternative means to analyze the complex mixtures without having to resort to isolation of the products of the reactions—provided, of course, that references of suspect degradation products are available. Thus the groundwork has been laid for analyzing the components of very small complex molecules.

5. EXPECTATIONS AND NEEDS FOR THE FUTURE

With the rapid technological advances of instrumentation, it would seem that we should have leaped ahead or at least kept pace with our sample preparation techniques. Unfortunately, this has not been the case. If it were not for methods developed by and for microscopists (both light and electron), we would still be in the dark ages (pre-1985).

As I reviewed these past few pages, I see a great need for imaginative approaches to sample preparation. Squashing a specimen to make it broader and thinner, pyrolyzing a small sample in a small tube rather than a larger sample in a large tube, or hydrolyzing a specimen in a capillary rather than a flask are not the results of flashes of genius.

I have no doubt that in the future, molecular spectroscopy will achieve picogram sensitivities. If we fail to prepare for this eventuality by investigating new and better sample preparation techniques, we will find ourselves playing a catch-up game just as we are today.

REFERENCES

1. Harms, D. L. (August 1953). Identification of complex organic materials, *Anal. Chem., 25* (8): 1140–1155.

2. McCrone, W. C., and Delly, J. G. (1973). Instrumentation and techniques, *The Particle Atlas,* 2nd ed., Vol. VI, Ann Arbor Science Publishers, Ann Arbor, Mich., p. 228.
3. Humecki, H. J. (1988). Polymers and contaminants by infrared microspectroscopy, in *Infrared Microspectroscopy* (R. G. Messerschmidt and M. A. Harthcock, eds.) Marcel Dekker, New York, pp. 68–72.
4. Tyndall, G. W., et al. (1991). Preparation of polyester samples for composition analysis, *Anal. Chem.*, *63*: 1251–1256.

Index